CIM SYSTEMS
An Introduction
to Computer-Integrated
Manufacturing

F. H. Mitchell, Jr.
Gonzaga University

Prentice Hall, Englewood Cliffs, New Jersey   07632

Library of Congress Cataloging-in-Publication Data

Mitchell, F. H. (Ferdinand Haverman) (date)
 CIM systems : An introduction to computer-integrated manufacturing
 F. H. Mitchell, Jr.
   p.   cm.
 Includes bibliographical references and index.
 ISBN 0-13-133299-6
 1. Computer integrated manufacturing systems.   I. Title.
TS155.6.M58   1991
670′.285—dc20   90-36106
                                                      CIP

Editorial/production supervision
  and interior design: Serena Hoffman
Cover design: Ben Santora

The author and publisher have used their best efforts in preparing this book. These efforts include the development and evaluation of the theories, models, problem-solving strategies, and computer programs. The author and publisher make no warranty of any kind, expressed or implied, with regard to these methods and programs or the documentation contained in this book. The author and publisher shall not be liable in any event for incidental or consequential damages in connection with, or arising out of, the furnishing, performance, or use of these theories, models, and programs.

Printed in the United States of America
10 9 8 7 6 5 4 3 2 1

ISBN 0-13-133299-6

PRENTICE-HALL INTERNATIONAL (UK) LIMITED, *London*
PRENTICE-HALL OF AUSTRALIA PTY. LIMITED, *Sydney*
PRENTICE-HALL CANADA INC., *Toronto*
PRENTICE-HALL HISPANOAMERICANA, S. A., *Mexico*
PRENTICE-HALL OF INDIA PRIVATE LIMITED, *New Delhi*
PRENTICE-HALL OF JAPAN, INC., *Tokyo*
SIMON & SCHUSTER ASIA PTE. LTD., *Singapore*
EDITORA PRENTICE-HALL DO BRASIL, LTDA., *Rio de Janeiro*

*to my mother and father . . .
with thanks*

CONTENTS

5 SOFTWARE FOR MODELING SUPPORT 164

6 MANUFACTURING EQUIPMENT AND AUTOMATION 221

7 INFORMATION FLOW AND COMPUTER NETWORKS 339

8 ORGANIZATION AND MANAGEMENT 392

9   PRODUCT DESIGN AND MANUFACTURING   428

10 SYSTEM DESIGN AND IMPLEMENTATION   462

PREFACE

Manufacturing organizations in the United States are under intense competitive pressures. Major changes are being experienced with respect to resources, markets, manufacturing processes, and product strategies. As a result of international competition, only the most productive and cost-effective industries will survive.

Manufacturing organizations are thus faced with the need to optimize the way in which they function in order to achieve the best possible performance within necessary constraints. This is a difficult task, both in terms of understanding the nature of the problem and the most effective solution strategies, and in forming and implementing plans that develop from this understanding.

Many of the efforts in this direction are being carried forth under the banner of *computer-integrated manufacturing* (CIM). CIM is a conceptual approach to helping manufacturing organizations respond to the difficult environment in which they operate. However, in the same way that there are many different perceptions of the problem, there are many different viewpoints regarding the nature of the CIM solution.

Although a definition of CIM is pursued in detail in Chapter 3, it is important to have an ''up-front'' understanding of the meaning presented in this text. As described here, CIM refers to the application of system concepts to produce a manufacturing enterprise that can best achieve performance objectives in today's market setting.

Many different types of planning and learning efforts for system change are required to satisfy enterprise objectives, and all of these activities are considered essential aspects of CIM. In most systems of interest, the system design will require an *integrated* information flow, which, in turn, depends on *computer* networks, thus giving rise to the CIM label. However, these aspects of system design are typically a necessary but not sufficient contribution toward satisfying performance objectives. There are many other important concerns. The overall system must be rationalized, requiring that the work flow, organizational structure, and management methods must be redesigned to obtain performance objectives. The entire meaning of product design must be assessed and modified as necessary to optimize system performance. The most appropriate use of technology can then be selected within this context.

> *CIM is not a product that can be purchased and installed. CIM is a way of thinking about and solving problems. The emphasis is on understanding how to create effective manufacturing enterprises. The determination to apply this understanding must come from personal commitment.*

CIM is thus taken here to involve the design or redesign of an entire manufacturing enterprise so that *all* aspects of the system work together effectively. In most cases of interest, *integrated* information flow, the widespread application of *computers*, and high levels of *automation* result from such design efforts. However, these technologies are identified here as a means to an end; the emphasis is on understanding how to most appropriately use all available resources to achieve enterprise objectives.

Given the different perspectives regarding CIM, experiences by companies that are struggling to redesign their manufacturing systems must be viewed somewhat cautiously until the context is understood. But the potential for improvement seems to be substantial. For example, based on widely quoted estimates by the Manufacturing Studies Board, the advantages of conversion to CIM-oriented operations for five companies were found as follows (National Research Council 1986, p. 119):

Reduction in engineering design cost	25–30 percent
Reduction in overall lead time	30–60 percent
Increase in product quality	2–5 times
Increase in capability of engineers	3–35 times
Increase in productivity of production operations	40–70 percent
Increase in productivity of capital equipment	2–3 times
Reduction in work-in-progress	30–60 percent
Reduction in personnel costs	5–20 percent

Whether or not these somewhat anecdotal data turn out to be typical, they do indicate the range of potential impacts that are now being associated with CIM.

Companies (and countries) that emphasize CIM are likely to have a significant competitive advantage over those that do not. Because of changes in the competitive environment, the nature of manufacturing is evolving in fundamental ways. The future standard of living and social integrity of the United States may well depend on the widespread implementation of CIM concepts by industry. Manufacturing continues to be es-

sential to a strong economy, and CIM concepts provide the pathway toward competitive manufacturing systems (Jonas 1987; Cohen and Zysman 1988).

The management of a manufacturing system to best use all resources, and the management of technology to best achieve system objectives, are both essential aspects of problem solving. As defined by the Task Force on Management of Technology (National Research Council 1987): "[The] . . . management of technology links engineering, science, and management disciplines to plan, develop and implement *technological* capabilities to shape and accomplish the strategic and operational objectives of an organization." If the effective management of technology is made a part of effective system management, competitive enterprises will result.

Skilled management is required to make the best use of available resources. In today's manufacturing setting, technology must be viewed as a potential resource. Thus, the management of technology—including selection of the most appropriate technology for use and its optimum application—is an essential management task.

The conversion of U.S. manufacturing to a CIM-oriented capability can be regarded as a challenge relating to appropriate system design and management, applied to what has become a knowledge-based field of endeavor. The types of problems that must be addressed and the characteristics of the solutions that are required are in many ways unique, because they involve an integrated approach to linking state-of-the-art engineering and science and management capabilities to address the needs of pragmatic organizations in competitive environments.

As described by those with experience in the field, many of the basic concepts of the field are still under development. As noted by Chiantella (1986), "The initial problem management faces is the lack of a methodology to explore a CIM initiative" (p. 5), and "At this writing there is not a general set of management procedures to address the CIM opportunity" (p. 10). In a similar interpretation, Lardner (1986) has noted that ". . . the essential nature of what we call manufacturing is not widely understood . . ." (p. 10). Further, the importance of continuing efforts to improve the "theoretical substructure" for manufacturing engineering has been noted by the National Academy of Engineering (1985, p. 5).

Educational institutions have a critical role to play in helping industrial organizations understand their choices and make the best possible use of their opportunities. Any effort to develop an educational strategy for CIM must begin by proposing a method for problem definition and problem solving. This text draws on the efforts of a number of investigators and open system concepts to introduce a step-by-step approach to problem definition and solution approaches. System design is treated as an iterative learning process rather than as an effort to predict a "final state solution" in advance. The *learning system* concept provides an important viewpoint of manufacturing system evolution. A general strategy is combined with in-depth studies of the management and technological opportunities and constraints that must be anticipated.

The text makes use of a simple system-environment simulation (SES) model to illustrate the application of key concepts. A copy of the software for this model is provided with the text. The text also discusses several commercially available software packages that can support system design, implementation, and evaluation activity.

Chapters 1 to 3 of the text introduce manufacturing systems and their environments. The purpose of this discussion is to explore the requirements that are placed on manufacturing systems by the market environment and to consider how the optimum internal structure and function of the system can be linked to external constraints.

This material is approached by summarizing open system concepts that are useful for analysis of U.S. manufacturing systems (Chapter 1), providing an historical overview of manufacturing in the United States (Chapter 2), and defining the nature of the problem that is faced in manufacturing system design and proposing an appropriate planning strategy (Chapter 3). The emphasis is on systems that are designed for high performance in markets characterized by intense competition and rapid change. Computer-integrated manufacturing is introduced as an effective strategy for systems in such settings.

Chapter 4 introduces step-by-step approaches that can be used for the detailed design of CIM systems; Chapter 5 explores alternative computer-modeling methods that can support planning and design activity. Chapters 6 to 9 provide detailed discussions of the strategies and technologies that can be used to implement competitive systems and the opportunities and constraints associated with these strategies. Various problem-solving approaches, issues, and trade-offs are introduced and evaluated against the requirements set by the system-market interaction. These chapters discuss manufacturing equipment and automation (Chapter 6), information concepts and computer networks (Chapter 7), organization and management (Chapter 8), and design and manufacturing (Chapter 9)—concepts that are essential to the design and implementation of effective CIM systems.

Chapter 10 draws on all previous development to consider the properties of integrated system design and implementation. The previously introduced strategies and technologies are discussed from a system-wide perspective. Step-by-step procedures illustrate the transformation of system performance objectives into implementation plans.

In overview, the text follows this sequence:

- Assessing the problem
- Determining a solution approach
- Developing an understanding of the relevant areas of engineering, science, and management
- Integrating the above information to achieve system design and implementation

The text is intended for interdisciplinary use. Successful experience has been gained in both single-discipline settings and in settings with students from diverse technical backgrounds. A system-oriented approach to manufacturing is introduced and developed. This system framework provides an approach that can be understood and appreciated by all disciplines. CIM system design must be multidisciplinary by nature, and a common knowledge base must be available to all participants in the planning process to enable communications and cooperative effort. The system framework provided here is common to all disciplines and can be linked to the specific interests of all participants. The examples and assignments throughout the chapters reinforce and extend this emphasis on a common understanding of CIM systems.

In order to gain maximum insight into the design and operation of CIM systems, it is helpful to consider the characteristics of such systems in general and to focus on particular types of manufacturing in order to provide a more targeted focus for understanding. In this text, the two manufacturing sectors selected for emphasis are associated with *machine tools for metalworking* and the *production of small electronics assemblies*. Historically, the machine tool industry has been a major focus for advanced manufacturing strategies owing to the broad economic impact associated with this sector. Many of the significant advances in computer integration have been led by new equipment concepts and products developed in this area. Electronics manufacturing is also appropriate for consideration because electronics-related products give rise to one of the largest industries today in the United States, and their manufacture is essential to national competitive capabilities. It is essential that the United States maintain a strong manufacturing capability in the electronics sector as it relates to computers, commercial products, state-of-the-art aircraft, space satellites, and numerous other applications.

The two manufacturing areas emphasized in this text relate to machine tools for metalwork-
ing and the production of small electronics assemblies. The CIM concepts and problem-solving strategies used in these applications can be generalized to many other settings.

The machine tool metalworking and electronics manufacturing areas have been chosen for focus here because they are important in their own right and because, taken together, they illustrate many of the principles and issues associated with industry at large. By focusing on these particular areas, the text addresses some of the detailed issues that are experienced in designing a manufacturing system; at the same time, the processes and equipment involved are sufficiently representative that the application of CIM to other industry sectors can be related directly to the particular focus that is provided.

The educational strategy developed for this text makes use of narrative material and examples in each chapter, student assignments at the end of each chapter, supportive software, and an accompanying Instructor's Manual. The text narrative and examples provide students with fundamental insights regarding the design, implementation, and operation of manufacturing systems. This includes an understanding of fundamental concepts, methods for applying these concepts, and issues associated with these applications. The examples further define and integrate the material in each chapter.

The assignments at the end of each chapter are designed to help students understand the issues discussed in the text, develop experience and skills, and participate more actively in learning activities. The assignments require students to take an active role in understanding the concepts and issues discussed. In addition, the assignments allow students to pursue their own specific interest areas. Many of the assignments require out-of-class interaction with research and industry resources; the intent is to stimulate and enrich in-class discussions among all course participants. In this way, participants can extend the total range of the materials being discussed. This strategy also helps students learn to work in teams in which different disciplines are represented. In-class presentations and written reports are required to strengthen communications skills.

The end-of-chapter assignments are flexible and are intended to allow students to pursue their individual interests. The assignments help students develop an understanding of CIM concepts and problem-solving strategies through active involvement in the learning process.

The system-environment simulation (SES) software provided with the text gives students a means to explore the implications associated with many CIM design principles. The software also allows students to extend their ability to use computer models as part of the system design process. The Instructor's Manual includes a discussion of alternative teaching strategies for courses based on the text and provides selected viewgraph masters for classroom use.

The academic core of the text centers on alternative design strategies that can be applied to complex, multidisciplinary systems. The objective is to extend student understanding of specific discipline areas, so that the students can appreciate how to cross-relate between these specific discipline areas and address realistic strategies for complex system design. Figure A provides a conceptual description of cross-relating among various disciplinary "strands."

The text provides an appropriate means for allowing many different disciplines to work together to achieve effective manufacturing system design and imple-

mentation. It can provide a linking mechanism for a wide range of other courses that relate to specific aspects of such systems. The material is accessible to many different disciplines because of its dependence on system design concepts and levels of math and problem-solving skills that are common to different curricula at the senior/graduate level. At the same time, the materials can serve as a bridge between different interest groups, to support exploration of the specific disciplinary areas that contribute to manufacturing system design and implementation.

Figure B illustrates how a course based on this text can link related courses in specific disciplines. The CIM course presented here provides an overview of the complex systems that are encountered in advanced manufacturing and in many other settings, and the course enables the representatives of various disciplines to learn to relate to one another and to engage in team problem-solving efforts that extend specific disciplinary capabilities to more general settings.

The proposed educational process for a class is as follows:

1. Students read assigned text material.
2. The class discusses the material, and the instructor contributes his or her

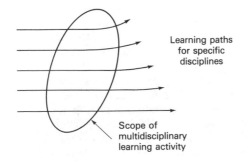

Figure A

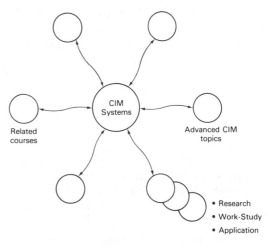

Figure B

own personal experiences and references.

3. Students are given assignments from the end-of-the-chapter materials.

4. Assignments are performed out of class.

5. The assignments are prepared, presented to the class, and discussed to obtain maximum insight.

6. The cycle repeats with new material.

The end-of-chapter assignments can be applied in several different ways. The same assignments can be assigned to all or several students, or they can be assigned individually or to groups. Several assignments or all assignments may be used when any particular chapter is being studied.

Written papers may be used selectively when appropriate, as part of these assignments, or not included when the instructor wishes to focus on classroom presentations. It may be useful to have each student prepare four or five papers as part of the course in order to obtain experience in the types of synthesis, analysis, and writing necessary to produce such materials. Part of the class time may be used to discuss the text and assignments at the end of each chapter, permitting interaction among all class members. Further discussion of teaching options is included in the Instructor's Manual.

The structure of the class is intended to develop and strengthen skills that students need to function in realistic advanced manufacturing settings and in other complex systems. The course requires of each student an ability to perform background research, gather information to interpret collected materials, synthesize a personal understanding of the topic, communicate these results verbally, discuss the topic with other members of the class, make presentations to the class and respond to questions, and prepare written reports as assigned.

The course design provides the flexibility to integrate different viewpoints and insights regarding the design of complex systems; at the same time, the material allows each individual to emphasize specific interest and discipline areas. The course thus provides a means for combining individual interests and emphases with the types of integration activities and team discussions that are necessary to understand the full nature of the system in which specific discipline activities take place.

ACKNOWLEDGMENTS

Many organizations and individuals have provided assistance to the author during the development of this text. Externally sponsored educational and research and development projects have provided a basis for enriched discussion throughout. The insight gained through these sponsored activities has been essential to the material presented here.

The National Science Foundation (NSF) College Science Instrumentation Program (CSIP) provided two equipment grants in support of laboratory development related to microelectronics manufacturing and the application of CIM concepts. Grants by the International Society for Hybrid Microelectronics (ISHM) assisted both with laboratory development and system studies.

Contracts sponsored by Hughes Aircraft Company, Teledyne Industries, and Tektronix, Inc., have related to the automation and integration of manufacturing equipment and systems. Particular appreciation is expressed to Peter Bullock, Bill Garrigues, Ray San Vicente, and Dr. Kevin Shambrook of Hughes Aircraft Company; Phil Migdal, Ralph Redemske, and Jay Allen of Teledyne Industries; and Bob Stanton, Bob Holmes, Raj Garj, and Mark McPherson of Tektronix, Inc., for the many ways in which they have shared their understanding of the field.

Performance of these grants and contracts has been supported by a dedicated team of seniors, graduate students, and research associates, including Scott Anderson, Vance Blakely, Darice Brayton, Darryl Brayton, Fred DeCaro, Tim Dorsett, Stacey Flynn, Paul Graham, Eric Hanson, Mike Hoffman, Heather Jewett, Jim Noble, Corey Peterson, Tim Pollock, Mike Reitcheck, Ron Rowbotham, Carl Sunderman, and Todd Sundsted.

Appreciation is expressed to the equipment manufacturers who have assisted by providing materials and photographs describing their products and by giving permission for discussion of their products in the text. The names and addresses of these companies are provided in the text.

Several companies have allowed descriptions of their software products to be included in the text. Appreciation is expressed to Pritsker Corporation, CACI Products Company, Meta Software Corporation, and Network Dynamics, Inc., for discussion of their system modeling software; to MiCAPP, E-Z Systems, and Manufacturers Technologies for descriptions of their cost modeling software; to International Business Machines and Teledyne Industries for discussion of their factory automation software; and to Swanson Analysis Systems, Inc., CNC Software, Inc., Battelle, and CADKEY, Inc., for an introduction to their computer-aided design and engineering products. Addresses for these companies are provided in the text.

Assistance in the evaluation and application of software tools was received from Ron Rowbotham with respect to MANUPLAN II and from Darice Brayton with respect to Design/IDEF. This support included both evaluation and application of the software for the examples in this text. The system-environment simulation model introduced in Chapter 3 was developed with the support of Tim Pollock, Todd Sundsted, and Eric Hanson, who assisted in both program development and application of the model.

The CIM system design strategy presented in this text has evolved over an extended period of study and application. Assistance during a recent project to develop this strategy in further detail was received from Mike Reitcheck (project coordinator), Jeri Moore (project editor), and the following contributors: Scott Anderson, Vance Blakely, Darice Brayton, Darryl Brayton, Fred DeCaro, Stacey Flynn, Lyle Hatfield, Mike McGreevy, Chris Moyer, Ron Rowbotham, and Jim Sever. Additional support was provided by several faculty members who served as consultants to this group, particularly Professors Gail Allwine of the Gonzaga School of Engineering, and Jerry Monks and Will Terpening of the Gonzaga School of Business.

Helpful reviews of this text were conducted during its evolution by Professors George Harhalakis of the University of Maryland, Michel A. Melkanoff of the University of California at Los Angeles, Shimon Y. Nof of Purdue University, and Hamid R. Parsaei of the University of Louisville. Additional feedback and comments were received from Professors Gail Allwine, Jerry Monks, Carlos Tavora, and Will Terpening of Gonzaga University.

The preparation of draft materials has benefited by editing and developmental support from Jeri Moore. Critical secretarial support has been provided by Mrs. Kelly Rowse, Lara Mihelich, Kadi Bence, David Leslie, and Courtney Flynn. The preparation of this text has been dependent on continuing administrative support from Gonzaga University. This assistance has been both essential and appreciated.

The author appreciates the assistance he has received from the above groups and individuals during preparation of this text. As is always the case, this final product represents the decisions made by the author, who accepts full responsibility for the materials as presented.

F. H. Mitchell, Jr.

1

APPLYING
OPEN SYSTEM CONCEPTS

This chapter provides a framework in which complex open systems can be understood and improved. The discussion addresses many of the underlying concepts that affect the design of appropriate manufacturing systems.

It is important to consider various types of problem-solving strategies that can be applied to manufacturing system design and to understand the limitations associated with these approaches. The design of manufacturing systems involves problem solving applied to complex nonlinear settings, and only partial information is known about the initial states for systems of interest. As a result, a wide range of linear modeling approaches can support such efforts in limited ways but cannot provide *solutions* to the issues associated with such systems.

In most circumstances, future states cannot be predicted based on present states. The future states of a system that evolves through nonlinear interactions may often be bounded, but the particular states typically cannot be predicted analytically based on a limited understanding of dynamics and initial conditions. Given the extreme complexity of such systems and the nonlinearities involved, it is essentially impossible to ensure that a particular intervention will lead to a predicted future state.

In such a setting it is more appropriate to explore how the organization may be turned into a "learning environment" in which planning both guides and supports the learning activities conducted by the organization. In order to remain competitive, an organization must constantly engage in learning activities and must evaluate the effects of those learning activities in order to modify and improve the planning process.

The evolution of a manufacturing system from one state to another thus requires a fundamental commitment to system learning, to undertaking step-by-step change efforts and evaluating each change effort with care. As discussed in this text, a number of useful modeling approaches can support such a learning activ-

Given the complexity of the problems in manufacturing system design, it is not useful to seek final "solutions." Rather, managers must establish a learning environment that will continually guide the organization toward improved system designs. An emphasis on unique solutions (that do not exist) can divert managers from the critical issues of learning and adaptation.

1

ity. However, it is important to remember that modeling approaches are limited in accuracy and relevance, and must be viewed as only one resource among many to help select the changes and learning procedures that are appropriate to the organization.

The design and implementation of manufacturing systems involves a problem-solving activity that is different from the linearized, well-defined problems encountered in many disciplines. Manufacturing enterprises are complex, partially understood systems to which a wide range of problem-solving approaches can be applied; it is appropriate to consider how best to use these problem-solving approaches to achieve the type of system evolution that is desired.

In order to study manufacturing enterprises and understand the problems they face and potential solution strategies, it is first necessary to have a way to "think about" or conceptualize such enterprises. It is important to develop an explicit understanding of the concepts that are being used, along with their strengths and limits.

This chapter provides a foundation for the application of system concepts to manufacturing organizations and their settings. System concepts are often used in the manufacturing and management literatures without a discussion of the underlying assumptions associated with this way of thinking about the problem. Such efforts run the risk of confusion and misunderstanding. Therefore, this text begins with an overview of the system approach to addressing organizational functioning. This overview is drawn upon both to structure problems and to interpret solutions.

1-1. Problem Solving

A typical manufacturing enterprise is a complex organization that depends on and interacts with a wide range of external organizations and activities. Thus, managers of such enterprises are faced with difficult problem-solving situations as they attempt to optimize performance. Recognizing the complexity of the situation, it is important that managers adopt solution concepts that will be productive. To this end, it is useful to consider first some of the general features associated with problem solving.

In order to solve a problem, it is necessary to develop a description or *model* of the features of the problem and to engage in activity that will produce a solution. In dealing with simple, everyday problems, the features of the model can often be readily identified, and experience suggests appropriate solution approaches.

People and organizations engage in a continuous flow of simple problem-solving activities. However, as the problems become more complex, difficulties are encountered in adequately describing the problems and engaging in appropriate solution efforts. One of the underlying reasons for this difficulty is that people are limited by the information they can receive and process and in their capacity for analysis to produce decisions (Simon 1976).

To cope with this situation, a common approach is to limit the problem description to a workable scope by "defining out" those features that are not amenable to consideration. The problem is replaced by a simpler, workable model. Solution action is then taken based on model prediction.

The dangers associated with this strategy are obvious. If the simplified model is not a good approximation of the actual problem, the solution effort will be inappropriate and may even worsen the problem.

The development of powerful theories can provide a valuable asset in deciding how to model a complex problem. The theories associated with science and engineering have provided the means to make rapid, large-scale improvements in problem solving and have contributed in essential ways to the manufacturing technology of today. However, even here, many problems are too complex for exact solution.

Whenever problem definition is extended to include organizational settings, the situation becomes more difficult. Organizational theory is based on a fragmented and partially developed body of knowledge, and can only provide limited guidance in the formulation of problem models. Managers commonly use their experience and instinct in dealing with complex problems that include organizational aspects. As a result, the creation of a competitive manufacturing enterprise involving both technological and organizational aspects is a task that requires an understanding of both the opportunities available and the limitations associated with related problem-solving techniques.

It is often easier to work on those problems that are readily understood and apply established problem-solving approaches than to address fundamental problems with new approaches to problem solving. The former is a comfortable activity, whereas the latter is unsettling. Unfortunately, traveling the familiar road often leads to a predetermined destination; new strategies and designs require a willingness to engage in new ways of thinking and managing.

EXAMPLE 1-1. PROBLEM SIMPLIFICATION

A manufacturing manager is faced with the following problem: Product A is no longer selling well in a competitive market.

What actions can be taken to make product A more competitive? In order to produce a workable scope, the problem is redefined as follows: What actions can be taken to reduce the factory production costs associated with product A?

Discuss some of the difficulties that may be experienced with this redefinition of the problem.

Results

The redefinition of the problem is based on a number of assumptions that may or may not be true. The degree to which the manager will take appropriate actions will depend on the assumptions that are made with respect to the situation. The assumptions implied by this redefinition are as follows:

1. The best way to meet the competition is to reduce product costs. This approach neglects other issues, such as improved product quality or development of a new market segment, that might be more effective approaches to producing a competitive product A.

2. Product costs are determined primarily by production costs. An assumption is made that other costs related to design, inventory and distribution, and general and administrative costs are not important with respect to establishing a competitive position in the market.

3. Product A results in the best use of resources to meet enterprise objectives. Not addressed is the general problem of the use of the manufacturing system resources to meet objectives and the role that product A plays within that setting.

Unless these assumptions are validated, it may turn out that solving the redefined problem will not necessarily improve the situation. It is necessary for the manager to consider all of the assumptions implicit in the problem definition that is taking place and examine whether a particular problem-solving approach is likely to be effective in this setting.

1-2. Open System Concepts

In order to frame the nature of the problems facing enterprise managers, it is useful to draw on *open system concepts* for insight into the problem at hand. The open system approach to understanding can provide concepts and definitions that will be useful throughout the problem-solving effort. The manufacturing enterprise of interest is considered to be an *open system*. The enterprise is assumed to function in an *environment*. The manufacturing system and the environment interact together in many complex relationships.

There are numerous ways in which to define a system. As proposed by Van Gigch (1978), "a system is an assembly or set of related elements. The elements can be concepts, objects or people" (p. 3). The world in which the system exists is assumed to be made up of an assembly of additional systems with their own related elements. The environment thus describes all organizations and activities outside the system that can affect system performance (Fig. 1-1).

These definitions are somewhat deceptive, in that they imply the existence of a *boundary* between the system and its environment but do not explicitly address many of the problems encountered in defining this boundary. For example, for a manufacturing enterprise, are suppliers part of the system or of the environment? Dividing the world into a system-environment "split" can require difficult decisions, and will affect the appropriateness of the problem-solving activity.

Open system theory emphasizes the close relationships between a system and its supporting environment (Katz and Kahn 1978). Thus, system-environment relationships are considered essential in effective system decision making. As noted by Porter (1980), "The essence of formulating competitive strategy is relating a company to its environment" (p. 3).

A *closed system* is one for which it is assumed that system-environment interactions can reasonably be neglected in problem solving. Given the limitations of human problem-solving capabilities, the closed system approach is often used to formulate problems and decide on

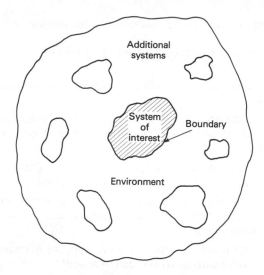

Figure 1-1.
A system viewpoint of a manufacturing enterprise and its environment. Difficulty may be experienced in defining an appropriate boundary between the system of interest and the environment.

solution strategies. In some cases, a very limited range of environmental interactions may be considered in an attempt to improve decision making. The focus on closed systems and limited environmental interactions is considered necessary because the actual problem is too complex to be workable.

The open system concept provides an alternative way of thinking about a problem and approaching problem definition and solution efforts that is reasonably matched to the manufacturing setting.

As noted by Kuhn (1970), the way of thinking about a problem can be described as a *paradigm*. As will be discussed, authors who write about changes in manufacturing often refer to shifts in "paradigms" and "mind-sets" (Clark, Hayes, and Lorenz 1985), "culture" and "business beliefs" (Krooss and Gilbert 1972), and "ideologies" (Perrow 1972). In each case, the author is referring to a basic set of concepts that defines a way to view a problem.

It is important to gain an awareness of the paradigm that is in use for problem solving, as it defines the ranges of response strategies that are considered. As is discussed in Chap. 2, the history of manufacturing in the United States can be associated with several different paradigms, and the competitive problems now affecting U.S. manufacturing are likely associated with the widespread use of paradigms that provide an inappropriate understanding of the problems and of responsive solution strategies.

As discussed by many authors, including Lardner (1986) and White (Solberg 1987), open system complexities must be considered when attempting to understand or design advanced manufacturing systems. Manufacturing must be developed as a "uniform, integrated, indivisible whole" that is matched to the environment (Lardner, p. 10). Throughout this text, the open system paradigm will be used as a way to view manufacturing enterprises and their settings and to explain past efforts and offer effective solution strategies for today.

If misused, the application of open system concepts to manufacturing can lead to a sense of overwhelming complexity and consequent management paralysis. It is essential that managers learn how to combine the open system perspective with commitment to a learning-based organization. Realistic action then results, with continuing learning and adaptation.

An open system can be associated with the following properties:

- A system is considered to consist of building-block *elements* or *subsystems*.
- The elements can be described in terms of attributes or characteristics.
- The elements are related to one another through these attributes. The structure and function of the system depend on these relationships.
- The system engages in purposeful activity. The goals and objectives of the system motivate system function.
- The system takes in resources from the environment (input), engages in conversion or transformation, and produces an output to the environment.
- The system takes in information to support decision making and provides information to the environment.
- The relationships among elements typically involve feedback loops. For each such loop, the output from a given element is linked back to other elements in the system that affect input to the given element. See Fig. 1-2.
- The system-environment interactions produce feedback loops in which the system acts on the environment, changing the ways in which the environment acts on the system. An action-reaction circular flow of cause and effect thus results (Mitchell and Mitchell 1980).

Figure 1-3 indicates how the development of an open system modeling effort can be implemented. The study objectives drive the definitions of system and environmental attributes that will be used to describe the problem. Based on these attributes, the system is divided into subsystems, with relationships formed among the subsystems. The

Figure 1-2.
System elements and interactions illustrating feedback loops. The outputs from several system elements are linked back to the inputs of other elements in the system, producing a coupling between output and input.

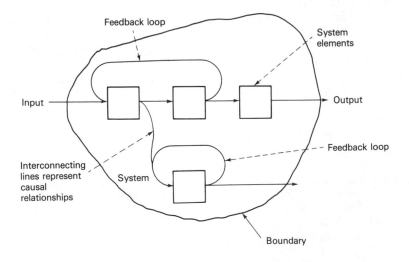

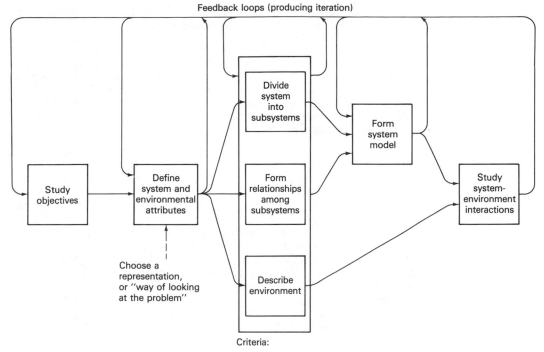

Feedback loops (producing iteration)

Divide system into subsystems

Form system model

Study objectives

Define system and environmental attributes

Form relationships among subsystems

Study system-environment interactions

Choose a representation, or "way of looking at the problem"

Describe environment

Criteria:
- Minimize complexity within each subsystem.
- Minimize complexity of relationships among subsystems.
- Minimize complexity of the environmental description.

Figure 1-3.
A common approach to system study, illustrating how system and environmental attributes can be used to form a representation, or "way of looking at the problem." This problem definition can be used to explore subsystems and their interactions, as well as the nature of the system-environment interaction.

combination of subsystems and relationships is used to create a system model. Interactions between the system and environment are then explored to achieve study objectives. Feedback takes place among all phases of the effort.

There are many difficulties with this strategy. In complex settings, there are many more possible attributes than can be reasonably considered, so a selection process must be conducted, thus limiting the validity of the study. Then, in order to achieve a useful system model, the subsystems must be defined to have a relatively simple internal structure and relatively simple relationships (for the given attributes). Otherwise, the subsystems and relationships are not amenable to analysis. Finally, the means must be found to study system-environment interactions and to draw conclusions regarding study objectives.

EXAMPLE 1-2.   CLOSED AND OPEN SYSTEM PERSPECTIVES

A manufacturing system problem can be described in the following alternative ways, making use of closed and open system viewpoints. Discuss the features of each description.

Closed System Perspective: A manufacturing system is experiencing quality problems that lead to extensive rework and high product costs. A plan must be developed to improve manufacturing system process control in order to reduce the rework and enable a reduction of product cost.

Open System Perspective: A manufacturing system is experiencing quality problems that lead to extensive rework and high product costs. A decision is made to evaluate the quality of the components and materials being used in the product and the manufacturing processes being applied. In addition, decisions are made to evaluate the product design to determine whether the manufacturing requirements are consistent with available technology, and to study further the market in which the product is being sold to determine if it is the most appropriate design for the market.

Results

The closed system perspective focuses on internal parameters alone. It is assumed that the problem is caused by flaws in internal process control, and that all other factors that extend beyond the immediate manufacturing system can be neglected. This approach is misleading and may not produce effective problem solving if other factors are important in the situation.

The open system perspective considers both the aspects of the manufacturing system itself and its environment. Input components and materials are to be evaluated because they may cause process failure. This leads to an examination of vendor quality, the kind of input testing that is required, and other system aspects to deal with the interface between the manufacturing system and the vendors. Then, once the input components and materials are evaluated, attention turns to the manufacturing processes themselves and those aspects of the problem that are addressed in the closed system viewpoint.

The next effort is to determine if product design is such that the available manufacturing processes and technology can effectively produce the desired product without rework. It may be that the design is inherently flawed with respect to the available technologies, so that rework is implicit in the product design, and no amount of effort can compensate for this design flaw.

Finally, the open system perspective leads to a review of the market to see if the design of this product can and should be changed in order to better address market opportunities.

The open system perspective provides a much more extended approach to problem definition and solution options associated with the extensive rework and high production costs. It may turn out that a very different understanding of the problems will emerge from the open system perspective, leading to different management actions and outcomes. The manufacturing system manager must constantly try to ensure that the problem definition includes all possible relevant aspects of the problem, as part of an effort to maximize production.

1-3. Application to Manufacturing Systems

Given the open system perspective of a problem, combined with the difficulties associated with system modeling, how should problem-solving activity be initiated? This is an essential question, and unless care is used in producing a strategy, misleading results can be obtained.

One common problem-solving approach is to assume that a valid model of the system and environment can be formulated. Then, starting with the present system state and the anticipated environment, the features of various future states can be evaluated. An optimum future state can be selected, based on enterprise objectives, and an implementation process can be used to guide the system from the present to the desired future state.

Numerous modeling efforts have followed this problem-solving strategy. In a major effort, complex simulations of national and world policy implications were developed by Forrester (1969, 1973). In a broader sense, this viewpoint has been extended as the proposed foundation for a *general systems theory* that could relate to a wide variety of system types (von Bertalanffy 1968; Van Gigch 1978; Bowler 1981). In all of these activities, it is assumed that realistic system models can be developed to deterministically relate present and future states of the system. (Attributes that must be described statistically are used to produce statistical distributions of future states, with the basic assumption of determinism still held valid.)

This approach to problem solving is based on an assumption that differences between the actual system and the system model can be treated as perturbations or correction terms to the solutions that are obtained. Thus, predictive errors are assumed to be small, or at least not critical to the validity of the results. The advantage of this approach is that it is convenient and well understood. It follows the pathway associated with the solution of linear systems of equations.

Unfortunately, the basic premises of such an approach can be questioned. Complex systems with feedback typically will be described using nonlinear differential equations (or nonlinear partial differential equations). Such networks of equations are often not well behaved. As has been demonstrated (Gleick 1987), such nonlinear relationships can produce non-steady-state solutions that are highly sensitive to initial conditions. Given that initial conditions for the selected attributes will be known with only limited accuracy, the future state of the system cannot be predicted with any confidence (Fig. 1-4). At any future time, the system attributes can have what appears to be a random relationship to one another, based on the errors associated with establishing initial

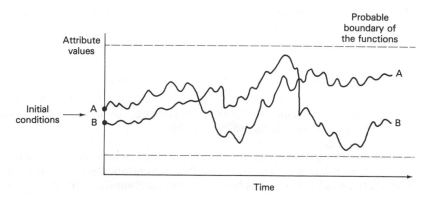

Figure 1-4.
Non-steady-state solutions to nonlinear equations, showing sensitivity to initial conditions. Given small differences in initial conditions and nonlinear interactions, the system attributes appear to have random relationships with one another.

conditions. It is also true that for many such systems, the attribute values will often tend to stay within an envelope that bounds the properties of future states. Thus, from this perspective, the future state cannot be predicted but may possibly be bounded.

If the system model cannot be used to predict future states, but only to develop a general understanding of likely boundaries for those future states, what type of planning and problem-solving strategy should be implemented in a manufacturing setting?

As proposed here, the value of system modeling is to force the organization to think about the system, the constituent subsystems and relationships, and interactions with the environment. This qualitative overview and understanding can be used as a context in which to evaluate and make use of more limited quantitative modeling techniques. The methods of operations research and economic analysis thus become inputs into a system-oriented qualitative overview of the enterprise.

Analytic models can be used to effectively support the design of manufacturing systems. However, given the complexity of open systems, such models must be used to stimulate organizational learning, not to establish the "solution" to the "problem." Managers cannot avoid the need to learn and adapt.

An open system understanding of the enterprise can be used to develop an appreciation of the likely nature of future circumstances, within an envelope or boundary of uncertainty. From this generalized appreciation of the likely kinds of futures to be encountered, conclusions can be drawn about the general characteristics that will be associated with the competitive enterprises of the future. Management objectives can then be directed toward these general characteristics.

Within the open system context, a series of short-term transition efforts is typically required, based on both an understanding of the present system, with its strengths and weaknesses, and on insight into the general characteristics that should be sought for survival in the future. These efforts can be used as reasonable risk learning experiences. Each transition effort should be associated with constant evaluation to ensure that progress is being made in the desired direction and to help formulate the next step.

Based on the open system paradigm and on an understanding of realistic system modeling complexity, the following approach to problem solving is thus developed:

1. Study the system and environment to learn as much as possible. Use limited-scope modeling wherever helpful to understand the present system and environment.

2. Estimate the future attribute boundaries that will likely be associated with competitive enterprises.

3. Draw conclusions regarding the desirable characteristics of the specific enterprise.

4. Plan a sequence of transition steps with learning activities.

5. Conduct the transition steps with a cycle of try-evaluate-learn repeated over and over. Use limited-scope modeling wherever helpful to design these transition steps.

This strategy is a common-sense one that is intuitively satisfactory to many managers. The best possible understanding of the general characteristics of the future is based on organizational experience, supported by data gathering and modeling. Many subtle nuances enter into the qualitative understanding that results. The management team members develop a fuzzy picture of the types of future enterprise performance characteristics that are likely to be competitive and can describe the characteristics that will be sought for their enterprise.

The essence of this paradigm is to emphasize organizational learning to achieve well-understood objectives. Organizational learning as the pathway to competitive enterprise performance is often mentioned by those examining the present condition of U.S. manufacturing. As noted by Clark, Hayes, and Lorenz (1985), a key theme of creating more competitive U.S. manufacturing enterprises is "the need for more effective management of organizational learning and ways to encourage it" (p. 279). They also noted that "the way we manage learning . . . (is) of fundamental importance" (p. 281). Similarly, Abernathy, Clark, and Kantrow (1983) have noted that in the new competitive environment, "more than ever before, to manage will be to learn" (p. 120).

Based on the paradigm developed here, manufacturing system evolution must take place within a continual learning process (Fig. 1-5). Through experience and evaluation, the system will learn how to improve itself. The type of planning that should take place in such circumstances should then be targeted toward the creation of a highly efficient and effective learning activity. Planning becomes a process that takes the system through a constant change procedure, learning from step to step.

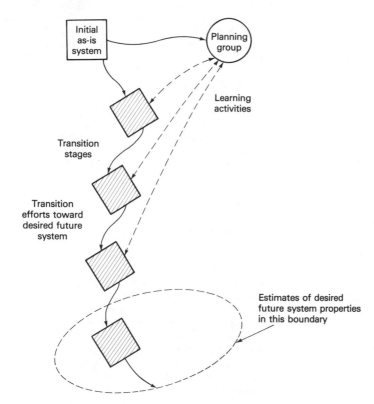

Figure 1-5.
A heuristic model that indicates how manufacturing system evolution must take place through a continual learning process. During evolution from the initial *as-is* system toward a desired future system, a series of transition stages will drive learning activities.

Within this framework, internal planning groups play an essential role: they must act as a focal point for understanding of the required enterprise characteristics for the future, the design of incremental evolution and learning activities, and evaluation of learning outcomes. All available forms of modeling and analysis should be used to support decision making in this context. The essential ingredients are that all sources of information be integrated by the planning group, and that a learning emphasis be maintained.

EXAMPLE 1-3. LIMITATIONS ON PREDICTIONS

The graphs in Fig. 1-6 describe the projected deterministic performance of a manufacturing system for the next five years. Discuss the ways, and reasons why, these predictions may turn out to be invalid.

Results

The graphs in Fig. 1-6 indicate a range of increases and decreases in the costs associated with a particular manufacturing system, based on projections related to current circumstances. Such trend lines extrapolate forward into the future based on current circumstances and cannot consider unexpected environmental changes and system-environment interactions for which no knowledge base exists.

Many aspects of the system and environment are not subject to system control, so these trend lines may not represent the future that actually occurs. There are always limitations in attempting to determine the present values of system parameters, based on past trends associated with these parameters, and the limits associated with estimating where the environment is today and the directions in which it is evolving.

The design costs are shown to decline over the time period. However, these costs may change due to new types of products, new types of computer-aided design tools, the availability of design engineers, and other factors. Any of these effects could drive design costs up unexpectedly.

A modest growth rate in component and materials costs is expected, based on recent experience. Certainly, shortages or surpluses of components and materials, or the introduction of new, unexpected components and materials on the market can extensively affect this growth both positively and negatively.

A decline is expected in manufacturing costs. Neglected, though, is the fact that additional requirements may be placed on the system by new designs, market demand, and competition, resulting in higher levels of manufacturing system investment, and higher costs.

The inventory and distribution costs are shown in decline. However, these costs are based on the planned and expected system-market interactions. These interactions are subject to rapid change if the market shifts, and thus can produce variances in the output.

The general and administrative (G and A) costs are flat, without change, over this period. This assumes no organizational change that will force an increase in this cost area. However, any of the other items considered above could significantly affect the G and A costs.

Finally, profit is a function of all of these cost areas and the selling price of the products from the system. The profit figure may be substantially "off the mark," depending on what happens in the future.

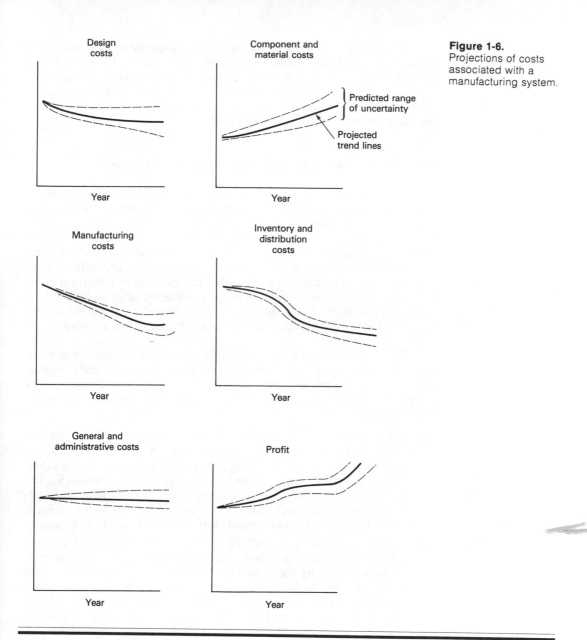

Figure 1-6.
Projections of costs associated with a manufacturing system.

The effort in Example 1-3 to predict the future of the system in detail may be a useful trial method, but it should always be remembered that deterministic, final state planning may be either reasonably accurate or off-target substantially. A more effective approach is to apply a learning cycle in which step-by-step efforts are made to optimize the manufacturing system by repeating the try-evaluate-learn cycle. The decision-making strategies should constantly be changed in order to adapt to the realities of the experience base that is developed.

At this point in the discussion, it may be useful to examine some of the graphs in Chap. 2 (see particularly Figs. 2-12, 2-13, and 2-17) to see how sudden changes in trends can occur in essential environmental aspects affecting a business. The types of unexpected changes shown would have a significant effect on the performance of any business operating in the indicated environment.

1-4. Developing Models of Manufacturing Systems

In order to make effective use of open system concepts, it is often helpful to develop simple, intuitive models that describe the subsystem elements and the relationships among the elements. This approach to visualizing problems has become a commonly applied method for gaining an understanding of complex systems on a *heuristic* basis. A heuristic understanding of a system is associated with a preliminary perception of the nature of the system that can be used as a starting point to stimulate further investigation of the system. In other words, a heuristic understanding "sets the stage" for exploration of the system to discover further information and insights. Figure 1-5 is an example of a heuristic model.

Efforts have also been made to formalize the development of more detailed system models to improve communication, enhance learning, and study system performance. It is useful to categorize these models as relating to one of the following three ways for viewing a manufacturing system:

1. *Physical models:* describing the visible aspects of the manufacturing enterprise. For convenience, the physical models can be grouped into those involving material transformation (including manufacturing equipment and operators) and those involving the hardware for information flow and system control (including computers and operators). Subsystems are defined in terms of hardware units and the movement of material through the system.

2. *Functional models:* describing the manufacturing system in terms of the functions it performs. The emphasis is on defining subsystem functions that can be well understood individually and then relating the functions to one another. This type of model does not represent the visible aspect of the system, but rather describes what the system is doing. For convenience, functional models can be grouped into those that describe the material transformation processes achieved by the product equipment and those that describe the information flow throughout the system.

3. *Organizational models:* describing the manufacturing system in terms of the organizational relationships that exist among the people in the system. The organizational relationships can be linked to the physical models to determine if workable machine-human interfaces exist, and to the desired system functions to determine if a functional congruence exists. These models can be used to consider how the formal and informal organizations relate to one another and to the overall manufacturing system activities.

Models of manufacturing systems generally fall into one of these three categories. Physical models are used to decide mechanical design issues and describe hardware performance, thus creating the physically observable manufacturing system. Functional models are used to describe how the physical equipment will operate to achieve production objectives, indicating how the observable aspects of the system are functionally related to one another, and thus prescribe the knowledge that is required to operate the system. Organizational models describe how people relate to one another to achieve system operations and how people function in the overall system setting. These categories are further expanded upon and applied in Chap. 3.

A physical model describes the visible aspects of the manufacturing system. A functional model describes how the system works. An organizational model describes how people are integrated into the system.

Simple heuristic models are commonly used in all areas of manufacturing to decide on the physical management of equipment, to describe a process, or to plan a reorganization. This type of system modeling is an intuitive aid to understanding and communication. More formal system models have both advantages and disadvantages. They can provide more in-depth system understanding, but they can also mislead if the limitations of the model are not kept in mind.

EXAMPLE 1-4. MULTIPLE SYSTEM PERSPECTIVES

Figure 1-7 describes a manufacturing process from the physical, functional, and organizational perspectives. Discuss the linkages that exist among the three perspectives.

Results

Figure 1-7a describes the manufacturing system from the physical perspective. Shown here is a *work cell* in which several different units of semi-automated equipment are grouped together for a particular manufacturing purpose, and operators are used to support the operations of the work cell. (Work cells of this type are introduced and discussed in Chap. 6.) This drawing shows the physical location of the manufacturing equipment, the operator stations, the conveyor belt that connects the units, and a computer network.

Figure 1-7b provides a functional perspective of the manufacturing system. Boxes and arrows are used to indicate the various operations that must be performed by the physical system. The physical equipment and operators provide the means for producing these functions, and the knowledge base controls the performance of these functions. One approach for producing such functional drawings, which makes use of the IDEF formalism, is discussed in Example 1-5.

Figure 1-7c shows a simple way of indicating how the team for this particular manufacturing work cell relates to the larger organization. Certainly, the illustration does not provide detailed insight into the nature of the internal team interactions or the nature of the interaction between the team and the larger organization.

These three points of view provide quite different insights into the manufacturing process. An understanding of all three is needed to obtain a composite view of how the system operates and of the strengths and weaknesses in the system. (Much more detail regarding these different perspectives of a manufacturing system is provided in Chap. 3.)

Figure 1-7.
Descriptions of a man-
ufacturing system in
terms of physical,
functional, and organi-
zational perspectives.
(a) Physical view of
the system, showing
work cell arrangement.
(b) Functional descrip-
tion of the system,
using the IDEF formal-
ism. (c) Formal (bu-
reaucratic) organiza-
tional description of
the system (discussed
in Chap. 8).

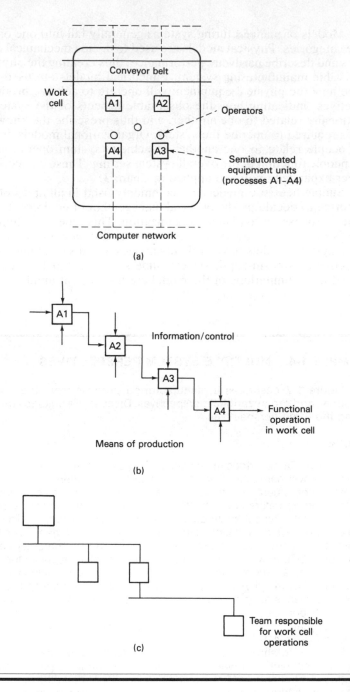

(a)

(b)

(c)

1-5. Graphical Modeling Methods

When dealing with complex manufacturing systems, it is often
difficult to find the most effective way to describe the functions that

must be performed and the relationships among these functions. Because of system complexities, many different functions and complex relationships among these functions are typically encountered.

A variety of graphical modeling methods can be used to aid the designer in visualizing the system and its properties. The development of general representations of this type is explored in a branch of mathematics relating to *graph theory* (Harary 1972; Ralston 1983). A graph consists of a collection of points and lines that connect some or all of these points together. A wide variety of graphical types can be developed. Graph theory can be used in system modeling in order to describe the components of a system and their relationships and to develop an understanding of the fundamental relationships that may be associated with graphical representations. Various mathematical tools, such as matrices, can be used to describe particular graphical forms and to aid in manipulation of these forms.

Algorithms have been developed to describe the requirements associated with searching certain types of graphical arrangements called *trees*. Such algorithms are of importance whenever a search is being performed to find selected data elements within a large data structure.

Graph theory can provide insight into the properties of large, complex systems. Once a set of points and lines is associated with the functional operation of a system, an understanding of the characteristics of graphs enables one to appreciate some of the features of the system that results.

Manufacturing managers encounter many types of graphs. Broad use is made of the Project Evaluation and Review Technique (PERT) and Critical Path Method (CPM) project management techniques, both examples of applied graph theory. For these techniques, connecting lines represent tasks, and nodes represent completed tasks; a path through such a network is any set of successive activities leading from the beginning to the end of the network. The *critical path* is the one with activity times that are longer than any other path and that places the maximum time constraint on the project completion. The development of such planning tools can be used to estimate the time required to complete a project, to monitor project implementation, and to evaluate whether resources should be shifted from other paths to the critical path in order to achieve overall project completion more quickly.

Another graphical application involves *Petri nets,* which can be used to model complex systems (Peterson 1981; Ralston 1983). The nodes of such a network are used to represent system states; linking arrows describe the necessary transitions that must take place to move among system states. Petri nets are studied by considering the circumstances under which particular transitions will occur to allow specific states to be reached. Petri nets can be used to model information flow and decision making in a system. They are especially useful for systems that include parallel operations. Petri nets can be used effectively to model computer hardware and software and provide many insights in this context. The general results of graph theory make up a framework in which the specific characteristics of Petri nets can be pursued. As noted by Bermond and Memmi (1985), "Petri nets were introduced in

1962 to model the dynamic evolution of discrete systems. Since then they have been widely used and they constitute now one of the most complete and advanced formal models of parallel computation.''

PERT/CPM methods and Petri nets are tools that can be effectively applied to selected classes of problems (for example, relating the sequence of events and times required to produce a product from component operations, or determining the conditions under which certain system states will or will not occur).

Other applications of graph theory are often encountered. The use of the common tree formalism to describe formal organizations is a graphical representation, as are other combinations of boxes and lines that describe the functional operations of a system. Graphical methods to study complex systems have been found effective in day-to-day approaches to system description. It is often most advantageous to describe complex systems in terms of ''boxes and arrows''; less often is use made of the literature to understand some of the properties associated with these representations.

One of the most ambitious modeling effects for manufacturing systems was developed by the U.S. Air Force Integrated Computer-Aided Manufacturing (ICAM) program in 1981 (Air Force Systems Command 1981a, 1981b; D. Appleton Co. 1985). The objective of the ICAM program was to ''develop structured methods for applying computer technology to manufacturing and to use those methods to better understand how best to improve manufacturing productivity.'' The ICAM program developed three well-documented modeling methodologies around the ICAM Definition (IDEF) approach to system study: (1) a functional model of a manufacturing system and environment (called the $IDEF_0$ model); (2) an information model of the system and environment (called the $IDEF_1$ model); and (3) a dynamics model to describe time-varying system behavior (called the $IDEF_2$ model). In the IDEF formalism, nodes represent planning stages or system processes, and several different meanings are assigned to the lines that relate these processes to one another.

Because of the development of a number of related software support tools, the IDEF methodology is one of the most widely used functional planning methods in manufacturing today. This text makes use of an IDEF formalism in several places. However, note that other modeling methods, such as PERT/CPM and Petri nets, also can be used effectively by the system design team in support of planning and implementation efforts.

As noted in the IDEF documentation, the methodology can be applied as a *descriptive* tool or an *analysis* tool. The former application can be readily understood as an extension of the heuristic study of systems; the latter application calls for a more careful evaluation. To the extent that the IDEF methodology is used to study localized subsystem issues defined to have a deterministic, nonadaptive nature, such models represent an extension of the standard methods of operations research (Fig. 1-8). Thus, the value of the methodology is to establish a common language to enhance communications and evaluation. As the methodology is applied to less well-defined, more adaptive portions

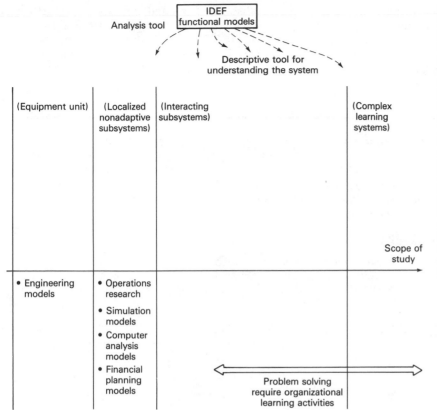

Figure 1-8.
Description of a framework that can be used to discuss and compare various problem-solving strategies as a function of the scope and type of study. The scope of study can range from individual equipment units, with a focus on engineering models, to complex systems that require organizational learning activities. The IDEF functional models can be used as an analysis tool for more narrowly defined problems and as a descriptive tool for complex problems.

of the system, less useful information will be obtained, and the potential for error will grow.

Thus, the IDEF methodology can be most usefully viewed as a structured way to think about system problems and as an analysis tool that can be used in a careful way to study specific design issues of interest. The dominant usefulness of IDEF-based models will be to enhance the organizational learning process. These models can be viewed as tools to be drawn upon by the organization as it cycles through iterative learning experiences. However, such models cannot be used in final state planning to define the system configuration that will most effectively meet enterprise objectives. As noted, the types of systems involved are not amenable to this "endpoint" modeling.

In December 1987, a commercial software product called Design/IDEF™ (Meta Software Corporation) became available as a means for developing functional models of manufacturing systems and their environments.* Design/IDEF is applied in Example 1-5 to describe the functional operation of a system, as a descriptive support to organi-

The IDEF methodology can be used to stimulate organizational learning. However, such models can represent only portions of complex, open manufacturing systems. Care must be taken to avoid misapplication of the method in efforts to define "problems" and "solutions" that will not require learning and adaptation.

---

* Addresses for all companies mentioned in the text are listed in the References. For the software products mentioned here, refer to Chap. 5 references.

zational learning activity; it is discussed in detail in Chap. 5 and further applied to system design in Chap. 10. Another commercial IDEF-related software product, called IDEF/LEVERAGE℠, is offered by D. Appleton Company (DACOM). This product is discussed further in Chap. 5. Other types of modeling are described in the following section and in Chaps. 4 and 5.

EXAMPLE 1-5.   APPLICATION OF A DESIGN/IDEF MODEL TO A MANUFACTURING SYSTEM OVERVIEW

Figure 1-9.
Description of a manufacturing system using Design/IDEF (courtesy of Meta Software Corporation). (a) Definition of the IDEF format showing how the system inputs, outputs, constraints, and resources are used to describe an activity.

Figures 1-9 and 1-10 show an IDEF model (the model is further explored in subsequent chapters). Discuss the meaning of this model and its usefulness and limitations.

Results

A manufacturing system is described by IDEF in terms of functions (represented by boxes) and relationships (represented by arrows). The position of any arrow with respect to a box will determine if the information flow

USED AT:	DESIGN COMMITTEE	DATE: 1/10/90 REV:	X	WORKING	READER	DATE	CONTEXT:
				DRAFT			
				RECOMMENDED			
	NOTES: 1 2 3 4 5 6 7 8 9 10			PUBLICATION			

System constraints

System inputs → Activity name A0 → System outputs

System resources

Purpose: To identify and solve system problems using graphics.

NODE: A-0	TITLE: The IDEF Format	NUMBER:

(a)

is an input, output, constraint, or resource. A sample IDEF module is illustrated in Fig. 1-9a with the position definitions identified.

The diagramming method associated with IDEF utilizes a top-down design technique. The top level of the model provides a general overview of the system, and the lower levels describe the subdivisions of the system. A major advantage of this design tool is that inputs and outputs of higher layers are passed down as the IDEF model is extended in further detail.

Figure 1-9b shows the manufacturing system in overview, with the manufacturing activity itself in the center of the drawing (box A3). The system is supplied by vendor kits that enable production to take place. The vendor kits are obtained from vendor companies, and a vendor kit inventory is maintained as an interface between the vendor output and the factory input. The output of the factory is divided into finished (good) product, which goes to product inventory, and rework/scrap product, which goes to rework/scrap operations. The finished product inventory is delivered to the market according to the sales that are achieved.

The IDEF subsystem areas shown in each of the blocks in Fig. 1-9b can be subdivided to define further each of these areas. For example, the manufacturing system can be treated as the "parent" of a more detailed "child" description of the system's elements. A further definition of the manufacturing system is shown in Fig. 1-10, in terms of three operations: fabrication, assembly, and test and packaging. An information network provides the control

Figure 1-9. continued
(b) Application of this format to a manufacturing system and its environment. (The system constraints and resources are not shown.)

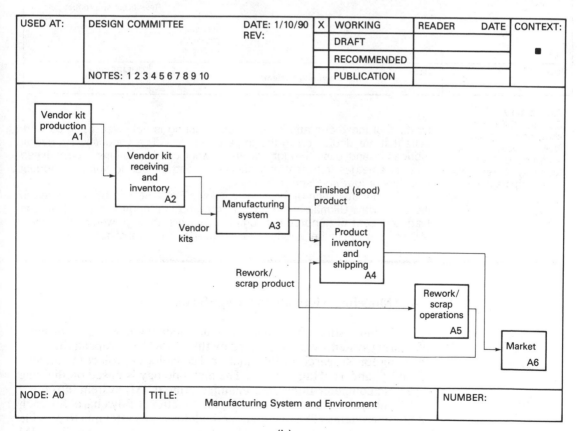

USED AT:	DESIGN COMMITTEE	DATE: 1/10/90 REV:	X	WORKING	READER	DATE	CONTEXT:
				DRAFT			
				RECOMMENDED			■
	NOTES: 1 2 3 4 5 6 7 8 9 10			PUBLICATION			

NODE: A0	TITLE: Manufacturing System and Environment	NUMBER:

(b)

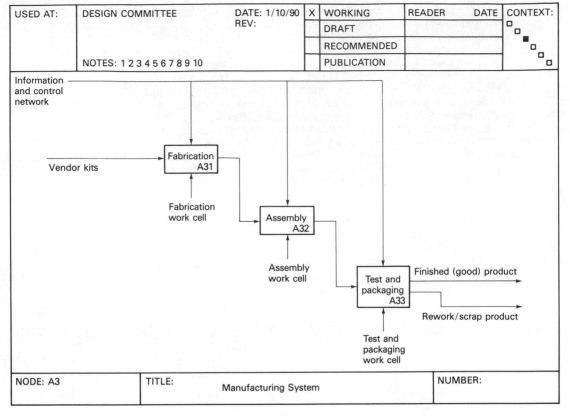

USED AT:	DESIGN COMMITTEE	DATE: 1/10/90 REV:	X	WORKING	READER	DATE	CONTEXT:
				DRAFT			
				RECOMMENDED			
	NOTES: 1 2 3 4 5 6 7 8 9 10			PUBLICATION			

| NODE: A3 | TITLE: Manufacturing System | NUMBER: |

Figure 1-10.
Design/IDEF chart showing the "child" level of system block A3 from Fig. 1-9b (courtesy of Meta Software Corporation).

for each of these operations, and manufacturing is achieved by three work cells that are dedicated to the three operations. The vendor kits arrive at fabrication and pass through the three work cell operations. The finished product emerges from test and packaging and passes to finished product inventory or to rework/scrap.

Further definition of the other blocks in Fig. 1-9b can be developed in the same way. Ultimately, the system drawing that begins in Fig. 1-9b can lead to dozens or hundreds of "child" sublayers, through which all aspects of the system are further subdivided until the desired level of detail is reached.

1-6. Modeling Methods for Manufacturing Systems

An interesting effort to develop a "modeling language" for manufacturing systems was conducted by the ESPRIT (European Strategic Planning for Research in Information Technology) project (Yeomans, Choudry, and TenHagen 1985). The methodology is based on dividing a generic computer-integrated manufacturing (CIM) system into functional subsystems and then developing functional flowcharts for each of the subsystems. The resultant methodology thus attempts to provide generalized flowcharts for a CIM system. The effort to generalize CIM operations in detail and codify the nature of CIM systems is largely an exercise in final state planning, and thus is probably not particularly useful from this perspective. On the other hand, the flowcharts, if viewed as a descriptive method for supporting learning efforts, might help stimulate useful discussion and decision making.

A number of additional modeling methods for CIM systems are useful for quantitative subsystem study. The most useful have been implemented on computers to enhance their support of learning efforts. Chapter 3 introduces the system-environment simulation model provided with this text and used in examples and problems to illustrate key concepts. Chapter 5 introduces several commercially available computer-based models that can be used to support the design and implementation of CIM systems. These models are applied to illustrate a range of considerations that enter into CIM system design.

ASSIGNMENTS

The assignments in the text often require that students gather information regarding a specific manufacturing setting—by firsthand experience, by contacting a manufacturing site of interest and obtaining documentation on the setting, by talking with someone who has had detailed experience with a specific manufacturing setting, or by using available reference resources. In doing the assignments, students should identify several possible strategies for obtaining the desired information. This practice allows students to gain experience in applying the concepts developed in the text to specific areas of application and interest.

See Appendix A for suggestions regarding class presentations and written reports.

1-1. In order to achieve maximum educational insight from Chap. 1, it is helpful to stimulate class discussion of the concepts that have been included, develop critical reviews that probe the advantages and limitations of the concepts, and explore alternative ways for approaching the topic. Select one or several topics from the bold-faced headings that introduce sections and examples in this chapter. Come to class prepared to explain this topic, critically describing the text materials and suggesting ways the topic discussion might be strengthened.

1-2. Experience in comparing the closed and open system approaches to defining a problem can provide essential insights. By drawing on reference materials, personal communications, or personal experience, describe a specific manufacturing problem in each of these frameworks. Then discuss the ways in which using the closed system point of view might not appropriately address some aspects of the problem that are considered when using the open system approach. Discuss difficulties that might be associated with applying an open system problem-solving approach, as well as oversimplifications that would be experienced as part of the closed system approach.
(a) Prepare a class report in the format shown in Fig. 1-11. Be prepared to present your materials and to discuss the implications of your findings with other members of the class.
(b) After the class discussion, prepare a brief written report describing your findings and insights.

1-3. It is useful to obtain experience in developing system models for the types of complex settings that are encountered in manufacturing and related fields. Identify a manufacturing-oriented setting that will enable you to gather the information that is necessary to develop a preliminary system description. Visit this location and gather sufficient information to identify some of the key system elements and the relationships among them that determine system operation.
(a) Prepare a brief list of some of the main features of system behavior. Come to class prepared to display a system drawing and to discuss some of the basic features of the operations associated with the system. Also describe those aspects of the system for which you were not

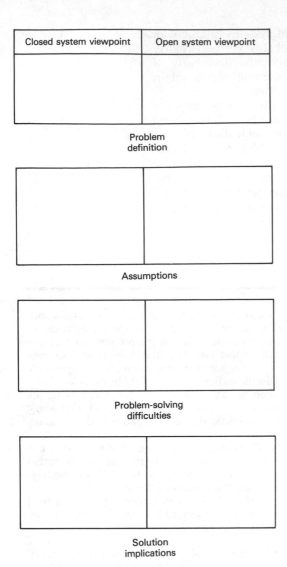

Closed system viewpoint	Open system viewpoint

Problem
definition

Assumptions

Problem-solving
difficulties

Solution
implications

Figure 1-11.
The format to be used for Assignment 1-2.

able to gather data and some of the difficulties in developing a complete system description.
(b) After class discussion, write a brief paper that includes your insights regarding the system you have explored, and describe how you might complete this project for a full system description.

1-4. Making use of the reference material authored by Forrester, study the complex simulations that he has developed. Discuss the nature of the system descriptions and the types of system behavior that he predicted as a result of his models. Consider both the strengths and weaknesses of his modeling approach. To what degree do you believe it would be reasonable to make real-life policy decisions based on this model? What boundaries would you place on the application of this model? What do you think are the major learning experiences that can be obtained from this effort by Forrester?
(a) Come to class prepared to present and discuss your findings.
(b) After class discussion, write a brief written report on your observations and insights.

1-5. Based on the references by von Bertalanffy and by Bowler, describe the nature of general system theory in some detail. Describe the rationale for general system concepts and the significance of the developments in this area. Develop a critique of general system theory as presented in the references. Would you feel comfortable with making real-life system decisions based on these concepts? What might you learn from these efforts?
(a) Come to class prepared to describe and discuss your findings.
(b) Following the class discussion, write a short paper describing your reaction to general system theory and its possible application for organization decision making.

1-6. Develop a description of a manufacturing system. Obtain references that describe the nature of such a manufacturing setting. Based on the references, describe how the manufacturing setting can be described in terms of physical models, functional models, and organizational models. Describe how these three models might relate to one another and the insights that might be developed as a result of these models.
(a) Be prepared to explain the manufacturing setting you have selected and to describe briefly the physical, functional, and organizational models you developed for this setting. Explain why your models take the form they have and some of the limitations that you have experienced in trying to develop these models.
(b) Following the discussion, write a brief paper describing the insights you have gained from this activity.

1-7. Obtain information about a manufacturing system in sufficient detail to describe it using the $IDEF_0$ methodology. Based on the information, give the functions and relationships of the system using this methodology. Develop several parent and child layers to indicate selected aspects of the manufacturing system. What conclusions can you draw regarding the utility of the $IDEF_0$ method? What strengths and weaknesses can you identify? Is the manual preparation of these models difficult, and would it be advantageous to have computer-supported modeling software (of the type discussed in Chap. 3)? Define the limitations associated with your application of this method.

(a) Come to class prepared to display your drawings and to discuss them.

(b) Following class discussion, write a short paper describing your experience with the $IDEF_0$ model and its use. Do you believe that this would be a useful method for describing more complex systems?

2

THE EVOLUTION OF MANUFACTURING SYSTEMS IN THE UNITED STATES

It is essential to gain an understanding of the system development strategies that have succeeded and failed in the past. However, it is difficult to draw conclusions from the past that will prove valid in a very different present. An effort must be made to extract from past experience an understanding of those general concepts and strategies that have succeeded previously and can help the managers of today continue to succeed.

In order to understand the present status of manufacturing systems in the United States, and develop insights into potentially useful planning strategies for the future, it is helpful to reexamine the past states of the system and the ways in which evolution has taken place. This chapter provides an historical overview of system environments that have existed for U.S. manufacturing and an evaluation of the problem-solving approaches that have been applied during different periods of national development.

Together, Chaps. 1 and 2 provide insight into the type of problem solving that must be employed in manufacturing system design and the past evolutionary states of the system to be considered. A strong relationship is thus established between the evolution of individual manufacturing systems and the evolution of the total environment or context in which these individual systems exist.

Insight into the present-day circumstances affecting U.S. industry can be gained by reviewing briefly the historical development that has led to the present situation. As noted by Abernathy, Clark, and Kantrow (1983),

> . . . to understand the critical, managerially relevant linkages among production competence, technology and competition, we need to know something about the general history of American manufacturing, as well as about the history of particular industries. (p. 31)

Since the nature of manufacturing clearly depends on the technologies in use, an historical review of manufacturing must also address the evolution of technology. As noted by Guile and Brooks (1987), "technological trajectories or life cycles" can be linked to industrial development, and therefore can help shape the focus and performance of manufacturing systems. The importance of technological change and productivity growth to the overall economic well-being of the United States has been emphasized by many groups, including the National Academy of Sciences (Cyert and Mowery 1987). If, as is often suggested (National Academy of Sciences 1987), U.S. dominance in the area of high technology is declining, then significant changes will occur in manufacturing system performance, as well as in relationships among countries.

The discussion in this chapter is brief, intended only to provide a framework in which current problems can be considered. The emphasis is on understanding the evolution of the U.S. industrial environment and perceiving how manufacturing system responses are linked to this evolution. To illustrate major changes in the dominant business paradigm, U.S. history has been subdivided into six periods, from colonialism (circa 1607) to the present (1990). It should not be interpreted that discontinuities occurred between periods, but rather that rapid (and sometimes cumulative) change led to evolutions in dominant paradigms from one period to another. The method used here for historical interpretation is only one of many possible choices. Other authors have suggested alternative ways for interpreting the evolution of U.S. manufacturing and the effect of technology on this evolution (Hawke 1988; Marcus and Segal 1989).

2-1. Applying Open System Theory to Manufacturing

Based on the concepts introduced in Chap. 1, a modern manufacturing enterprise can be considered as a complex system that consists of many subsystems interacting in multiple feedback loops (Fig. 2-1). The system is part of a larger environment that affects all aspects of system performance. Many different parameters can be used to describe the system and system-environment interactions. As discussed in Chap. 1, alternative approaches to system definition can be based on descriptions of the *physical, functional,* and *organizational* aspects of the system. The various periods of history and the associated paradigms can then be approached as shown in Fig. 2-2.

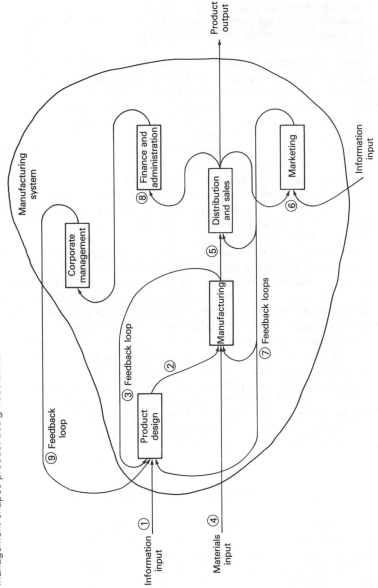

Figure 2-1. An extension of the open system concepts introduced in Chap. 1 (Fig. 1-2) indicating some of the elements and interactions in a simplified manufacturing system. (1) Product design is based on externally obtained information (providing a knowledge base). (2) Product design drives manufacturing efforts. (3) Manufacturing capabilities constrain and shape product design. (4) Manufacturing requires materials input. (5) Manufacturing output is delivered to the environment through distribution and sales. (6) Marketing gathers internal and external information. (7) Marketing feedback helps shape product design, manufacturing, and sales. (8) Sales results are fed through finance to corporate management. (9) Corporate management shapes product design decisions.

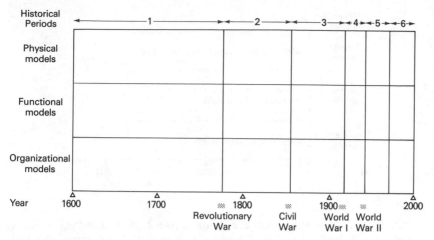

Figure 2-2.
A framework for
describing manu-
facturing systems and
system-environment
interactions during six
periods in U.S. history.

The following sections view the historical evolution of U.S. manufacturing from a system perspective. In each section, the chart in Fig. 2-2 is used as a method of summary and comparison. U.S. Department of Commerce data (1975, 1987) are drawn upon to help track this evolutionary growth.

2-2. Manufacturing System-Environment Relationships from Colonial Settlement to the Revolutionary War

The settlement of the North American continent was fed by many factors, including fundamental changes in European economics (Fig. 2-3). A growing commercial revolution, caused by increases in population and growth in market-oriented economic activity, resulted in a

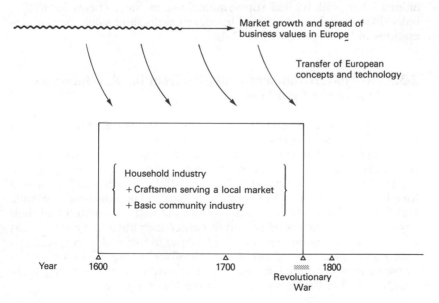

Figure 2-3.
Manufacturing during
the colonial years,
showing the transfer of
European concepts
and technology to the
colonies.

Figure 2-4.
System-manufacturing characteristics during the colonial years, described in terms of a framework that relates to physical, functional, and organizational models.

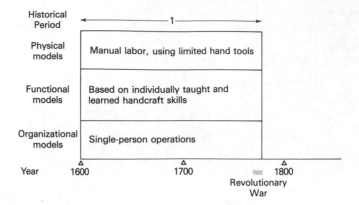

shift from a bare subsistence economy to a higher standard of living and growth in the need for the production and distribution of goods (Ratner, Soltow, and Sylla 1979). Trade grew rapidly, and colonial development was seen as a method for enhancing the resources available to the colonizing countries, companies, and individuals.

As settlements in the new world were established and began to expand, household fabrication of goods for personal consumption was the dominant form of manufacturing. Commercialized industry was gradually initiated to support community needs, and sawmills and gristmills became more common. Further expansion of manufacturing then extended to textiles, shipbuilding, and metalworking (Groner 1972). The type of manufacturing was determined by the environment in which the settlers existed and their needs for survival.

To a large degree, the technologies were imported from Europe, adapted as necessary to meet colonial needs. From the forming of the colonies to the Revolutionary War, the dominant manufacturing paradigm in both Europe and North America was based on human and animal labor (with limited supplemental use of waterwheels for mills), individually taught and learned handcraft skills, and single-person operations in local markets (Fig. 2-4).

2-3. System-Environment Evolution from the Revolutionary War to the Civil War

From 1750 to the end of the century, an industrial revolution took place in England. This era of change was marked by "a period of stable governments, a confident middle class, wars fought by professionals with little damage to the economy and, above all, of expanding European trade . . ." (Derry and Williams 1961). In this setting began a long history of effort that eventually culminated in industrially feasible technology. The inventions often associated with this period had their beginnings in an extended period of earlier learning and were the start of a long process of improvement. The dates associated with inventions thus represent a time of rapid growth of technology. Significant improvements were noted in the textile industry, in ironworking technologies, and in the development of the steam engine.

One of the most important activities during this period was the effort to find a source of power to replace human and animal labor. As early as 1690–1712, the combination of scientific study and trial-and-error learning had resulted in the first useful steam engine. Over the following half-century, improvements in materials, manufacturing technology, and skilled labor resulted in continuing improvements in steam engine performance. In 1765, James Watt added the condenser to the engine configuration that had evolved in England, and in 1776 his first two engines were set to work (Derry and Williams 1961). By the time of the Revolutionary War, steam power was close to becoming commercially available, and manufacturing methods for fundamental goods (such as textiles and iron) were evolving rapidly.

After the Revolutionary War (1775–83) and the Constitutional Convention (1787), the environment and perceived needs of the new country began to change. In 1791, Alexander Hamilton wrote: "It seems not always to be recollected that nations who have neither mines nor manufactures can only obtain the manufactured articles of which they stand in need, by an exchange of the products of their soils . . ." (Groner 1972, p. 50). The importance of manufacturing was thus recognized.

In addition, the concept of the division of labor to enable mass production was already being proposed, with Adam Smith as one of

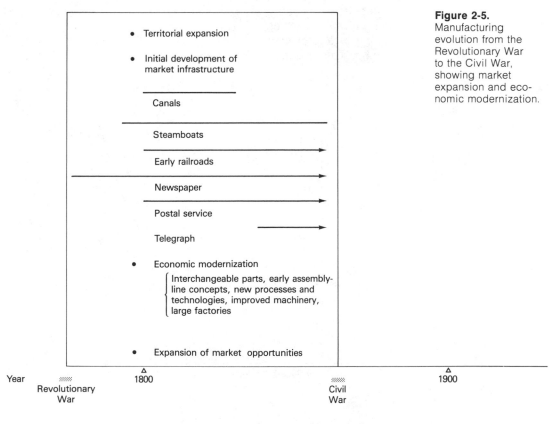

Figure 2-5. Manufacturing evolution from the Revolutionary War to the Civil War, showing market expansion and economic modernization.

the influential contributors to this business paradigm. In *The Wealth of Nations* (published in 1776), Smith argued that the standard of living of any society would grow rapidly with manufacturing based on the division of labor. As he explained, the efficiencies associated with having each worker perform a specific task, with high productivity, would result in lowered manufacturing costs and higher levels of goods for all:

> It is the great multiplication of the productions of all the different arts, in consequence of the division of labour, which occasions, in a well-governed society, that universal opulence which extends itself to the lowest ranks of the people. (p. 8)

Adam Smith thus helped provide the rationale for a shift from craftsman-oriented manufacturing systems, with individual responsibility for an entire product, to mass production–oriented systems, with individual responsibility for only a small part of a product. The growth of factories and mass production in the United States was strongly influenced by his work.

The United States began to develop the infrastructure required for market and industrial growth (Fig. 2-5). New means of transportation and communication were central to this evolution. The use of canals gave way to steamboats and then to railroads, which became the core of the transportation system. The telegraph provided a matching capability in communication, resulting in the networking of the U.S. population into a market.

Figures 2-6 through 2-8 illustrate some of the changes in the national infrastructure that were taking place during this period. Figure

Figure 2-6.
U.S. population, 1790–1880 (U.S. Department of Commerce).

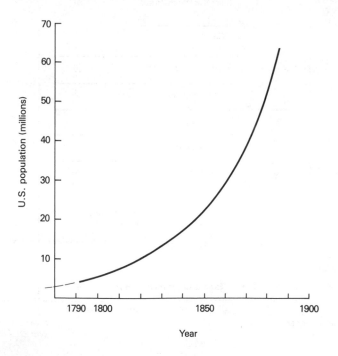

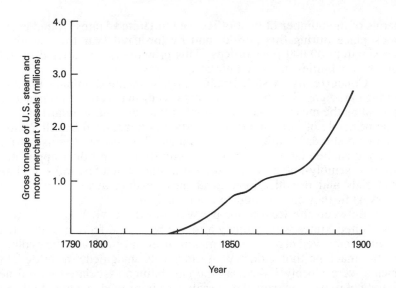

Figure 2-7.
Gross tonnage of U.S. steam and motor merchant vessels, 1790–1900. Note the temporary change in growth rate caused by the Civil War (U.S. Department of Commerce).

2-6 shows how the U.S. population grew from the Revolutionary War to the Civil War—by approximately a factor of 10, from about 4 million to about 40 million. This increase in population made possible a broadened, more complex economic structure and expanded market opportunities.

Figure 2-7 indicates how the number of steam and motor merchant vessels increased during this same period. Prior to 1830, there were few vessels in this category, but by the Civil War a rapid growth had begun, providing new transportation opportunities. This type of growth was paralleled by the other transportation areas discussed earlier.

Figure 2-8 illustrates an aspect of communications growth in

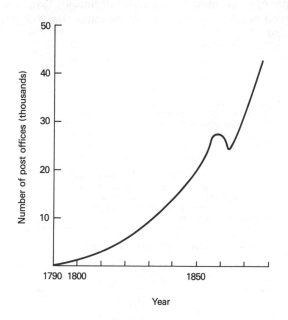

Figure 2-8.
Number of U.S. post offices, 1790–1880. Observe the drop in the number of post offices associated with the Civil War (U.S. Department of Commerce).

terms of the number of post offices in the United States. Rapid growth took place during this period, and by the Civil War there were approximately 30,000 post offices. This growth was paralleled by telegraph capabilities, as noted earlier.

Concurrently, a shift began to occur in the dominant manufacturing paradigm. Steam-driven machines began to replace human labor, providing the means to increase productivity and the standard of living. The principle of interchangeable parts was introduced by Eli Whitney to create standardization in gun making. This allowed the use of relatively unskilled labor and increased control over product application. Early assembly-line concepts were introduced in a few mills, and new materials and organizational strategies (leading away from the solo worker) further contributed to the paradigm.

Between the Revolutionary War and the Civil War, industry in the United States was strongly affected by the perfecting of the steam engine, improved processes for making iron and steel, and the evolution of the machine-tool industry. As might be imagined, these developments were closely linked, as the capabilities associated with a new source of power, improved materials, and improved machine tools contributed to a resultant advance in manufacturing technology. The use of electricity began to enter into the development of a communications infrastructure. With reliable batteries came telegraph technology, the widespread availability of telegraph offices, and a rapid communications capability.

During the first half of the nineteenth century, manufacturing became the fastest growing segment of the U.S. economy, paced by the iron industry and a rapidly growing array of products. The introduction of new processes and technologies was combined with large factories, a new type of business leadership for these factories, and a national interest in inventions (Groner 1972). Contributing to this growth were the concepts of interchangeable parts, continuous flow manufacturing, and improved machinery. Figure 2-9 indicates the dominant paradigm of this period.

During this period, U.S. manufacturing was characterized by the rapid exploration of ideas. A new country and new technologies required experimentation and learning for effective growth. Lacking established concepts, there was a willingness to explore, learn, and change.

Figure 2-9.
System-manufacturing characteristics from the Revolutionary War to the Civil War, applying the physical, functional, and organizational model framework.

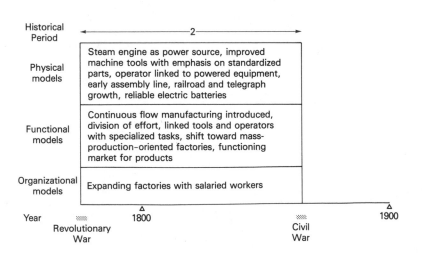

Americans were used to the idea of change leading to betterment, and thus largely endorsed these changes (Groner 1972; Lebergott 1984). There was a materialistic interest in using new products to improve the standard of living (Groner 1972), so a market developed once the manufacturing means and infrastructure were present. However, worker resistance to factory employment became apparent during these early formative years (Clark, Hayes, and Lorenz 1985). The breakup of social relationships, externally imposed discipline, and the required subordination of self were not easily accommodated. Thus, the dominant acceptance of change for new technologies and products did not easily extend to the new workplace.

EXAMPLE 2-1. INFRASTRUCTURE GROWTH

Discuss how the growth of the U.S. population, transportation, and communication during the period from the Revolutionary War to the Civil War likely affected the growth of U.S. manufacturing.

Results

The rapid growth of the population, as illustrated in Fig. 2-6, provided an expanded work force and a growing market to support the development of manufacturing systems, as well as the potential for developing a new type of manufacturing economy. Paralleling population growth were improvements in the transportation and communications systems that provided the means for the population to be integrated into a more cohesive manufacturing environment.

The new types of transportation (see Fig. 2-7) provided the means for the increased population to become more mobile and for meeting the basic transportation needs of commerce. Complex manufacturing systems could not be developed unless materials and products could be conveniently moved from the plants to the various markets. This rapid growth in the transportation infrastructure provided a further stimulus.

Another element necessary for a growing manufacturing environment is the ability for all aspects of the economic system to communicate with one another in order to exchange information necessary for manufacturing and marketing. Figure 2-8 illustrates one aspect of the typical growth in communication systems taking place during this period.

2-4. System-Environment Evolution from the Civil War to World War I

During the period between the Civil War (1861–65) and World War I (1914–18), improved technology brought significant paradigm changes to U.S. industry (Fig. 2-10). Rapid changes occurred in metallurgy, power engines, machine tools, managerial enterprise, and mass production. Electricity was further harnessed for communication by means of national telegraph and telephone networks and for indus-

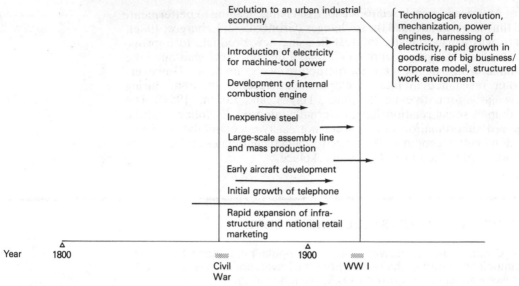

Evolution to an urban industrial economy

Technological revolution, mechanization, power engines, harnessing of electricity, rapid growth in goods, rise of big business/corporate model, structured work environment

Introduction of electricity for machine-tool power

Development of internal combustion engine

Inexpensive steel

Large-scale assembly line and mass production

Early aircraft development

Initial growth of telephone

Rapid expansion of infrastructure and national retail marketing

Year   1800   1900

                Civil   WW I
                War

Figure 2-10.
Manufacturing changes during the period between the Civil War and World War I, describing evolution toward an urban industrial economy.

trial use with the introduction of dynamos and power distribution systems, the incandescent light bulb, and electric motors.

Electric-powered machine tools began to produce rapid change in factory design and capability. The development of the internal combustion engine took place in the decades before and after 1900, and led to the mass-produced automobile. The combination of new power

Figure 2-11.
U.S. patents issued, 1790–1890. The growth in number of patents issued indicates the rapid changes in technology taking place (U.S. Department of Commerce).

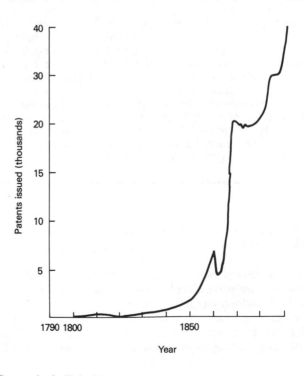

sources, new materials, improved process technology, and advanced machine tools was the basis for a new type of production. National and world markets were formed using extensive railroad and communications capacities combined with multinational shipping. The establishment of major steel and auto industries accompanied continuing growth of the national infrastructure.

Figures 2-11 through 2-13 provide further insight into the infrastructure growth during this period. The networking capabilities and new products began to interact to create a stronger economic framework. Figure 2-11 shows the rapid growth in the number of patents issued between the Civil War and World War I. Such growth indicates the introduction of new products and the application of new technologies in the manufacturing system.

Figure 2-12 shows the railroad mileage during this interval. By

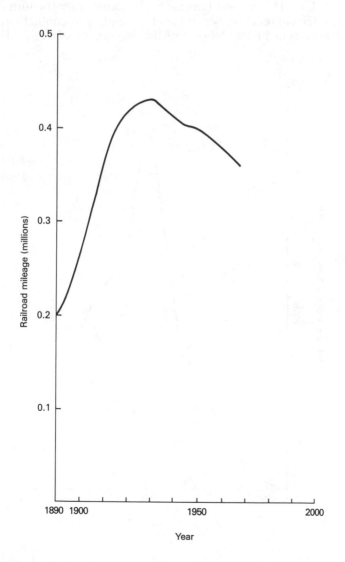

Figure 2-12.
U.S. railroad mileage, 1890–1970. The rapid growth of railroad mileage continued until the early 1900s, then began a steady decline (U.S. Department of Commerce).

World War I, the total mileage of railroad track was nearing its maximum. This rapid growth in railroad transport provided the key means for manufacturing systems to obtain the raw materials and input products required and to distribute the output products to market.

Figure 2-13 illustrates how the number of post offices continued to grow in this period, reaching an all-time peak around the turn of the century. Other areas of communication grew as rapidly during the same time, so that a combination of technology, transportation, and communication was in place to stimulate improved manufacturing systems.

The early part of this period was associated with uninhibited economic exploitation and competition. Herbert Spencer contributed to the dominant paradigm by introducing "social Darwinism" as an argument against government regulation and unionism (Groner 1972). Continuing pressure was placed on workers to fit into the factory-based industry (Clark, Hayes, and Lorenz 1985). Later, after the turn of the century, a reform reaction began to set in, leading to conflict over the roles of management and labor and the emergence of national labor unions.

Figure 2-13.
Number of U.S. post offices, 1790–1980. A rapid drop in the number of post offices was associated with improved transportation systems and increased mobility (U.S. Department of Commerce).

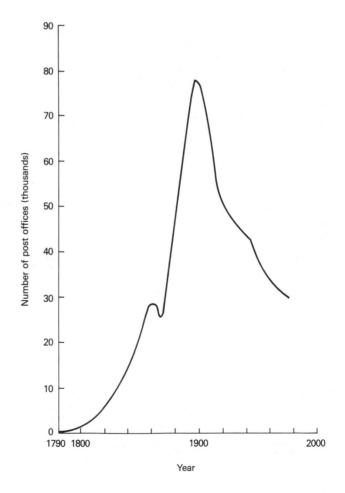

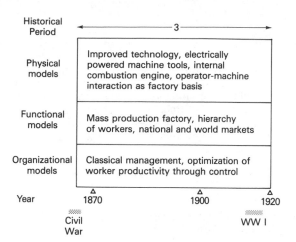

Figure 2-14.
System-manufacturing characteristics from the Civil War to World War I, applying the physical, functional, and organizational models.

As already discussed, Adam Smith's concepts about the division of labor strongly affected manufacturing in the United States. Also highly influential in the school of classical management were the writings and speeches of Frederick W. Taylor (1916). He introduced the idea of *scientific management,* based on having each worker optimize productivity by most effectively applying the principle of the *division of work.* The time and motion study was proposed as a way to study scientifically each worker action and to teach the worker how to perform the assigned task most efficiently.

According to Taylor, all society would benefit from such efforts. The increased earnings from higher productivity would be shared by management and labor, improving both profits and wages. His was an argument that the "larger productivity pie" would benefit all. He called for a "mental revolution" (or paradigm shift) that would lead both employers and workers to see the advantages of such a system.

Taylor provided an in-depth rationale for speeding up the mass production processes of the day. He saw that higher levels of productivity could be obtained if workers would only perform more efficiently. From his perspective, the result would be beneficial to everyone. The continuing management paradigm of this period thus emphasized survival of the fittest and the use of workers as specialized machines, optimized for the job. Workers' resistance to such a framework was assumed to be shortsighted, not in their best interest, and subject to change through pressure and education regarding system advantages. Figure 2-14 summarizes the paradigm of this period.

EXAMPLE 2-2. COMPARISON OF GROWTH STAGES

Contrast the growth graphs in Figs. 2-6 to 2-8 with those in Figs. 2-11 to 2-13, and discuss the relative effects on U.S. manufacturing that might reasonably be associated with these different periods.

Results

Figures 2-6 through 2-8 show the initial growth of the infrastructure necessary to support an advanced manufacturing economy (between the Revolutionary War and the Civil War). These and many other types of data from this period illustrate the changes in the fundamental nature of the economy. The graphs represent the introduction of new ideas and new capabilities.

Figures 2-11 through 2-13 show the continuing expansion in related infrastructure areas between the Civil War and World War I. During this period, the emphasis was on taking maximum advantage of the new paradigms to achieve national growth.

2-5. System-Environment Evolution from World War I through World War II

The United States became a world industrial power between World War I (1914–18) and World War II (1939–45). A paradigm shift brought recognition of research and development (R&D) and science-based industry as potential driving forces for economic growth (Fig. 2-15). The resultant changes in energy sources, materials, chemistry, and electronics established a new industrial framework. Matching these expanded production capabilities, increased sophistication in retailing brought products to consumers more efficiently.

Figures 2-16 through 2-19 indicate some of the important trends between the world wars. As noted in Fig. 2-16, there was a rapid change in the transportation system in the increased amount of surfaced roads and streets. At the time of World War I, there were many miles of roads and streets in the United States, but most of them were not

Figure 2-15.
Manufacturing evolution from World War I through World War II, showing the rise of the United States as a world industrial power.

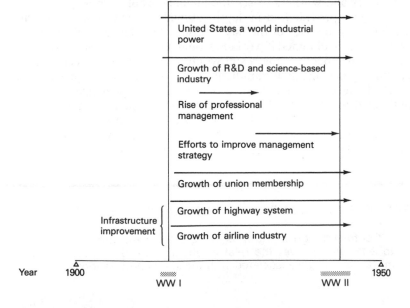

surfaced. By the end of World War II, the majority of roads and streets were surfaced.

This shift was driven by the growth of automobile production. At the same time, as may be noted in Fig. 2-12, growth in automobile transportation began to reduce the use of railroads, and railroad mileage began to decrease. Further, as shown in Fig. 2-13, a rapid drop in the number of post offices occurred during this period. With the advent of the automobile and surfaced roads and streets, a single post office could service a larger area, so fewer post offices were required.

Figure 2-17 illustrates a basic technology shift in terms of the application of electric energy to the economy. Between World War I and World War II, the foundation was prepared for growth in electric energy consumption. The means for rapid change were in place by the time World War II was over, and the resulting growth is shown.

Figure 2-18 shows the steady growth in number of telephones between the two world wars. By the end of World War II, the economy was poised for even further expansion. Accompanying this change was the initial growth in the application of electronics to commercial radio and television. Figure 2-19 shows the number of radio sets produced during this period and the number of households with radio sets. A similar increase in communications took place with the introduction

Formal research and development (R&D) activities provided a new strategy for harnessing technology to drive economic progress. Success would depend on maintaining a spirit of innovation and change within a structured environment. Failure would result if the formal approach stifled the learning focus.

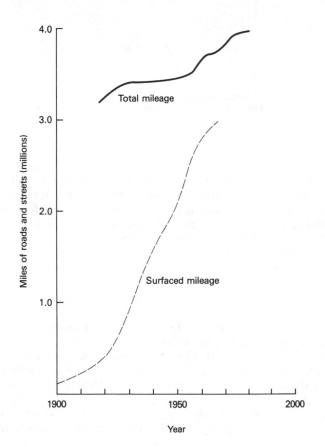

Figure 2-16.
Miles of U.S. roads and streets surfaced and total, 1900–80. A rapid shift from unsurfaced to surfaced mileage took place during this period (U.S. Department of Commerce).

Figure 2-17.
Use of electric energy
in the United States,
1900–70. The explo-
sive increase in elec-
tric energy consump-
tion indicates the
availability of new
technical products
(U.S. Department of
Commerce).

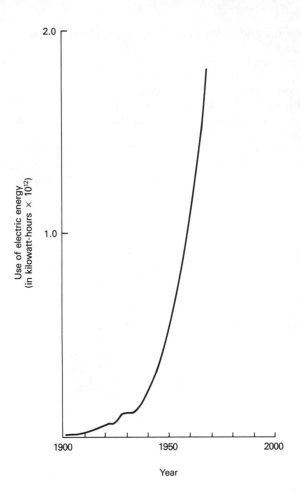

of the television set immediately following World War II, with a con-comitant increase in the number of households with television sets.

The typical data shown in Figs. 2-16 through 2-19 illustrate how the combination of technological change and the development of a transportation and communication infrastructure combined to produce a rapid shift from one set of paradigms to another as the United States grew in sophistication of the technology used and developed a more complex manufacturing system.

This rapid growth, with the application of science and technology as the driving force, led to the exploration of alternative management strategies. Worker resistance to mass production factories remained a problem, leading to efforts to better understand the situation and how to deal with it. By the 1920–30 time period, a new management paradigm began to emerge from the work of several research groups. Worker resistance to the mass production setting characterized by a division of labor and optimized through time and motion studies caused investigators to look more closely at the workplace in order to both understand what was happening and to find improved ways to increase

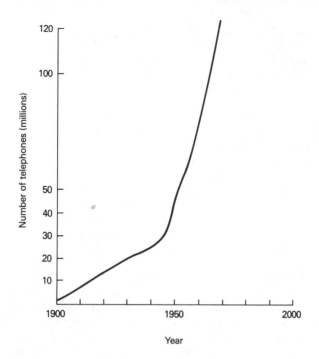

Figure 2-18.
Number of telephones in the United States, 1900–70. A rapid increase in the number of telephones took place after World War II (U.S. Department of Commerce).

productivity. Much of this effort was concerned with worker motivation.

A series of famous experiments was conducted by F. J. Roethlisberger and others (1941) at the Hawthorne plant of the Western Electric Company to identify those elements of the work environment that

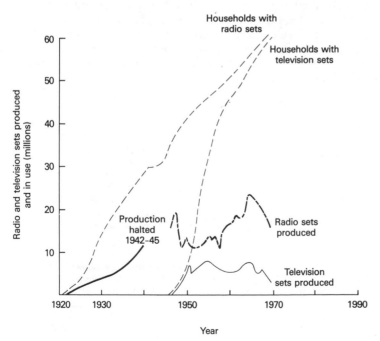

Figure 2-19.
Radio and television sets produced and used in the United States, 1920–70. The number of households with radio and television sets grew rapidly as soon as the technology became available. Levels of production generally increased to satisfy this demand. However, during World War II, commercial radio set production was halted (U.S. Department of Commerce).

had the most influence on worker productivity. Studies were performed of worker production as a function of technical factors (such as lighting levels) and social factors (such as team grouping).

The experimenters concluded that social factors dominated productivity, and that "the worker is a social animal and should be treated as such." They noted that all people want to have a skill that is "socially recognized as useful" and that "most of us want the satisfaction that comes from being accepted and recognized as people of worth. . . ." They concluded that a new management approach was required to better motivate workers. It should be noted that subsequent reviews of the Hawthorne experiment data have suggested that other economic factors entered into the studies and may have affected the results (Carey 1967). However, during the last half of the twentieth century, the original descriptions of these experiments have created an interest in improving worker production by changing social factors.

In the 1940s, Maslow (1943) extended this concept to explore a "hierarchy of human needs" that could be used to understand motivation. He identified those needs as physiological, safety, love, affection and acceptance, esteem, and self-actualization. Once a given need is satisfied, motivation rises to the next level; for example, once the basic physiological and safety needs are met, satisfaction of the needs for acceptance, esteem, and self-actualization will dominate motivation. This theory further contributed to concepts regarding social incentives as a link to productivity.

The management activities of this period included R&D related to people as integral elements of manufacturing systems. A focus on the control of workers led to the definition of structured, monitored work environments in which learning and new procedures were unwanted side effects.

As social motivation studies continued into the World War II period, McGregor (1957) identified the classic control-oriented management style as being associated with a "Theory X" view of workers (people must be forced to work and be closely supervised) and the new self-actualization management style as being associated with a "Theory Y" view of workers (people want to work if given a supportive setting). Courses were developed to teach managers how to master Theory Y concepts and to blend the two styles in an optimum way.

Figure 2-20 illustrates how the business approaches of this period fit into the evolutionary view of history.

Figure 2-20.
System-manufacturing characteristics of the United States from World War I to World War II, showing the growth of new business foundations and concepts.

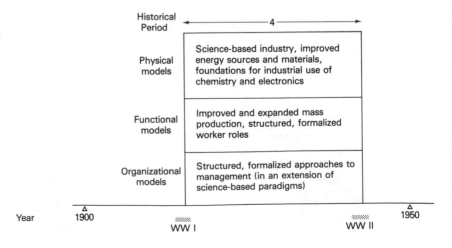

EXAMPLE 2-3. ALTERNATIVE PARADIGMS

A manual, operator-based production system is achieving much less efficiency than expected. System managers review the situation and determine that the reason is inadequate productivity. Discuss how management might address this problem based on the works of (1) Smith, (2) Taylor, (3) Roethlisberger, (4) Maslow, and (5) McGregor.

Results

A brief look at this problem from several different viewpoints can be revealing in terms of the alternative paradigms available to managers. Adam Smith was an advocate of the division of labor. Thus, he might view a manufacturing system with inadequate productivity in terms of the need for further subdivision of labor in order to improve the efficiency of the overall manufacturing process. This point of view emphasizes shifting toward a mass production–oriented system, with each individual having responsibility for only a small part of the production process.

Frederick W. Taylor might be inclined to apply time and motion studies to each worker action in order to teach the worker how to perform the assigned tasks most efficiently. From his point of view, inadequate productivity is probably due to the inadequate training of workers to achieve their maximum potential. From Taylor's perspective, an emphasis on individual performance is good for both the worker and the management system. The concepts of Smith and Taylor fit together smoothly in terms of a management problem-solving approach.

A quite different approach is associated with the work of F. J. Roethlisberger and the Hawthorne experiments. As generally described in literature, the orientation would be toward modifying social factors to improve productivity. Thus, the inclination would be to engage in social change to produce an environment in which the workers feel accepted and socially rewarded, with an accompanying increase in productivity.

In a similar vein, Abraham Maslow would likely try to determine where on the hierarchy of needs the workers are presently being motivated and would seek to move the workers toward self-actualization. This involves creating an environment in which the workers could fulfill themselves to achieve higher levels of satisfaction and productivity.

Douglas McGregor might say that the best approach would be a synthesis of the Theory X approach (associated with Smith and Taylor) and the Theory Y approach (associated with Roethlisberger and Maslow). Thus, for McGregor, it would be best to teach managers how to synthesize these concepts in an optimum way to achieve the desired level of productivity.

Note that for each of these approaches, the emphasis is on motivating the worker to achieve the maximum productivity, not on improving the technology that the worker is using. These solutions thus represent only one pathway toward improving manufacturing system performance.

2-6. The Post–World War II Years

At the end of World War II, American industry dominated the world. Given manufacturing superiority, management's attention turned to optimizing performance through financial and marketing

Figure 2-21.
Manufacturing evolution after World War II, describing the rapid growth of a technology-based economy.

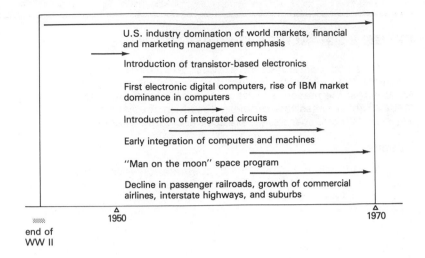

U.S. industry domination of world markets, financial and marketing management emphasis

Introduction of transistor-based electronics

First electronic digital computers, rise of IBM market dominance in computers

Introduction of integrated circuits

Early integration of computers and machines

"Man on the moon" space program

Decline in passenger railroads, growth of commercial airlines, interstate highways, and suburbs

1950   1970

end of
WW II

means (Fig. 2-21). Manufacturing moved further down the organization and received less management focus. The factory came to be seen as a financial resource, to be controlled and applied in support of larger (financial and marketing) considerations. Manufacturing was carefully monitored and directed to serve the larger corporate purposes.

During the 1950s and 1960s, American industrial and management know-how became the envy of the world. It has been suggested that 1967 was the peak year for admiration of U.S. strategy (Hayes and Wheelwright 1984). U.S. industry developed a sense of "having arrived" (Abernathy, Clark, and Kantrow 1983). The system paradigm for this period has been well described by Clark, Hayes, and Lorenz (1985):

> Technological change is assumed to originate outside the company and is available to all competitors on an equal footing. No long-term commitments are required; investments are reversible, and the factors of production—including human skills and relationships—can be acquired (and divested) through market transactions. In this environment the firm's only problem is to choose and assemble the combination of factors hat will produce the right level of output given the prices it faces. (p. 2)

U.S. manufacturing enterprises viewed themselves as having reached the maximum possible level of success. The objective of management was to control, stabilize, and exploit. Business had "grown up" and had all of the answers; any further learning and change could only reduce this optimal situation.

Chandler (1977) has discussed the effects of this type of enterprise on the organization structure and market environment. As he suggested, business enterprises of the time could be seen as having many distinct operating units and being managed by a hierarchy of salaried executives, in contrast with the traditional single-unit business enterprise. In this modern enterprise, the processes of production and distribution became combined in complex organizations that took over many coordination and integration functions from the market. Chandler concluded that the transition to the modern business enterprise occurred when the new configuration became the most effective way to deal with the market. Thus, he saw the enterprises of this period as an appropriate response to the environment they faced.

This was the prevailing view of industry for 20 years following World War II. United States business was internationally successful, and it seemed inevitable that the situation would continue indefinitely. Industry had finally achieved its peak design and strategy.

Beneath this dominant management strategy, however, significant changes in technology were taking place. New materials and processes, more sophisticated machinery, and rapid improvements in electronics contributed to manufacturing capabilities and options.

The transistor, invented and marketed in the late 1940s, was followed by integrated circuits in the 1950s. The first electronic digital computers became available in the early 1950s, and by the late 1950s and early 1960s International Business Machines (IBM) was growing toward market dominance. By the late 1960s, the large mainframe computer was well established as an essential element of the business enterprise.

The first steps were also being taken to link manufacturing machines and computers. A numerically controlled (NC) milling machine was created in 1952, with commercial applications following shortly thereafter. The NC machine used a punched-tape input and an electronic control system to drive mechanical operations. By 1967, direct numerical control (DNC) of computer numerically controlled (CNC) machines became a reality. A central computer provided control messages to a production machine, and the production machine provided status information to the central computer (Harrington 1973). This growing technical capability was employed by management only so long as it served the financial strategy of the corporation.

The 1960s also became the era of *final state* planning. Based on the economic boom and world leadership, both public and private sectors turned their attention to meshing organizational objectives, systems, and the environment in a well-understood planning or policy-making activity that provided the ultimate in directed control. Computer data bases and modeling were introduced in large-scale applications, increasing the sense of power and success. So long as the environment remained benign, this situation was able to continue.

Figures 2-22 through 2-25 provide additional information on the nature of the manufacturing environment that existed during these years. As Fig. 2-22 shows, the U.S. population continued to grow rapidly, reaching 115 million people in the post–World War II years.

Figure 2-23 illustrates one aspect of the major shifts in transportation. Rapid growth in commercial air transportation began immediately after World War II and continued over the next two decades. A comparison of these results with those in Fig. 2-12 shows that the growth in air travel was accompanied by a continuing decrease in the amount of railroad track mileage. As noted in Fig. 2-18, the use of telephones grew rapidly as communication requirements became more demanding. The intensity of the interaction required among the elements of the manufacturing system continued to grow in this period; air transportation and telephones provided an essential evolutionary step in infrastructure evolution necessary to achieve the desired level of systems operations.

Figure 2-24 shows the large growth in R&D funds that began in the post–World War II years. The United States began to turn toward manufacturing systems driven by both technological means of production and focused on new products based on technology. Figure 2-25 shows how the resulting infrastructure produced a rapid expansion in retail sales in the post–World War II years. Between 1950 and 1970, major changes occurred in the nature and scope of retail sales, providing a basis for the rapid sales growth in the following decades.

A few voices were raised in concern, but these were treated more as points of interest rather than issues of serious concern. Simon's studies (1946) pointed out that there are finite limits to human abilities to gather and assimilate information and to make decisions. His principle of "bounded rationality" suggested limits to the macro-planning optimization studies being conducted by business. He also showed clearly that many of the "laws" of administration were really "proverbs," reflecting current world perceptions. He thus challenged the "invariant right" way to management success.

Figure 2-22.
U.S. population growth, 1800–1975, demonstrating the continued exponential growth in population throughout this period (U.S. Department of Commerce).

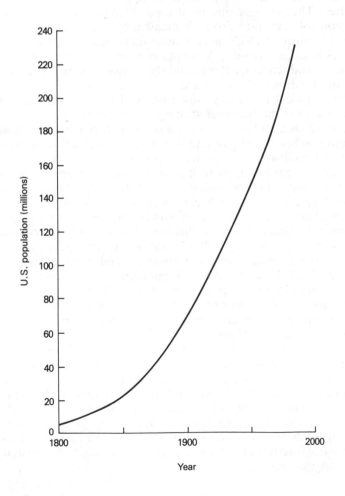

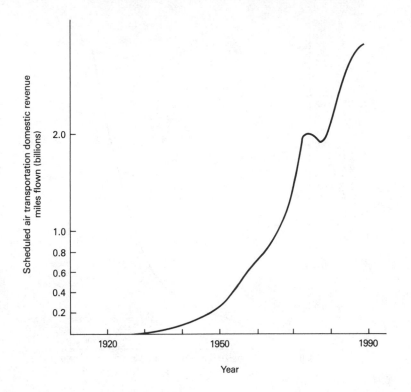

Figure 2-23.
U.S. scheduled air transportation, 1920–85. The growth in air transportation shown here and the growth in surfaced roads and streets shown in Fig. 2-16 may be linked to the decline in U.S. railroad mileage shown in Fig. 2-12. (U.S. Department of Commerce).

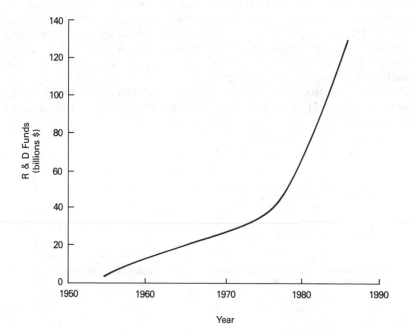

Figure 2-24.
U.S. research and development funds 1955–85. The impact of rapidly growing new technologies is shown in the increased growth rate of R&D funds over this period (U.S. Department of Commerce).

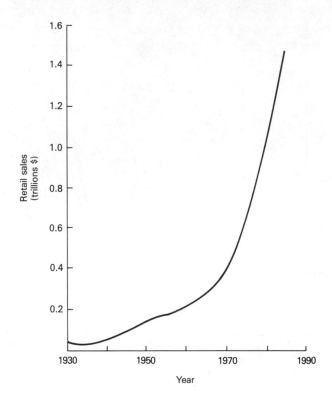

Figure 2-25.
U.S. retail sales, 1930–85, showing growth of the retail market (U.S. Department of Commerce).

Simultaneously, Lindblom raised another challenge to final state planning (1959). He noted that such efforts assume "intellectual capabilities and sources of information that [people] simply do not possess." He recommended an alternative strategy of "muddling through," suggesting that administration (should and does) take place through a series of successive limited actions and evaluations. He thus challenged the ability to perform final state planning, but went to another extreme, that of muddling through. "Learning planning" (introduced in Chap. 1) might be regarded as a compromise between these two concepts.

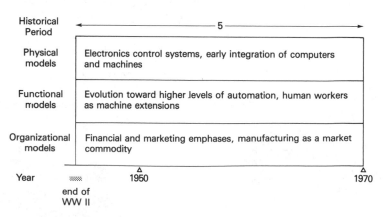

Figure 2-26.
System-manufacturing characteristics in the years after World War II, indicating the rapid growth of electronics and automation.

During the two decades, the international environment changed in many ways not perceived by U.S. industry (Fig. 2-26). The growing inappropriateness between U.S. internal management practices and the changing external environment was summarized by Hayes and Wheelwright (1984):

> Since World War II, the United States [has] taken for granted the superiority of its technology and management practices. This complacency (one might go so far as to call it arrogance) [has] tended to blind U.S. managers, at least until the late 1970s, to the rapid improvements taking place elsewhere. (p. 3)

EXAMPLE 2-4. CHANGES IN THE BUSINESS ENVIRONMENT, 1945–70

Discuss how changes in the U.S. business environment in the period 1945–70 might reasonably have affected manufacturing systems.

Results

In the post–World War II years, the U.S. business environment continued to evolve rapidly in terms of complexity and the intensity of the interactions taking place among the elements of the system. A rapid growth in population provided the means for achieving both manufacturing and market efficiencies. There was also a rapid introduction of new technologies, leading to new approaches in transportation and communication. Communication through radio and television began to mature, and growth in research and development programs provided the technological force for the evolving system (Fig. 2-21).

As a result, the market expanded rapidly (Fig. 2-25). The business environment continued to evolve toward higher levels of interaction and an increased level of networking to link all elements of the manufacturing system and the market.

2-7. The Period 1970–90

At the beginning of the 1970s, a dichotomy began to develop between the innovative technologies available for manufacturing and the use of such technologies by U.S. industry (Fig. 2-27).

In 1973, Harrington coined the phrase *computer-integrated manufacturing** as the logical direction for growth of the manufacturing enterprise (Shrensker 1985). In his book (1973) he described the potential associated with the introduction of high levels of automation to the factory, and took a strong advocacy position for corporate evolution in this direction. Harrington observed that the division of labor

---

* Harrington purposely avoided developing an acronym for this phrase. However, others were quick to do so, producing the abbreviation CIM (pronounced sim).

Figure 2-27.
Manufacturing evolution, 1970–90, associated with a rise in intense international competition and a reduction in U.S. industrial performance.

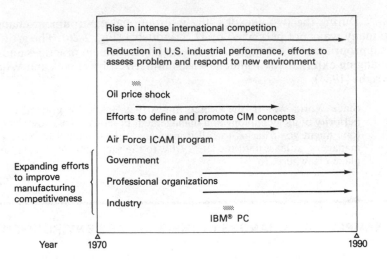

trend of the past might be reversed by the introduction of high levels of automated production, noting that

> the conventional organizational pattern of manufacturing has been evolving for centuries. Until recently it has been characterized by its growing compartmentalization, doubtless a result of the increasing size of organizations, specialization of skills, and formalized boundaries of authority and responsibility. This was not always so, and the future will probably bring more changes. There is some evidence that the wheel has come full turn, and that the organizational pattern of the future may embody some of the characteristics of earlier forms in which specialization and formal organization were less emphasized. An historical perspective will remind us that these patterns are not static and may forecast the direction of their future evolution. (p. 27)

Harrington also noted that optimization of a manufacturing system could not be achieved by maximizing the productivity, efficiency, and output of the various organizational sectors. He concluded that

> while management structure can be fragmented as much as one wishes, the manufacturing function carried out under that management remains a monolithic unit. Everything depends on everything else—"interdependence absolute" as Kipling once described it. If such a structure is to be optimized, it must be as a totality and not piece-by-piece. (p. 33)

Harrington thus argued against the highly fragmented structure of manufacturing organizations, which could only result in localized optimization, and urged a more system-oriented approach to operations. He saw that the introduction of computers to factories might have the potential for the creation of improved, more competitive systems. However, he also pointed out the resistance to change that should be expected. Regarding the current factory paradigm, he concluded that "The system seems well entrenched. . . . It is built into every corner of our industrial world" (p. 34). He thus saw the potential for both change and resistance, concluding that change would be inevitable and predicting the continuing growth of computer-integrated manufacturing.

Management response to the CIM opportunity developed only slowly, due to a variety of factors. If this new technology was to require a redesign of the entire organizational structure, numerous vested interests would be involved, and corporate management typically did not have the background to understand the technologies being used. The concept of a very expensive major corporate redesign around technology that was not well understood by management led to an understandable slowness in change. In fact, the imperative of technical evolution as seen by Harrington was not strong enough to overcome management resistance in most cases.

As the 1970s unfolded, however, the world began to change. Europe and Japan established a completely rebuilt industrial base and engaged in intense efforts to compete with the United States. New demands were placed on U.S. industries. Productivity increases and market share began to suffer, and U.S. management was no longer perceived as the role model for success.

Management in the United States was slow to understand the changing environment, and a "business as usual" attitude still prevailed for the most part. Upper management of the large corporations often knew little about technology or manufacturing and were preoccupied with the financial aspects of the enterprise. Little attention was paid to the broader implications of a static manufacturing capability.

In 1985, Clark, Hayes, and Lorenz (1985) noted: "A recognition that manufacturing prowess is central to the success of an industrial firm has, until recently, escaped many managers" (p. 139). This building pressure began to make itself felt by the start of the 1980s. Management was finally forced to begin considering that a new way of doing business might now be required. Prescriptive advice began to surface from many directions.

A variety of observers concluded that in the future, international success would require the use of production capability as a competitive weapon. No longer could the factory be treated as a static resource to be manipulated by other aspects of the corporation; rather, it would have to become an essential focus of competitive performance.

Several insightful books have helped define the situation. The essential problem has been diagnosed in several ways:

> The pursuit of . . . modern management approaches has led American manufacturing companies to . . . emphasize analytical detachment and strategic elegance . . . , focus on short-term results . . . , and emphasize the management of marketing and financial resources. (Hayes and Wheelwright 1984, p. 8)

> Our starting point in the discussion that follows is a single harsh but inescapable fact: the nation's lackluster industrial performance in recent years (the early 1980s) is, in part, the result of failure of many of its traditional manufacturing industries to adjust to a troubling new set of competitive realities. (Abernathy, Clark, and Kantrow 1983, p. xi)

> If managers continue to view a production system as a limited, neutral piece of apparatus, which is to be blindly run according to fixed design

By this time, management emphasis on maximizing the productivity of stabilized systems had led to widespread resistance to change. The learning of new and improved ways to operate had been actively discouraged in favor of maximized short-term productivity. Organizations emphasized the financial performance of component business units, without attempting system-level improvements.

specifications and used by rote to meet budgeted cost objectives and volume levels, they will enjoy no greater benefit from the new technology than if they had mistaken it for a coat rack. (Abernathy, Clark, and Kantrow 1983, p. 124)

Clark, Hayes, and Lorenz (1985) further noted:

Just as economic research has failed to unravel the [declining] productivity enigma, so has most recent managerial research. As a result, both managers and managerial theorists are beginning to recognize that their conventional "business strategy" and "corporate portfolio" frameworks are unable to explain satisfactorily the dynamics of productivity and competitive behavior at the level of the operating manager. (p. 1)

In 1983, Abernathy, Clark, and Kantrow emphasized the importance of industry learning processes to achieve a solution strategy:

Thus, only when grafted onto a production system dedicated to on-going learning and communication, only when used in tandem with a skilled and responsible work force, can new technologies realize their potential as competitive weapons. Only when such a work force is truly engaged in the enterprise and encouraged to learn and excel, can a company hope to introduce competitively new products in a timely fashion. Innovation is no substitute for competent work-force management or for a mastery of production. The future does not belong to firms that try to make up for poor work-force management or sloppy plants with cutting-edge technology. Nor does it belong to firms that do not bother with new advances in technology because their factories are well run by present standards. The future belongs only to those firms—and managers—who eagerly seize opportunities on both fronts. (p. 125)

Summarized in another way:

The solution to improving one's manufacturing competitiveness . . . [involves] continually putting . . . [the] best talent and resources to work doing the basic things a little better, every day, over a long period of time. It is that simple—and that difficult. (Hayes and Wheelwright 1984, p. 390)

During the early 1980s, several related paradigms began to develop in response to external pressures for change. Two major theories involved the management of the enterprise to optimize manufacturing performance and the management of technology in this context.

In the post–World War II years, Japan had to learn and adapt to survive. As might be expected, the learning focus developed around established social patterns. Efforts to understand Japanese methods from the perspective of U.S. social patterns have often resulted in substantial confusion for all parties.

Interest began to grow in the Japanese manufacturing paradigm as U.S. managers found themselves trying to understand why their methods of manufacturing were no longer competitive. In 1981, Ouchi coined the term "Theory Z" to describe the Japanese approach to management. Involved workers were identified as the key to productivity. He indicated that productivity is basically a problem in managerial organization. Theory Z emphasizes the importance of mutual trust, subtle relationships, and intimacy of close social relations. Ouchi summarized the need for U.S. industry to move away from existing management paradigms to the concept of an "industrial clan" as the key to productivity.

Concurrently, a growing emphasis emerged on the appropriate management of technology to enhance enterprise performance. The movement toward more automated equipment and wider use of computers developed images of the "factory of the future." (As discussed here, the computer-integrated manufacturing [CIM] concept involves the creation of competitive manufacturing enterprises using the most appropriate strategies available. Changes in both management and in the use of technology are included. CIM is thus both a management and technology paradigm.)

During the early 1980s, the system-oriented application of manufacturing technology to optimize enterprise performance, as described by the CIM concept, came to be recognized as having major potential for responding to the competitive situations. Still, progress has been slow, due to a variety of factors:

- The present corporate leadership is often unfamiliar with technology and manufacturing and with the types of management changes required for effective use of this technology.
- The proposed solution strategies require major changes in corporate culture.
- Present accounting methods include built-in biases that make it difficult to justify the required investment.
- The required investment is often associated with a reduction in short-term profits.
- The perceived risk is high.

Caught between competitive pressures and the unpalatable aspects of the CIM solution, national progress has been slow. While a few companies have moved rapidly to substantial success and many are involved in limited study and experimentation, CIM nationwide still remains an elusive target.

Given the inhibiting factors, various efforts have been made to encourage corporate change:

- The Society of Manufacturing Engineers (SME) has become an advocate of CIM, and conducts a wide range of educational programs and conferences to encourage action in this direction. LEAD awards (Leadership and Excellence in the Application and Development of Computer Integrated Manufacturing) have been presented to industry leaders beginning in 1981.
- Concerned with the noncompetitive nature of U.S. manufacturing, the Department of Defense (DOD) has been sponsoring efforts by industry to produce more competitive operations. The Air Force Integrated Computer-Aided Manufacturing (ICAM) program of 1979–84 (introduced in Chap. 1) was a large-scale effort to achieve technology development and transfer for "factories of the future." The ICAM Definition (IDEF) formalism for designing manufacturing systems was created by this program (further discussed in Chaps. 3 and 5). The Manufacturing Technology (MANTECH) and Technology Modernization (TECH-

MOD) DOD programs have extended this same pattern in support of industry change.

- The National Institute of Standards and Technology (previously the National Bureau of Standards) has created an Advanced Manufacturing Research Facility (AMRF) to function as an R&D and demonstration site for U.S. industry (Simpson, Hocken, and Albus 1982).

- Efforts have also been developed at state and local levels. For example, in 1982, the Industrial Technology Institute was established in Michigan by the governor's High Technology Task Force as a publicly supported, not-for-profit corporation to conduct contract research and development in automated manufacturing technologies. An Advanced Manufacturing Center has been established to serve as an R&D and demonstration facility.

- Companies have created cooperative efforts to perform the advanced R&D efforts necessary to establish leadership in manufacturing technology. An example is the SEMATECH consortium of semiconductor manufacturing companies, planning on an intense automation development activity with both business and government support.

- Many universities and consultants have become involved in the effort to develop the skills and insights required to strengthen the competitive manufacturing capabilities of the United States (Owen and Entorf 1989). Several large companies are taking a leadership role in implementing CIM concepts and guiding others in the same direction.

Progress is being made. An improved understanding of the problem is being gained, and trial solutions are being tried. Capabilities and experimentation are growing. What is required now is the creation of an educational framework in which progress can continue as rapidly as possible, combined with continuing experimentation and evaluation by industries across the country. The means must be found to apply effectively limited resources to meet competitive demands.

The historical description of this chapter, leading to Fig. 2-28, illustrates that, until recently, U.S. manufacturing has engaged in practice-based learning activity. The outpouring of inventions in the nine-

Figure 2-28.
System-manufacturing characteristics, 1970–90, with growing attempts to employ manufacturing as a competitive strategy.

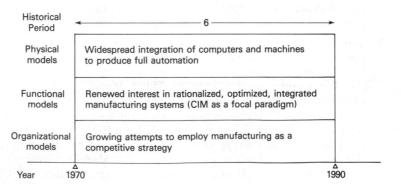

teenth and twentieth centuries, the creation and application of experimental technology, and the harnessing of scientific R&D have all served this purpose. However, the scientific-engineering paradigm, emphasizing completely known, closed systems, has extended into organizational planning and management with unforeseen and undesirable outcomes. This shift in paradigm has enabled the rapid growth of the easily quantifiable financial aspects of corporations as the central planning mechanism.

The training of managers in the technologies of final state planning has produced a paradigm shift that is not effective in the open system environment. Competition today requires the reestablishment of a learning setting that can lead to future success by U.S. industry.

EXAMPLE 2-5. PRODUCTIVITY

In order to understand the historical evolution of manufacturing in the United States and to develop effective plans for the future, it is necessary to collect and make use of data that adequately describe the system of interest. However, a constant effort must be exerted to guard against overdependence on such data. Small errors always exist in any statistics, and the use of data can vary widely, leading to possible misinterpretations. For example, based on official U.S. government data, conclusions have been drawn that "except for business cycle movements, the shares [of the economy reflected by] real manufacturing output and final goods output have been remarkably stable for 25 years" (Kuttner 1988; Pennar 1988). Thus, based on these data, it has been assumed that manufacturing has been doing acceptably well in a competitive environment.

Lack of concern for the status of U.S. manufacturing can be based on such findings, if they are accepted as correct. However, if such data are reviewed in detail for accuracy and completeness, can significant issues be raised regarding these conclusions?

Results

Recent reexamination of government data has resulted in proposed revisions. Based on a number of detailed factors, it has been estimated that the productivity increases over the period shown have been substantially smaller than originally interpreted. As concluded by Mishel (1988a, 1988b), the errors in measurement are attributable to failing to account properly for imports used in the production process, incorrect use of an index to account for services purchased in the production process, and an overestimate of the growth of the computer industry. When these adjustments are made, Mishel concludes the following:

> Manufacturing's share of output (has) declined by three or four percentage points since 1973, most of which occurred between 1979 and 1987.

> This recent shrinkage of manufacturing's share of output is a sharp break from historical trends. . . . Had the output shrinkage not occurred, my estimate is that we would have at least three million more manufacturing jobs today.

> This substantial downward revision in the manufacturing output and productivity growth statistics reveals a substantial weakness in our international competitiveness.

The debate over productivity growth in the United States and the stability of the manufacturing sector illustrates how difficult it is to measure the present state of a system, to understand the interaction of the system with the environment, and to project forward to future system and environment states. In this environment, every possible type of supporting analysis should be performed. However, it should also be understood that the evaluation and interpretation of these data must include qualitative overviews of the nuances of the system and not just rote manipulation of data.

Those who come to depend on such data often tend to assume that the numbers are exactly correct and that interpretations of past performance and projections into the future can adequately be performed based on such information. When the possible errors associated with such information are adequately evaluated, projections into the future are likely to be inaccurate.

Often, the best that can be said is that the parameters associated with likely economic performance may be bounded in terms of the most optimistic and the most pessimistic performance limits, based on the ability of the system to engage in self-correction. However, the prediction of particular future states based on past and current data is not likely to be successful.

ASSIGNMENTS

2-1. In order to achieve maximum educational insight from Chap. 2, it is helpful to stimulate class discussion of the concepts that have been included, develop critical reviews that probe the advantages and limitations of the concepts, and explore alternative ways for approaching the topic. Select one or several topics from the bold-faced headings that introduce sections and examples in this chapter. Come to class prepared to explain this topic, critically describing the text materials and suggesting ways the topic discussion might be strengthened.

2-2. Based on the data in this chapter, develop brief descriptions of each of the paradigms that existed during the defined evolutionary periods. Then link these paradigms in the format shown in Fig. 2-29. By considering the shifts from one paradigm to another, briefly describe the changes that took place between each of the periods. To what do you attribute the changes? If you were to develop a system description of the entire U.S. socioeconomic system, in order to explain the changes in these paradigms, what system elements and what interactions or rela-

Figure 2-29.
The format to be used for Assignment 2-2.

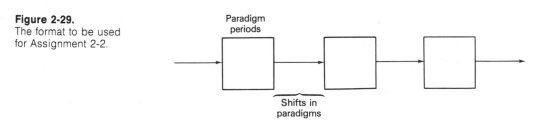

Paradigm periods

Shifts in paradigms

tionships might you include? Can you suggest a partial system description that would explain some of the changes? Can you use this historical interpretation to draw any conclusions about likely trends for the future? What are the opportunities and constraints associated with anticipating the future based on such modeling efforts?

(a) Come to class prepared to describe your interpretive analysis of the questions you have raised and present your version of Fig. 2-29.

(b) Following class discussion, write a brief paper summarizing your learning experiences.

2-3. Using reference materials, explore some of the technical and organizational changes in the United States that took place during one or several of the periods illustrated in Fig. 2-2. Graph the data you select as shown in this chapter. Do the data support the evolutionary paradigms that have been described, or are they in conflict? How do you interpret the meaning of these data within the larger context that has been set in this chapter? What changes would you make in this chapter based on your data?

(a) Come to class prepared to present your graphs and describe your interpretations.

(b) Following class discussion, write a brief paper on your interpretation, insights, and observations regarding the historical evolution of manufacturing in the United States.

2-4. Read the relevant parts of Adam Smith's *The Wealth of Nations* and gather other information regarding his point of view about the development of society. What other comments can you make about the issues raised in this chapter? Locate one or more analyses of Smith's book and come to class prepared to discuss both his point of view and those of the other authors regarding his work. What is your own interpretation of this work and its meaning in terms of the evolution of the United States?

(a) Be prepared to share your results and participate in discussion of the issues raised.

(b) After discussion, prepare a brief paper on your insights.

2-5. Read several references on Eli Whitney and his contributions to manufacturing concepts. What is your own interpretation of his contributions? How would you describe the relationship between his activities and the evolution of the dominant paradigms of the times? Which other individuals would you consider as

having an important role in the contributions made by Whitney?

(a) Come to class prepared to share your results and participate in discussion of the issues raised.

(b) After this discussion, prepare a brief paper on your insights.

2-6. Review some of the writings of Herbert Spencer and his concepts of social Darwinism. What conclusions do you draw about his work? How do you believe he meant for his discussion to be related to the organizational system of the time? Read one or more commentaries regarding his work and decide what conclusions you can draw from these other points of view. Given these combined resources, consider the issues raised by these discussions, how you believe they affected the dominant paradigm of the time, and how they might still be affecting current paradigms in the United States.

(a) Be prepared to share your results in class and participate in discussion of the issues raised.

(b) After this discussion, prepare a brief paper on your insights.

2-7. Apply Assignment 2-6 to Frederick W. Taylor.

2-8. Apply Assignment 2-6 to F. J. Rothlesberger and the Hawthorne experiments.

2-9. Discuss your insights into the manufacturing system paradigm that exists in the United States today. To what degree do you believe this paradigm is related to earlier paradigms and evolutionary forces? To what degree is this paradigm driven by technology, and to what degree is it driven by social and economic issues? Do you see these two aspects of a sociotechnical system feeding back in a supportive or a conflicting way with one another?

(a) Come to class prepared to share your results and participate in discussion of the issues raised.

(b) After this discussion, prepare a brief paper on your insights.

2-10. Review one or more textbooks on manufacturing that were written in the post–World War II years (1945–70). Use these resources to understand the ways in which the world was viewed at the time. Can you paraphrase the paradigm of the manufacturing system of this time by viewing the world through the eyes of these authors?

(a) Come to class prepared to share your results and participate in discussion of the issues raised.

(b) After this discussion, prepare a brief paper on your insights.

2-11. Gather information on one of the university or government programs that supports the development of advanced manufacturing strategies for the United States. Explain how the program is set up and its activities. Do you believe that the program is making a major contribution to the United States? If so, in what areas? If not, what critique would you raise?

(a) Come to class prepared to share your results and participate in discussion of the issues raised.

(b) After this discussion, prepare a brief paper on your insights.

MANUFACTURING SYSTEM DESIGN

3

Based on the insights into complex, nonlinear systems presented in Chap. 1 and the overview of the historical evolution of U.S. manufacturing systems in Chap. 2, this chapter addresses problem-solving approaches that have a reasonable potential for producing successful system designs. Emphasis is on the most effective design strategies, given the types of problems faced and the supporting methods of analysis available. At issue is achieving a system design process that is appropriate for complex systems that are constantly evolving and changing in response to internal and external adaptation.

Chapter 3 approaches this problem from several different points of view. An overview is presented of a planning and learning strategy that can be effective for problem solving in the types of environments that are being considered. A system-environment simulation (SES) software model, included with this text, is presented as a simple way to help understand some of the important interactions between a manufacturing system and its environment.

A manufacturing system must link effectively to its environment if it is to remain financially viable. The roles of information gathering from the environment and decision-making algorithms used by the system are explicitly addressed.

Also considered are a number of design principles that can support manufacturing system design. These principles suggest directions in which system change may take place and may be regarded as resources for use in problem-solving efforts. These idealized principles are rarely instituted by themselves or completely; rather, they represent suggestions for ways in which improved system operations may be obtained. In any planning setting, the objective is to draw on *combinations* of these principles as most appropriate for the particular system and its environment.

Software and documentation for the system-environment simulation (SES) model are provided with this text. The SES model can be used to study the relationships between manufacturing systems and their environments for a limited number of situations, and can also be modified for extended applications.

An introduction to open system concepts and manufacturing enterprises was provided in Chaps. 1 and 2. This chapter considers how viable enterprises can be developed to function in a specified environment. This is a difficult task in which the external market setting must be effectively linked to the internal structure and function of the manufacturing system.

It is important to recognize that decision making regarding future enterprise development is of critical interest to the organizational leadership. Therefore, any strategy for enterprise change must involve this leadership. The resultant actions will combine system modeling insights *and* a political negotiation process among those exercising organizational power, affected by personal and institutional objectives, individual perceptions, and the individual propensity for risk taking.

Within this framework, decision making will typically be shaped partly by the quantitative predictions of various types of limited-scope modeling techniques and partly by technological and management opportunities and constraints. The modeling techniques and technological and management factors provide input to the development process that is defined. Design efforts for a manufacturing enterprise will thus center around human decision making, typically in a group setting, supported by quantitative methods and an understanding of technology and management options.

This chapter describes how the concepts introduced in Chaps. 1 and 2 can be linked to management action for manufacturing system design. Chapter 3 provides a link between the concepts and the detailed problem-solving methods of Chap. 4.

This chapter probes the nature of the development activity for manufacturing systems. The discussion begins with a description of the problems encountered in planning the evolutionary development of any such enterprise. It is then extended to describe the solution approaches that promise to be most effective.

The importance of a system approach to this design task has been noted by several authors. For example, Hutchinson (1985) has observed that design must reflect a "continual awareness of system status" in order to avoid a situation in which "the parts are efficient but [still result in a whole] that is ineffective" (pp. 8–9). Without a system overview, development efforts may result in locally optimized portions of the manufacturing activity that, in combination, do not result in an optimized system. This difficulty sometimes results in "islands of automation" that are separately effective but poorly integrated, and therefore lead to inefficient system operation.

The Manufacturing Studies Board (National Research Council 1986) has concluded that "the systems approach is a key principle . . . ," but that "such a concept has been foreign to most U.S. managers. . . ." As envisioned by the Board, "Managing manufacturing as a unified system will profoundly affect every activity involved (in the system) . . ." (p. 50).

A system approach must be concerned with more than the technical aspects of a manufacturing facility. For example, in a statement delivered to the U.S. House of Representatives, Alic (1984), of the Office of Technology Assessment (OTA), observed that "it makes . . . sense to view integrated production systems as integrating people and machines. Improving the efficiency of such systems . . . will take a better sense of how to allocate tasks and responsibility among people and their machines." Thus, the system must include both technical and organizational/management aspects of the facility being studied.

3-1. Problem Definition and Design Approach

In order to achieve directed change within an organizational framework, the design or redesign of a manufacturing system requires the establishment of a planning group. This planning group must act to collect information, evaluate alternatives, and make decisions with respect to the perceived choices. And to be effective, the group must be endorsed by the organizational leadership. This emphasis on a politically endorsed group as the focus for system change is considered further below.

The following discussion refers to the present (*as-is*) and future (*to-be*) environments and to the present (*as-is*) and future (*to-be*) system configurations. The planning activity, conducted by a planning group, must consider the present environment and the present (*as-is*) manufacturing system. In addition, the group must also attempt to estimate the future environment and alternative possible future (*to-be*) system configurations.

Figure 3-1 builds on Fig. 1-5 to illustrate how these various aspects of system design can be conceptualized, using a heuristic visual model.* The planning group receives limited information regarding the present environment and even more limited estimates of probable future environments. The environment is in constant change and can only be partly known. The planning group also must develop a detailed understanding of the present (*as-is*) manufacturing system. This understanding is limited by available information and the perception limits of the group's members.

The planning group becomes the focal point for CIM system design and implementation. The group can draw on supporting modeling efforts and a variety of CIM design principles to guide the enterprise along productive directions of change.

Finally, the planning group is assumed to have access to a set of manufacturing design principles that provide general guidance into the types of *to-be* system designs that have a reasonable likelihood of providing a competitive manufacturing system. (The latter part of this chapter explores such principles for computer-integrated manufacturing.) These principles may be used to produce idealized "reference system" configurations.

The reference systems are developed by drawing on one or more design principles. They typically represent extreme, idealized configurations with one or several concepts carried to a limiting case. The reference systems typically do not reflect the many subtleties and dimensions of the *to-be* systems, nor the many trade-offs involved in system design, but they do represent limiting cases for selected features of a CIM-oriented system. The reference systems then emphasize particular design factors in a way that demonstrates the relevance of the principles being drawn upon. These reference systems, based on design principles, may then be considered for idea generation when a realistic *to-be* system is configured.

* Heuristic models are introduced in Chap. 1.

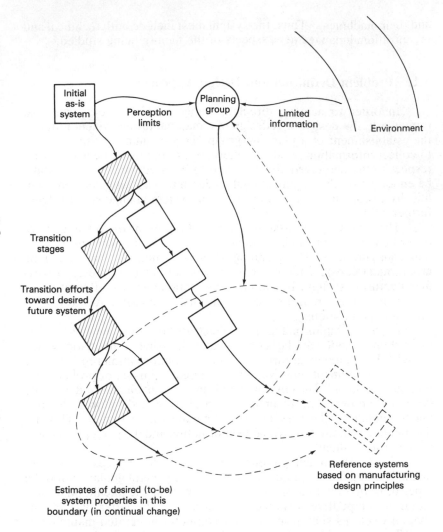

The planning group thus has a sense of the environment, a knowledge of the present system, and some general guidelines into the types of reference system configurations that have promise. The tasks of the planning group are: (1) to produce estimates of desired future (*to-be*) system concepts, and (2) to define viable transition stages that lead from the *as-is* system to a desired *to-be* system.

Since the environment and reference systems are consistently changing (due to new information and insights), and the *as-is* system is continually evolving, the preferred *to-be* system concepts will also change with time. The desired future state for the system will not be fixed but will evolve in time. Thus it is reasonable to conclude that in many cases, the *to-be* system must be regarded as a temporary concept that will never be achieved. Rather, the *to-be* system provides a direction for change and serves to specify or constrain the near-term transition stages. As a transition stage is achieved, the *to-be* system concept will probably have changed, so that the planning group must develop a new track to the future, from the transition state to a new *to-be* system.

The resultant evolutionary pathway is likely to be a series of transition stages that become the bases for new evolutionary paths. The definition of flexible, robust transition stages thus becomes a key element of the design process. Each transition stage must be a stand-alone, viable configuration that does not excessively limit future *to-be* system concepts. It may often be necessary to use a transition stage as a starting point toward a revised final configuration that was not originally anticipated. Major difficulties are thus encountered, since the planning group is working with a moving target.

There are significant problems in designing a *to-be* system, even if static assumptions are made. The means are not available to predict a final, optimized system configuration for a given starting *as-is* system and environment; as discussed in Chap. 1, the problem is too complex and rich in mixes of qualitative and quantitative considerations. An alternative viewpoint of the design process must be considered. Based on the type of system-environment configuration that is encountered and on current experience, the design of a manufacturing enterprise must be regarded as a "search and learning" procedure rather than as an "optimize in advance" procedure. A different kind of design activity thus ensues, which is quite distinct from typical engineering product

The enterprise will experience a series of transition stages as it evolves toward the future. These transitions must encourage the needed learning experiences for the organization and must result in robust system designs that can be maintained for an indeterminant period of time.

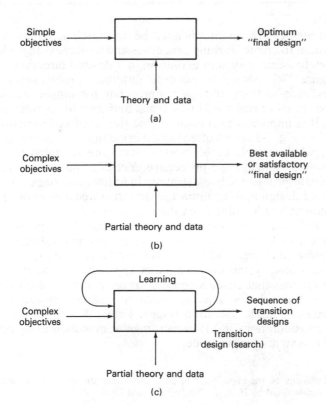

Figure 3-2.
A comparison of alternative problem-solving strategies. (a) Simple, well-defined problem (of the type often taught to students). (b) Extension to more complex problems, without feedback. (c) Extension to real-world problems with feedback.

Figure 3-3.
Evolutionary growth cycle for a manufacturing system, applying the concepts of Fig. 3-1. Information regarding the enterprise objectives and knowledge of the *as-is* system is merged with insights regarding reference systems based on CIM principles and an understanding of the environment to create *to-be* system concepts. Viable transition states may then be developed, and feedback/learning will take place to drive evolution of the transition states.

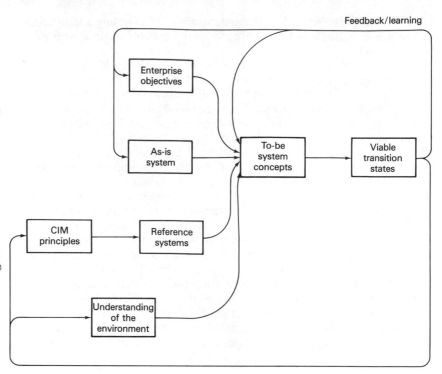

design strategies. The emphasis must be on search activities that have the potential to create learning processes and to harness these learning processes to achieve system evolution in a desired direction.

Figure 3-2 compares several different problem-solving approaches, ranging from strategies appropriate for simple, well-defined problems, to more real-world problems, and then to complex problem solving. It is important to recognize that the iterative problem-solving approach of Fig. 3-2c is not often taught in engineering and management courses, where simple problem statements are intended to produce unique solutions in a single procedure. Yet this approach is becoming the dominant problem-solving strategy in industry settings, using computer-aided design (CAD) support as an integrated aspect of searching for a solution and learning from the experience.

Manufacturing system design thus cannot be approached as a static, linear endeavor with a unique, fixed, optimum solution. Rather, due to constant change and adaptation, the system follows a sequence of trial solutions. System design must be approached as an evolutionary learning process that enables continuous adaptation to a changing environment.* The design objective must be a gradual system evolution using robust, flexible transition stages. Figure 3-3 illustrates how the various considerations result in incremental evolution using reference and *to-be* systems for guidance.

* A similar focus on the importance of continuous learning to create *dynamic manufacturing* is discussed by Hayes, Wheelwright, and Clark (1988).

3-2. Computer-Integrated Manufacturing

The type of business environment that is of principal concern today consists of a highly competitive, rapidly evolving market. International competition continues to place intense pressure on all manufacturing systems and to place the most demanding requirements on market participants. It is this type of setting that provides maximum stress on system design. The computer-integrated manufacturing (CIM) concept has evolved in response to this situation. Given the complexity of the issues involved, there are many different definitions of CIM. The definitions proposed by Harrington, who suggested the phrase *computer-integrated manufacturing,* are discussed in Chap. 2.

Alternatively, as concluded by the Committee on the CAD/CAM Interface (CCCI) of the National Research Council (1984), computer-integrated manufacturing in a manufacturing enterprise occurs when:

- All the processing functions and related managerial functions are expressed in the form of data.
- These data are in a form that may be generated, transformed, used, moved and stored by computer technology.
- These data move freely between functions in the system throughout the life of the product, with the objective that the enterprise as a whole have the information needed to operate at maximum effectiveness. (p. 9)

Warndorf and Merchant (1986) provide an alternative definition: "The integration and automation of data, information and the control of product, from perception through production, to shipment and support, is referred to as Computer-Integrated Manufacturing (CIM)" (p. 162).

These definitions are useful and provide insight into the meaning of CIM. However, the definition proposed in this text starts with an emphasis on enterprise objectives, then defines CIM as a strategy that can contribute to obtaining these objectives:

> The purpose of manufacturing system design is to produce an enterprise that best meets stated performance objectives. A computer-integrated manufacturing (CIM) system results when the design effort includes the use of *computers* to achieve an *integrated* flow of manufacturing activities, based on *integrated* information flow that links together all organizational activities.

This definition identifies the meeting of enterprise objectives as the motivating force for all manufacturing design, and defines CIM within this context. However, given the information-intensive nature of modern manufacturing operations, it may be argued that when efforts are made to meet performance objectives for most settings of interest, integrated computer systems will be part of the solution, so that CIM design is required. Thus, the objective of producing the most effective manufacturing enterprises is viewed predominantly as an issue in producing the preferred CIM system design.

However, there is a danger that the "CIM equals integrated computer networks" identification will be formed. This can lead to unfortunate results in terms of enterprise objectives. The overall system must be rationalized, requiring that the work flow, organizational structure, and management methods be redesigned to best obtain performance objectives. The entire meaning of product design must be assessed and modified as necessary to optimize system performance. The most appropriate use of technology, including information flow and computers, can then be selected in this context.

Figure 3-4 illustrates how all of these areas are part of CIM. As considered here, CIM-oriented manufacturing systems will include information integration obtained through computer networks. The uses of this technology, however, will be based on system rationalization and a product design strategy that is an integrated aspect of the system. The definition in this text is useful because it will continue to force attention on the actual problems in manufacturing system design and will orient all solution strategies toward evaluation in this framework.

In typical cases, manufacturing operations are data intensive, and the costs associated with information handling are a large percentage of total production costs. In such a setting, integrated computer systems turn out to be an essential element of a CIM strategy. Similarly, most manufacturing systems can benefit by automation. The issue is how much automation and of what type is appropriate to a given enterprise. The design of a CIM-based enterprise thus requires a careful look at the specific market and manufacturing system and an evaluation of the most appropriate use of automation and computers.

Obtaining the most effective system operation through worker involvement is also critical to CIM design. In many circumstances, the competitiveness of a CIM system will ultimately depend on how well human resources function within the system. Technology and management options must work together to produce an effective production capability.

Experience has indicated that the preparation of a functional flowchart of current operations is an appropriate way to begin study of the *as-is* system (Chiantella 1986). This flowchart can then represent one source of information for design of the *to-be* system. As has been noted by those with experience in the field (Chiantella 1986), the design of a CIM system can then emphasize the formulation of a "conceptual specification" of the functional operation of the "new (*to-be*) way of doing business." This specification will be part of the "search and learn" procedure that is to follow (p. 37).

The planning process of Fig. 3-1 can be applied to CIM system design when a highly competitive, rapidly changing environment is assumed, and an effort is made to optimize system performance in this

Figure 3-4.
Illustration of how the CIM system definition used in this text is a response to enterprise objectives. Objectives drive the system design or redesign, producing transition stages for the system.

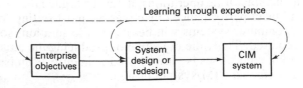

setting. Sections 3-6 through 3-12 provide an introduction to a set of CIM principles that can be used in support of the system planning activity.

It is essential to recognize that moving toward a CIM system does not mean simply introducing higher levels of automation into an existing system. Adding automation to an inefficient system will likely produce a highly automated, inefficient system (Chiantella 1986).

The challenge is to make the most appropriate use of technology and management to optimize the business enterprise. Because the core of the concept is a strong emphasis on systemwide integration of activities, CIM will often require a change in emphasis and management attitudes and developing new ways of thinking about the business enterprise.

As noted by Chiantella (1986), "CIM by its very nature requires a unique collaborative effort by the entire management team involved. The . . . team . . . must learn to focus collectively on a single system view across the entire business cycle of manufacturing" (preface). This point has been reinforced by Appleton (1984), who notes that "integration is a management issue, not (only) a technological issue. . . . It is a managed infrastructure. . . . For CIM to become a reality, it must become a management style."

The importance of the planning group in manufacturing design was discussed earlier. Within the context of management responsibility in CIM systems, it is useful to return to this subject. As noted by a number of companies and investigators, the role of the chief executive officer (CEO) in CIM is of critical importance:

> The first step in establishing . . . (an) enterprise-wide perspective is to define the integration control structure within which planning . . . will be conducted. (Appleton 1985, p. 4)
>
> The CEO directs the development of a company-wide, long-term plan . . . (and) . . . assembles an interdisciplinary team . . . (to perform the necessary planning). (National Research Council 1987, p. 163)

The importance of the planning process itself has been reinforced by Warndorf and Merchant (1986). In order to accomplish CIM, they state, "There first must be a plan . . . which allows everyone to become a part of this new concept or goal" (p. 163). Thus, strong political support from the CEO within a controlling structure must be combined with an adaptive management strategy and a capable planning process to begin the evolution toward CIM.

The design of any effective manufacturing system is a difficult task, and CIM applications might be regarded as the most difficult. To achieve the desired integration objective, it is necessary to incorporate top-down planning that considers how all aspects of the enterprise will eventually function together to optimize business performance. At the same time, realism regarding budgets, technology, risk, the search process, learning time, and the present *as-is* system requires an incremental evolution from the bottom-up (implemented through transition stages).

The challenge is to choose an evolutionary, modular growth

Since the application of CIM concepts can require the introduction of new methods for enterprise planning and management, the role of the CEO is critical. Strong leadership and support for planning and implementation teams is required to achieve the desired learning environment.

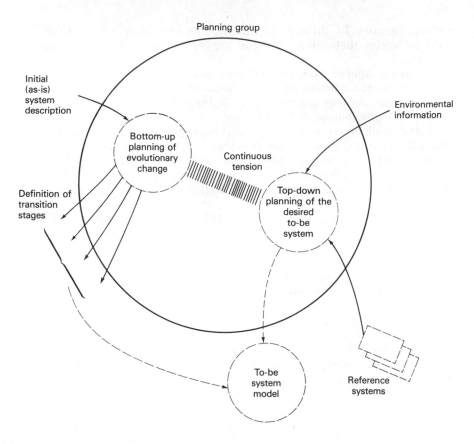

Planning group

Initial (as-is) system description

Bottom-up planning of evolutionary change

Continuous tension

Top-down planning of the desired to-be system

Environmental information

Definition of transition stages

To-be system model

Reference systems

Figure 3-5.
A heuristic model illustrating the inherent tension between bottom-up and top-down planning. Tension will always exist between the bottom-up definition of transition stages (based on evolution of the *as-is* system) and top-down planning of the desired *to-be* system (based on reference system concepts).

mechanism that reflects the search and learn strategy and will ultimately lead the system toward the desired system design objectives. This requires the definition of a sequence of transition steps to extend the system of today toward the desired future state of tomorrow. Continuous tension will exist between those attempting to guide the enterprise upward to the next stage, within financial, technological, and organizational constraints, and those who are attempting to guide the enterprise toward a final configuration (Fig. 3-5).

Both design and implementation strategies must reflect this dual approach and the requirements associated with combined top-down and bottom-up change. If the top-down portion of the approach becomes too strong, it can produce high-risk transition stages, leading to organizational collapse. On the other hand, if the top-down portion becomes too weak, the resultant growth will not lead to the desired system configuration. Similarly, if the bottom-up portion of the approach becomes too weak, high-risk transition stages can again result. If the bottom-up portion is too strong, overriding strategic guidance, the risk is that the final system will reflect an updated version of the *as-is* capability, not growth toward the integration objectives. And given a continuing environmental change, the world will not remain the same from either perspective.

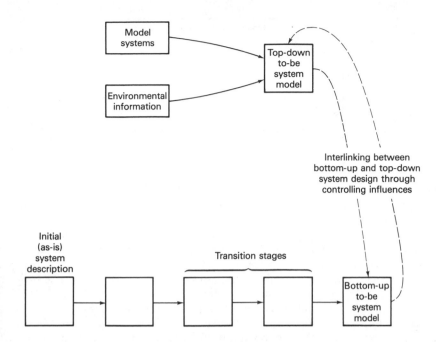

Figure 3-6.
Top-down system design reflecting model systems and the environment. The top-down system model is linked to the bottom-up model shown in Fig. 3-7.

Figure 3-7.
Bottom-up system design reflecting technology and management opportunities and constraints. The two approaches shown in Figs. 3-6 and 3-7 indicate the types of interlinking that develop between these alternative design approaches.

Figures 3-6 and 3-7 illustrate how the design concepts of Figs. 3-1 through 3-5 can be described in further detail to emphasize the top-down and bottom-up features of the design process. Figure 3-7 shows how technology and management opportunities and constraints produce a (bottom → up) aspect of system design. Figure 3-6 illustrates how the (top → down) required descriptions of the environment and system are interlinked by complex feedback loops. Refer to Fig. 3-5 to see the tension that can arise between these two emphases.

The planning group has a difficult task. There is no "cookbook" approach to making the necessary decisions. The design will never be complete, but will always be in a state of transition as the environment changes. However, a number of methods and techniques can help the planning group in its task. When used with care, these methods can provide assistance for appropriate action and decision making. This text defines the design problem facing planning groups, introduces the reader to some of the technical and management resources available for problem solving, and describes how these resources can be harnessed to the design and implementation processes.

3-3.   A Simple Introductory Model for Studying System-Environment Interactions

Many different types of models are available to assist in the design of manufacturing systems. Some are based on analytic equations; others make use of simulation strategies. Such models are often integrated into computer software to provide maximum access and utility. In general, these models all have strengths and weaknesses, and they can

Figure 3-8.
Input/output relation-
ships for the SES
model. The model pro-
vides a link between
vendor kits arriving
from the environment
and product sold to
the environment.

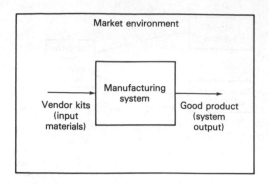

support the system design activity if they are used with understanding and care.

Chapter 5 provides a detailed introduction to several computer-based models that are useful in this context. Included are descriptions of the concepts that are incorporated into the various software programs and outlines of possible program application strategies. The programs are then illustrated in examples.

This chapter makes use of a simple introductory model for studying system-environment interactions. The effects of various design choices on manufacturing system performance can be summarized by means of a system-environment simulation (SES) model. The SES model used below does not deal with the many detailed aspects of modeling and simulation that are discussed in Chap. 5. However, it does provide an overview that can be used effectively to illustrate key principles and design concepts. The structure of the SES model software is discussed in Appendix B. A floppy disk that includes the software is provided with the text.

The SES model is a simple planning and evaluation tool. This model provides a *rough-cut, approximate method** for use in considering some of the performance aspects of a manufacturing system within its environment. As illustrated in Fig. 3-8, the model is driven by interactions between the system and its environment.

The SES model tracks work-in-progress from vendor kit availability, through the complete manufacturing process, to delivery of the final product to the market. By using this model, it is possible to explore interactions between unexpected shifts in market behavior and system decision making. Various management and decision strategies can be explored in a wide variety of situations in order to select those best matched to the enterprise and its market.

Figure 3-9a illustrates a functional flowchart of the SES model. The model is based on tracking work-in-progress (WIP) from the input of the manufacturing system through the system itself and into the market. System status is determined at a series of times that are identified by the parameter J, which takes on integer values 1, 2, 3, . . . N. Manufacturing activities are assumed to take place through a series of steps that are sequentially associated with the J values. Several counters describe the system performance for each activity and time, as defined by the value of J. For each value of J, the operation of the system depends on the previous system states and on current decision-making processes. As the J values increment, the work-in-progress will move through the system according to the defined sequence of activities.

Figure 3-9a makes use of the IDEF model that was introduced in Example 1-5. This redrawing of the system and environment indicates

* These terms are discussed further in Chap. 5.

the specific definition of parameters that are used in the SES model. Figures 3-9b through 3-9e show the various "child" levels that can be associated with this overview. The manufacturing system is related to the vendor suppliers and to the market as shown. The vendor kit* orders VT(J) and the desired level of factory production ET(J) control the manufacturing system and the vendor.

Figure 3-9b indicates that once vendor kit orders are received, a number of steps may take place, producing a delay before vendor kits are shipped to the manufacturing system, represented by N1(J). Figure 3-9c shows how the vendor kit inventory "box" in Fig. 3-9a can be further divided into a receiving function, inventory function, and transfer-to-system function, thus further subdividing the system. Figure 3-9c shows the relationship of N1(J), Q1(J), and N2(J). Figure 3-9d shows how the manufacturing system itself can consist of a set of processes, which can cause a delay between the arrival of the vendor kits N2(J), the shipment of good product N3(J), and the shipment of product to rework or scrap R(J). Figure 3-9e shows how the product inventory function can be divided into product receiving, product inventory, and product transfer-to-market. It also indicates how N3(J), Q2(J), and N4(J) are related to one another. Figures 3-9a through 3-9e provide another indication of the usefulness of the IDEF methodology in understanding the step-by-step nature of manufacturing system operations.

The flowchart in Fig. 3-9a begins with the input of components and parts into the system. Each vendor kit provides the building-block

Figure 3-9.
Illustration of the SES model using Design/IDEF, introduced in Fig. 1-9. The version shown here has been modified to indicate definition of the SES model parameters. (Design/IDEF courtesy of Meta Software Corporation.) (a) Overview of the manufacturing system and its interaction with the environment. (b) Development of a "child level" for box A1 to show the steps involved in vendor kit production. (c) "Child level" associated with box A2, showing vendor kit receiving and inventory. (d) "Child level" associated with the manufacturing system (box A3), indicating the sequence of steps that can be required for such a system. (e) "Child level" shown for box A4, indicating product inventory and shipping to link manufacturing to the market. (f) Parameters used to define the decision-making process. (g) Definition of terms used in Fig. 3-9.

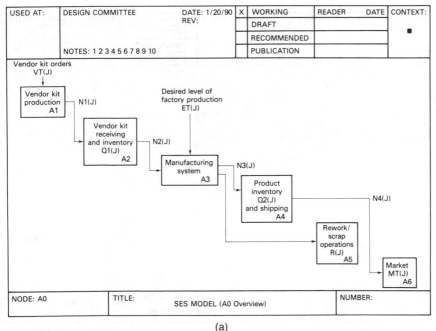

(a)

* Some of the advantages of kitting are described by Votapka (1988).

Figure 3-9. continued

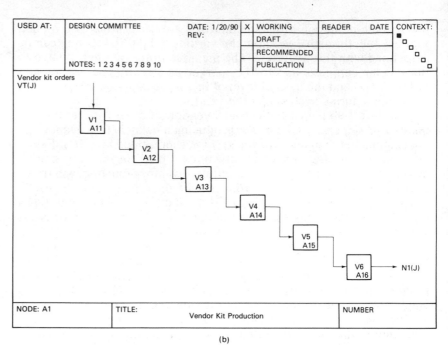

USED AT:	DESIGN COMMITTEE	DATE: 1/20/90 REV:	X	WORKING	READER	DATE	CONTEXT:
				DRAFT			
				RECOMMENDED			
	NOTES: 1 2 3 4 5 6 7 8 9 10			PUBLICATION			

Vendor kit orders
VT(J)

V1 A11

V2 A12

V3 A13

V4 A14

V5 A15

V6 A16 → N1(J)

NODE: A1	TITLE: Vendor Kit Production	NUMBER

(b)

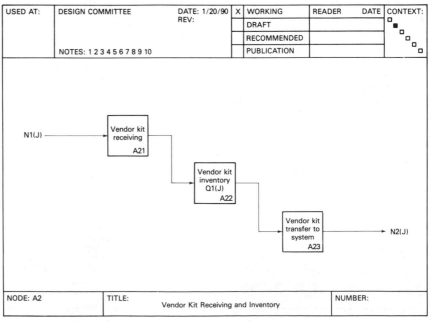

USED AT:	DESIGN COMMITTEE	DATE: 1/20/90 REV:	X	WORKING	READER	DATE	CONTEXT:
				DRAFT			
				RECOMMENDED			
	NOTES: 1 2 3 4 5 6 7 8 9 10			PUBLICATION			

N1(J) → Vendor kit receiving A21

Vendor kit inventory Q1(J) A22

Vendor kit transfer to system A23 → N2(J)

NODE: A2	TITLE: Vendor Kit Receiving and Inventory	NUMBER:

(c)

Figure 3-9. continued

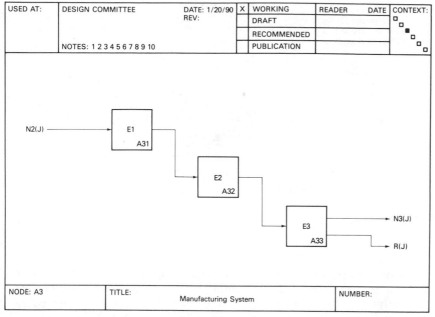

USED AT:	DESIGN COMMITTEE	DATE: 1/20/90 REV:	X	WORKING	READER	DATE	CONTEXT:
				DRAFT			
				RECOMMENDED			
	NOTES: 1 2 3 4 5 6 7 8 9 10			PUBLICATION			

NODE: A3 | TITLE: Manufacturing System | NUMBER:

(d)

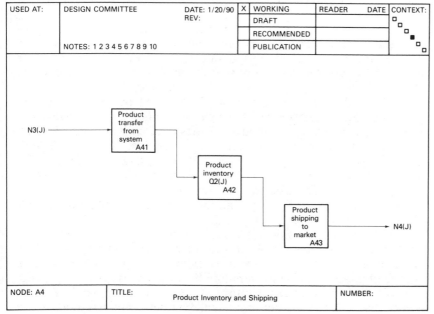

USED AT:	DESIGN COMMITTEE	DATE: 1/20/90 REV:	X	WORKING	READER	DATE	CONTEXT:
				DRAFT			
				RECOMMENDED			
	NOTES: 1 2 3 4 5 6 7 8 9 10			PUBLICATION			

NODE: A4 | TITLE: Product Inventory and Shipping | NUMBER:

(e)

Figure 3-9. continued

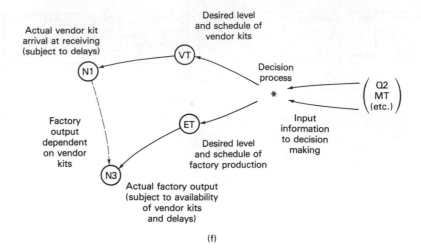

(f)

elements required for one output product. The number of vendor kits ordered at a given time is represented by VT. Once a vendor kit order is received by the vendor, the variable V1 will increase to the value represented by the order. This represents the receipt of the order by the vendor and the beginning of preparation to fill the order. The series of counters V1, V2, V3, . . . then represents the delays associated with filling this order. The order being prepared advances from V1 to V2, then from V2 to V3 as J increments forward. At the end of a predetermined number of steps, the vendor kit order will be filled and the vendor kit will be received by the manufacturing system, as represented by the variable N1.

Variable N2 represents the movement of kit inventory from receiving into the factory. During any time increment, if N2 is less than N1, then the excess vendor kits go into kit inventory Q1. On the other hand, if N2 is greater than N1 during any increment, kit inventory will be reduced to the degree that inventory is available.

After the vendor kit inventory is received by the factory, the work-in-progress moves through the various manufacturing stages. The factory itself can be represented as a series of processes, each of which is associated with one time increment. Therefore, work-in-progress advances from E1 to E2, E2 to E3, E3 to E4, and so forth, as J increments forward. At the end of a chosen number of steps, the work-in-progress will be released from the factory. N3 describes the number of good units from the factory, and the variable R counts the number of units sent to rework or scrap.

The arrival of the finished work in shipping is represented by N3, and N4 represents the actual sales to the market from shipping. If, in any time increment, the volume of sales to the market is less than the volume of goods arriving in shipping, then the excess factory production will go into product inventory Q2. On the other hand, if N4 is greater than N3, the market orders will be filled from inventory to the degree that inventory is available.

A decision-making loop determines the level of purchasing from vendors (VT) and the desired level of production in the factory (ET) (see Fig. 3-9f). In general, the feedback loop uses information on the product inventory (Q2) and the market (MT) as basis for placing vendor orders (VT) and for determining the desired level of factory production (ET). The variable MT determines the market sales that are possible. The level of market sales during any particular time increment can vary up or down, depending on many influences outside the manufacturing system.

Information regarding market behavior (MT) is fed into the model. Various algorithms for manufacturing system decision making can be then explored to determine system operation within the assumed market environment. The SES model shown in Figs. 3-9a–9e thus describes the relationships that exist in a given market environment between the number of vendor kits ordered from suppliers, the level of manufacturing operations for the factory, the levels of inventory that result, and sales to the market. The parameters associated with Fig. 3-9a are summarized in Fig. 3-9g.

The counter VT represents the number of vendor kits ordered during each time increment. These values must be related to system decision making on an a priori basis before exercising the model operation. Q1 describes the number of vendor kits in inventory at any time. This value will change, depending on the relationships between the arrival rate of vendor kits N1 and the use of vendor kits N2 by the factory. The parameter EMAX represents the maximum production capability of the factory. The values for factory production (ET) are limited by this upper boundary. The variable U represents the efficiency rate of the equipment, and is used in conjunction with the EMAX

Figure 3-9. continued

MT(J)	Maximum number of finished products that can be sold by the factory to the market at the time defined by J
VT(J)	Number of vendor kits ordered at the time defined by J
NI(J)	Number of vendor kits delivered to the receiving unit of the factory at the time defined by J
N2(J)	Number of vendor kits delivered to the factory from receiving at the time defined by J
Q1(J)	Number of vendor kits stored in receiving inventory at the time defined by J
ET(J)	Level of desired factory production that is set by the decision-making loop (actual production may be less than ET if an insufficient number of vendor kits, N2, is available)
EMAX	Maximum number of products that can be produced by the factory at the time defined by J
U	Efficiency rate of the factory equipment
R(J)	Number of vendor kits lost to rework or scrap at the time defined by J
N3(J)	Number of (good) products transferred from the factory to shipping at the time defined by J
N4(J)	Number of products sold to the market at the time defined by J
Q2(J)	Number of products in inventory at the time defined by J

factor to determine the maximum possible production level of the factory. The variable Y is the rework/scrap percentage. This parameter determines the percentage of vendor kits entering the factory that will be lost to rework/scrap during each time increment.

The parameter R represents the number of vendor kits lost to rework or scrap for each time unit, and N3 represents the number of good products transferred from the factory to shipping in any time unit. Q2 is the number of final products in inventory and will vary, depending on the relationships between the factory output N3 and the sales to the market N4. MT describes the market purchasing characteristics and is provided as an initial model assumption. Decisions regarding each value to be used for ET and VT are linked to a decision-making loop (generally a function of Q2 and MT) inside the manufacturing system.

The model shown in Fig. 3-9 is a simple one that enables many important aspects of manufacturing systems to be considered. Relationships among suppliers, factory operations, and the market can be specifically illustrated. The impact of decision making on kit inventory and product inventory can also be specifically studied. The effects of different types of decision-making feedback loops in the system can be demonstrated directly. Delays associated with the provision of vendor kits and with factory production can also be illustrated. This simple model of a series of counters linked through defined logical operations can provide many useful insights into manufacturing systems.

The following sections use this SES model in a number of ways to illustrate specific issues and strategies being discussed. Further details of the model design are included in Appendix B.

The discussion so far indicates how the SES model can be used to describe the operations of a manufacturing system as the system interacts with its environment. It is also important to be able to describe the financial performance of such a system; the ultimate measure of system performance is often related to financial parameters.

One useful measure for such a system is the production cost associated with each unit sold into the market. Figure 3-10a defines financial parameters that are useful for such a calculation. As shown, $C is used to represent, on a cash basis, the cost of units sold during any time increment J. (This increment is often taken to be one day or one month for the examples presented in this text.) The cost per unit sold $C depends on the fixed daily costs ($FC), the variable costs per unit ($VC), the vendor kit daily storage costs ($VD), and the output product daily storage cost per unit ($ST). These four financial parameters can be combined with the SES system operational parameters to produce Eq. (3-1) in Fig. 3-10b.

Equation (3-1) indicates how the cost per unit sold $C during any time increment J is related to fixed costs, variable costs, vendor kit storage costs, and product storage costs. The variable costs ($VC) are multiplied times the total factory production (N3 + R) for each time increment. The vendor kit storage costs and product storage costs are multiplied times the inventory levels for vendor kits (Q1) and products (Q2), respectively. The denominator of this equation is given by N4, the total number of units sold for the time increment selected. This

Figure 3-10.
Cost analysis linked to
the SES model. (a)
Definition of costing
parameters that can
be used in conjunction
with the SES model.
(b–e) Equations that
can be used to de-
velop costing esti-
mates based on the
parameters defined in
Fig. 3-10a.

$C(J)   Total cost of products sold at the time defined by J
$FC(J)   Fixed cost per time unit (or day) of products sold at the time defined
           by J
$VC(J)   Variable cost of products sold at the time defined by J
$VD(J)   Vendor kit storage cost at the time defined by J
$ST(J)   Output product storage cost at the time defined by J

(a)

FIGURE 3-10B, FIGURE 3-10C, FIGURE 3-10D, FIGURE 3-10E

$$\$C(J) = \frac{\$FC + \$VC * (N3 + R) + \$VD * Q1 + \$ST * Q2}{N4} \quad (3\text{-}1)$$

(b)

$$(\$C)_{AV} = \frac{\sum\limits_{J\,min}^{J\,max} \$C(J)}{\sum\limits_{J\,min}^{J\,max} N4(J)} \quad (3\text{-}2)$$

(c)

$$J\,max - J\,min = \Delta T$$

$$(\$C)_{AV} = \frac{\$FC(\Delta T) + \$VC * [(N3)_{AV} + (R)_{AV}]\Delta T + \$VD * (Q1)_{AV}\Delta T + \$ST * (Q2)_{AV}\Delta T}{(N4)_{AV}\Delta T}$$

$$= \frac{\$FC + \$VC * [(N3)_{AV} + (R)_{AV}] + \$VD * (Q1)_{AV} + \$ST * (Q2)_{AV}}{(N4)_{AV}}$$

(d) $\qquad (3\text{-}3)$

$$\begin{aligned}\frac{\text{Income}}{\text{over cycle}} &= \$P_{AV} \sum\limits_{J\,min}^{J\,max} N4(J) - (\$C)_{AV} \sum\limits_{J\,min}^{J\,max} N4(J) \\ &= [(\$P)_{AV} - (\$C)_{AV}](N4)_{AV}(\Delta T + 1)\end{aligned} \quad (3\text{-}4)$$

(e)

equation can be used to develop a value for the cost per unit sold for each time increment.

It also turns out to be useful to calculate the average cost per unit over a period of time. When system operation is in a repetitive cycle (as is illustrated in Examples 3-4 and 3-5), then the average value should be calculated over one of these cycles. If the system is not varying in a periodic cycle, the average value must be calculated over a reference period that is defined.

Equation (3-2) in Fig. 3-10c shows how the average cost per unit can be calculated by knowing the costs associated with each time increment (or day of operations) and the number of units sold during each day of operations. The average cost per unit is obtained by choosing the period of time over which the averaging is to be done, then adding all costs associated with this period and dividing by the total units sold during this period. If the cost parameters \$FC, \$VC, \$VD, and \$ST are constants, then Eq. (3-2) may be rewritten as shown in Eq. (3-3) in Fig. 3-10d. Therefore, the average cost per unit can be calculated given the (constant) cost values (\$FC, \$VC, \$VD, and \$ST) and the average values of N3, R, Q1, Q2, and N4 over the specified period of time.

For several of the following examples, these average values are provided over a defined period of time or cycle. Thus, it is possible to use the results of the SES model not only to study the relationship between the system and the environment in terms of product volumes but also in terms of the cost impact. From the above information, it is straightforward to calculate the factory income over the defined period of time. If \$P is the average price for the product that is manufactured during this time, then the income produced is given by Eq. (3-4) in Fig. 3-10e.

The estimation of fixed and variable costs for existing or proposed manufacturing products and systems is a difficult but essential task. It is necessary to understand all aspects of system operation in terms of related costs and to estimate the total volume of sales expected. System costing can draw on a variety of modeling strategies linked to cost analysis, and commercial software can assist in product costing.

Equations (3-1) through (3-4) (Fig. 3-10) describe product costs in terms of the fixed and variable costs that apply to a given manufacturing system. The values for these costs (such as those in Example 3-6) are difficult to estimate and require the definition of many component elements. Chapter 5 (Sec. 5-7) discusses some of the commercial software programs that are available to assist in product costing. Chapter 6 provides further discussion of issues involved in costing for capital equipment. Chapter 10 presents a system design for which costing insights from previous chapters are considered together.

The SES model, in conjunction with the financial equations in Fig. 3-10, is a useful tool for the study of manufacturing systems and their environments and the financial operations of such systems. The system operational parameters and financial performance parameters together can provide a useful description of the manufacturing system and its financial operations. Alternative factory designs and alternative levels of automation and integration can be evaluated by considering the effects of these alternatives on the factory system-environment interactions and on the cost parameters associated with factory operations.

The SES model thus can be used as a link between factory design considerations and the financial performance associated with the factory. This relationship will enable trade-offs to be explored among alternative factory designs, alternative factory-market interactions, and the financial performance of the system. The model can be used to provide a link between specific CIM design considerations and the performance advantages associated with these operations.

EXAMPLE 3-1. OPTIMAL SYSTEM OPERATION

The objective of this example is to provide a reference system description for comparison with the more complex situations to follow. Consider an ideal setting with an optimal environment for the factory. Market purchases are always greater than the maximum factory output, and the entire system operates in a stable setting without any internal or external changes. The market demand is at a constant level, above the maximum production level for the factory.

This is the simplest situation that can exist for the factory, because with such stable operations, no adaptive decision making is required. The same kit order rates are placed to vendors as a function of time, the factory always operates at full capacity, and the market always buys the product produced at the prices that have been established.

In this type of environment, what types of issues are faced by the manufacturing system, how can the SES model describe this system-market relationship, and what are the issues associated with such a system?

Results

The results are illustrated in Fig. 3-11. The market (MT) is constant at a level of 500 products per time unit. For simplicity, assume that each time increment in this model represents one day; then, as illustrated, the market demand is sufficient to ensure that 500 products can be purchased by the market during each day of operations.

For the particular system shown here, EMAX, the maximum factory production, multiplied times U, the efficiency rate of the equipment, is set at 450 products per day. The maximum factory output thus will always be less than the available market. Vendor kit orders (VT) are constant at 450 per day. The input to receiving (N1) and kit release into the factory (N2) are also constant at 450 a day. For this optimized system, Q1 = 0 for all J.

All of the vendor kits pass directly into the factory, so ET is constant at 450 units per day. From this total, 45 units (10 percent) are routed to rework/scrap (R) during each day of operations. N3 shows the output of good product from the factory, which is constant at 405 per day. These products go directly into the market at the rate of 405 per day, as shown by N4. For the optimized system Q2 = 0 for all J.

Thus, for this particular arrangement, all parameters within the factory are held at a constant level. The sole functions of management are to continue

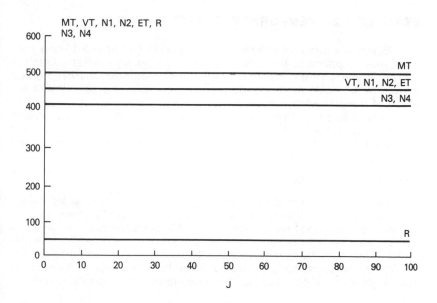

Figure 3-11.
SES model data for Example 3-1.

to make sure the market will absorb all product that is being produced, to try to reduce the amount of rework/scrap, and possibly to try to increase the total manufacturing capability of the factory (EMAX). This stable environment allows management to take a closed system view of operations. In such a stable, benign market environment, the tendency is to begin to neglect the market and market forces and to focus completely on internal operations. Since the market will always buy any products that are produced, it is easy to stop being concerned about the effects of the market. Effort turns to optimizing the long-term, steady-state operations of the manufacturing system. Costs are trimmed at every point to produce a highly optimized factory for the particular product and the particular market.

The user of the SES model must decide on the types of market behavior that are most important for consideration and on decision-making algorithms that are most likely. Then the quantitative results must be interpreted in terms of potential impact on enterprise objectives and operating strategies.

A number of dangers are obviously present in such a system. If any shifts in the market begin to occur, the system may be poorly suited to dealing with them. The decision-making loops may be inactive and unable to evaluate market changes adequately and produce adaptive responses. The type of system represented in Fig. 3-11 can be misleading. Although the system will optimize profits over the period of time illustrated, system operations also set the stage for a major disruption in the future when market change occurs. Long stable periods of operations can be dangerous for a manufacturing system, because the system can lose its ability to adapt rapidly. When change comes, the manufacturing system is poorly prepared to cope.

As discussed in Sec. 2-6, this condition is somewhat like that experienced by U.S. industry during the post–World War II years (1945–70). Assumptions about long-term market behavior were based on the market during the period. It was assumed that the market was stable and benign and that all products produced could be sold into that market. Since the manufacturing system was stable and did not have to change rapidly, corporate focus on the factory could be reduced, resulting in a major concern with fine tuning and with using financial and administrative controls to optimize the profit of the system. This type of simple model effectively sets the stage for what happens when the market environment begins to change.

EXAMPLE 3-2. TEMPORARY MARKET DISRUPTION

Given the frame of reference developed in Example 3-1, now consider what happens to this system when a temporary market disruption occurs. This example considers the effect of a temporary market drop on an otherwise stable factory system environment. It is assumed that because of the preceding period of stability, the factory is not prepared to make any changes in operations during this temporary interruption and continues to run at its full maximum capability (EMAX). However, when the market shifts, the relationships between the manufacturing system and the market will change.

Results

Figure 3-12 shows the results of applying the SES model for such a situation. Market purchases (MT) stay constant until day 50. From days 51 through 75, the market drops from buying 500 products per day (which is greater than the manufacturing system capacity) to 350 products per day (which is less than the manufacturing system capacity). The variable VT does not change, since the factory does not adapt to the situation and vendor kit orders continue without change. The stable relationship continues between

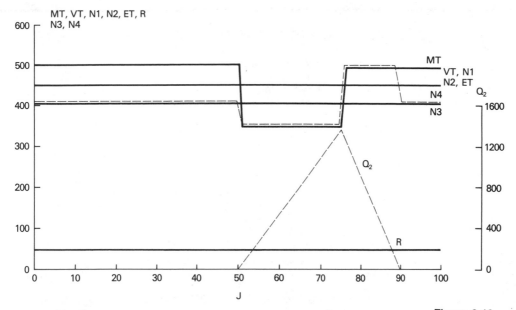

Figure 3-12.
SES model data for
Example 3-2.

the vendor kit arrivals and factory operations; there is no buildup of kit inventory Q1 because N1 = N2. ET also remains constant because it is assumed that the factory has had a long preceding period of stability and is not able to respond to this temporary market shift.

During this period of time the amount of output product going to rework or scrap R stays constant since system operations are constant. However, because the market drops temporarily below factory output N3, market sales (N4) will experience a temporary drop. A buildup of product inventory (Q2) results during the period of reduced market sales.

On day 76, the market MT again becomes greater than the factory output N3. Sales N4 rise above the factory capacity EMAX so long as inventory Q2 is available to meet the full market demand. The product inventory is gradually worked down since N4 is now greater than N3. During days 88–90, the final inventory is sold, and all effects of the transient market drop are eliminated.

Thus, N1 = N2, N3 stays the same since factory operations don't change, and N4 varies. On day 51, N4 drops below the stable level as sales to the market drop. Then on day 76, N4 rises above market capacity to allow the working down of product inventory at full product capacity. By day 90, the inventory has been worked off, and a stable situation returns.

This example illustrates what can happen to a system that has been stable for a long period of time before experiencing a temporary market drop. The management response to such a situation can take several forms. Management can view the market change as a one-time, temporary phenomenon and ignore it, since within a reasonable length of time product inventory has been worked off. On the other hand, such an experience may cause substantial alarm among management and engender a concern about what will happen if future market changes take place. Such a temporary shift in market demand may be viewed as a warning to force management to ask itself many difficult questions. On the other hand, it may also be convenient for management to neglect the experience of this market shift and continue business as usual.

EXAMPLE 3-3. PERMANENT MARKET SHIFT

This example illustrates what happens when a permanent market drop reduces the sales from the factory to the market. It is assumed that the drop is sufficiently large that the factory system begins to produce more than the market will absorb.

Results

Figure 3-13 shows the assumed market performance (MT). On day 51, market demand drops from 500 parts per day to 350 parts per day, and this drop continues throughout the 100-day period. Shipping of vendor kits (VT) is assumed to remain the same because during this period the factory is not adapting. The factory is assuming that the drop will be temporary and that the situation will again stabilize. The output N3 of the factory also remains constant, because it is assumed that this is a temporary market disruption; once the market returns to normal, the inventory will be reduced and all will be well. The factory continues normal operations (ET = constant), vendor orders continue normally, N1 = N2, and there is no buildup of kit inventory (Q1 = 0).

The rework/scrap level (R) remains constant since factory operations are not being adjusted. Factory output N3 is constant and shipments to the market (N4) drop when market demand decreases from greater than factory output to less than factory output. The product inventory (Q2) begins to rise on day 51. There is a continuing rise in the product inventory level that is unbounded so long as factory output continues greater than market demand.

This example illustrates how a manufacturing system in a stable environment can suddenly be significantly endangered. As the product inventory rises rapidly, so will storage costs, and as sales drop, an acute cash flow problem can quickly develop. This simple application of the SES model illustrates the dangers associated with the transition from a stable, benign

Figure 3-13.
SES model data for Example 3-3.

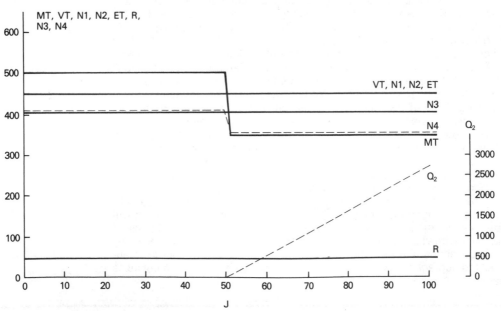

market to a market in which suddenly all factory product will not be purchased.

Again, this situation can reflect the experience of U.S. industry during the years 1945 to 1970. A long period of benign market demand resulted in manufacturing systems with poor ability to adapt to market changes. As the market for one type of product was reduced, perhaps in favor of a new market product, the manufacturing system continued to turn out the same volume of product without change, resulting in financial stress.

Examples 3-1, 3-2, and 3-3 provide a simple introduction to the use of the SES model to illustrate some of the important relationships among vendors, manufacturing systems, markets, and decision-making procedures. In the sections that follow, more specific issues involved in manufacturing system operations are illustrated through use of the SES model.

EXAMPLE 3-4. INTRODUCING A SIMPLE ADAPTIVE MARKET RESPONSE

This example builds on the results of Examples 3-1 to 3-3 to illustrate how management can use a decision-making algorithm to relate product inventory and factory output parameters to vendor orders and factory operations in order to "balance" the factory to the market. Instead of taking a completely closed system view of manufacturing system operations, a simple adaptation to market changes is introduced. A specific algorithm is assumed to relate the factory to its environment. It is assumed that the (daily) production level of the factory will be cut in half if the (daily) product inventory plus factory output exceeds twice the market demand. Thus, if the product inventory plus factory production level rises too high with respect to the market, the production level of the system will be cut in half. The production rate of the factory will stay at 50 percent until the product inventory reaches zero. At this time, factory operations will return to the maximum production level.

It is assumed that this adaptive system response is limited by the delay times associated with the decision-making loop. The production rate of the factory can only be changed at the end of three time units, or three days. Therefore, once market information is known, a delay is experienced in making the factory changes. Further, the orders of vendor kits are assumed to be delayed by five time units, or five days from the time that the market and product inventory information is available. These constraints can be associated with the time required to reduce staffing through layoffs, or to hire the necessary personnel, and to continue with contract commitments to vendors.

Results

A detailed analysis of Example 3-4 is included in Appendix B. The following discussion provides an overview of the important features illustrated by this example.

Consider Fig. 3-14. As soon as the market (MT) drops from above to below the existing level of production (ET), the product inventory (Q2) begins to build. The factory is overproducing for the market, resulting in inventory buildup.

This situation continues until the combined factory output and product inventory exceed the critical level. Once this condition is reached, a decision is made to reduce the factory production by 50 percent. Orders are given to

Figure 3-14.
SES model data for
Example 3-4.

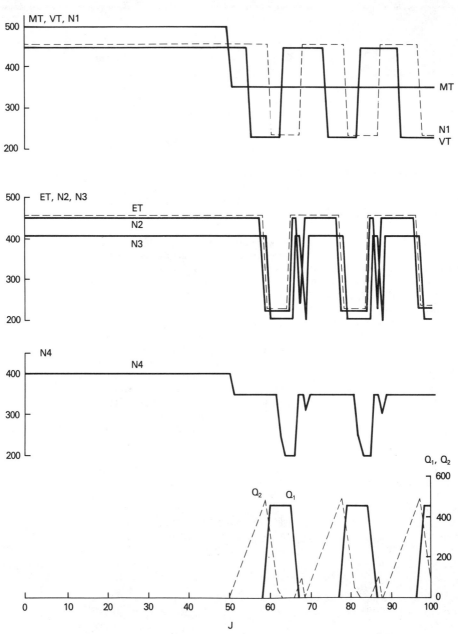

reduce the level of factory production (ET) and the vendor kit orders (VT). The implementation of these decisions is delayed by requirements associated with the partial closing down of suppliers and the factory. The arrival of vendor kits at receiving (N1) and the factory production level (N3) also drop once the decisions are implemented.

As soon as the production level drops, product inventory (Q2) begins to decline. The factory has reduced its output in order to prevent further inventory buildup and is now producing at a level that is below the market. After several days at this reduced level the product inventory Q2 is reduced to zero. At this time, a decision is made to resume the higher level of manufacturing, as reflected in the changes in VT and ET. Following turn-on delays, the arrival of vendor kits (N1) and the factory output (N3) increase.

At this point, an unexpected result appears. The factory output (N3) drops temporarily after the management decision is in effect for a higher level of production. The delay in the buildup of vendor kits (VT) has caused the vendor kit inventory to be reduced to zero. Without adequate vendor kits available, the factory cannot maintain its full production level. Thus, even though a higher output is desired, production slips temporarily due to an inadequate supply of vendor kits. As soon as the vendor kit arrivals increase to the higher level, production is able to resume as desired.

The resulting delivery of product to the market (N4) and the increases and decreases associated with product inventory (Q2) are illustrated. The temporary buildup and decrease of inventory that may be noted (between the major peaks) is due to start-up followed by a temporary cutback in factory output associated with inadequate kit inventory. As kit deliveries increase, product inventory (Q2) then begins a steady increase due to overproduction.

It is immediately clear that this type of decision-making algorithm leads to operational problems. The factory seems unable to match itself to the new market environment but rather alternates between turning out too few products and too many products. The average productivity of the manufacturing system can be determined from N4, which shows actual shipments to the market, by averaging values over a cycle. The average number of products provided to the market can be calculated. At the same time, Q2 shows the product inventory buildup, and the average inventory can be calculated from this figure. Thus, these two figures provide a summary of the number of parts being delivered to market on the average and the average inventory levels of inventory of completed parts. The graph of Q1 can be used to calculate the average inventory of vendor kits Q1. These three parameters provide a means for evaluating the average inventory levels and the average number of good parts being shipped to the market.

Although it may seem obvious from looking at these data that the algorithm needs to be changed, it is nonetheless true that many manufacturing systems find themselves in situations in which they alternate between production levels that are too high and too low for the market. This places a great stress on manufacturing operations and personnel.

The strategy being used by management in this example is to bracket the market by overproducing and underproducing. Such a strategy might be used for many reasons. The factory may be designed in such a way that flexibility only exists for a 50 percent reduction in operations. It may not be feasible to reduce operations by any other percentage rate. For such a case, the factory is slave to market variations, as shown here. It should be expected that such operations will be costly because of the start-up and shutdown costs and because of the costs associated with the storage of kit inventory and product inventory. It is clear that an operation of this type will produce products that are expensive with respect to the optimum production costs associated with Example 3-1.

The SES model can illustrate the problems associated with oversimplified decision-making algorithms. At the same time, the model forces planners to identify specifically the information base required to implement various algorithms and to evaluate the availability of the information in a timely manner. A significant advantage can accrue when such decision strategies are subjected to careful analysis and criticism.

This approach to matching a factory to an environment may be viewed as arising from limited factory capabilities and unsophisticated decision-making loops. Thus, this type of adaptation might develop during a shift from the benign environment of Example 3-1 to a much more complex and demanding environment. This approach to matching a manufacturing system to its environment might be viewed as a temporary learning process in which experience is gained with respect to the effects of certain decisions. Certainly, based on this type of performance, factory management would quickly begin to understand the need for a wider range of factory capabilities in terms of partial production levels. Also, the costs associated with kit inventory and product inventory would suddenly have a major effect on overall financial operations and would become of concern to factory management.

EXAMPLE 3-5. MODIFIED ADAPTIVE RESPONSE

For this example, management attempts to match factory operation to the market environment by relating factory production directly to the product inventory level, without attempting to track market demand. The product inventory will be allowed to vary between 20 and 200 percent of the factory production level. If the upper limit of 200 percent is reached, the factory cuts production in half; if the lower limit is reached, the factory increases production to 100 percent. The factory can still only operate at 100 or 50 percent of capacity. Once one of the trigger levels is reached, the factory takes three time units (or three days) to make the changes in production, and again it is assumed to take five time units (or five days) to change the number of vendor kits ordered. These constraints are similar to those in Example 3-4, to allow for personnel changes and to conform with subcontract requirements.

Results

A detailed analysis of Example 3-5 appears in Appendix B. The following discussion provides an overview of the important features illustrated by this example.

Figure 3-15 illustrates the impact of this modified adaptive response on factory operations. In general, the system response is similar to that illustrated in Example 3-4. As soon as the market (MT) drops, product inventory (Q2) starts to build. When the inventory reaches a critical level, a decision is made to cut back on vendor kit delivery and factory production. Delays in the implementation of these decisions are reflected in the vendor kit arrival and the factory output parameters (N1 and N3).

The same general output result is observed here, although some of the specific features of the system performance have changed. There is still a periodic increase and decrease in both the kit inventory (Q1) and the product inventory (Q2). The resulting product sales to the market (N4) take the form shown. A change in the specific parameters and threshold levels between Examples 3-4 and 3-5 has altered some of the details of the system performance. Yet, at the same time, the fundamental inability to match the factory to the market continues to dominate system behavior.

Figure 3-15.
SES model data for Example 3-5.

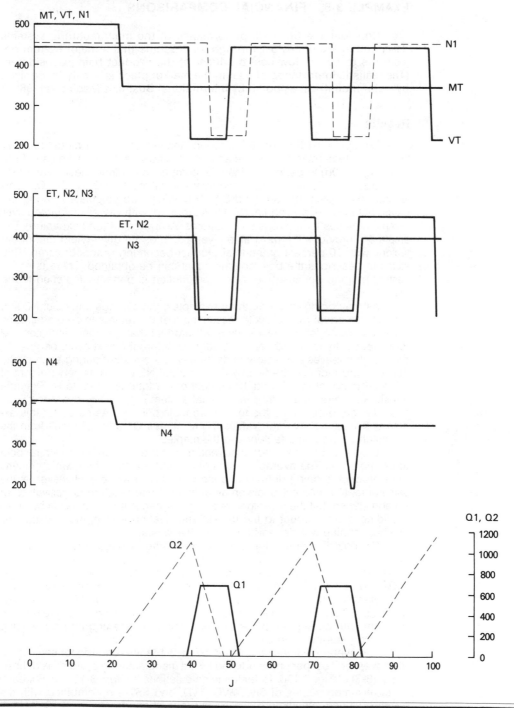

EXAMPLE 3-6. FINANCIAL COMPARISONS

Consider the financial performance of the manufacturing systems described in Examples 3-1 through 3-5. Assume that a major system objective is to be the low-cost producer of the product being considered. (The realistic importance of cost in the market place is widely recognized by major industries, as noted by International Business Machines [1988].)

Results

Examples 3-1 through 3-5 introduce the use of the SES model to study types of system-environment interactions. It is now useful to make use of Eq. (3-3) (Fig. 3-10d) to perform a financial comparison among these examples.

Figure 3-16a summarizes the average values for N3, R, Q1, Q2, and N4 from Examples 3-1, 3-4, and 3-5. The values are calculated during the period of time after the market (MT) has dropped, with values calculated over complete cycles for the repeating patterns. As indicated (and expected), Example 3-1 provides the most attractive option for the given yield rate of the factory, with 10 percent of the total product becoming rework or scrap. This example describes the best operation that can be obtained. There is no inventory buildup because the factory production is perfectly matched to the market.

As may be noted for all three examples, the average values of N3 and N4 must be equal over a cycle (assuming that Q2 returns to zero after each cycle, as shown for the cases here). All product that goes into inventory Q2 is subsequently removed from inventory Q2 before the next cycle begins. All product that leaves the factory finds its way to the market during each cycle. The total product into the factory is the sum of N3 plus R. Ninety percent of this total is equal to N3, and 10 percent of the total is equal to R. Example 3-4 shows a significant drop in the total system output. The average output of the factory drops from 405 to 318 products per day. As noted for this example, substantial average values of inventories Q1 and Q2 result from the mismatch between the factory and the market.

Example 3-5 shows some advantages and disadvantages with respect to Example 3-4. The average values of N3 and N4 are larger, which means more product is being delivered to the market over a cycle. However, at the same time, the different decision algorithms in use result in larger values of Q1 and Q2, so that the inventory levels are higher. A trade-off exists between providing more product to the market and dealing with higher inventories. The best choice will depend on the relative costs.

For both Examples 3-4 and 3-5, the average values of Q1 and Q2 are not the same. This is because if both Q1 and Q2 return to zero during each cycle, as shown for the cases here, differences between the average values of Q1 and Q2 will result due to the different holding times in inventory, leading to different lengths of time it takes the product to be removed from inventory. If Q1 and Q2 do not return to zero after each cycle, then a cumulative buildup over time will be superimposed on the cyclical increasing and decreasing of inventory during each cycle.

The average values calculated in Fig. 3-16a are required for calculation of an average cost per unit produced by the manufacturing system, as defined in Eq. (3-3) of Fig. 3-10d. Refer to the calculations in Figs. 3-16b–e. For each case, assumed values of $FC, $VC, $VD, and $ST are combined with the average values associated with system operations from Fig. 3-16a to cal-

	Example 3-1	Example 3-4	Example 3-5
$(N3)_{AV}$	405	318	340
$(R)_{AV}$	45	35	38
$(Q1)_{AV}$	0	166	225
$(Q2)_{AV}$	0	180	513
$(N4)_{AV}$	405	318	340

(a)

$C

	Ex. 3-1	Ex. 3-4	Ex. 3-5
$FC = $1000 $VC = $10 $VD = $1 $ST = $1	$13.60	$15.30	$16.23

(b)

$C

	Ex. 3-1	Ex. 3-4	Ex. 3-5
$FC = $5000 $VC = $5 $VD = $1 $ST = $1	$18.52	$22.20	$22.44

(c)

$C

	Ex. 3-1	Ex. 3-4	Ex. 3-5
$FC = $5000 $VC = $1 $VD = $1 $ST = $1	$13.58	$17.92	$17.99

(d)

$C

	Ex. 3-1	Ex. 3-4	Ex. 3-5
$FC = $5000 $VC = $1 $VD = $0.50 $ST = $0.50	$13.58	$17.38	$16.90

(e)

Figure 3-16.
Cost analysis data for Example 3-6. (a) Data from Examples 3-1, 3-4, and 3-5. (b–e) The results from Example 3-6.

culate average costs for the products produced. The fixed and variable costs* are used in combination with operational parameters to provide a financial estimate of the cost per unit for each of the cases under consideration.

Case A involves relatively low fixed costs ($FC = $1000), high variable costs ($VC = $10), and storage costs of $VD = $1, $ST = $1 for each kit or product per time unit. Figure 3-16b shows the results that are obtained for the product costs for Examples 3-1, 3-4, and 3-5 under this set of assumptions.

---

* See Wittry (1987) for a discussion of the appropriateness of unit costs as a measure of manufacturing system performance, and Gustavson (1986) for a discussion of the characteristics of fixed and variable costs.

Note that, as expected, Example 3-1 gives the lowest cost because of the ideal match between the factory and the market. Example 3-4 provides a higher cost, and Example 3-5 the highest cost for the particular set of assumptions used here.

Figure 3-16c shows an alternative set of assumptions, with the fixed costs increased substantially, which might be associated with investment in capital equipment. The variable costs are reduced by a factor of 2, and the inventory storage costs remain the same. The baseline cost per unit for Example 3-1 increases significantly, and the costs for Examples 3-4 and 3-5 are similar.

In Fig. 3-16d, the fixed costs allocated to each period of time remain at $5000, the variable costs are reduced to $1, and the storage of inventory cost for kits and products remains the same. In this case, the production cost per unit is about the same for the reference case of Example 3-1. Example 3-4 and Example 3-5 are substantially higher but are almost equal to one another again.

Finally, in Fig. 3-16e, the fixed and variable costs remain the same as in Fig. 3-16c, but the inventory cost is reduced in half. The reference case of Example 3-1 remains the same, since the change in inventory cost has no effect here. However, with reduction in inventory cost, the approach in Example 3-5 now has lower cost than in Example 3-4.

The SES model illustrates how a significant change in market behavior can alter the relative advantages and disadvantages of a specific manufacturing system. A changed environment can cause a successful enterprise to rapidly reach an endangered status. In rapidly changing, turbulent environments, all enterprises are constantly being challenged for survival.

As seen here, operational performance and cost structures must be combined in order to select an optimum strategy for factory performance. It is possible to change operational approaches and change cost structures in such a way that significant differences exist in production costs. Analyses of the type shown here can be helpful in exploring operational strategies to select the one that will be most effective within the anticipated market setting.

This example also indicates how the advantages of one manufacturing system may change radically when the market changes. The lowest cost solution for product manufacturing depends on the behavior of the market and on the algorithms that are used to link the market and inventory levels to production and vendor orders. This example points out the importance of considering trade-offs between fixed costs and variable costs. It also indicates the importance of inventory cost, as changes in this cost can change the nature of the preferred solution.

3-4. A Planning Strategy

The design (or redesign) of a manufacturing system and its implementation are formidable tasks. Figure 3-17 illustrates one approach to viewing the challenges faced by the planning group.

The first step in the planning strategy is to *define the planning tasks* that must take place as part of the design effort. The tasks define what has to be done to obtain a (design) solution, and task linkages indicate the relationships among tasks that are expected. The creation of this "functional planning" flowchart can be the first step toward a solution. The tasks are defined, although the methods for accomplishing these tasks are not yet considered. Preparation of a task flowchart can be regarded as the first step in system modification, as shown in Fig. 3-17. The object of step 1 is the definition of the tasks that must be performed.

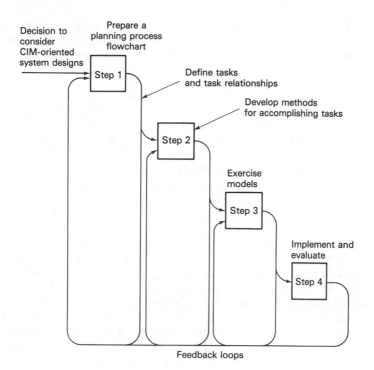

Decision to consider CIM-oriented system designs

Prepare a planning process flowchart

Step 1

Define tasks and task relationships

Develop methods for accomplishing tasks

Step 2

Exercise models

Step 3

Implement and evaluate

Step 4

Feedback loops

Figure 3-17.
Overview of a planning strategy for manufacturing system design, showing how each step feeds back to previous steps to create an ongoing learning process. The steps relate to preparing a planning process flowchart (step 1), developing methods for accomplishing the task (step 2), exercising models that are called for by the methods (step 3), and implementing the modeling results, with ongoing evaluation (step 4).

The second step in the problem-solving process is to *develop methods* for accomplishing each of the defined tasks and for linking the tasks together in the indicated manner. The focus is now on *how* the planning tasks are to be performed. Specific means must be introduced for performing each defined task and dealing with relationships. For complex problems and processes, many different approximate models must be introduced and related at this stage. Care must be taken that the methods selected are appropriate to the overall problem-solving objectives and constraints. The planning group provides the integration mechanism.

The third step is to *exercise the models and methods*, thereby performing the indicated tasks, in order to produce information. Many iterations will be required, and trade-offs considered, in order to balance off all of the constraints.

The last stage is *implementation and evaluation* (to support the learning cycle). The implementation tasks must be defined, and the implementation must be combined with ongoing evaluation throughout the process.

The four problem-solving steps of Fig. 3-17 are closely interrelated through feedback loops. The planning process flowchart (step 1) will often have to be modified to accommodate the methods available for accomplishing tasks (step 2). As the models are exercised (step 3), changes may be required in the task requirements (step 1) and methods for accomplishing tasks (step 2). Finally, implementation efforts will require constant adaptation of the entire cycle.

The four steps are useful because they emphasize that the planning group should first examine the tasks that *should* be completed, then ask *how* they can be completed. Any compromises due to the feedback process are then noted and evaluated as part of the design activity.

Figure 3-17 can be interpreted as a first-level system overview making use of the $IDEF_0$ model (introduced in Chap. 1). This methodology has been used for manufacturing system design by a variety of groups, including the Society of Manufacturing Engineers (Savage 1986; Bertain and Hales 1987).

The planning strategy in the flowchart of Fig. 3-17 is a complex one, due to the numerous feedback loops. It is not a sequential, linear activity, with a readily identifiable start and conclusion. Rather, Fig. 3-17 illustrates an iterative, system-oriented process without specific start or finish. Each activity can affect not only the "downstream" tasks, but can also feed back to the "upstream" tasks. All tasks are interrelated. A planning strategy must depend on "search and learn."

Planning for manufacturing systems presents many challenges. The best plan cannot be obtained by developing a matrix of equations and "turning the crank" to produce an analytic solution. Efforts must be made to "work away" continually at the problem, making maximum use of a search, learn, and adapt strategy.

Given such a complex process, how can a company proceed with planning for a CIM-oriented system? Certainly the complete problem is unworkable, in the sense that a unique, optimum answer cannot be produced from a matrix of equations. Too much is unknown, and the relationships are too complex. On the other hand, doing nothing is an unsatisfactory decision, since the competitive environment requires a response. Lack of action means that no experience will be gained, so the situation will only be less tractable in the future.

What is required is an approach to producing workable problems and solutions that have a reasonable probability of success. Efforts must be made to "work away" at the complete problems, learning all the way, in order to develop the ability to produce a revised system that more closely attains enterprise objectives. An effective approach is thus to work on selected portions of the full problem, while concurrently cycling and recycling through the flowchart of Fig. 3-17 to maintain perspective. The combined top-down and bottom-up strategy is a useful approach in this context.

3-5. The Planning Group

The successful development of a CIM-oriented manufacturing system depends on the establishment of an effective planning group. Group members must be able to develop the needed planning process and work productively as a team. They must have the ability to operationalize the planning process by accomplishing each task or relationship and to implement the plan that is developed.

Many different types of modeling efforts may be required in support of the planning activities (as reviewed in Chap. 4 and in subsequent chapters). These can range from calculations relative to a specific item of equipment to complex simulations that consider functional system operations or system-environment interactions. Group members must have the ability to understand the quantitative results of supportive modeling efforts and the technical and management aspects of present and future systems. However, since no single model encompasses all aspects of the system and environment, the planning group must ultimately perform the integration function.

As more *expert systems* are developed, growing levels of model sophistication will become available (as is discussed below). However, treatment of the nonlinear aspects of the system design and coping with nonquantifiable decision-making factors will likely still require a human integration process. And it must be remembered that political aspects of decision making make it unlikely that critical system decisions will be made completely on the basis of modeling predictions.

From an implementation perspective, the planning group is essential. An organizational mechanism must be in place to understand what has to be done, to direct the performance of the necessary tasks, and to accept accountability for program outcomes.

3-6. A Multilayer Model for the Study of Design Principles

As introduced in Fig. 3-1, the design of a CIM system will often take advantage of a set of design principles that are incorporated into idealized reference systems. These reference systems help guide the formation of the *to-be* system concepts and the resultant transition stages.

The design principles in use today reflect a combination of historical experience and theoretical constructs that seem to be generalizable to a wide range of systems. The resultant reference systems can be based on the specific activities of several enterprises and their learning conclusions, or on seemingly attractive constructs resulting from theoretical analysis and modeling, or from a combination of both resources.

Design principles often emphasize a particular aspect of a system, resulting in a reference system that does not reflect the full complexity of a *to-be* system. In applying such principles, it is important to avoid building a one-dimensional system and a single-reference concept. The mechanism of application must still emphasize an integrated systems approach for *to-be* system design.

The most useful way to discuss design principles is to develop a model of the internal aspects of a manufacturing system. In order to do this, however, it is necessary to recognize the many different ways in which such a system can be viewed, in terms of: (1) its physical and observable properties; (2) the functional relationships that exist in the system; and (3) the management/organizational relationships that exist in the system. (This three-level approach was introduced in Chap. 1 and applied in Chap. 2.)

This section describes a multilayer model of the internal manufacturing system (Fig. 3-18). The model expands on the three perspectives for viewing a system to create a five-layer model. Each layer provides a different way of describing the system. Strong interrelationships exist among the layers, ensuring that they provide complementary descriptions of the same system. The five layers result from an extension of the concepts introduced.

The multilayer model presented here describes a manufacturing system in terms of its physical, functional, and organizational aspects. These three ways of viewing the system relate to the visible (physical) aspects of the system, the knowledge (functional) information base included in the system, and the human (organizational) relationships of the system. The five-layer model is an extension of these three system aspects; the model separates "touch labor" and supporting manufacturing functions, and thus provides a more detailed structure to guide system design.

Figure 3-18.
Multilayer model of the internal manufacturing system showing how the physical, functional, and organizational system perspectives have been used to define a five-layer model. Resource materials are provided as input to layer 1, the *material processing* layer. Equipment and human workers create the output product. Layer 2 is the *process function* layer, which contains the knowledge and information base that describes the functions that must be performed to achieve the desired process. Layer 3, the *organizational/management* layer, describes the functioning of people within the organizational setting. Layer 4, the *information function* layer, receives input information, processes the information, and produces output information. Layer 5, the *control* layer, consists of equipment and human workers to provide control over the processes taking place. Strong coupling and interrelationships exist among these five layers.

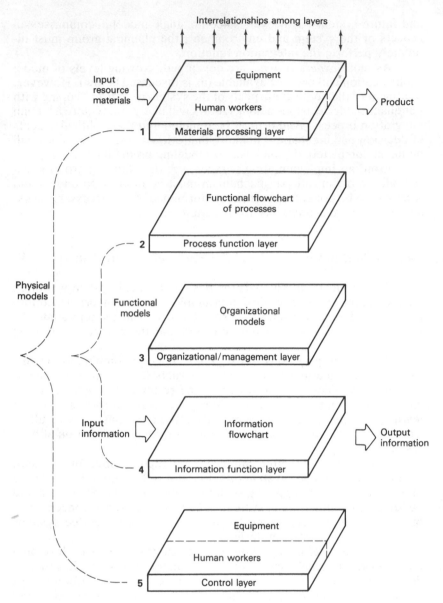

The parallel emphasis on materials transformation and information flow in Fig. 3-18 has also been suggested by Wozny of the National Science Foundation (Piszczalski 1987) and others, including Lardner (1986), Brower (Burggraaf 1985), and Kutcher (1983).

Viewing the manufacturing system, an observer will see equipment and human workers that are transforming input resources into a product (layer 1), and equipment and human workers that are providing the required control activities to operate the resource processing system (layer 5). The equipment and workers in layer 1 process the resources by direct contact, whereas the equipment and workers in layer

5 are not in direct contact with resource processing but function in a supporting and supervisory capacity. Control is exerted by having layer 5 provide information to and direct the processing workers or operating equipment (layer 1). Together layers 1 and 5 constitute the observable portion of the manufacturing system and determine the performance boundaries associated with the system.

It is important to distinguish between the *direct processing* system and the *indirect supporting and supervisory* system. In many manufacturing facilities, the bulk of the employees work at the indirect level, often involved with handling the flow of information that is required to maintain the process flow. Thus, the design of more efficient manufacturing systems must deal with both the most competitive strategies for layer 1, the observable material and product transformation activity, and the strategies for layer 5, the indirect support structure that enables the transformation activity.

For a highly manual production facility, the material transformation and indirect support layers are formed by human workers. In a semiautomated facility, these layers are based on equipment/worker interactions. For a highly integrated and automated facility, the role of the human worker can be significantly reduced in both the direct production and control processes. (In the latter case, human involvement in the system may be limited to securing the required system input, maintenance, and system improvement.)

Layers 1 and 5 describe the observable part of the system and determine the physical constraints on the manufacturing activities. However, the intelligence of the system, although not visible, is clearly a crucial aspect of facility operation. Layer 2 may be thought of as a functional flowchart of the processes being performed by layer 1. Layer 2 describes how the processes of layer 1 must relate to one another to transform the input resources to the desired product. Layer 2 is thus a functional model of the desired processes, whereas layer 1 is the visible implementation of the related activities.

The functional model of layer 2 can reside in the memory of human workers, where it is used to guide the activities of layer 1. For partly or highly automated systems, portions of the functional model are built into the equipment itself. The combined human worker/equipment combination must contain all the required capabilities to complete the processes according to externally imposed requirements.

Where layer 5 includes the physical equipment and human workers used to control the materials processing activity, layer 4 may be thought of as an information control flowchart for the system. Where layer 2 describes the process functions necessary to manufacture a desired product, layer 4 describes the information flow that must exist to direct the desired process combination. Layers 1 and 5 provide a complete description of the observable system; layers 2 and 4 provide a complete model of the functions and information that form the system knowledge base.

For all manufacturing settings, the activities of workers depend on human behavior characteristics and on the organizational/management structure for the facility (layer 3). This structure is also a nonobservable determinant of behavior but is of critical importance in system operation. Unless the organizational/management layer is matched to the defined system, worker activities may not be consistent with the desired flowchart models. The result will be a system with all of the physical requirements in place but lacking satisfactory system function. Similarly, if the knowledge base models of layers 2 and 4 are inadequate, decisions by human workers and equipment will not satisfy desired performance standards, and the product output will not be satisfactory.

To obtain acceptable system operation, the necessary physical system (workers and equipment) must be in place, an adequate knowledge base must be incorporated into the system, and the organizational/management structure must be appropriate to the system. The preferred organizational/management structure can change significantly as manual and semiautomated systems evolve to fully automated systems.

Figure 3-19.
CIM design principles associated with various layers of the multilayer model. The design principles are linked to one another by interactions among the modeling layers. Integrated operation is obtained through system management that develops an overview perspective based on the design principles.

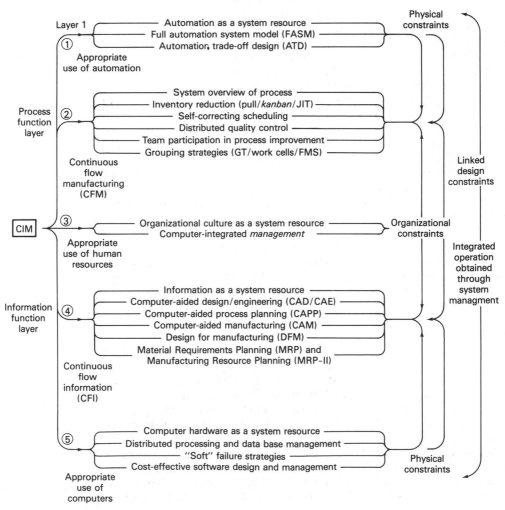

Design Principles

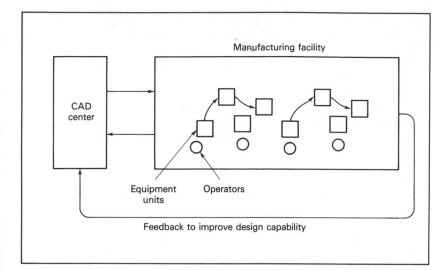

Figure 3-20.
Functional flowchart corresponding to layer 2 (the functional layer) of the multilayer model. This drawing illustrates how a computer-aided design (CAD) center can be linked to a manufacturing facility represented by equipment units and operators. The CAD design center interacts closely with the manufacturing facility, and feedback from the facility is used to improve design capability.

The five-layer model can provide a useful entry into the internal study and design of manufacturing systems. The model provides a representation that can be used to examine the system from several perspectives and to identify some of the layer trade-offs and areas of complication that can be encountered.

The design principles available to the planning group (Fig. 3-1) can be conveniently explained and related to one another in the context of the multilayer model.* A range of reference systems for use in the planning/design activity can then be defined through various idealized combinations of these principles. In turn, these reference systems will help drive the learning and improvement processes.

In Secs. 3-7 to 3-12, the multilayer model is used as a framework in which to present some of the design principles that contribute to CIM development and to illustrate some of the interrelationships that exist among these principles. Figure 3-19 can be used as a road map for the discussion.

3-7. Design Principles for the Process Function Layer (Layer 2)

Many of the CIM design principles are formulated around the process function and information function aspects of the system (layers 2 and 4 of the multilayer model in Figs. 3-18 and 3-19). Guidelines are proposed for optimal functional relationships. The physical aspects of the process and control layers (layers 1 and 5) can then be viewed as real-world constraints that must be incorporated into the functional design.

Figure 3-20 illustrates a functional flowchart corresponding to layer 2 of the multilayer model. The manufacturing operations include

---

* Alternative considerations for CIM system design are discussed by Cousins (1988) and Fuchs (1988).

a computer-aided design (CAD) capability that is linked to a CIM facility. The facility consists of a variety of equipment units and operators that provide the necessary material transformation process. One of the more or less obvious issues in facility design is to decide how the various processing stations can be linked most effectively. Several design principles have been presented to address this issue.

When looking at the facility in overview, *continuous flow manufacturing* (CFM) has been suggested as an appropriate emphasis. CFM is a system-oriented concept that expresses the need to achieve internal balance among the multiple manufacturing processes in a CIM facility. An effort should be made to optimize the relationships among all tasks in the facility in order to maximize all *value-added* activities and remove all *non-value-added* activities. One implication of the CFM principle is that there should be no buildup of work-in-progress (WIP) among the processing stations. Each process station should produce enough output to satisfy the needs of the following process stations, without building up inventory (which is expensive to finance and store).

A *pull-oriented* CIM system sets process equipment performance levels by starting with the desired output flow at the final production stage, then working backwards through the process flow to set performance levels for every station to ensure the needed input, but without excess (Fig. 3-21).

The production process can be controlled by using a series of *tickets* or *tokens* for communication among process stations. This type of pull emphasis is also called *kanban,* a Japanese word that means "ticket" (Suri and DeTreville 1986). Each station can control the availability of input to itself by passing a ticket to a previous station each time an input product is required. In this way, the pull system is generated, with the output at each process station controlled by the needs of the following stations. The fewer the number of tickets in the entire system, the more in-process inventory is reduced.

This same concept can be extended to control input inventory to the manufacturing system from all suppliers. If the tickets are passed between the manufacturing company and those providing the required resources, both input and in-process inventory can be reduced. The result is called a *just-in-time (JIT)* approach to managing input inventory (Fuchs 1988).

Suri and DeTreville (1986) have emphasized how the kanban concept can be used to drive the search and learn efforts conducted by the system to improve itself. In a typical job-shop or assembly-line-oriented facility, external input and in-process inventory can be used to smooth out production problems and cover up inefficiencies. If the *to-be* system design includes transition stages toward CFM and a pull system, then a major learning program will be required. The transition stages can involve a gradual reduction in the effective number of tickets in circulation, forcing a search and learn activity to improve process efficiency and flow. The CFM principle, as reflected in model system concepts, can be used to help drive system evolution.

The pull system discourages the buildup of inventory levels at intermediate manufacturing process stations. In turn, this requires facility-wide coordination and quality control at each process station (to

Simple concepts can have widespread impact on manufacturing system design and operation. An effort to maximize all value-added activities and remove all non-valued-added activities in a manufacturing system can result in significant change. Again, however, the design must be evaluated against a variety of market conditions to prevent "brittle" solutions that are highly effective for the moment but are limited as environmental changes are encountered.

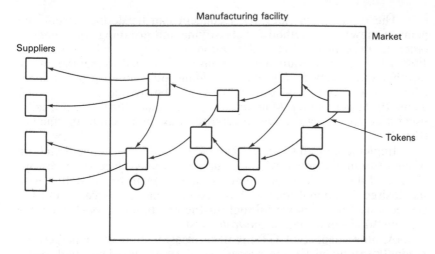

Manufacturing facility

Suppliers

Market

Tokens

Figure 3-21.
The use of tokens to establish a pull-oriented manufacturing system. As shown in this functional flow-chart, the tokens, which stimulate production, are driven by the market. For each output product that is required by the market, a token is provided to the final processing stage of the manufacturing facility. Each manufacturing stage then passes the token to the prior stage in order to achieve the required input. This passing of tokens continues throughout the manufacturing facility and then to the suppliers to stimulate delivery of the vendor kits or components and materials required.

prevent rework and the development of insufficient WIP). A complex monitoring process must be in place, and a self-correcting scheduling system is required.

The CFM concept requires active operator/worker participation in producing a process flow that is optimized at each station and throughout the facility, combined with system-oriented leadership by managers. Extensive interaction is required among the operators and managers to produce a coordinated team effort. Figure 3-20 illustrates how work at the individual equipment level must interact with a system overview to produce the required CFM functions.

CFM thus appears to be a reasonable design principle, but one that can be difficult to implement. The requirements for system overview, coordination, high quality levels, and close teamwork may not be compatible with equipment capabilities (constraints from layer 1) or with the existing organizational/management structure (layer 3). A reference system based on the CFM principle has many attractive aspects, but also may not be completely realistic. As noted in Sec. 3-1, the reference system provides an idealized, extreme perspective that can help in the configuration of *to-be* systems. The CFM concept is very broad and does not specifically indicate *how* the processing functions can best be arranged to achieve the desired levels of effectiveness and efficiency. Supplementary design principles are useful to provide more immediate guidance in this area.

In viewing a manufacturing system of the type shown in Fig. 3-20, a reasonable question is how the various equipment stations should be functionally related to one another. Should some type of grouping be introduced to optimize the facility? In the past, the functional relationships among equipment units were often achieved by: (1) placing units where they would "fit" (drawing on the physical constraints of layer 1 to determine placement); (2) grouping by class or type of equipment; or (3) grouping to achieve a linear assembly-line flow.

The issue here is how to achieve a grouping strategy that will best serve system performance objectives. Several approaches to this issue have been embodied in design principles.

The *group technology* (GT) concept can be defined (from one point of view) as a method for classifying and grouping equipment in order to maximize functional relationships (Hyer and Wemmerlov 1984, 1988). That is, equipment stations are arranged so that they work together effectively as a group, providing interrelated process capabilities. These combined stations can then be treated as a *work cell* within the larger facility. Manufacturing takes place through work cells, using a cellular organization to arrange equipment within the facility (Fig. 3-22).

Implementing this concept can be difficult because the definition of the attributes through which grouping will take place is not always clear, and the physical constraints (of layer 1) can be in conflict with the desired functional relationships. Nevertheless, this version of GT can be a useful and powerful tool for the creation of model reference systems to aid in design of the *to-be* system.

Another aspect of GT involves classification of manufactured products to emphasize the process similarities associated with each, thereby shifting functional emphasis from equipment to products. GT applied to products will have a coupled effect on the formation of appropriate work cells and will also have a major impact on new-product design approaches. Given an appropriate set of attributes, GT can simplify and expedite the design process. (This aspect of GT is further discussed below with respect to computer-aided design.)

A further evolutionary step in work cell design is to integrate physically several equipment stations into a single *flexible manufacturing system* (FMS). As defined by Jaikumar (1986), "[an FMS] is a computer-controlled grouping of semi-independent work stations linked by automated material-handling systems" (p. 70). An FMS is thus a "miniature automated factory" within the larger CIM facility, which may consist of a number of such work cells. From another viewpoint, an FMS involves integrating individual process equipment into a single subsystem capable of multiple functions. The FMS concept is a powerful approach to maximizing work cell automation and integration and to CIM system design.

Figure 3-22.
Grouping manufacturing equipment into work cells. As illustrated, manufacturing tasks can be linked in various ways into work cells that consist of both equipment and operators. The objective is to choose grouping principles for the work cells that will optimize production.

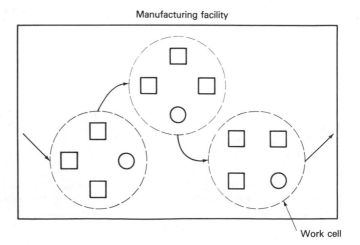

Manufacturing facility

Work cell

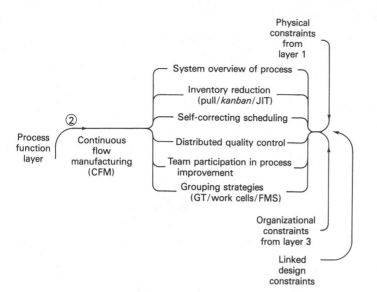

However, as is true of all design principles, many problems may be experienced in implementation. Physical design constraints may have to be overcome to achieve the desired levels of automation in a cost-effective manner, and the management of the FMS can stretch organizational capabilities. As Jaikumar (1986) noted, the effectiveness of an FMS is very management-dependent. The required organizational impact will include expanded team-building efforts and a broadening of engineering management roles to include manufacturing.

Figure 3-23 shows how the general framework of Fig. 3-19 describes design principles associated with layer 2 of the multilayer model.

3-8. Design Principles for the Information Function Layer (Layer 4)

In the same way that continuous flow manufacturing (CFM) can be taken as a design principle for the process function layer (layer 2), continuous flow information (CFI) can be regarded as a design principle for the information function layer (layer 4). The flow of information should be optimized to best achieve performance objectives for the facility (Fig. 3-24).

Classically, manufacturing information flow has been characterized by a "paper trail" that results in duplication, errors, and limited access. Many organizations are built around such a paper flow, and an attempt to implement CFI may be painful. The objective is to treat information as a manufacturing resource, to be handled as appropriately as possible. Duplication should be minimized, access provided to all who need it, and costs reduced as much as possible.

It has been estimated that more than 50 percent of the costs of a manufacturing facility can be associated with an information overhead.

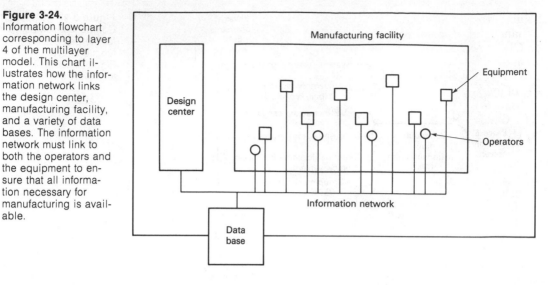

Figure 3-24.
Information flowchart corresponding to layer 4 of the multilayer model. This chart illustrates how the information network links the design center, manufacturing facility, and a variety of data bases. The information network must link to both the operators and the equipment to ensure that all information necessary for manufacturing is available.

Many companies have come to understand that the paper-based handling of information can drive total product cost. Significant savings can result through elimination of the multiple organizational layers and departments that are required to develop and process the paper flow in a given manufacturing setting. However, such change can strike at the core of the organizational structure and may be difficult to implement with success.

Thus, large potential savings can be associated with CFI. On the other hand, such a large cost area will be associated with a wide range of vested interests and resistance to change. Successful application of the CFI design principle (layer 4) will depend on the willingness of the organization (layer 3) to evolve in significant ways, combined with an understanding of the principal implementation constraints (layer 5). CFI implies the integration of the information flow, the use of computer networks, and a common computerized data base.

The information layer encompasses the entire product design-manufacturing cycle, including information flow related to system management. The information flow is the major unifying aspect of the system. Part of this information flow will be data from all of the computer-aided activities taking place, as related to design, process planning, manufacturing, and feedback from the manufacturing system (and to allow improved design and planning activities). In addition, a continuous information flow must take place between the management sector and the factory. Management decision making regarding factory operations must be continually passed to the factory, and data regarding factory performance must be passed back to management. The design, management, and manufacturing sectors of the system must be linked by a continuous information exchange that enables each part of the system to function effectively as part of the whole. (Information flow among the various parts of a manufacturing system is discussed in detail in Example 9-3.)

The name *computer-integrated manufacturing* is used to describe the type of systems and environments being discussed here because the development of a CFI-based system will require the extensive use of computers and computer networks. Where layer 4 is concerned with the functional information flow, layer 5 considers the physical (hardware) computer requirements. The types of CIM systems described in this chapter can exist only because of the computer capabilities that are now available on a cost-effective basis. Providing the desired information capability and adaptability to achieve CFI requires the ability to handle large quantities of data very accurately and at low cost. For all CIM-oriented systems, then, it can be expected that computers will form a vital aspect of the systems.

Within the larger umbrella concept represented by CFI are a number of additional computer-integrated design principles that are helpful in configuring model systems.

Computer-aided design (CAD) and computer-aided engineering (CAE) are used to improve the creation of product design information and the flow of information from design to manufacturing. CAD involves the use of computers to support the design activity. In typical settings, products are designed through use of a graphics terminal. The designer provides data to a computer regarding the desired product geometry, and the computer creates a drawing of the resulting product on the terminal. The designer can modify the proposed geometry and study the changes as they appear in the screen of the terminal. Once the design is complete, the computer can maintain a record of all product input specifications and can re-create the product drawing at any time.

CAD can be a supportive tool for the version of GT that involves the standardization of products. If all product designs are stored in a computer, then the computer can be used as a mechanism to define attributes and develop standardized product design strategies. When a product is to be developed, maximum use can be made of past design efforts as building blocks for the needed configuration. A more rapid product development cycle can result by using linked designs to avoid completely "reinventing" the type of product line of interest.

Limitations arising from the process flow (layer 2) and equipment constraints (layer 1) will restrict the scope of products that can be produced in any manufacturing system and will thus limit the range of CAD flexibility. In addition, defining a finite set of (useful) attributes will limit the ability to spin off one product from another. However, despite these restrictions, CAD will typically become a vital aspect of any CIM system. In addition to providing design support through GT linkages, CAD can also be used to reflect the constraints associated with the manufacturing system itself (layer 1 and layer 2), as discussed below in conjunction with design for manufacturing (DFM).

Computer-aided engineering (CAE) can be used to strengthen CAD by linking the geometric design process with modeling and simulation tools that can predict how a given product design will actually function. The models use the geometric design and materials specifications to explore how such a product would actually perform. An important design aid, CAE is widely used for both mechanical and electronics products.

Computer-aided process planning (CAPP) uses computer capability to help plan an optimized process flow to manufacture a specific product (Fuchs 1988). Thus, the information capability of layer 4 can be used to help design the process flow of layer 2. In a reciprocal sense, the selected process flow configuration of layer 2 (implementing CFM concepts) will help determine the required information flow (implementing CFI concepts) of layer 4.

CAPP can be used as a means of controlling the selection of subsystem processes and for integrating these processes to achieve the most effective process flow. CAPP can consist of many constituent subsystems linked together, and can make use of expert systems to capture the best historical insights regarding process planning for the specific industry and processes required. Such computer-aided planning can provide an improved, integrated information base that will achieve a more optimum process flow than will a collection of individuals who are less "networked" together. The best use can be made of all available knowledge regarding the processes to be employed.

However, the application of CAPP is based on obtaining the needed information base, incorporating the information base into an appropriate computer system, and developing needed software to achieve the required level of planning proficiency. In many realistic applications, the CAPP design principles are used to create reference systems that affect *to-be* system design. Computers are often used effectively to support process planning. However, a fully integrated CAPP strategy can be beyond the capability of available technology and may not adequately reflect the organizational/management interests expressed through layer 3.

Computer-aided manufacturing (CAM) involves the use of computers to plan and conduct the production of a product. Computers can be used to collect and process data (layer 4), control the manufacturing process (layer 2), and provide decision-making information (for layer 3). CAM can involve distributed quality control and test and inspection built into the manufacturing process to support the layer 2 functional relationships. CAM thus involves using computer information-handling capabilities associated with CFI to achieve the functional objectives expressed for CFM.

The design for manufacturing (DFM) concept combines CAD/CAE/CAPP/CAM in order to customize the product design process for the particular manufacturing facility. The capabilities of the manufacturing system are modeled and fed back into the design activity in order to produce product designs that are most effective for the available manufacturing system, thereby maximizing effective facility use and minimizing product development costs.

The implementation of design for manufacturing (DFM) concepts requires breaking down walls between the design and manufacturing aspects of an organization. Changes of this nature can result in significant competitive advantage for an enterprise. At the same time, these barriers are often so strongly established that major, continuous efforts may be required to achieve the desired level of integration.

DFM is an attractive concept. However, difficulties occur in providing the required system models, providing the needed design-support software, linking the resultant design to manufacturing, and training designers in system use. A trade-off exists between DFM complexity and utility. Each setting must determine the best solution, depending on the search and learn activity that is introduced.

Information flow must also be used to support management operations and decision making. The MRP (Material Requirements Planning) and MRP-II (Manufacturing Resource Planning) software systems support this purpose (Fuchs 1988). An MRP software program provides a mechanism for using computers to manage materials and resources in the factory. A manufacturing data base links schedules, bills of material, and inventory data in order to make the most effective use of the CFM system. Feedback must be used continually in a learning process to improve program design and application. An MRP-II soft-

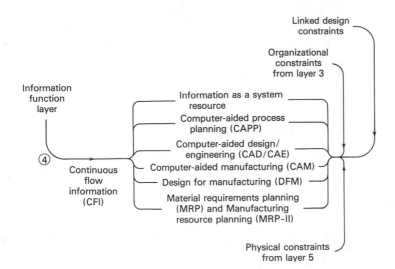

ware program extends this concept to add business planning and operations evaluation to the information flow. Management can thus use the same data base to form business decisions and study their impact on a continuing basis.

The information layer can make use of *expert systems,* which are software programs that incorporate the decision rules of human experts. Such programs can then be used to support automated decision making. Expert systems are often configured to deal with incomplete data and form supportive decision patterns for the allocation and control of resources. They can enable the automated portion of the manufacturing system to participate actively in the system search and learn procedure that is necessary to produce an improved configuration. Expert systems also are a useful way to standardize the knowledge base of a manufacturing process in order to maximize process efficiency.

The information layer can be used to support modeling and simulation in both system planning and design for manufacturing (DFM). The modeling and simulation can be associated with tactical issues (involving the internal structure and function of equipment stations) or with strategic issues (involving the entire manufacturing system or system-environment interactions). Automated equipment simulations can also be used to design and select equipment and to help generate subsystem software. These aspects must be combined for a complete subsystem design.

Figure 3-25 shows how the framework of Fig. 3-19 describes design principles associated with layer 4 of the multilayer model.

3-9. Design Principles for the Physical Layer (Layer 1)

The physical layer (layer 1) describes how the process function flowchart of layer 2 is implemented in hardware. The properties of the equipment used in layer 1 will restrict the range of functional operations, and the functional flowcharts will help specify the equipment

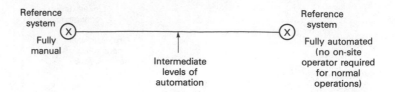

Figure 3-26.
The automation of pro-
cess and materials
transport equipment
may be viewed as fall-
ing on a continuum.
Two reference systems
can be defined, asso-
ciated with fully man-
ual operations and
fully automated opera-
tions. Many intermedi-
ate levels of automa-
tion can be defined
between these refer-
ence system extremes.

properties needed for system implementation. A close interrelationship thus exists between these two layers.

In the same way, layers 3 to 5 are also linked to layer 1. The equipment must operate with the types of machine/human interfaces associated with layer 3, and the information software and hardware of layers 4 and 5 must be consistent with the equipment characteristics. Layer 1 provides the most visible aspect of the manufacturing system, since it is often the materials transformation equipment that can be most readily viewed. Layer 5, involving computer hardware, is the next most visible aspect of the system.

For layer 1, important design principles relate to the concepts of fully automated processing equipment and fully automated materials transport equipment. These two concepts contribute toward a reference system in which full factory operations can take place without direct human intervention. This is a limiting case, as expected for a reference system, that may not be the most appropriate strategy for a given manufacturing system and setting. It is useful to view the automation of process and materials transport equipment as part of a continuum, as shown in Fig. 3-26.

The level of automation should be selected as most appropriate, leading to automation trade-off design (ATD). In many situations, the hardware requirements of fully automated processing equipment may be beyond the technological state of the art or be too expensive for use. If a manufacturing system is designed around state-of-the-art levels of automation for a given industry, scheduling and cost allowances should be made for tuning up the technology for commercial application. The principle of full automation for process and transport equipment is a limiting case or reference system to be used in considering system design.

Figure 3-27 shows a generic flowchart (associated with layer 2) that is an appropriate way to view fully automated processing equipment. Work-in-progress (WIP) must be automatically fed into and out of the station under positive identification and control. Optical bar coding is often used to label and identify each item in the assembly process. The process equipment station must then be able to complete all transformation processes without direct human intervention. This requires that the station be able to receive the WIP from the materials transport subsystem; place the WIP in the correct location, using *fixturing* methods to hold it in place; perform the required operations while ensuring continuous quality control; and transfer the WIP to the materials transport subsystem.

The elements of the system are internal and or/external sensors, actuators to drive effectors, which then act on the product, and a controller that links sensors to the actuators/effectors. If internal sensors

In order to understand the available system design options, it is essential to explore the levels of equipment automation and integration that are possible and desirable. Figure 3-27 provides a starting point for understanding the key elements of automated manufacturing equipment. Examples of families of equipment units that can achieve a given manufacturing function with various levels of automation are discussed in Chap. 6.

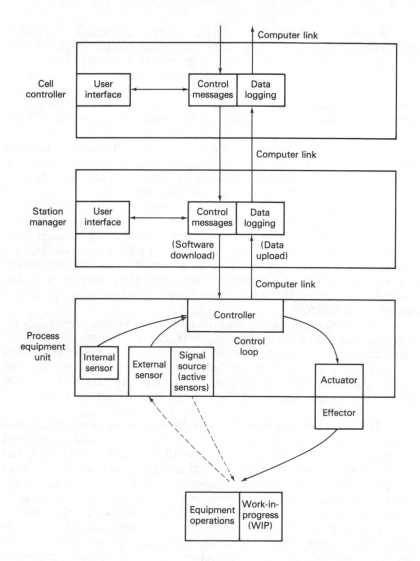

Figure 3-27.
A generic approach to describing fully automated equipment. Shown here are a cell controller, station manager, and process equipment unit that direct equipment operations. The process equipment unit typically consists of sensors to gather information regarding the equipment operations and the work-in-progress, a controller to receive the sensor signals and develop control signals, and actuators and effectors to change the nature of the operations on the work-in-progress. The sensors, controller, and actuators/effectors form a control loop. The process equipment unit can be linked to a station manager that passes control messages "down" through the network and data logs information "up" through the network. A user interface may be available for operator intervention at this point. At a higher level, the computer link can extend to the cell controller and from there to the factory controller.

are used (for example, a position indicator associated with a robot arm), the system cannot perceive the outside world but can only make measurements on its internal condition. This configuration requires rigid moving parts so that the controller at all times can know exactly where the effector is placed. Such systems are typically not very adaptable and are constrained in the number of ways in which they may be used. Most robots in use in the United States fall into this category.

An alternative approach is to make use of external sensors that can gather information from the equipment operations and work-in-progress on an ongoing basis and to link these external sensors through the controller to the actuator and effector. Passive sensors make use of sensory information that is available in the environment, and active sensors require a signal source to produce the desired sensor input.

A typical example of an external sensor is a machine vision system, with the lighting forming the signal source and the reflected light from the product being used to drive the external machine vision system. The machine vision sensor provides information to the controller, which can appropriately guide the effector to achieve the desired action. Control loops of this type can produce smart equipment operations, but can also lead to a number of difficulties (as discussed in Chap. 6).

The controller can perform a wide range of functions, including *data logging* (to provide information to the cell and factory regarding the processes that are taking place) and *controlling* (downloading software from the factory to tell the controller the types of actions that need to be taken). In both types of systems, integrated information flow with the larger environment can improve controller performance, both in terms of allowing the controller to accept information from other locations and to provide data logging information that allows statistical analysis and statistical quality control. A variety of different types of manufacturing equipment units and their properties is discussed in Chap. 6; the information flow related to such systems is discussed in Chap. 7.

Information can be fed into the equipment station from other locations, and information regarding WIP can be communicated back into the computer network. A dedicated station manager (associated with the equipment station) can link to a work cell controller and then to the systemwide factory controller. The implementation of the computer network portion of Fig. 3-27 is considered part of layers 4 and 5.

The flowchart of Fig. 3-27 can be viewed as the generalized flowchart for a *robot*. A robot may be defined as any software-controlled mechanical device. Introduced in the 1950s, early robots were numerically controlled (NC) machines that used a punched-type control system. Computerized numerical control (CNC) was introduced in the 1960s. Growing levels of machine intelligence (in terms of sensors, control loops, and processing capability) have produced highly flexible robots that can be completely operated by remotely downloaded software. Such robots can exercise their full range of operations by responding to the downloaded instructions provided through a computer network.

Many (perhaps most) of the robots in use in the United States today do not make use of closed-loop control systems but have only rudimentary sensor and control capabilities. The result is a nonadaptable response that severely limits equipment performance. As might be expected, ensuring quality production without adequate fixturing and process control can be a difficult task. Thus, equipment limitations can restrict the types of flowcharts that are developed for layer 2.

A close relationship must exist among automated equipment capability, the information flow of layer 4, and the computer capability of layer 5. A matched balance must be obtained among these levels for effective system operation.

In a fully automated system, the process equipment must be interconnected by automated materials handling subsystems. Transport among process stations can be handled by conveyor belts, tracked subsystems, or automatic guided vehicles (AGVs). The subsystem must also be able to place WIP into storage and retrieve WIP from storage based on computer-controlled instructions, and similarly handle input materials distribution throughout the manufacturing system. Optical bar coding may be a necessity for maintaining positive identification control over all resources and WIP. The entire materials transport subsystem must be closely tied to the information flow and computer network layers (layers 4 and 5). In many cases, a partially automated materials handling subsystem will result in the most appropriate system design. For this situation, the organizational/management structure of layer 3 must also link closely to the subsystem.

The full automation system model (FASM) provides a limiting design principle for the physical layer. It is important that this principle be linked to automation trade-off design (ATD) to obtain the most appropriate use of automation for the design of specific manufacturing systems (refer to Fig. 3-19).

3-10. Design Principles for the Computer Layer (Layer 5)

The computer stations and networks of layer 5 provide the hardware implementation required to achieve information flow. (The computer hardware that is incorporated internally into process equipment is defined here to be a part of the equipment, and is thus discussed as part of layer 1.)

As noted in Sec. 3-8, sophisticated computer capability is usually required to achieve CIM-oriented system design. Thus, CIM facilities have become possible only in the 1980s, as computer performance/cost ratios have grown very rapidly. The implementation requirements associated with the system computer network (layer 5) can still provide significant constraints on the information flow (layer 4) that determines how close to a desired *to-be* system the planning group can actually approach. Limits are encountered with respect to processing speed, memory requirements, and costs.

To achieve a CIM-oriented manufacturing facility, it is necessary to obtain a free exchange of information among many system users. This leads to a need for distributed processing and data base management. In a limiting reference case, all data will be available to all users, making use of hardware that degrades "gracefully" or "softly" under load or various possible failure modes.

The computer implementation requirements are so specialized that the costs associated with hiring and maintaining the necessary personnel and retraining the existing work force can be substantial. The success or failure of a highly automated manufacturing system will often be associated with the skill in designing and implementing the (level 5) computer network.

The implementation of computer networks is made more complex by the multiplicity of languages (or protocols) that now exists for communication links. Despite efforts to standardize these protocols (discussed in Chap. 7), the situation is still unsatisfactory from the user

The advantages of eliminating the manual, paper-flow-oriented management of information from an enterprise have been discussed previously. The elimination of paper will require the introduction of computer-based information handling to the system. Thus, more competitive manufacturing systems often result in widespread use of computers and integration. The result is CIM. However, the difficulties associated with the transition should not be minimized, as many new design and operational problems will be encountered.

viewpoint. Significant improvements are still required in this area before true building-block components can exist for the required subsystems.

The necessary skills must be obtained through consultants, permanent staff, or combinations. The difficulty in obtaining and operating computer hardware and software for a highly automated manufacturing system is an important issue to be addressed by the planning group. However, computer technology is at the center of modern manufacturing technology and must be assimilated into the organization at all levels. Organizations that successfully master and apply computer technology will be able to completely dominate those that do not.

Figure 3-19 summarizes the design principles for layer 5.

3-11. Design Principles for the Organizational/Management Layer (Layer 3)

The success or failure of CIM system implementation depends on the ways in which human resource issues are approached (refer to Fig. 3-19). If the employees do not try to make the system work, it won't. A can-do team spirit must pervade the organization.

Successful implementation requires a major change in the organizational culture, not easily achieved. Leadership must come from the top, and extensive changes in organizational communications, worker retraining, team building, and revised employment conditions may be required. These changes often take place slowly, and can cause delays that must be incorporated into the planning process. Certainly an active concern with all aspects of organization and management must be an essential aspect of any CIM effort.

It has been observed that the management of technology is a critical aspect of responding to competitive demands. Thus, CIM could appropriately stand for computer-integrated *management*. The feasibility of any CIM plan depends on how well management is able to motivate employee participation and cope with problems as they develop. This aspect of CIM system design is discussed in detail in Chap. 8.

3-12. Design Principles and the SES Model

Figures 3-28a and 3-28b provide a helpful comparison of an inventory-oriented production system and a pull-oriented system. Figure 3-28a, for the former case, shows how the market for a new product drives design and manufacturing setup, after which inventory production begins. The inventory production level is based on an algorithm that relates to the factory's past experience and estimates of the nature of the market, and can be either too low, about right, or above the actual sales that can be achieved.

Once the market drops, a delay time is experienced in the ability of the manufacturing system to sense the market change and to reduce production levels. An evaluation process must take place in order to

Figure 3-28.
Various strategies that can be used to define the levels of factory production: (a) A build-to-inventory system. Production is driven by an increase in the market for the new product. System and vendor design and setup plus time required to achieve production output result in a delay before an increase in production can begin to fill product inventory. Once inventory is available, sales of the new product can begin. A potential mismatch can exist between production levels and inventory if the market is not correctly estimated. Once a decrease in market starts to take place, a delay time is experienced in closing down production and halting factory output. For this type of system, it is necessary to estimate continually the level of appropriate inventory, attempt to build up that level of inventory, and then adapt as the need for inventory is reduced. (b) A pull system response based on build to order. An increase in the market for a new product is used to create market orders and tickets (or *kanban*) that will stimulate factory output. System and vendor design and setup delays are experienced before the factory output can rise. Based completely on a build-to-order strategy, production will be matched to the market orders. Once market orders begin to drop, factory production will drop, but all products will be sold since they were built to order.

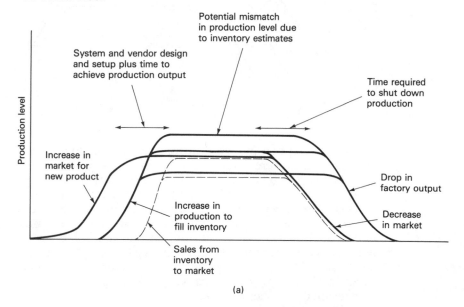

(a)

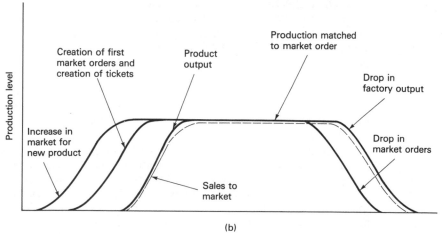

(b)

Figure 3-28. continued

(c) Combined system response. For many applications, both inventory building and build to order are used to drive the manufacturing system. The products delivered to the market will be based on orders and the available inventory. Delays are experienced in developing the product and in detecting a market decrease that will drive the need to reduce factory output.

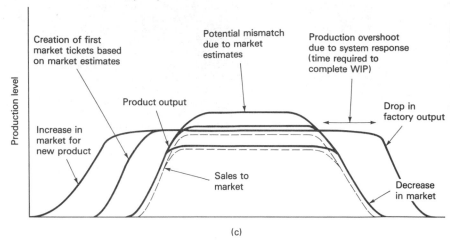

(c)

estimate how much and when the production should be reduced. (Several such cases were discussed in the previous examples.)

Figure 3-28b contrasts the inventory system with a pull system based on "build to order." Once the market increases, there follows a period of design and manufacturing setup for both the manufacturing system and the vendor. Following this period, the product is announced. Orders are received from the market. Tickets created at the output pass through the system, beginning to create the pull production, and tickets also are passed to the vendor to create just-in-time delivery. In an idealized system, the passage of tickets is sufficiently rapid to be neglected as a cause of delay.

Once the first ticket reaches the vendor, production can begin. A delay time is experienced in passing the work-in-progress through the system; then the product emerges and is available to the market. The combined design and setup time, plus the time to produce, results in the composite response delay. When the market drops, the number of tickets entering the system from the market drops shortly thereafter, so that no more product is created. In an idealized pull system, product is only manufactured as the market causes tickets to be generated at the system output; therefore, in this idealized system, there is no buildup of inventory.

As has been indicated above, the buildup of inventory is an expensive aspect of a system, and mismatches between the actual market and production levels can rapidly lead to such a buildup. An idealized pull system prevents such a result, and therefore allows for a much lower-priced product. On the other hand, it is difficult in a realistic setting to achieve an idealized pull system without any inventory at

Examples 3-7 through 3-9 illustrate how the SES model can be used to explore the application of various reference concepts for manufacturing systems. As illustrated, rapid system response to market change is a powerful system advantage. When rapid market response is combined with other design principles (such as minimizing product cost through elimination of non-value-added activities, DFM to achieve systemwide integration of product development and manufacturing, and elimination of paper-based information in favor of computer-based information), the result can be a powerful manufacturing system that can survive intense competitive pressures.

any level of manufacturing. It may often be that such a limiting arrangement will prevent the most effective operation of the system and the most effective addressing of the market. Therefore, a compromise may be the most attractive solution, in which a basic pull system is combined with a modest inventory system that provides some slack but still avoids a full focus on inventory production (Fig. 3-28c).

The following examples show the performance of a pull manufacturing system for three different cases in which there is a long response delay, a medium response delay, and a very short response delay built into the system. These different response delays are associated with alternative applications of CIM principles within the manufacturing system. In each case, the objective is to illustrate how a pull system, when combined with other CIM design principles, can lead to more optimized manufacturing performance. The impact of the manufacturing system principles is illustrated by making use of a variation of the SES model.

EXAMPLE 3-7. MODELING PULL SYSTEM PERFORMANCE

It is assumed that a manufacturing system is designed to make effective use of pull system concepts, with just-in-time vendor deliveries. The system is slow to adapt to market changes because of manual design and setup, minimum automation, and manually oriented information transfer. These aspects of the system require a 12-time unit (month) delay in shifting from one product to another. It is further assumed that because of this manually oriented system, rework of scrap is at the 10 percent level for the manufacturing system. It is assumed that the factory can discontinue a product within 1 time unit (or 1 month) after the market drops to zero.

The system thus achieves: (1) production matched to a market increase with a 12-month delay and matched to a market decrease with a 1-month delay; (2) use of a just-in-time system so that vendor kit arrivals are matched to factory production; and (3) a minimum work-in-progress inventory, which will reduce manufacturing costs.

What insights into this type of manufacturing system can be obtained from the SES model?

Results

Figure 3-29 shows the market (MT) that is assumed to exist for products 1, 2, and 3 for the particular environment being discussed. Note that each product market exists for 20 time units (or 20 months, as applied here), and that the market level is flat at 400 parts per month. The actual shape of these market curves is less important than for prior examples because the pull system tends to create a near-match between the market and the manufacturing system.

The vendor kit orders (VT) begin 12 months after the market develops. This 12-month delay is associated with product redesign and tooling, as provided for in the system description. After the market becomes established, vendor kits start to arrive at a sufficient level to allow for a 10 percent scrap/rework rate. The flow of vendor kit arrivals directly into the factory (N2) and the result of manufacturing system operations (ET) are also illustrated. Again,

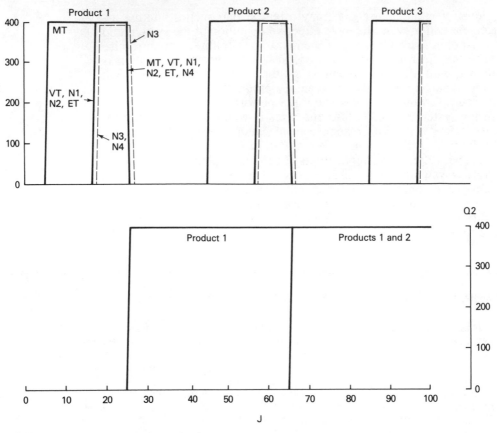

Figure 3-29.
SES model data for
Example 3-7.

a 12-month delay can be noted, along with a level of production that is matched to the market.

Because of the just-in-time delivery system, there is no buildup of vendor kit inventory (Q1). Therefore, Q1 is not shown. Factory output (N3) starts after a delay of 12 months for design and setup plus 1 month for production and ends 1 month after the market drops. Product inventory (Q2) will develop as the market changes from one product to another, as the factory continues to create products the market will not consume for the one month delay associated with shutdown. The inventory that results from the final month of in-progress production is shown as Q2.

As can be immediately noted, the total sales of the system have been severely restricted because of the delay caused by new product development and tooling up the factory to produce a new product. Of a potential 20 months of production for which a market was available, only 8 months of production are actually obtained. The final month of production goes into inventory, since the market no longer exists. Thus, the cost to the manufacturing system must be spread across this much-reduced level of sales. Although the system achieves a match between the environment and the manufacturing system due to the application of pull and just-in-time concepts, the system still performs poorly because of the time delays associated with the product adaptation.

EXAMPLE 3-8. IMPROVED DESIGN AND MANUFACTURING CAPABILITIES

It is assumed that the system has improved computer-aided design (CAD) capabilities and improved manufacturing capabilities due to automation and integration. However, the linkage between design and manufacturing is still primarily manual. It is assumed that under these conditions, the 12-month delay associated with new product introduction can be reduced to 4 months. It is also assumed that for this improved manufacturing facility, rework can be reduced from 10 to 2 percent. What insight can the SES model reveal for this situation?

Results

Figure 3-30 shows the same market variation. The vendor kit arrivals (N1) and the flow of vendor kits directly into the factory (N2) are illustrated. The factory output of good product (N3) begins after a turn-on delay of 4 months for design and setup and 1 month for production. The output decreases 1 month after the market drops. The buildup of inventory is the same as in Example 3-7 because of the 1-month delay associated with closing down the manufacturing system. The delivery to the market is given by N4.

Note that the number of months during which delivery is made to the market has increased from (20 − 12 − 1 = 7) of Example 3-7 to (20 − 4

Figure 3-30.
SES model data for Example 3-8.

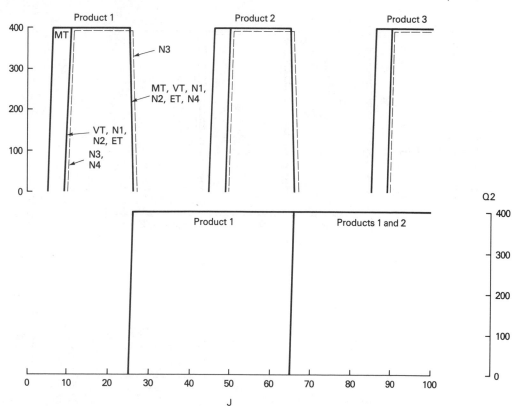

− 1 = 15) for this example. The production is produced over 16 months; the last month of production goes into inventory because of the 1-month delay in recognizing the market drop and implementing an appropriate shutdown. The increased design and manufacturing capability has resulted in significantly improved duration of production and therefore a substantially higher level of sales for this manufacturing system.

EXAMPLE 3-9. DESIGN-MANUFACTURING LINKAGE

It is assumed that the manufacturing system is further improved by maximizing the design-manufacturing linkage and incorporating a design-for-manufacturing (DFM) philosophy. Rework is assumed negligible. The result is a system that can very rapidly respond to market changes with integrated information flow throughout the entire system. What can the SES model reveal about this case?

Results

Figure 3-31.
SES model data for Example 3-9.

Figure 3-31 shows the market demand for a sequence of products. The vendor responses (VT) and the desired production levels are shown, as is

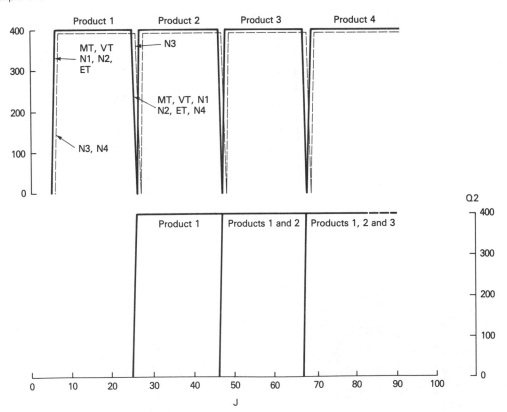

the good product output (N3). The resulting inventory is Q2, and output to the market is N4. For this example, a 1-month delay is experienced in introducing the new product and in achieving production, and a 1-month delay is experienced in ending production of the product. The result is an excellent match between the market and the manufacturing system.

In comparing the results of Examples 3-7, 3-8, and 3-9, it can be seen that the total sales to the market increase rapidly as the adaptability of the system increases and the response time is reduced due to the application of CIM principles. It is essential in the design of a manufacturing system to study such interactions in order to determine the response time that is required for a given facility in order to optimize its operations. Additional costs are experienced as the operations delay times are reduced, so that a trade-off will result.

However, it must be remembered that there are other advantages to rapid-response manufacturing systems. The reduction in delay times is not the only advantage of the improved system; there has also been a reduction in rework and scrap. Further, this type of system will likely lead to a higher-quality product with reduced service requirements, another definite advantage for the corporation.

3-13. MIS Integration

As illustrated in Fig. 3-19, the various aspects of a manufacturing system must be integrated through the system management function. System management must understand all activities that are taking place in the system and must be able to achieve appropriate decisions and control functions.

Information is received from all system layers. The most formal information flow will be received through the functional information system (layer 4). This layer provides one of the primary mechanisms through which system information is guided into the management process. The content of the information will relate to the physical, functional, and organizational layers, and thus, the information relates to the entire operation. The information will flow to system management in a defined manner, and control actions taken by system management will affect all five layers of the system.

In many manufacturing settings, the formal information base originates as part of a management information system (MIS). The objective of the MIS is to provide information to management that will enable appropriate actions to be taken. Unfortunately, it is often the case that such management information systems do not function in a way that allows effective management of the manufacturing system. MIS reports are often late and of questionable validity or do not provide adequate insight into the type of decision making that needs to take place.

It is essential that an MIS be designed as an integral part of the manufacturing system. The type of information that is presented, the timing on which this information is available, and the mechanisms that exist for analyzing this information must be appropriate to the type of manufacturing system being implemented. In addition, the control structure that enables system managers to make changes within the system must be closely linked to the MIS and must provide for appropriate decision-making mechanisms. Unless management can understand exactly how the system is operating and intervene with appropriate decisions and control, the system will not achieve its intended effectiveness.

3-14.   Statistical Quality Control

A key component of the management of any CIM system is the ability to ensure the high quality of the product. The dominant approach to such assurance is often through final product testing, at the end of the manufacturing process. Those products that do not pass test are routed to rework or scrap. Such an approach is expensive and inadequate. Rather, the system should operate in such a way that it is constantly determining its own performance and ensuring that rework and scrap are of insignificant quantity. The objective is not to produce, then test to find the bad product, but to design the system in such a way that only good product is produced. Not only does this avoid the high cost of rework and scrap, but it also ensures satisfaction and a higher level of product performance in the market.

The use of statistical quality control (SQC) to make only good product and to eliminate rework and scrap is another important design principle. By combining many different design principles, a "vision" of the successful "enterprise of the future" can be developed and used to guide in system planning.

The application of statistical quality control (SQC) is a key ingredient in any effort to ensure the production of only good product (Klippel 1984; Gould 1986; Ryan 1989). SQC basically involves the collection and use of timely information on the operation of the entire system so that only good product is being produced at any time. The data and analysis should provide guidance to management when problems begin to develop and enable management to intervene and correct those problems before bad product results (Gould 1986). In this way, statistical quality control is a monitor that enables management to achieve a system that functions as desired. Efficient information gathering and analysis, combined with control of the manufacturing system on a continuing basis, results in only good product being made. The final (and often very expensive) test procedures can be discontinued.

The ultimate objective of any manufacturing system should be to produce only good product and to eliminate the manufacture of bad product. However, this objective is difficult to achieve, and in many cases no significant effort is even made in this direction. Rather, an alternative approach is used, with an emphasis on "make it the best way we can, and then use rework to repair those products that do not meet standards."

Figures 3-32a and 3-32b compare these two very different strategies. For the former, ongoing quality control ensures that only good product is produced; for the latter, bad product is accepted, and rework is included at the end of the manufacturing system. One of the reasons that the second method is so prevalent is because it is difficult to achieve the first. It is a major challenge to operate a manufacturing system in such a way as to eliminate all bad product and produce only

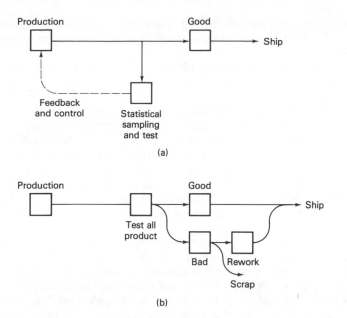

(a)

(b)

Figure 3-32.
Alternative approaches to quality control. (a) Production is sufficiently under control to produce only good product. Statistical sampling and testing of the product are used for feedback and control to enable this level of good production to be maintained. (b) All product is tested and divided into "good" and "bad." The bad passes to either rework or scrap. The product that is successfully reworked is combined with the good product and shipped.

good output. On the other hand, there are tremendous incentives to work toward the high-quality system, since the elimination of testing for every product has the potential for sharply reducing manufacturing cost.

Figure 3-33 illustrates the nature of the problem. In order to produce only good product, it is necessary first to identify those parameters or variables that control whether a product is good or bad. If the manufacturing process operates within each of these parameter ranges, a good product will result; if any of the parameter values fall outside these ranges, a bad product may result.

The foundation of SQC is a complete understanding of all aspects of the manufacturing processes. This type of in-depth knowledge is required in order to identify the parameters that can determine a good product and to establish the appropriate ranges for these parameters. The major handicap to implementing SQC will often be the lack of understanding and control over many processes that exists in many industries. If a satisfactory degree of understanding and control does

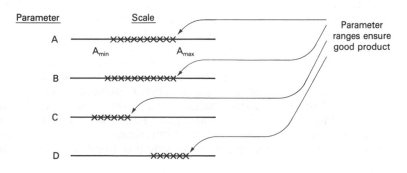

Figure 3-33.
Parameter control used to obtain good product. A number of parameters can be defined and parameter ranges assigned to each parameter, so that measured values of these parameters within the defined ranges will ensure that only good product results.

Manufacturing System Design **121**

not exist, then the manufacturing system must fall back on testing each product, separating the good output from the bad and reworking the bad product.

Another major concern relates to the manufacturing equipment. Even when the knowledge exists on how to design such a manufacturing system, the necessary equipment may not be available. Many types of equipment do not enable the manufacturing system to collect data adequately on all the necessary parameters to obtain only good product. Even though the process is sufficiently well understood, it may not be possible to implement the desired system, oriented toward high-quality product, if the original equipment manufacturers (OEMs) do not provide equipment with the necessary sensors. These sensors are necessary to monitor constantly the effect of the manufacturing equipment on the product and to monitor the activities of the equipment unit itself.

Although the manufacturing strategy in Fig. 3-32a appears much more attractive than the method in Fig. 3-32b, there are significant handicaps to achieving such a simple manufacturing system. The restriction goes much deeper than not using certain mathematical techniques. The difficulties extend to the basic understanding of how the manufacturing system processes its products and to the capability to obtain the data required for a complete tracking of the system. Once the knowledge base has been developed and the equipment capability has been obtained, then an effort may be mounted to implement the system option shown in Fig. 3-32a.

Monitoring of this type of system can be achieved by sampling tests that are performed on selected products passing through the system. The result is statistical quality control. The objective is to use sampling methods to obtain product parameters that do not drift outside the acceptable limits. If the sensors on the equipment are sufficiently accurate and self-monitoring to ensure that all processes are within parameter limits, and if these parameters are adequate to define good product, no type of sampling testing is required. However, due to variations in processes, areas in which knowledge is not complete, or areas in which adequate information cannot be fully collected, it is often useful to maintain a statistical sampling process to confirm that the system is operating as desired.

The complexity of SQC is associated with monitoring the manufacturing process, based on the samples that are drawn and tested. The ideal objective would be that if the process is satisfactory and under statistical control (behaving according to the statistical variations), then the observer will know that all product output is of acceptable quality. On the other hand, it would be desirable to ascertain immediately if the process is not performing according to expectations.

Since only samples are being drawn from the full product line, it is not possible to obtain such a clear-cut resolution; rather, it is necessary to draw on statistical insight to draw conclusions. The application of SQC involves setting certain *control levels* associated with each of the parameters to be measured and examining whether or not the measurements on the samples fall within these control levels. There is a trade-off between the average run length (number of samples to be drawn) before a *false alarm* occurs versus the average run length to reveal that a process is *out of control*. If an effort is made to eliminate false alarms as much as possible, then it will take longer to reveal that the process is out of control. On the other hand, if false alarms are accepted, then a more rapid determination can be made that a process is out of control. Thus, the key ingredient to applying SQC is to choose control levels that will provide the desired solution for the particular manufacturing system.

Figure 3-34 shows a simple type of control system applied to parameter A. Each time a sample is drawn, the mean value of parameter

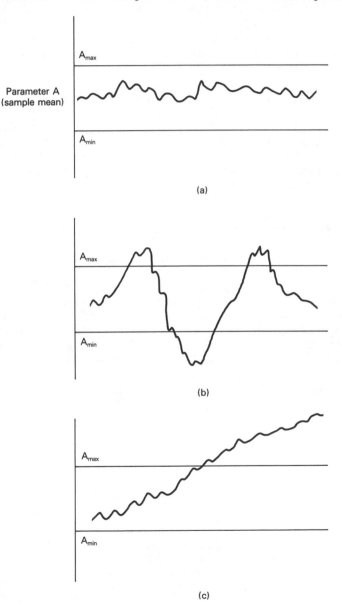

Parameter A (sample mean)

A_{max}

A_{min}

(a)

A_{max}

A_{min}

(b)

A_{max}

A_{min}

(c)

Figure 3-34.
The use of control charts for statistical quality control (SQC). (a) A parameter that stays within control limits so that only good product is produced. (b) A parameter that is periodically outside control limits so that a mix of good and bad product is produced. (c) a manufacturing system that is drifting out of compliance and steadily shifting from good to bad product.

A is calculated for the sample. By inspection, the mean value of A must lie between A_{max} and A_{min} (refer to Fig. 3-33) if good product is to result. (These outer limits will significantly tighten most applications that draw on statistical trade-offs.)

As shown in Fig. 3-34, a control chart can be developed through which the means of the sample group are plotted as a function of time on a chart that indicates the minimum and maximum acceptable values for this mean. The measurements of the sample cannot fall outside this range and still meet the parameter requirements. The result is a control chart through which the characteristics of the samples are plotted as a function of time and compared with control levels.

This general approach to SQC can be applied to many different parameters and many different methods of analysis. In each case, it is necessary to define the procedure to measure a property of the sample group, plot the measurement results as a function of time, and then compare these results with control levels that are contained on the charts. If the resulting parameter values extend beyond the control levels as shown in Fig. 3-34b, the process is out of control; it can be concluded immediately that both good and bad product are being produced. Figure 3-34c shows another type of system, in which there is a gradual drift toward being out of control due to equipment failure or wearing of the equipment unit. This type of process drift can alert the system to shut down and call for repair.

Many complex statistical methods can be used to choose the most appropriate control limits and to enable the manufacturing system to draw appropriate conclusions about the nature of system performance from the data that are being collected. The discussion here has provided the framework in which all of these more specific methods can be applied.

If a complete understanding of the manufacturing processes exists and if the equipment used can gather the needed information regarding all these processes, then an effective SQC program can be implemented by drawing on consultant support to decide how to set the measurement control limits and how to achieve decision making based on the samples being tested. On the other hand, if a lack of process knowledge exists and/or if the equipment units cannot collect the needed data, no amount of mathematical analysis will enable the manufacturing system to achieve a high-quality production capability.

ASSIGNMENTS

3-1. In order to achieve maximum educational insight from Chap. 3, it is helpful to stimulate class discussion of the concepts that have been included, develop critical reviews that probe the advantages and limitations of the concepts, and explore alternative ways for approaching the topic. Select one or several topics from the bold-faced headings that introduce sections or examples in this chapter. Come to class prepared to explain this topic, critically describing the text materials and suggesting ways the topic discussion might be strengthened.

3-2. Discuss your concept of how a planning group should be developed. What disciplines should be included in this group, and what procedures should the group follow in the planning process? How should the planning group be related to the rest of the enterprise? What guidance would you provide to the planning group to accomplish its task most effectively? Consider a specific manufacturing setting that is facing many of the issues raised by computer-integrated manufacturing; discuss the type of planning group that you would find effective in this setting.

(a) Come to class prepared to share your results and participate in discussion of the issues raised.

(b) After this discussion, prepare a brief paper on your insights.

3-3. Develop a description of computer-integrated manufacturing and compare it with the definitions provided in this text and with any other definitions from reference materials. Discuss the strengths and weaknesses of your definition, contrasting them with those of the other available definitions. Why do you prefer your definition over others? What are the implications of your definition in terms of designing and implementing CIM systems?

(a) Be prepared to share your results in class and discuss the issues raised.

(b) After this discussion, prepare a brief paper on your insights.

3-4. Assume you are the chief executive officer (CEO) of a manufacturing enterprise. You are concerned that your company may not be making the most effective use of technology. What actions might you take to begin to address the issues facing your company? Relate your proposed actions to the particular enterprise that you are discussing. Identify the strengths and weaknesses of your strategy. Can you identify other problem areas that might develop, and ways in which you can modify your strategy in response to these problem areas?

(a) Come to class prepared to share your results and participate in discussion of the issues raised.

(b) After this discussion, prepare a brief paper on your insights.

3-5. (a) The SES model allows a wide range of studies to be conducted regarding interaction between a manufacturing system and its environment. Study the material provided in this chapter and in Appendix B, and practice your interaction with the software provided with this text in order to become capable of applying the model. Define a system-environment setting that you believe may reveal issues facing a manufacturing system. Relate your discussion to one or several of the design principles provided in this chapter or

to other design principles that would provide an important reference model for the planning group. Consider the type of data that should be fed into the SES model in order to study the situation you have described. Consider how the nature of the manufacturing system and its environment, and the design principles in use, will lead to assumptions regarding data fed into the SES model. Carefully define and describe these input assumptions. Then, by using the model, simulate the interaction between your system and the environment and observe the results that are obtained. Provide a step-by-step interpretation of these results and present them to the class.

(b) The data to be provided for the SES model can take the form of individual data points, equations to generate data points, or output from a random number generator. Using these various methods to simulate market environments, try other environments for the case you developed in part (a). What insights to your situation can be obtained by trying different market setting excursions for your case of interest?

(c) Be prepared to discuss your findings in class.

(d) After class discussion, write a brief paper on your assumptions and results for parts (a) and (b).

3-6. Develop an SES model application by repeating one of the examples provided in this chapter. Then change the nature of the assumed market either by using individual data points, equations to generate data points, or the random number generator. Describe how the results of your simulation might affect the conclusions drawn by the example. To what degree can further insight be obtained?

(a) Come to class prepared to discuss your findings.

(b) After class discussion, write up your assumptions and results in a brief paper.

3-7. Consider the various types of markets that may be generated for a system environment. What type of market characteristics might be most difficult for the system to accommodate and still maintain its effectiveness and efficiency? For this kind of environment, what would be the most effective system design? Use the SES model to simulate the type of market environment you have considered and the type of system design that might be most effective, and explore the qualities of this system-environment interaction. Try other variations on your original assumptions to see whether the situation

changes in ways you would predict. Consider the meaning of your effort.
(a) Come to class prepared to discuss your findings.
(b) After class discussion, write up your assumptions and results in a brief paper.

3-8. Modify Example 3-6 in order to use different financial data for your analysis. Then conduct an analysis that is similar to the one shown in this example. How are your conclusions changed as a result of the different data you have assumed? Can you identify specific ranges of financial data that produce different types of conclusions regarding the most desirable system design?
(a) Come to class prepared to discuss your findings.
(b) After discussion, write up your assumptions and results in a brief paper.

3-9. Describe in your own words the characteristics of each of the layers of the multilayer model. Consider a specific setting that might be experienced, and describe this setting in terms of the multilayer model. What strengths and weaknesses in the model do you note? What insights regarding the manufacturing system can be obtained by describing it in terms of this modeling approach?
(a) Be prepared to share your results in class and participate in discussion of the issues raised.
(b) After the discussion, prepare a brief paper on your insights.

3-10. Develop a concise description of each of the following design principles: CFM, pull-oriented system, *kanban* system, just in time, group technology, and flexible manufacturing system. What additional information regarding these design principles would be useful for their use by a planning group? What are the strengths and weaknesses of each of these design principles? Can you think of other related design principles or aspects that might be useful for the planning group?
(a) Come to class prepared to share your results and participate in discussion of the issues raised.
(b) After this discussion, prepare a brief paper on your insights.

3-11. Discuss the aspects of the following design principles: CFI, CAD, CAE, CAM, CAPP, and design for manufacturing. Discuss the insights associated with each of the design principles and how they might be used by the planning group.
(a) Come to class prepared to share your results and participate in discussion of the issues raised.
(b) After this discussion, prepare a brief paper on your insights.

3-12. By exploring the references at the end of the book for this chapter or by drawing on other reference material, develop further insights into the design principles that are available for CIM systems. (Your discussion may involve exploration of the principles mentioned here in further depth, as well as the introduction of other design principles.) What conclusions can you draw from your research in these areas? Do you have additional insights to how design principles might be used and some of the concerns that should be expressed with their application?
(a) Prepare a presentation for class to describe your insight based on your studies.
(b) After class discussion, write a brief paper describing your findings.

3-13. Drawing on the references for this chapter or other reference materials, review in detail the characteristics of an MRP-II system. Develop an understanding of the flow of information that takes place in this particular system and prepare one or several flowcharts to describe system operations. Critique the system in terms of its ability to support the type of operations needed for various CIM settings. In what settings will the information system be most effective and least effective?
(a) Come to class prepared to discuss your findings and explain the aspects of the particular product.
(b) Based on the above discussions, provide a brief analysis of this product in terms of its strengths and weaknesses and write a paper on this subject.

3-14. Consider a particular manufacturing setting that might be the subject of efforts for improved manufacturing capabilities. Discuss the type of equipment that is or might be available for use in this setting. How will the physical characteristics of this equipment limit the CIM design options that are available to the planning group? Based on the initial first cut, can you estimate the strengths and weaknesses of some of the various equipment available? What are some of the factors that you believe should enter into automation trade-off design? Based on the ex-

perience and data for this setting, what would be your estimate as to the type of technological applications that might be most appropriate?
(a) Be prepared to share your results in class and discuss the issues raised.
(b) After this discussion, prepare a brief paper on your insights.

3-15. Discuss the types of sensors that are available for automated equipment (including machine vision, pressure sensors, or other types that produce high levels of automation). Consider a specific type of sensor, its strengths and weaknesses, its applications, and the state of the art in technology. Discuss how this particular sensor might strengthen the ability to design cost-effective CIM systems.
(a) Come to class prepared to share your results and participate in discussion of the issues raised.
(b) After this discussion, prepare a brief paper on your insights.

3-16. Repeat Assignment 3-15 for actuators.

3-17. Repeat Assignment 3-15 for control loops and controllers.

3-18. Based on the references provided for this chapter, collect information on a computer numerically controlled machine. Consider the specific capabilities of the machine and describe its strengths and weaknesses. What constraints on the physical manufacturing system would be introduced by the use of such a system? Discuss how such a system might limit the ability to produce the most cost-effective system behavior.

(a) Come to class prepared to share your results and participate in discussion of the issues raised.
(b) After this discussion, prepare a brief paper on your insights.

3-19. Discuss some of the aspects of modern-day computer capabilities as they relate to CIM facilities. What computer capabilities are available that would help the design and implementation and operation of a CIM system? What limitations of these computer systems should be noted? Can you find information regarding the evolution of performance-cost ratios for computer design the past ten years? (Make use of the references provided for this chapter and other resource material you can find.)
(a) Be prepared to explain the type of computer hardware you are discussing, the reasons for your selection, and the strengths and weaknesses.
(b) Prepare a brief paper describing your findings.

3-20. Drawing from this text and other resources, discuss a method of quality control and how this method can be applied in realistic manufacturing settings. Work out a simple problem and explain the problem to the class. Describe the type of data collection that takes place and the type of analysis. Display the types of control charts that might be used and their significance. Come to class prepared to explain the reasons for your selection. Provide an adequate description of the methods being used and of the interpretation of the results.

4

IMPLEMENTING SYSTEM DESIGN CONCEPTS

The objective of Chap. 4 is to operationalize the concepts and insights developed in Chaps. 1 to 3. Given the foundation that has been developed, this chapter describes the various ways in which a planning group can proceed with system design or redesign.

Once the basic insights into manufacturing design and implementation have been established, the next step is to decide on specific methods that can be used by the planning group to guide system design.

This chapter discusses several different types of design solutions that can be applied, along with some of the strengths and weaknesses associated with each. One of the main issues is to decide on the degree to which concurrent planning activities and iterative planning activities can be most effectively used for a particular planning environment. After various strategies are discussed, a compromise approach is selected for further study and application. This compromise approach is discussed in detail throughout the text and is used as a means for showing how the important elements of a system design can be performed.

The methods of Chap. 4 provide a link between the problem definition given in earlier chapters and the more specific design techniques of later chapters. The system design concepts discussed here provide an action framework for a planning group. The effort is to develop a step-by-step understanding of the considerations that enter into system design and evolution, and to present ways in which planning efforts may be used to support the desired change activities.

Chapters 1 through 3 provide the conceptual foundation for a system-oriented approach to the design and implementation of computer-integrated manufacturing (CIM) systems. As emphasized in these chapters, final state, optimizing planning cannot be used in such a context; rather, a search and learn procedure must be established. As a result of the system design process, the manufacturing setting passes through a series of robust transition stages as it evolves into a more competitive enterprise. The task of the planning group is to guide this overall evolutionary process by producing estimates of desirable *to-be* system concepts and by defining viable transition stages that can lead in the general direction that has been established. As noted in the earlier chapters, it should be understood that the target *to-be* system is constantly in a state of change, so that each transition step can become a new starting point for system development.

Section 3-4 and Fig. 3-17 provide the starting point for the planning approach that is suggested in this text. This approach involves the introduction of system design strategies, the introduction of a range of models in support of these tasks, and the use of these models to strengthen the overall planning process. These models can provide insight to the planning group but cannot result in a complete definition of the actions that should be taken. The group must integrate the predictions of these models with a broad understanding of the *as-is* environment and system and an appreciation for the parameter ranges for the desirable directions of evolution for the manufacturing system.

The flowchart in Fig. 3-17 is complex because of the numerous feedback loops. Therefore, modeling must be viewed as part of an iterative procedure in which the results of the models are used to support the design effort, and design efforts are used to further define the required modeling activities. During this process, a combined top-down and bottom-up approach must be maintained, as illustrated in Fig. 3-5. A continuous tension will exist between the bottom-up planning of evolutionary change and the top-down planning of the desired *to-be* system. In many ways, the models will contain elements of both of these viewpoints. It will be important to use these models so that they conform to the top-down objectives in the manufacturing system but are also compatible with the system's bottom-up capabilities.

The objective here is to apply the conceptual foundation developed in earlier chapters to the creation of an *operational approach* for the planning and design of computer-integrated manufacturing systems. The approach taken is to develop a step-by-step method that can guide the design process in general. Based on this framework, Chap. 5 introduces several computer models that can be used for modeling efforts in support of the design activity. Chapters 6 through 9 provide information that is required for application of this design method to CIM systems. Chapter 6 provides an in-depth look at the issues involved in the physical equipment layer, Chap. 7 examines computer network requirements, Chap. 8 reviews organizational opportunities and constraints, and Chap. 9 addresses the design-manufacturing interface. Chapter 10 draws on all of these chapters to apply the modeling framework that is described here in Chap. 4.

4-1. Design Strategies

In Figure 4-1 the system approach and the five-layer model of Fig. 3-18 are applied as planning mechanisms. Illustrated in Fig. 4-1a are inputs to the design process, a design effort with supportive modeling contained within the design process itself, and outputs from the design activity. An iterative learning activity takes place among the output, the input, and the design activity.

As indicated in previous chapters, many different factors must be included as input to the design process. There must be an understanding of and an appreciation for the enterprise objectives and the environment in which the enterprise functions. Further, in order to guide system evolution in the desired direction, there must be insight into the CIM design principles and the reference systems toward which the manufacturing system can evolve. The *as-is* system must be understood as the present manufacturing base, and a strategy must be in place to define learning processes that will support system evolution.

The design process itself will take place under the guidance of the planning group. As has been emphasized, no models exist that can completely predict the optimum system configuration for the future. Nonetheless, a wide range of limited-scope modeling can be used to assist in the design process. The center box in Fig. 4-1a is where the actual design, modeling, and decision-making efforts occur to produce the CIM system for the future. The conceptual descriptions discussed earlier provide a framework for studying the transactions inside this box. The design output includes both the *to-be* system and the definition of proposed transition stages. Continual iteration must exist throughout the design process in order to obtain a learning cycle.

In developing a system design and implementation strategy, the planning group must function in a complex environment and address realistic opportunities and capabilities. The intent must be to establish a learning pathway that will strengthen the competitiveness of the enterprise on a continuing basis. The result will be an ongoing change process and an effective adaptation to the environment. A number of modeling tools can be harnessed in support of this effort. When applied as part of a balanced activity, limited-scope modeling can be an effective resource.

Figure 4-1a provides a means for coupling the concepts introduced in earlier chapters to a specific design approach. This figure illustrates several important aspects of the design process, including a system approach with inputs and outputs, a design process controlled by a planning group, the use of supportive limited-scope modeling as part of the design process, and an iterative learning activity.

Figure 4-1b indicates how the five-layer model can be used to structure the design process. Within the central design process block, it is possible to begin the design activity by establishing an information base for all five layers or viewpoints of the manufacturing activity. Interactions can then be allowed to take place among these five layers, and an output obtained in terms of system definitions for each of the layers. Thus, the five-layer model provides a structure through which to describe the *as-is* system and the *to-be* system, and this structure can be used as a framework for study of interactions that take place among these layers. The model shown in Fig. 4-1b provides a paradigm or perceptual framework for considering the internal characteristics of the manufacturing system.

It is possible for such a design process to take place entirely on a qualitative and intuitive basis. However, decision making can be improved by performing limited-scope modeling to support the design activity.

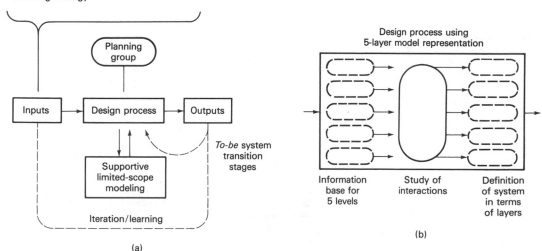

- Enterprise objectives
- Environment
- CIM design principles and reference systems
- *As-is* system
- Learning strategy

Planning group

Inputs → Design process → Outputs

To-be system transition stages

Supportive limited-scope modeling

Iteration/learning

(a)

Design process using 5-layer model representation

Information base for 5 levels

Study of interactions

Definition of system in terms of layers

(b)

A wide variety of techniques exists for performing a system study using the approach described in Fig. 4-1. One strategy might call for having support groups individually address each of the five layers of the system, having these groups interact simultaneously in order to consider the relationships among the layers, and having the groups work together in a synthesis process to produce a completely new system description in terms of the five layers (Fig. 4-2). The difficulty with this approach is that such a parallel process can become unwieldy and difficult to implement.

Another approach is to develop a more serial solution and use various types of iteration and feedback to ensure that all factors are allowed to interact with one another to produce a final configuration. It is often useful to begin the design process at the functional level, as expressed by layers 2 and 4 of the multilayer model. The physical system characteristics of layers 1 and 5 can be treated as inputs to the functional design, and the organizational/management considerations of layer 3 can be considered in evaluating the feasibility of the solutions that are obtained. If this process is followed, then the particular approach outlined in Fig. 4-3 can result. In this figure, the method of performing the design process using the five-layer model begins with modeling support in the two functional layers 2 and 4, using input constraints from layers 1 and 5. An evaluation of organizational/management impact for each of the various functional designs is considered in tandem with the financial evaluation. The supportive limited-scope modeling involves defining the input information to the models, concurrently applying functional models of the manufacturing system and

Figure 4-1.
Limited-scope modeling in support of the system design process. (a) How the design process itself relates to the planning group and supportive limited-scope modeling. (b) Application of the five-layer model to support the design process.

Figure 4-2.
Concurrent modeling efforts in support of the planning function. Planning associated with all layers is performed during the same period of time, with continuing interaction among layers. This is one possible approach to the design process. Significant operational difficulties may be experienced in attempting to perform all of the planning tasks at the same time, with continuing interaction among the task efforts.

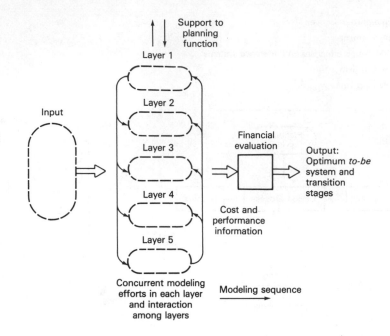

Figure 4-3.
Modeling effort that begins with the independent consideration of layers 2 and 4 using the physical system aspects (layers 1 and 5) as inputs and constraints. Layers 2 and 4 are then allowed to interact, and other design aspects follow sequentially, leading to iteration.

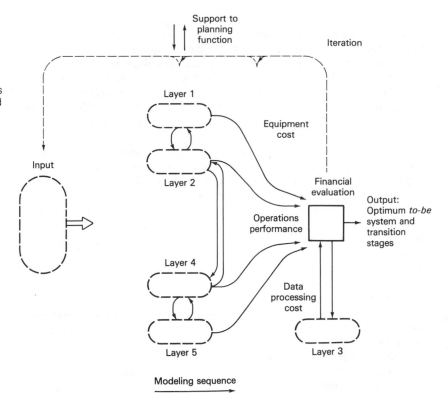

information system, and then allowing interaction to take place among the two model layers. The output can be used to drive an iterative process and improve on the proposed designs. This approach to structuring limited-scope modeling is often a useful one in realistic application.

Another, even more serial, approach to system design is indicated in Fig. 4-4. As indicated here, the design process begins with the functional model of layer 2, constraints from layer 1, and a set of specified inputs. The results of the layer 2 design are then fed into evaluations associated with layers 4 and 5, and a process of iteration results. The initial functional studies are simplified to emphasize the product flow and process functions within the manufacturing facility. All other aspects of the design are considered following this initial development. The advantage of this approach is the simplification that results. The disadvantage is that in producing this degree of serial development, a large number of iterations may have to be performed in order to begin to converge on an optimum solution.

The three study strategies shown in Figs. 4-2 to 4-4 represent a subset of all potential strategies. It is also possible to begin the design of a system with the information-related layers (4 and 5) or with the consideration of organizational opportunities and constraints of layer 3. However, in many realistic applications, the approaches illustrated in Figs. 4-3 and 4-4 turn out to be useful. Figure 4-4 is discussed in further detail below as a method for implementing the design of a manufacturing system.

To provide a more in-depth understanding of design process requirements, it is helpful to take the simplified model of Fig. 4-4 and

Figures 4-2 to 4-4 describe alternative planning procedures that can be adopted by the planning group. It is essential that the group consciously select and apply a strategy that is likely to bring the desired results. The preferred choice will depend on the nature of the enterprise, the orientation and preferences of the planning group, and the types of design efforts being considered.

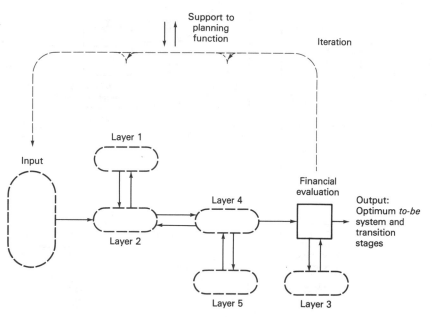

Figure 4-4.
Modeling effort that begins with process functional design (layer 2) using physical equipment aspects (layer 4) and inputs as constraints. The information functional design (layer 4) is then considered witih physical aspects (layer 5) as inputs and constraints. Other design aspects follow sequentially, leading to iteration.

Figure 4-5.
Extension of the modeling strategy of Fig. 4-4. (a) Input information for the four areas shown can be used to define a product line in terms of product categories.(b) Once the product line is defined, the manufacturing requirements associated with the products can be used to develop a description of the composite manufacturing operations required for all categories. Based on the definition of parameters that can describe these manufacturing operations, alternative manufacturing system configurations can be evaluated, making use of the information base for manufacturing equipment. (c) Once alternative manufacturing configurations have been formulated, they can be evaluated in terms of information flow and financial and organizational impact. Alternative strategies for product design and manufacturing are considered as complementary aspects of an integrated system. The result will be an implementation plan for an optimum *to-be* system, with associated transition stages. (d) The steps in Figs. 4-5a–4-5c can be combined, as shown here. Iteration among all aspects of the design activity can then take place. (e) Detailed view of a combined concurrent/sequential modeling process formed by combining Figs. 4-5a–4-5d. Application of this model is dis-

expand upon its use in a design effort. The result of such an expansion is indicated in Fig. 4-5. As shown in Fig. 4-5a, the design process begins with definition of the product line, which is a function of the market that exists, the objectives of the enterprise, and the technology that is available to create the selected products. The enterprise reviews these three areas and selects one or several initial or revised product lines for consideration.

The next sequence of tasks leads to descriptions of alternative manufacturing system configurations that are appropriate for the combined product lines. As shown in Fig. 4-5b, the products that are being considered are used as input to define the alternative manufacturing configurations. In turn, the configuration descriptions depend on the manufacturing equipment that is available (or can be produced) and on the application of CIM design principles and reference models. A number of computer modeling techniques can be applied to enhance the description of these configurations.

The final activities require evaluation of the configurations and selection of the best system design, using a series of selection *screens*. As shown in Fig. 4-5c, the information systems associated with each

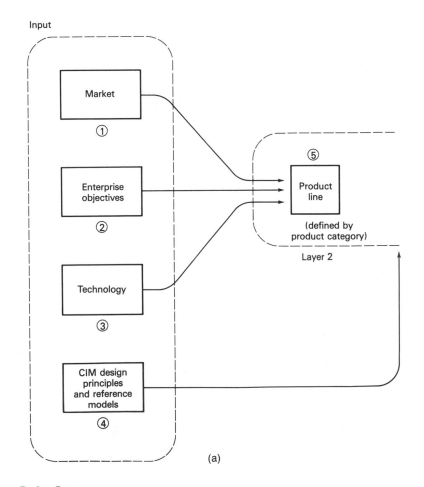

(a)

configuration must be evaluated in terms of the strategies to be used for product design and manufacturing and for the design-manufacturing interface. A financial evaluation is then necessary to establish the costs associated with performance of each system alternative, while the organizational evaluation is necessary to define the advantages and disadvantages that will be experienced in trying to achieve operations with each system alternative. Finally, having passed through these screens, the evaluation process can be completed, and an implementation plan can be selected.

This sequence initiates the design, evaluation, and learning activities that must take place. Once these tasks have been completed, it is necessary to start all over again and iterate through the cycle on a continuing basis (Fig. 4-5d). Once an implementation plan is selected, the iteration can be used to learn from each stage of the experience. The combined model in Fig. 4-5e provides a simple way of approaching manufacturing system design and represents a compromise between simultaneous and sequential planning activities.

Figure 4-5f shows the type of product design and manufacturing system implementation and operations that will follow. The imple-

Figure 4-5. continued
cussed further in this chapter and in Chap. 10. (f) Once the implementation plan has been developed, a series of transition stages can be performed, with continuing feedback and iteration from stage to stage to continue the learning process.

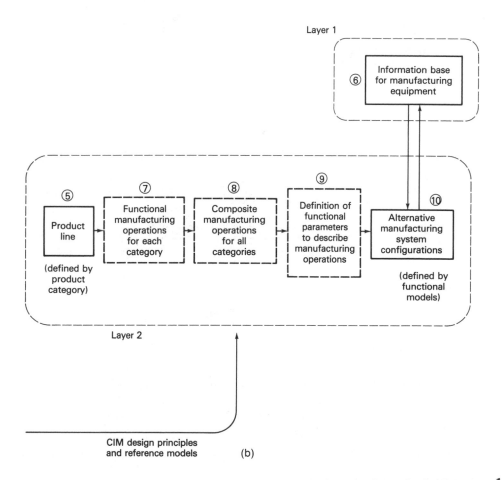

(b)

Figure 4-5. continued

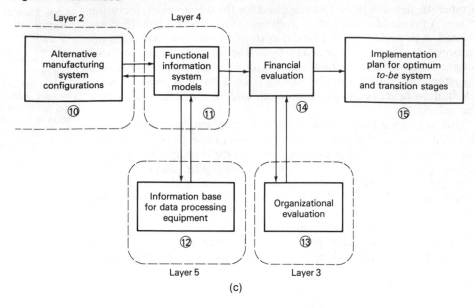

(c)

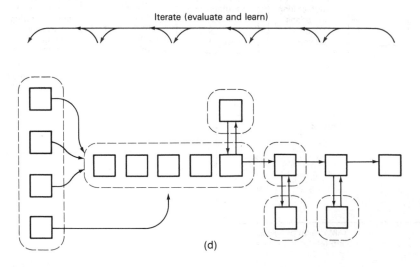

Iterate (evaluate and learn)

(d)

Figure 4-5. continued

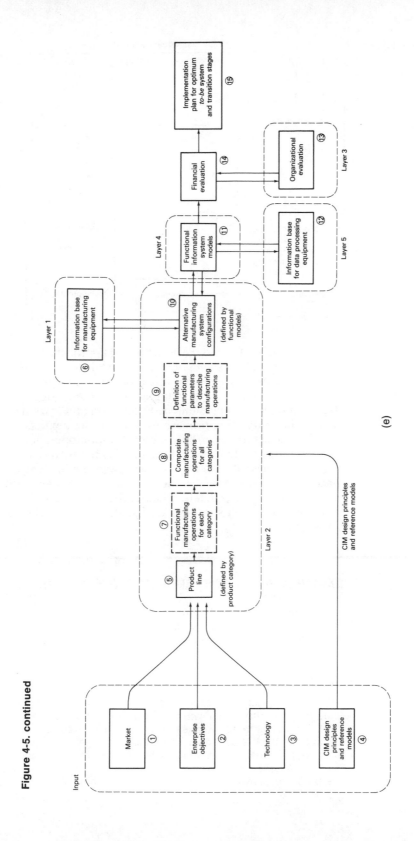

(e)

137

Figure 4-5. continued

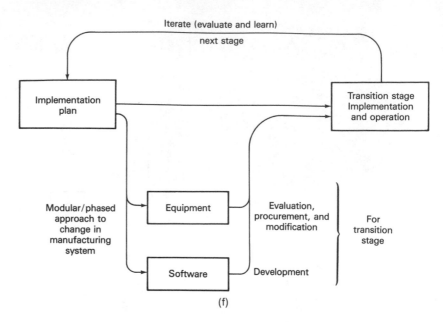

(f)

mentation plan is used to define a phased approach to change, involving a series of transition stages. For each transition stage, equipment evaluation, procurement, and modification must take place in parallel with the necessary software development for the system. The transition stage can then be implemented. Evaluation and learning activities can be used in an iterative process to develop the next stage.

The design/redesign and implementation/operations aspects of the manufacturing system cycle through this process on a continuing basis. **The strategy described in Fig. 4-5 for design and implementation of manufacturing systems provides the structure for this text and is discussed below in step-by-step detail.** Chapter 10 links these design considerations together by illustrating a design study for a particular setting and set of objectives.

EXAMPLE 4-1. PROBLEM SOLVING THROUGH LEARNING

Discuss some of the implications associated with setting enterprise objectives based on the perceived environment and then working to address these objectives, rather than continuing with established problem-solving strategies.

Results

One approach to problem solving in a manufacturing setting is to continue to apply problem-solving strategies that are familiar. This is certainly the first reaction in many cases, and is often the most comfortable approach because of past experience. The difficulty is that strategies that have been effective in the past may or may not be effective in addressing new problems.

An alternative approach involves setting objectives that are based on the perceived environment and system and then working to address those objectives through a learning experience. It may turn out that the given combination of objectives and environment will require an active learning experience oriented toward new ways of understanding problems and new problem-solving approaches. Enterprise objectives may result in the need for major change in ways of coping with the system environment. Such a shift can be difficult to achieve because changes of this nature may require a significant investment (in terms of time and energy) in learning how to adapt to the situation.

Thus, if the choice is between extending past problem-solving methods or engaging in a difficult learning experience to develop new problem-solving strategies, the former will tend to prevail over the latter. A decision to achieve problem solving through learning experiences is often a difficult (if essential) way to achieve the desired system change.

EXAMPLE 4-2. ALTERNATIVE SYSTEM DESIGN STRATEGIES

Discuss the relative advantages and disadvantages of the system design strategies shown in Fig. 4-2, 4-3, and 4-4.

Results

The modeling sequence in Fig. 4-2 requires a simultaneous interaction among all of the component study efforts. The advantage of this approach is the simultaneity and concurrent synthesis; if an effective means can be found to have all of these groups work on their own project areas and relate strongly to other areas at the same time, then this can be an effective design approach. On the other hand, this approach can often collapse under the pressures of the organizational and communications requirements. It may be impossible for the individuals involved in such an effort to simultaneously perform their own work and keep in mind all of the concurrent activities happening in the other groups, and then, on a continuing basis, modify personal efforts to adapt to other related activities. The degree of fluidity and the rate of change in such a planning process may result in requirements that exceed participant capabilities. Frustration may develop because of an inability to "get on with the job."

Figure 4-3 shows an alternative approach in which there has been some reduction in the concurrent nature of the study. The groups studying layers 1, 2, 4, and 5 perform concurrent planning efforts, but communications among these efforts are limited. Layer 1 primarily interacts with layer 2, so that the physical equipment constraints are incorporated into the functional design of layer 2. In the same way, the layer 5 team interacts primarily with layer 4, so that the computer constraints are incorporated into the information functional layer. Layers 2 and 4 then interact so that the two functional study areas can be related. All four of these areas feed into the financial evaluation, and the incorporation of the organizational constraints is considered at this time.

The advantage of the method in Fig. 4-3 is that the communications and simultaneity requirements associated with the study have been reduced. At the same time, a disadvantage is that a number of iterations may be required throughout the planning process, which will slow down the overall planning activity and extend the duration of the study. Further, it is always possible that in such an effort closure will never be obtained to the degree desired; all of the aspects of the study may not be incorporated adequately in the final design.

Figure 4-4 shows an alternative approach in which the simultaneity and communications requirements are further reduced. The interaction between layers 1 and 2 now takes place prior to the interaction between layers 4 and 5. Once the layer 2 and 4 efforts have been completed, an interaction is allowed to take place between these layers. Results of these four study areas are then fed into the financial evaluation, at which time the organizational constraints are considered. Again, iterations must take place in order to produce convergence among the various aspects of the planning effort.

The advantage of this system is its simplicity and the ability of the participants in the study to work within a manageable scope at any one time. The difficulty is that the communications requirements and the sequential nature may limit the convergence of the study and may also place management constraints on the planning team.

4-2. Design Inputs

A number of inputs are required for the design strategy in Fig. 4-5. The market environment and selected market sector to be addressed by the factory must be specified (step 1), and documentation of the enterprise objectives must be compiled (step 2). These information bases, combined with an understanding of state-of-the-art technology in the field (step 3), will enable the formation of product definitions in terms of manufacturing *categories* (step 5). The functional manufacturing operations for each product category can then be defined (step 7), making use of insights regarding the CIM design principles to be applied (step 4).

As discussed here, CIM system design is based on defining the range of products to be produced by the planned factory (step 5). The product definition must reflect the market environment, the enterprise objectives, and an understanding of the relevant state-of-the-art technologies. In turn, the product definition will drive the manufacturing processes required for the facility, the nature of the supporting information system, and the overall cost-benefit of the resulting system.

To support system design, it is necessary to develop a method for categorizing the markets to be addressed and products to be manufactured. This method must provide a means for *mapping out* the entire product range that is of potential interest and must enable the defining of categories within the overall market or product area. These categories can be used to define which portions of the market can be most effectively addressed by the factory and which portions should be excluded from factory operations. An appropriate method for categorizing the market will enable the planning group to look at the entire range of possible product lines and to weigh these in terms of the enterprise objectives and technology strengths and opportunities. Based on an understanding of all product opportunities, specific categories can be selected for inclusion or exclusion in the factory scope of operations.

It is important to select a means for categorizing the market and product line that will be specifically useful for the planning effort to follow. As defined and used here, each product *category* includes a variety of different specific product *types*. A category is defined so that all product types within a category require similar manufacturing processes. Categories should be formed in anticipation that equipment changes will not be required in shifting production among product types. Within a given category, product types can be comingled to achieve an effective lot size that includes all types within the category. The CIM facility will be able to produce all of these products by making changes in the operator and software instructions provided to the equipment. Various product types within a category can be manufactured by changes in operator actions or by downloading different software programs to the manufacturing equipment. All product types within a category can thus be treated as a single lot size for purposes of economy of scale.

Figure 4-6 illustrates the category concept as defined here. Production within a given category takes place between defined *setup* times. Setup is assumed to involve hardware changes or significant servicing changes required to extend the adaptable range of existing equipment. Between setup times, any product type can be produced based on providing the appropriate information associated with that type.

In dealing with CIM systems, it is useful to develop idealized reference models that can be used for comparison and evaluation (this concept is introduced in Sec. 3-1). An idealized fully automated CIM reference system is defined here as being able to operate between setup times without operator intervention and produce 100 percent yield. The full variety of products can be manufactured during this period, where each category consists of multiple product types. Software control can be used to shift equipment operations among product types. This reference system thus represents a limiting case in which all equipment can operate w1thout operator intervention (Fig. 4-7). At another extreme is a system in which manufacturing requires maximum use of continuing operator support. For this second reference case, equipment is manual or minimally automated and requires ongoing operator servicing.

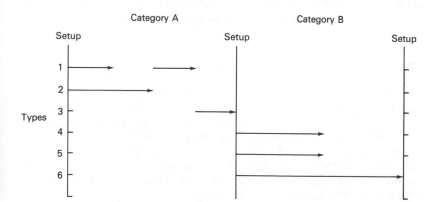

Figure 4-6.
Illustration of the product category concept. Within each category, various types of products can be manufactured by the factory without operator intervention. Each product category extends between manual/operator equipment setups. A number of product types can be produced within each category without basic change in the equipment or manufacturing system. If many product types can be produced within a category, requirements for setup will be reduced.

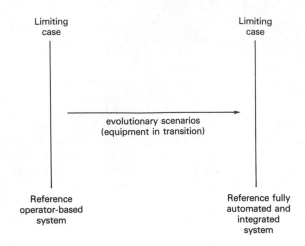

*Introduction of the
mean time between
operator interventions
(MTOI) as an equip-
ment or process pa-
rameter is a simple
way to distinguish be-
tween levels of auto-
mation. The MTOI de-
scribes the relationship
between equipment
and operators. By ex-
ploring alternative sys-
tem design concepts
and allowing the
MTOI to vary among
members of equipment
families, the planning
group can gain an un-
derstanding of the
types of automation
that will be matched to
the enterprise.*

These two extremes represent limiting *scenarios* for a manufacturing system. In between can be defined a number of evolutionary or
transition scenarios for which the equipment has varying degrees of
additional automation capability over the operator-run system, but has
reduced capability when compared with fully automated versions.

The idealized fully automated reference system provides an important perspective from which to view CIM options. Such an extreme
system can produce 100 lots of quantity 1 in approximately the same
time and at the same cost as 1 lot of quantity 100. This reference system
thus can provide a useful framework in which to consider other evolutionary or transition scenarios. (As explained below, this reference
system will often *not* be the preferred factory design choice due to the
high costs associated with its implementation.)

The degree of operator intervention in any of the transition scenarios can be viewed as a way to contrast the scenarios. It is possible
to associate required operator services with a mean time between operator interventions (MTOI) parameter. The MTOI provides a means
for measuring how a transition system configuration (or equipment
unit) relates to the two reference CIM systems. In many cases the
MTOI will be much less than the mean time between failures (MTBF)
for the equipment being used, and will dominate process activities between setups. This framework, which allows comparison between various intermediate scenarios and limiting reference systems, is further
applied below.

The definitions of product categories and types given above
should be applied during the product definition stage of a planning
effort. Of course, it is not always clear at the beginning of the study
as to the most appropriate definition of categories. In this case, preliminary categories can be formed and used as a basis for starting the
study. Iteration will provide a means for improving the categories during development. As defined above, then, the design process begins
with the selection of a market and product area to be the focus of factory
operations and the subdividing of the market and product line into
manufacturing categories. These categories form the basis for developing a preferred manufacturing strategy.

4-3. Manufacturing Equipment and Functional Operations (Layers 1 and 2)

Once the categories have been established, the next step is to list the functional manufacturing operations that are associated with each product category (step 7 in Fig. 4-5). After these operations are defined for each category, they may be combined to produce a composite description of the manufacturing functions for the entire product line, including all categories being considered (step 8). As the planning process proceeds, it is possible to consider different combinations of categories and different composite manufacturing functions as input to the design process and thus perform cost-benefit comparisons of alternative factory configurations. In this way, the factory can be most effectively targeted toward the market segments with the highest payoff and with the highest compatibility with enterprise objectives.

In step 8, the manufacturing functions by category are combined to produce a composite functional description for the entire product line. Each equipment operation, including operation of any materials handling robots, must be included as a separate function for purposes of this description. The resulting functional description can be represented by a flowchart in which the boxes define the manufacturing functions that must take place. Linkages among these boxes are indicated to document the functional relationships that must exist. It is possible to prepare such a functional flowchart by hand. However, as discussed in Chaps. 1 and 5, use of IDEF-based software models can improve the efficiency of the activities required to produce this flowchart.

Some equipment will require changes in input materials/components to produce different product types. For the purposes of this text, equipment is described in terms of the number of different input conditions that the unit can accommodate at a given time. To the extent that equipment requires manual input changes during production of a given product category, an operator intervention results. Equipment units with highly automated multiple feeders and process capabilities will require fewer such interactions, whereas units with manual changes in input will require many such interactions. A trade-off exists between a few automated-input units and many manual-feed units to provide the same input flexibility. The best choice for a given factory system will be determined by studying the cost-benefits of alternative possible system configurations. The advantages of equipment upgrading will be clear in this context.

The next task is to introduce a set of parameters that can be used to describe all equipment in the factory (step 9), including both processing equipment and materials handling equipment. These parameters should provide an adequate description of each equipment unit in the factory for the purposes of deciding on a preferred factory configuration. For this text, the following five parameters have been selected, as appropriate for any item of equipment within a CIM factory.

1. *Scope of operations:* This parameter is used to describe the number of categories for which a given unit of equipment can be used

in the manufacturing process. If a unit can be used to support production of several different categories of product, then it has an enhanced capability. In an ultimate, limiting case, if all equipment is able to support all product categories, then no setup time will ever be required, and the factory will be completely flexible with software download compatibilities. In a limiting case at the other extreme, if each item of equipment can be used with only a single product category, then a complete equipment change-out is required in going from one category to another. In between these two extremes are a number of cases in which an individual unit can be used to produce several categories of product, but not all categories. This scope of operations parameter is thus a measure of the flexibility of the equipment in terms of addressing the desired market environment and product line.

2. *Mean time between operator interventions:* Between setups, it is desirable that the factory operate with a preferred level of operator intervention. This requires that the equipment be able to function with an optimal level of operator assistance from one setup to the next. The mean time between operator interventions (MTOI) for a given piece of equipment will define how often an operator must intervene in the system during production runs (Fig. 4-8). A long MTOI will allow the equipment to operate fully from software control between setups. If the MTOI is longer than the estimated time between setups, then this fully automated strategy can be realized.

3. *Mean time of intervention:* This parameter describes how long it will take, on the average, to provide the required servicing for the equipment each time an operator intervenes. To simplify modeling, an operator intervention can be considered as combining the impact due to both equipment failure and the need for normal equipment servicing.

4. *Product yield:* The fourth parameter is the percentage of the product produced by each item of equipment that is of acceptable quality. For a CIM facility, it is desirable that this percentage be high, so that rejects and rework not constrain the system from reaching its full performance capability.

When taken together, the MTI (mean time of intervention) and the MTOI (mean time between operator interventions) can completely define average operator-equipment interactions for each manufacturing function. By varying these parameters, a wide range of system configurations, levels of automation, and equipment performance can be reviewed by the planning group.

Figure 4-8.
Illustration of the MTOI concept. In general, between hardware setups, each manufacturing equipment unit will require periodic servicing by operators. If the MTOI is sufficiently long to prevent servicing requirements between setups, production can be optimized within a given category.

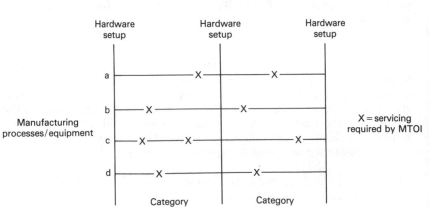

5. *Processing time:* The fifth and final parameter is the time required for each item of equipment to process the work-in-progress (WIP). The mean processing time is associated with the equipment design and product type. For continuous flow manufacturing (CFM), the processing time will be given by the time spacing between products as they exit from a manufacturing process.

These parameters can be used to describe each manufacturing unit and to develop functional models of the entire factory. Based on models that include these parameters, alternative combinations of equipment and routing strategies can be studied for cost-effectiveness.

The MTOI and MTI define how manufacturing equipment and operators relate to one another. The product yield and processing time then describe the combined equipment-operator performance.

EXAMPLE 4-3. INTERPRETING MTOI AND MTI DATA

Figure 4-9 provides a set of MTOI and MTI values for various equipment units. What can be concluded about the operational nature of these units?

Results

Equipment units A through E in Fig. 4-9 show a wide range of MTOI and MTI values. Unit A requires operator intervention in equipment operation every 0.1 minute, or every 6 seconds, and an average intervention lasts 0.1 minute. Unit A thus requires full-time operator interaction with the unit; the operator must be fully at attention and interacting with the equipment unit at all times.

For unit B, the MTOI of 1.0 minute and an MTI of 0.1 minute means that the operator must be on location, interacting and available to the equipment unit at all times since this interaction is required every minute. On the other hand, since the MTI is 0.1 minute, the operator typically has a rest period between interventions. For unit C, the MTOI is up to 10 minutes and the MTI is 1.0 minute. For this case, the operator must be in the vicinity and available, so that for 1 minute out of 10 the equipment can be serviced as required.

For unit D, the equipment can operate for 100 minutes, or 100/60 = 1.7 hours without servicing and the mean servicing time is 10.0 minutes. For this case the operator must be in the facility and available periodically for the required servicing, but the equipment basically operates on its own the rest of the time. Unit E can operate for 1000 hours without intervention and has an MTI of 10 minutes. This equipment unit can operate for 17 days without operator involvement, and might be viewed close to an ultimate unit for a "lights out" factory.

Equipment Unit	MTOI (min)	MTI (min)
A	0.1	0.1
B	1.0	0.1
C	10	1.0
D	100	10
E	1000	10

Figure 4-9.
MTOI and MTI data for Example 4-3.

To explore system performance fully, it is necessary to develop simulations that can accommodate a mix of product types and treat all equipment parameters as a function of product type. Statistical studies can then be performed to estimate the result of factory performance for a variety of product mixes. It is important that this method not be highly dependent on the product mix, as the mix can change with time. The performance of the system must be commercially attractive for a wide combination of product types. In many cases (as discussed in Chap. 5), this complete statistical simulation approach to solving the problem can often be viewed as providing too much detail during early design phases. More approximate methods may be useful during initial design. Simplifying assumptions regarding the mix of product types and the dependence of equipment parameters on the types are typically useful during a first-cut design.

The next step in the manufacturing process is to develop functional models of potential factory configurations (step 10) based on the required manufacturing functions (from step 8), the parameters of each function (from step 9), and design principles in use (step 4). These step 10 models should describe the number of equipment units necessary for each function, the preferred function placement, and alternative routing structures among the equipment units and the work cells.

The manufacturing equipment information base (step 6) can be used to set the parameter *constraints or limits* associated with the functional equipment descriptions in step 10. The results of the step 10 modeling can in turn be used to define the *preferred* parameter values for the materials processing equipment to be used (as defined in step 6). Thus, the modeling effort will ensure the most balanced and appropriate specification of equipment to be used.

In step 10, various types of models can be used to study the functional advantages and disadvantages of alternative system configurations, based on defining each equipment unit in terms of the five parameters described above. Obviously, the values given to these parameters depend on the physical manufacturing equipment that is being considered. Thus, a knowledge of the manufacturing equipment information base, step 6, can be used to set upper parameter limits for each of these five parameters. During the functional modeling, these parameters must not be allowed to take on values exceeding these limits. However, as a result of modeling, it may turn out that these limits are not required for all items of equipment and for all product categories. For this case, the functional process modeling can be used to define preferred parameter values that will often allow a cost reduction in the capital investment for equipment. The cycling between steps 6 and 10 allows a wide range of system configurations to be studied using the defined parameters and the product categories developed above, with consideration given to the constraints associated with the physical equipment. From this iterative process will result one or several preferred configurations for the factory.

Various system modeling techniques can be used to support the functional design associated with step 10. Once models have been constructed for a wide range of equipment configurations, the software can be run to determine the operational characteristics of each configuration. The modeling effort will result in a set of possible factory configurations described by related performance characteristics.

EXAMPLE 4-4. EQUIPMENT SELECTION BASED ON FUNCTIONAL PARAMETERS

Figure 4-10 lists values for various equipment parameters associated with various technologies and optimum parameter ranges determined by a first-cut system modeling effort. Which technologies are available for the system being designed and which technologies will limit available system configurations?

Results

Consider equipment type A, available in versions 1, 2, and 3. As illustrated, the optimum MTOI must be greater than 0.5 min. In comparing this

Equipment Function	Equipment Version	Available Parameter Values			
		MTOI (min)	MTI (min)	Yield (%)	Processing Time (min)
A	A1	0.1	0.1	90	12
	A2	1.0	0.1	85	10
	A3	10	1.0	80	8
B	B1	1.0	0.1	95	10
	B2	10	0.5	90	2
C	C1	0.1	0.1	98	3
	C2	5.0	1.0	98	2
	C3	8.0	2.0	96	1

(a)

Optimum Parameter Range			
MTOI (min)	MTI (min)	Yield (%)	Processing Time (min)
>0.5	<2.0	>70	<9.5
>0.5	<2.0	>85	<5.0
>2.0	<1.5	>90	<3.0

(b)

Figure 4-10.
System parameters for Example 4-4. (a) Available parameter values. (b) Optimum parameter range.

value with the available equipment parameters, note that only equipment units A2 and A3 are qualified. The MTI must be less than 2.0 min., so both A2 and A3 qualify. Similarly, both A2 and A3 meet the yield requirements. However, A2 does not satisfy the processing time requirement, so based on this particular modeling effort, equipment type A3 would be selected as appropriate for a given application. Equipment type B is available in versions 1 and 2. Only type B2 can meet the requirements associated with processing time. For equipment type C, available in versions 1, 2, and 3, only C2 can meet the combined requirements on the MTOI and MTI.

This type of analysis indicates how relationships may be developed between the available parameter values for equipment that is available on the market, and the optimum parameter ranges that are determined by system modeling efforts.

4-4. Scenario and Configuration Development

The objective of the modeling for step 10 must be to provide input that can be used to evaluate alternative factory configurations. A set of alternative scenarios must be developed involving different equipment capabilities and costs. The cost-benefits of each configuration can then be calculated in step 14, and the preferred solution selected.

Several ways exist for approaching scenario development. One simplifying approach that can be used initially is to assume a steady-state market in which the major objective is to produce the lowest cost per unit (cost/unit) product. The major issue at stake then is how to achieve the lowest cost/unit operations. A second, more sophisticated approach involves scenarios in which the manufacturing factory model is linked to a changing environment. This interaction can be used to study system response and adaptability and to consider advantages of alternative configurations. (As has been illustrated, the SES model can be used to show the advantages of a rapid product design and turn-around.)

For the following discussion, the initial steady-state market approach is considered. Given this approach to the problem, it is useful to bracket the potential design scenarios by the limiting highly manual and fully automated systems, as introduced in Fig. 4-7.

Figure 4-11 illustrates one way in which data from step 10 can be summarized. Alternative scenarios appear along the horizontal axis, as defined in terms of low to high levels of automated equipment operation. At the left end of this axis are reference operator-dependent systems and to the right are reference fully automated systems. The horizontal axis then represents the degree of automated performance. Along the vertical axis is the number of units that can be produced by the various factory configurations in a given time. Initially, it is assumed that a stable market exists, and the objective is to produce the lowest cost/unit product.

For each scenario, a variety of data points will be obtained, depending on the particular factory configurations being studied. For each scenario, a range of equipment mixes can be considered, producing

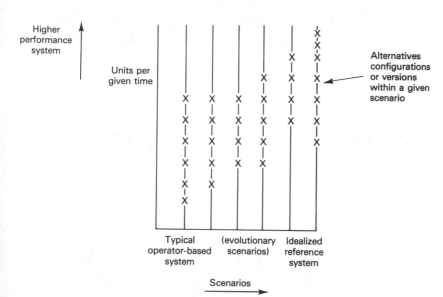

Higher performance system

Units per given time

X
X
X

X X
| |
X X X
| |
X X X X X X
| | | | |
X X X X X X
| | | | |
X X X X X
| | | |
X X X X
| |
X X
|
X

Alternatives configurations or versions within a given scenario

| Typical operator-based system | (evolutionary scenarios) | Idealized reference system |

Scenarios

More automated and idealized equipment

Figure 4-11.
Illustration of how data from step 10 (of Fig. 4-5) can be used to describe the productivity levels associated with various scenarios and versions. This chart indicates the number of product units that can be manufactured in a given time as a function of various scenarios that range from operator-based systems to fully automated systems. The multiple data points for each scenario are associated with alternative configurations or versions of the equipment for the system.

alternative configurations. Each configuration will have associated with it a performance level in terms of the number of units that can be produced in a given time. The idealized, fully automated scenario implies no operator involvement; the mean time between operator interventions is much greater than the time between setups. If 100 percent yield is assumed, the scenario shown to the far right might be viewed as the ultimate factory of the future.

Between the limiting scenarios, a variety of intermediate scenarios can be defined. These can follow one of two strategies:

1. In proceeding from the operator-driven to the fully automated systems, equipment modification may be assumed to take place starting with operator-driven equipment, then introducing fully automated equipment on a unit-by-unit basis. Thus, the intermediate scenarios start out with operator-driven scenarios and then move to the right as specific equipment units are assumed to have a higher performance capability. A unit-by-unit upgrading will then produce a spread of intermediate scenarios between limiting cases.

2. An alternative way to create intermediate scenarios is to consider equipment versions that are improvements over the operator-driven models, but are not fully automated. Essentially, any desired number of intermediate scenarios can be generated by varying the capabilities of the individual equipment units and looking at the associated performance levels.

In developing the data displayed in Fig. 4-11, it may be useful first to establish the limiting cases, then the more simple intermediate scenarios in which individual equipment units are fully automated in-

Figure 4-12.
One approach to developing alternative configurations within a scenario. The equipment unit with the highest utilization rate is duplicated, then the equipment unit with the next highest utilization rate is duplicated. This process continues until balance is obtained among the equipment utilization rates.

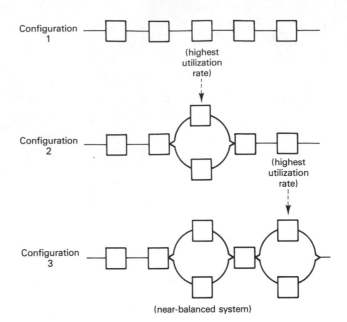

Configuration 1

(highest utilization rate)

Configuration 2

(highest utilization rate)

Configuration 3

(near-balanced system)

The scenario definitions allow the planning group to consider a sequence of system designs ranging from low to high degrees of automation. This sequence can be used to study the relative benefits and costs of system options as a function of the level of automation. Further, such studies can be used to determine which members of equipment families are most cost-effective for a given application.

dividually, and then finally the scenarios associated with subtle changes in equipment capability.

A potential strategy that can be used to develop factory configurations within a scenario is to start out with one equipment group in each type, as shown in Fig. 4-12, and then begin to add equipment units. One possibility is to add a new unit to replicate the manufacturing unit or function that has the highest utilization rate for the previous configuration. This process can continue until near-balanced utilization is obtained among all portions of the system.

Another strategy is to modify individual equipment units while maintaining one type of each unit. Combinations can also be considered in which upgrading is initially considered for individual units, and then multiple units of each type are introduced in order to achieve balanced utilization rates among all equipment types.

The preceding paragraphs describe methods that can be used to create multiple configurations for each scenario and multiple data points for each scenario, as shown in Fig. 4-11. The various scenarios are generated by considering the types of building-block equipment units that are being used in each case, going from the lesser to the more highly automated equipment. The result of the step 10 modeling can thus be summarized as shown in Fig. 4-11. This graph represents the relative performance levels of all factory configurations considered.

EXAMPLE 4-5.   AVAILABLE SYSTEM DESIGN OPTIONS

Figure 4-13 shows the predicted result of a system study, indicating the production rates that would be available for ten different scenarios for three different versions. Discuss the significance of this result.

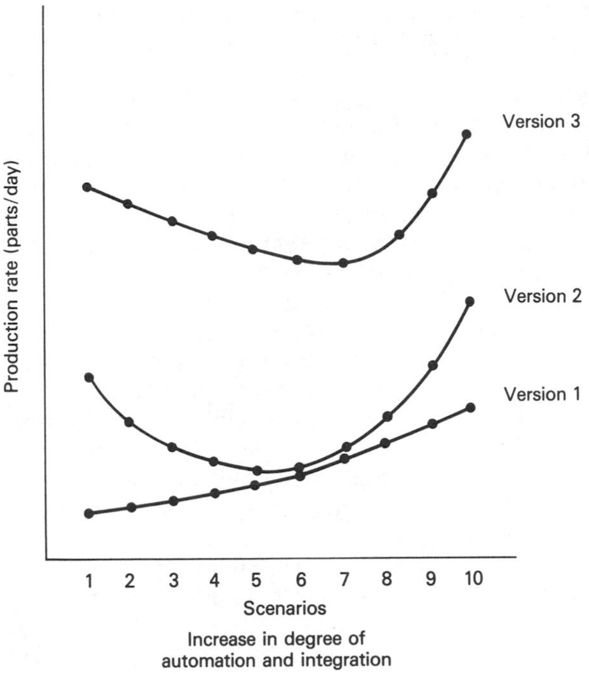

Figure 4-13.
Alternative system designs used for Example 4-5.

Results

For version 1, there is a steady increase in production rate for increasing levels of automation and integration. Thus, for this case, the preferred solution is to move steadily toward more CIM-oriented manufacturing systems. Version 2 indicates a situation in which a gradual movement toward automation and integration produces internal inefficiencies in the manufacturing system that result in a drop in production rate, until scenario 5. Beyond scenario 5, there is a steady and rapid increase in the production rate. Version 3 shows a similar drop that extends to scenario 7 before the overall production rate increases to its maximum value.

Based on the data shown here, the highest production rate for the manufacturing system will be scenario 10, version 3, followed by scenario 10, version 2, and scenario 10, version 1. If only the production rate is considered, then the best system design is clear. However, these data only address part of the problem. In the first place, there may be information system and organizational/management restrictions that will strongly affect the available scenarios and versions, and a cost-benefit analysis is necessary to determine which of these potential system designs will have the maximum benefit for the costs that are incurred. Therefore, the information in Fig. 4-13, which results from system modeling, is the starting point for determining the preferred system design, but the figure does not yet provide enough information on which to base a decision.

4-5. Functional Information Flow and Data Processing Equipment (Layers 4 and 5)

The next design activity involves development of a functional description of the information network (step 11) that will be required to support the defined factory processes (step 10). The data processing equipment information base (step 12) can be used to set initial parameter limits for the network, and network modeling can in turn be used to define the preferred parameter values for the information processing equipment (as defined in step 12).

The functional information models of step 11 are based on an understanding of the information flow requirements of the system, relating to both operators and the equipment units. It is necessary to assess the total data flow requirements of the configurations in terms of the quantity of data to be exchanged and the time available for this exchange. In addition for computer systems, the memory storage requirements must be anticipated. The dominant considerations in this step are thus the type and volume of information exchange that must take place, the time allowed for this exchange, and the total memory storage requirements with the necessary data base. The dominant computer specifications will be those associated with the automatic control and operation of the CIM factory, including the downloading and uploading of software as necessary, operation of scheduling systems, automated inventory tracking and control, and so on. These items are discussed further in Chap. 7. The management information system reporting from the information network can be considered an additional output requirement superimposed on the control and operations structure.

Figure 4-14.
Identification of configurations that must be reviewed for consideration because they exceed information flow capabilities. The configurations that are beyond information flow capabilities are associated with the highly automated and integrated systems, as noted by the dotted line.

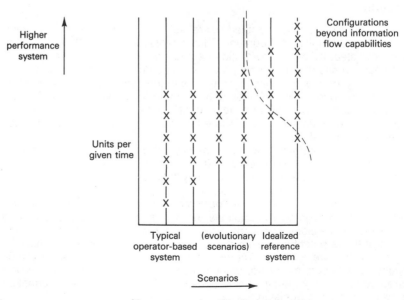

Once functional information models have been associated with the proposed factory, an effort must be made to determine the availability of and costs associated with the data processing equipment that will be required to implement the required information flow (step 12). If the requirements placed by the configuration cannot be accommodated within state-of-the-art technology or within budgeted resources, then the functional design of the factory must be modified to reduce the information flow requirements. An iteration between steps 10 and 11 must continue until it is determined that for each proposed factory configuration, the information flow requirements can be accommodated with available equipment and software.

As indicated in Fig. 4-14, some of the scenarios and configurations may be evaluated as beyond state-of-the-art information systems and must be excluded from the range of options considered. Obviously, if all data points have to be excluded on this basis, an iteration must be performed with different types of equipment and scenario strategies.

EXAMPLE 4-6. INFORMATION SYSTEM RESTRICTION

Figure 4-15 shows how the available scenarios and versions of Example 4-5 might be reduced because of information system restrictions. Discuss the significance of this result.

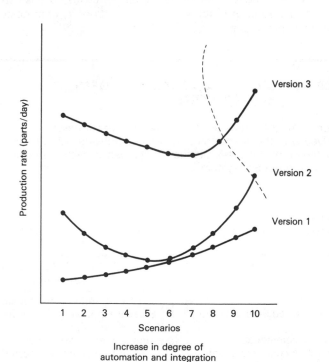

Figure 4-15. Limiting of design options due to information flow constraints. The scenarios and versions to the right of the dotted line must be excluded.

Results

A number of the more highly automated and integrated scenarios cannot be implemented because the information technology is not available. Before performing the cost-benefit analysis, it is appropriate to remove these possible solutions from the range of those being considered. The remaining scenario and version combinations may then be continued in the next step of the planning process.

4-6. Organizational/Management Issues (Layer 3)

At this stage, it is necessary to review the organizational requirements of each of the remaining data points in Fig. 4-15. Any scenario/configuration combination that cannot be effectively implemented due to organizational constraints must be removed from the set.

As indicated in step 13, each configuration must be evaluated in terms of organizational opportunities and constraints. This aspect of evaluation may include the costs and difficulties associated with reorganization and retraining, educational programs, hiring of new employees, and the development of an organizational culture that will provide effective use of the new factory. Whenever significant constraints are anticipated in any of these areas, the costs of accommodating the constraints must be included in the financial evaluation, or iteration must be used to modify the configurations to ease the anticipated organizational problems. These aspects of system design are discussed further in Chap. 8.

EXAMPLE 4-7. ORGANIZATIONAL/MANAGEMENT RESTRICTIONS

Figure 4-16 describes some of the organizational/management issues involved with the scenarios and versions introduced in Example 4-5 and restricted in Example 4-6. Discuss how these issues might change the selection of an optimum system.

Results

A number of issues might give rise to organizational/management problems. The types of employees who are required to operate the system might

Figure 4-16.
Potential causes of scenario restrictions as a result of information and organizational/management factors.

Information Restrictions	Organizational/Management Restrictions
Data rates too high Memory size requirements too large Does not accommodate presently available computer capabilities of company	Type of employee not available Present employees do not have necessary skills Management system would have to be changed, resulting in unrest

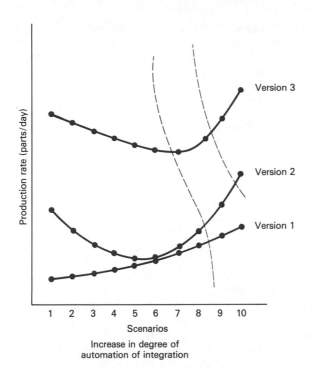

Figure 4-17.
Limiting of design options due to organizational/management constraints. The two dotted lines indicate the information flow restrictions and the organizational/management restrictions. Only those scenarios and configurations to the left of the dotted lines can be considered for implementation.

not be available. Or present employees do not have the required skills and are unwilling to undertake educational and training efforts to learn them. At the same time, the management of the company may have to be changed in essential ways, involving introduction of a new culture that would be technology centered as opposed to organization centered. All levels of management in this system may be unable to achieve this type of culture change. These and other types of related organizational/management factors might reduce the range of scenario and version options that are available to the planning group, as illustrated in Fig. 4-17.

4-7.   Financial Evaluation

The study process can be used to produce a variety of factory configurations that make best use of available manufacturing equipment and data processing equipment. The physical constraints associated with both types of equipment are reflected in the configurations. Chapters 6 and 7 provide additional information on the various constraints that will be experienced with respect to the manufacturing equipment information base and the data processing equipment information base.

At this stage of the study plan, factory configurations exist along with definitions of the manufacturing and data processing equipment required for each operation. In order to perform financial comparisons, it is now necessary to estimate the costs associated with implementing and operating each configuration (step 14). Such estimates must include allowance for manufacturing and data processing equipment costs (obtained from steps 6 and 12), personnel costs (obtained from consideration of the equipment design and organizational/management structure), and all other factory cost areas. The financial evaluations will result in cost-benefit analyses that examine the relative cost and performance properties of the alternative configurations that have been suggested. (Further detail on financial modeling is presented in Chap. 5, in Sec. 5-7. Equipment cost strategies are discussed in Chaps. 6 and 7.)

The financial evaluation must include not only an analysis of proposed CIM configurations using appropriate methods, but also comparative analysis with standard manufacturing systems that could be developed for the same product line. It will be necessary to compare the proposed CIM design with a typical facility based on current manufacturing norms in the industry to show the cost advantages of the CIM system. Thus, the financial evaluation must consider both the system under design and the costs of a typical manufacturing system achieving the same products (Fig. 4-18).

Cost-benefit ratios can be calculated for each facility configuration. Figure 4-19 shows the typical results that might be obtained. Two types of cost/unit responses are indicated. The lowest cost/unit configuration is shown for each scenario. Curve A indicates a product line, factory, and market that results in a continuing lower cost/unit as the CIM facility evolves toward more fully automated operation. In such a situation, the optimum design will be to move as rapidly as possible toward operator-free, fully automated operations.

Curve B indicates a different result in which there is an intermediate level of automation that produces the lowest cost/unit. For this case, the extreme, fully automated system is not the most cost-effective solution to CIM design. The additional features introduced into the fully automated equipment units have become so expensive that the improved performance does not warrant the required investment. For this case, the best design strategy is to move toward the lowest cost/unit solution and to stabilize at this point while continually testing the potential for higher levels of automation as equipment costs are reduced.

The strategy described here has assumed a steady-state environment in which the lowest cost/unit product will prevail. The evaluation

Figure 4-18.
Alternative combinations of people (operators) and equipment for system design. The alternatives can be used to compare manual, semiautomated, and fully automated factories with the same product line and product volumes.

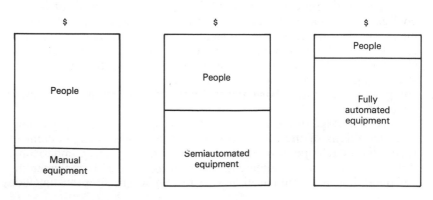

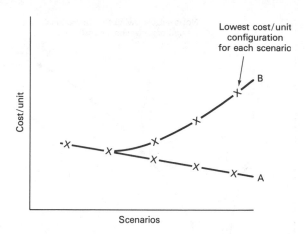

Figure 4-19.
Comparing the lowest cost/unit configurations for a set of scenarios, indicating two possible types of relationships. For curve A, the cost/unit declines steadily for scenarios associated with higher levels of automation and integration. Curve B shows a cost/unit that decreases with initial moves toward automation, then increases for more extended use of automation and integration.

has not dealt with other relative advantages of the system, including market responsiveness and adaptability and the advantages of producing a product that is more responsive to customer desires. Therefore, at this point, all of the configurations in Fig. 4-19 might be evaluated by allowing the various factory capabilities to interact with the SES model and performing additional ratings of the alternative configurations. Thus, the lowest cost/unit rating in Fig. 4-19 provides a start in comparing the various configurations, but in a complete evaluation, there must be additional criteria explored. A final design selection can be based on all of these criteria, evaluated in terms of the organizational objectives and in the larger context by the planning group.

EXAMPLE 4-8. COST-BENEFIT TRADE-OFFS

Figure 4-20 shows several graphical relationships that might result from system cost-benefit analysis. Describe the advantages and disadvantages associated with each set of scenarios described.

Results

Figure 4-20 shows how the remaining scenarios and versions of Figure 4-17 could be combined with cost data to produce cost/unit production estimates associated with each system configuration. As indicated, the consideration of cost data changes insight into which scenarios and versions would be preferable. Version 1, scenario 5, has the lowest cost/unit of all versions or scenarios. This solution indicates a compromise between manual and highly automated and integrated versions. If this solution were adopted, it would be important to continue to evaluate equipment availability and costs to determine if this lowest cost/unit solution might migrate over time toward higher levels of automation and integration.

Version 2 would reach almost the same low cost/unit solution available for version 1 if automation and integration were carried to scenario 10. However, other risks are associated with the maximum use of technology (as discussed in Chaps. 6 and 10). Therefore, even though version 2 might be

Figure 4-20.
Several types of graphs that might result from system cost-benefit analysis (see the discussion in Example 4-8).

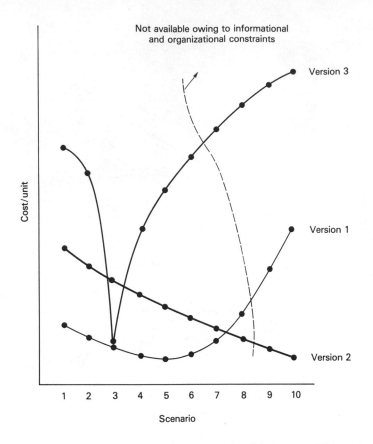

Not available owing to informational and organizational constraints

Cost/unit

Version 3

Version 1

Version 2

Scenario

1 2 3 4 5 6 7 8 9 10

a second choice, it is not as attractive as version 1 because of this potential risk.

Version 3 is in many ways a quite dangerous approach, because cost/unit rates are very high except for one optimized system around scenario 3. This version is an unattractive way to approach system implementation because it shows a high sensitivity to the exact scenario level of automation and integration being used and to the costs associated with equipment. Narrow optimum ranges are to be avoided because of the uncertainties associated with the data being applied. As noted in previous chapters, the ability to project into the future is sharply restricted by the quality of the data being used and the modeling assumptions. Very tightly optimized systems should often be viewed as unattractive because small changes in the assumptions regarding the input data might make large changes in the estimated cost/unit performance.

4-8. Study Output

The output of the study flow in Fig. 4-5 is a description of a preferred *to-be* system (step 15). As noted in Chap. 3, it is likely that this *to-be* system will never be completely realized, but that the system represents a direction toward which evolution will take place. The planning group must now define a series of robust transition stages leading to the *to-be* system, with the intent that each of these transition stages may, in the future, form a starting point for a new study to define the evolving system.

The modeling procedure described here provides a top-down overview of the manufacturing system and enables an evaluation of system components and relationships. During this modeling effort, it is necessary to maintain a bottom-up awareness of the building blocks of the system. In particular, the definition of equipment parameters and their limiting values (steps 6 and 12) depends on the physical equipment that is to be used in the facility. In commercial applications, it will often be necessary to restrict parameter values to those associated with readily available commercial equipment.

As the step 10 models are applied, the preferred parameter ranges for various equipment items can be established. Therefore, initially proposed parameter ranges can be relaxed or tightened to the maximum state-of-the-art value, as appropriate to the requirements of the system, as indicated by exercising the software. The system modeling and financial evaluations will enable the development of preferred system configurations, which in turn define the best parameter values for the system equipment units. Equipment specifications are closely related to the planned system application, so available equipment can be selected for most effective use. High-performance equipment will be used only in those parts of the system where the payoff is appropriate; otherwise, equipment performance parameters will be required to be satisfactory, but not above the required level.

Once the modeling is complete, detailed equipment specifications can be developed for the purchase of commercial equipment or production of custom state-of-the-art equipment. In addition, based on the information-handling specifications, computer networks can be designed to implement the proposed configuration. As a result of this system design study, the design team will produce a preferred manufacturing system configuration, reflecting the specifications required for the constituent equipment units and an appreciation for the cost benefit of the system.

Once the CIM system design is complete, an implementation plan must be developed, based on a series of incremental transition stages or experiments that will serve as learning experiences for the new facility. Implementing a radically new manufacturing system without an experimental validation of the strategy is obviously a high-risk approach. At this point, it is essential to define transition stages that can lead on a step-by-step basis toward the final factory configuration.

To make maximum use of these transition configurations, strategies must be carefully developed to evaluate the configurations as they become available. Conclusions can then be drawn regarding factory performance. The results of these transition stages will validate the ability to construct a full CIM factory.

4-9. Areas of Manufacturing Emphasis

The approach taken to advanced manufacturing in this text is generally applicable to a wide range of manufacturing settings. Given the diversity of industry sectors and the wide variety of enterprises within each sector, a discussion of computer-integrated manufacturing must of necessity provide tools and methods that are intended to work in a broad setting.

At the same time, it is often useful to combine general discussions with a more detailed exploration of specific industry applications. Specific settings can illustrate how the more general discussion can be applied and therefore serve as a role model for other applications. A number of general insights can be drawn if the industry sectors of emphasis are selected carefully.

The study of machine tools for metalworking provides insight into the mechanical product settings in which many fundamental CIM concepts have been developed. Study of the production of small electronics assemblies provides insight into the electronics manufacturing settings in which application of CIM concepts has been required to make effective use of rapidly changing technology to meet international competition. In combination, these two areas of emphasis provide a balanced viewpoint of the design and implementation issues associated with CIM.

The two manufacturing sectors selected for emphasis here are associated with machine tools for metalworking and the production of small electronics assemblies. The machine-tool industry historically has been a major focus for advanced manufacturing strategies, due to the broad economic impact associated with this sector. Many of the significant advances in computer integration have been led by new equipment concepts and products in this area. Electronics manufacturing is also appropriate for consideration, as electronics-related products give rise to one of the largest industries in the United States and the manufacture of electronics-related products is essential to the competitive capability of the United States. These two areas raise different and complementary issues regarding the diversity of mechanical processes involved in metalworking and the quite different set of processes required for electronics manufacturing. Taken together, these areas illustrate many principles and issues associated with industry at large.

Figure 4-21 indicates the areas of the text devoted to specific industry sectors. In Chap. 5, which discusses modeling and simulation, examples address both the machine-tool and electronics-products areas. Examples 5-1 and 5-2 discuss manufacturing enterprises in general. In addition, Example 5-2 discusses a flexible manufacturing system that performs machining operations on castings. The machining center consists of two lathes, a wash/load area, and ten horizontal milling machines linked together by automatic guided vehicle (AGV).

Example 5-3 discusses a general manufacturing center with identical machines producing a single product and then a more complex manufacturing system for printed circuit boards (an important area in electronics). Example 5-4 discusses a manufacturing system that performs a grinding operation on turbine blades, with both grinding and inspection processes. The example then turns to printed circuit manufacturing involving two types of boards and a variety of operations. Example 5-5 is concerned with a general product made in a flexible manufacturing system, and Examples 5-6 through 5-8 consider more complex versions of the same system. Costing studies for general manufacturing settings are discussed in Example 5-9. Example 5-10 introduces three cost-estimating software products that are specifically intended for use in machine shop settings.

	General Methods and Discussion	Application to Machine Tools for Metalworking	Application to Electronics Assemblies
Chap. 1	X		
Chap. 2	X		
Chap. 3	X		
Chap. 4	X		
Chap. 5	X	Examples 5-2 through 5-10	Examples 5-3 through 5-8
Chap. 6	X	Example 6-9(a–c)	Sec. 6-10 to end of Chap. 6
Chap. 7	X		Examples 7-1 and 7-8
Chap. 8	X		
Chap. 9	X	Example 9-1(a–c)	Examples 9-1(d) and 9-3
Chap. 10	X		Focus for chapter

Figure 4-21.
Identification of those portions of the text that apply specifically to machine tools for metalworking and to the manufacture of electronics assemblies.

Chapter 6 addresses manufacturing equipment and automation. Example 6-9 introduces machining centers, cells, and systems. This example discusses general concepts and describes two state-of-the-art machining center products on the market. Starting with Sec. 6-10, the remainder of Chap. 6 is concerned with a wide variety of processes used in electronics manufacturing. The objective is to review a range of related processes that can introduce the concept of a family of equipment units and to indicate the variety of processes that can be encountered in manufacturing.

Chapter 7 introduces information flow and computer networks. Examples 7-1 and 7-8 and Example 6-11 (previously introduced as part of the discussion of integrated equipment products) describe the types of software system products that are available for highly integrated electronics manufacturing settings. The software system product in Example 7-8 is now being applied to other types of settings.

Chapter 9 describes the properties of computer-aided design systems in general and includes examples for both metalworking and electronics-product design. Chapter 10 builds on the material in previous chapters to provide an in-depth discussion of manufacturing system design. The objective is to optimize the system to achieve the most cost-effective operations. The emphasis throughout Chap. 10 is on electronics assemblies, but the implications are general and can be readily extended to other manufacturing settings.

ASSIGNMENTS

4-1. In order to achieve maximum educational insight from Chap. 4, it is helpful to stimulate class discussion of the concepts that have been included, develop critical reviews that probe the advantages and limitations of the concepts, and explore alternative ways for approaching the topic. Select one or several topics from the bold-faced headings that introduce sections or examples in this chapter. Come to class prepared to explain this topic, critically describing the text materials and suggesting ways the topic discussion might be strengthened.

4-2. Figures 4-2, 4-3, and 4-4 describe three alternative strategies that may be used to design CIM systems. Develop and describe your own alternative method for such a purpose. Discuss the strengths and weaknesses of your approach as contrasted with the ones discussed in this chapter.
(a) Prepare a flowchart showing your strategy and be prepared to discuss it in class.
(b) Following class discussion, prepare a brief paper on the strengths and weaknesses of your strategy.

4-3. Figure 4-5 is an in-depth look at one possible CIM design process. Based on Chaps. 1 to 3, make a list of all the problems you believe might be encountered in the design of a CIM system. Then use this list as a means to critique the design in Fig. 4-5. In what ways does this design process meet the list of needs, and in what ways is it deficient? Do you have suggestions as to how the proposed design process might be improved? Are there classes of problems for which the design process would be more effective than others?
(a) Come to class with a list of objectives for any design process, a second list of strengths and weaknesses of the model in Fig. 4-5, and a third list describing possible changes that might be made in the design process and/or limitations that may apply to its use.
(b) After class discussion, prepare a brief paper that summarizes your findings.

4-4. The primary objective of all the design strategies introduced in this chapter is to support problem solving through learning. Describe how organizational learning cycles might be built around the strategies shown in Fig. 4-2 to 4-5. Make several lists describing: (1) the problems you think might be encountered in developing the desired learning cycle; (2) strategies that might be taken by the planning group to deal with these problems more effectively; and (3) modifications to the planning strategies themselves that might further accommodate a learning activity.
(a) Come to class prepared to share your results and participate in discussion of the issues raised.
(b) After this discussion, prepare a brief paper on your insights.

4-5. A number of design inputs are suggested for the design strategy in Fig. 4-5 (steps 1 to 4). Consider a manufacturing setting with which you are familiar and discuss some of the inputs that might be appropriate to this setting. Discuss the ways in which information might be collected and the problems that would exist in obtaining this information. How might you address these problems in order to obtain the necessary input information for the design process?
(a) Be prepared to share your results in class and participate in discussion of the issues raised.
(b) After this discussion, prepare a brief paper on your insights.

4-6. For a specific area of manufacturing, discuss the product types and categories that might be formed. What difficulties might you experience in developing these category definitions? Develop a first-cut list of the product categories. Describe how the planning group might further improve the category definitions. Select a market and product area as the focus of your design study and subdivide the market and product line into manufacturing categories. These categories can then form the basis for developing a preferred manufacturing strategy.
(a) Share your results in class and participate in discussion of the issues raised.
(b) After discussion, prepare a brief paper on your insights.

4-7. Consider the information you have available for a specific manufacturing setting. List the functional manufacturing operations associated with each project category (step 7 in Fig. 4-5). What difficulties do you experience in trying to define the appropriate operations? In what way could a planning group assist with obtaining the required information? Once the manufacturing functions by category are obtained, produce a composite functional description for the entire product line. You can prepare this functional flowchart by hand or use the Design/IDEF software if it is available.

(a) Come to class prepared to share your results and participate in discussion of the issues raised.

(b) After this discussion, write a brief paper on your insights.

4-8. This text lists five parameters that describe equipment performance in the factory. Use these five parameters to describe (to your best understanding) the available equipment that could be used in a factory setting of your choice. Use the parameters to describe the various types of equipment that are available or might be available for the setting you are describing. What problems do you foresee in obtaining the data required for these parameters, and what solution approaches might be followed to address these problems? Based on available information, can you develop a set of parameter values for the equipment available? If data are not available, how might the planning group for your particular topic address this problem? In order to decide on the most cost-effective design, it is necessary to develop a range of potential scenarios and configurations for your application. Based on your best understanding of the selected manufacturing setting, discuss the scenarios and configurations that you think would be of most use. Develop descriptions of possible scenarios and configurations and explain the rationale for your selections.

(a) Be prepared to define the alternative scenarios and configurations, to discuss the reasons for the selections, and to discuss any issues that might be raised about the cost-effectiveness of these choices.

(b) Prepare a brief paper summarizing your findings.

4-9. Discuss some of the issues that might be raised in a selected setting with respect to evaluating information flow and data processing equipment requirements (associated with steps 11 and 12 of Fig. 4-5), based on an initial rough cut. In what ways do you believe that the information flow and data processing requirements might affect the selection of a preferred system design?

(a) Come to class prepared to share your results and participate in discussion of the issues raised.

(b) After the discussion, prepare a brief paper on your insights.

4-10. Provide a rough-cut evaluation of some of the organizational and management issues that you might expect to arise for the setting you are considering. Discuss how the planning group might address these issues. Which issues might be amenable to problem solving by the planning group and which ones seem relatively intractable? Do you believe that any of your design strategies might be completely eliminated from consideration based on organizational/management issues? If so, identify those that you believe are at risk and explain the reasons for your concern.

(a) Come to class prepared to share your results and participate in discussion of the issues raised.

(b) After discussion, prepare a brief paper on your insights.

4-11. Once the range of possible scenarios and configurations for a manufacturing setting has been described and modeled, it is necessary to perform a financial evaluation of the alternatives. Consider some of the important issues that might enter into such a cost-benefit analysis. Can you identify problem areas that should be considered in advance by the planning group? What risks associated with the cost-benefit analysis might prevent you from obtaining the best solution? What types of criteria might you want to study in the cost-benefit analysis, in addition to the "lowest cost per unit price"? What approaches would be of most use in performing a financial analysis that would provide the best insight into the relative costs and benefits of the options available?

(a) Be prepared to share your results in class and participate in discussion of the issues raised.

(b) After discussion, write a brief paper on your insights.

4-12. Once an initial or trial *to-be* system has been described as a result of the study design, how should the planning group use this result? In what ways should the result of this activity be integrated into overall enterprise efforts? Can you suggest specific activities that could be mounted in order to make the most effective use of the study activity? In what ways should the *to-be* system configuration *not* be used? What dangers are associated with the definition of this system?

(a) Come to class prepared to share your results and participate in discussion of the issues raised.

(b) After this discussion, prepare a brief paper on your insights.

5

SOFTWARE
FOR MODELING SUPPORT

Commercial software modeling products can help stimulate the organizational learning process. New insights can challenge the existing order, and modeling results can support definition of the robust transition stages that are required for successful evolution of the enterprise for advanced manufacturing.

Within the problem-solving approaches developed in previous chapters, it is helpful to be able to perform various modeling efforts. Modeling can provide insight into limited aspects of the system that is being developed. The commercial software products discussed here are a useful resource for the system design team. At the same time, it is essential to recognize that no model can be used by itself as a basis for the system configuration that will be most competitive in a specific future environment. Rather, these software modeling efforts provide planning methods that can be used in support of the system design strategies described in Chap. 4.

The objective is to make use of limited-scope modeling methods to support the overall planning process. The planning activity must guide decision making, and the results of modeling must always be critiqued in view of the limitations associated with all such efforts. The results of the limited-scope modeling then can be interpreted by the planning group to consider the full scope of the system and environment under study.

A variety of software products can be used in support of the CIM planning process. As illustrated in Fig. 4-1, limited-scope modeling can give essential support to the planning group. Modeling efforts can assist in many different aspects of a typical design activity (for example, to assist in implementation of the procedure described in Fig. 4-5). Figure 5-1a illustrates the discussion emphasis of Chap. 5, and is related to the design model of Fig. 4-5. The software that is available for modeling (step 10) is used to support the study and evaluation of alternative functional system descriptions (associated with layer 2 of the five-layer model). These software models are driven by the manufacturing equipment information base and provide alternative capabilities for functional information modeling.

This chapter introduces several commercially available software products that can assist in the CIM design activity. A particular emphasis is placed on those products that are available in educational versions.

5-1. Overview of Software Products

Figure 5-1b lists the software products discussed in this chapter. Steps 7 and 8 of the design overview procedure in Fig. 4-5 involve models that can be used to provide a functional description of the elements and relationships of a manufacturing system. The Air Force IDEF model, Design/IDEF™ by Meta Software Corporation and IDEF/LEVERAGE™ by D. Appleton Co., Inc. fall into this category.

Step 10 of Fig. 4-5 can draw on software that is used to evaluate alternative system options and describe the operational nature of the manufacturing facility at a functional level. Several different fundamental strategies can be used for this type of modeling.* One approach is to develop a detailed simulation of the existing and/or proposed system. Another is to develop an analytic model. The SES model (introduced in Chap. 3) provides an overview of the nominal or average performance of a simplified manufacturing system and of the manufacturing system-environment interactions. The SES model can be used to extend a system study effort to include the system-environment interactions.

More complex simulations are required to produce a more detailed understanding of manufacturing systems and to study statistical variations in system performance. Such simulations can become quite complex because they require a detailed system specification and multiple runs of the model in order to produce statistically valid results. SLAM II® and SIMSCRIPT II.5® are examples of well-documented, widely used general simulation languages. XCELL+®, SLAMSYSTEM™, and SIMFACTORY II.5® are manufacturing system simulations for exploring alternative system configurations. (SLAM II and XCELL+ are registered trademarks of Pritsker Corp. SLAMSYSTEM

---

* The importance of modeling and simulation for manufacturing system design is described widely (see, for example, Grant 1988; Law and McComas 1988; Talavage and Hannam 1988; Vig, Dooley, and Starr 1989).

Figure 5-1.
Overview of Chap. 5.
(a) Discussion empha-
sis, related to the de-
sign model of Fig. 4-5.
(b) Summary of the
software products dis-
cussed. Addresses for
the companies offering
these products are in-
cluded in the refer-
ences.

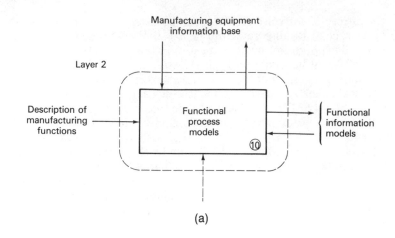

(a)

Model	Type of Application with Respect to Fig. 4-5	Model Name*
Functional description of manufacturing system elements and relationships	Steps 7, 8	Design/IDEF™ IDEF/LEVERAGE™
Functional simulations	Step 10	SLAM II®, XCELL + ®, and SLAMSYSTEM™ SIMSCRIPT II.5® and SIMFACTORY II.5® System-environment simulation (SES) model (see Chap. 3)
Analytic models	Step 10	MANUPLAN II™
Computer network models	Step 11	NETWORK II.5® and COMNET II.5™
Financial and cost-estimating models	Step 14	Extensions of the SES model (see Chap. 3) MiCAPP E-Z QUOTE™ COSTIMATOR® 3

* Trademark ownership is indicated in the text discussion of this chapter.

(b)

is a trademark of Pritsker Corp. SIMSCRIPT II.5 and SIMFACTORY II.5 are registered trademarks of CACI Products Company.)

Computer-based analytical models have been developed that provide a means for preliminary statistical studies of alternative system configurations and have the advantage of simplicity and speed. It is reasonable to view these analytic approaches as providing approximate, initial insights into proposed configurations, where simulations might be used for detailed confirmation of these approximate methods when required. One of the software products on the market today in the analytic first-cut category is MANUPLAN II™ by Network Dynamics, Inc.

The step 10 models produce estimates of the functional performance of alternative configurations of a manufacturing facility. These operational data may then be used to drive the financial impact data in step 14. NETWORK II.5® and COMNET II.5™ by CACI Products Company are computer network models, discussed below. Financial evaluations are important for CIM settings; however, many of the standard software tools are based on accounting strategies inappropriate for the CIM setting. Therefore, in order to perform financial evaluations, it is often necessary to consider both standard financial evaluation software and other tools more specifically targeted to the CIM setting. MiCAPP by MiCAPP, Inc., E-Z QUOTE™ by E-Z Systems, Inc., and COSTIMATOR®3 by Manufacturers Technologies, Inc. are discussed below.

The software products mentioned above are introduced in this chapter. These standard packages are particularly useful for educational purposes because they are well documented and are often available in academic versions at reduced cost.*

5-2. IDEF Models

As discussed in Chap. 1, the integrated computer-aided manufacturing (ICAM) functional definition modeling approach (IDEF) was developed by the U.S. Air Force. $IDEF_0$ can be used to provide a functional description of a manufacturing system in terms of manufacturing activities and relationships among these activities; $IDEF_1$, to produce an information model that represents the information flow to support the manufacturing functions; and $IDEF_2$, to produce a dynamics model to represent the time-varying behavior of the manufacturing system.

The $IDEF_0$ methodology is a graphical approach to describing the manufacturing functions of a system. A standard notation has been developed that includes boxes or blocks to represent activities or func-

---

* Product descriptions are based on information provided by the companies that offer these products and on limited evaluation by the author. Many other software products are also available. The author cannot guarantee the particular performance features of any product described here, and is not recommending any of these products for purchase.

tions and various connecting arrows that have specific representational purposes. A complete manufacturing system can be developed as a hierarchy of block diagrams in which the highest-order diagram shows the entire system and the component diagrams represent portions of the system with a parent/child relationship established between the functional representations. This pattern can be continued indefinitely to produce any desired level of functional detail. At the same time, this approach ensures that the detailed system levels are constantly being linked back to the full system overview.

As noted, the $IDEF_0$ methodology is useful as a means to describe the manufacturing functions that will be associated with a given product line that is specified for a manufacturing system (steps 7 and 8 in Fig. 4-5). The $IDEF_0$ model thus provides a means to document and communicate the nature of the manufacturing functions that are required for each product category and for the categories in composite. Software for the IDEF model is available from the U.S. Air Force under export control restrictions (Air Force Systems Command 1981a, 1981b; D. Appleton Co. 1985). Specific distribution control is required to make use of these IDEF products.

A commercial product that addresses a similar modeling technique is generally available. Named Design/IDEF™, this model provides software that can be used to implement IDEF-based planning and to create the necessary parent/child functional layers for a detailed description of a complex manufacturing system (Meta Software Corp. n.d.).

Design/IDEF is an automated system design tool that can facilitate the development of new manufacturing systems, enhancing the efforts of both individuals and teams. The software can be used to divide a large, complex problem into several subproblems and allow the design of these subproblems to proceed in separate teams while guiding integration of the final configuration. Extended features of Design/IDEF are under development to support additional applications, such as the modeling of behavioral system information. Future extensions of the method may include the functional use of color and the ability to trace pathways through animation.

Design/IDEF software automates many of the tasks involved in creating an IDEF model, including the development of a structure that supports creation of the IDEF drawings themselves (with boxes, arrows, and labeling), automatically links parent/child models, and provides a data dictionary that can be used as a reporting and maintenance feature as well as a cross-referencing resource for more complex Design/IDEF systems. This product also supports the decomposition of labels to child pages and the maintenance of consistency of labels across levels. The objective is to automate many of the labor-intensive, paper-oriented aspects of functional design and thus maximize support to the system designer.

Design/IDEF was developed for use on a Macintosh® and applies the user-friendly icons and menus that are available on this system. Subsequent versions of the software function with UNIX® and MS-DOS® operating systems, widening the application potential of the software.* The user interface and system behavior are essentially the same across a variety of platforms, and diagrams built on one platform can be applied and modified on another. Design/IDEF makes use of a proprietary graphics kernel that was developed specifically for this program.

Once the operating system software and Design/IDEF software are loaded, a straightforward step-by-step process can be used to create the types of IDEF drawings that are required. The software allows for

---

* Macintosh is a registered trademark of Apple Computer Company. UNIX is a registered trademark of AT&T Corp. MS-DOS is a registered trademark of Microsoft Corp.

the quick development of the required boxes (including features to expand and reduce box size as appropriate) along with methods for preparing the required labeling internal to the boxes. The connecting arrows can be drawn using an auto-routing program or can be directed by the planner. The labeling on the arrows can be added directly by using the available graphics support mechanisms.

Design/IDEF begins with the highest system description level (the A0 level) and automatically provides parent-child relationships between this level and all more detailed levels below this system level. The arrows into and out of each particular Design/IDEF layer can either be explicitly carried up and down between parent/child layers or suppressed, depending on the application. Each time text is to be added to a drawing, the software provides a field definition that can be used effectively to enter the necessary information. A number of different pointers are used to support the application of the software.

The diagramming method enforces strict rules for the representation of information and activities and for establishing the hierarchy of diagrams. An IDEF model may contain hundreds of pages from the top-level option description to a very detailed level that describes every process that must go into the product production. Design/IDEF provides one means by which to automate the modeling process.

Design/IDEF software has been applied in Chap. 1 to introduce a generic manufacturing system; see Fig. 1-9b. This same representation has been used as the basis for development of the SES model in Chap. 3; see Figs. 3-8 and 3-9. Chapter 10 draws on the concepts for specific application. The fabrication, assembly, and test and packaging work cells shown in Fig. 1-10 are analyzed to another IDEF level of definition to explore specific equipment units and their integration into FMS work cells. The particular equipment units in these work cell configurations are introduced in Chap. 6 and further discussed in Chap. 10.

EXAMPLE 5-1. DESIGN/IDEF

Figure 1-10 shows a Design/IDEF screen that can result from a planning effort. What type of interaction with Design/IDEF software will the user experience in producing such outputs?

Results

Figure 5-2a shows a typical screen that may be used to select the specific IDEF attributes to be employed. This screen illustrates the size and number of boxes to be drawn, the characteristics of the arrows, the numbering system, and the types of labels. Figure 5-2b shows how the various text characteristics may be defined in detail in order to achieve the desired output characteristics. Figure 5-2c shows the in-process development of the system flowchart; the editor is being used to place the boxes at the desired locations. Figure 5-2d shows how the boxes associated with the flowchart in process might be renumbered by the user. Figure 5-2e shows a master page template, illustrating the page structure; this screen can be used to audit work that has

Figure 5-2.
Typical Design/IDEF screens (courtesy of Meta Software Corporation). (a) Screen
used to select specific IDEF attributes. (b) Screen showing how various text characteristics may be defined in detail. (c) In-process development of a system flowchart.
(d) Illustration of how the boxes associated with the flowchart in process may be renumbered by the user. (e) Master page template illustrating the page structure. (f)
Illustration of the dictionary capability of Design/IDEF, which can be used for consistency checks.

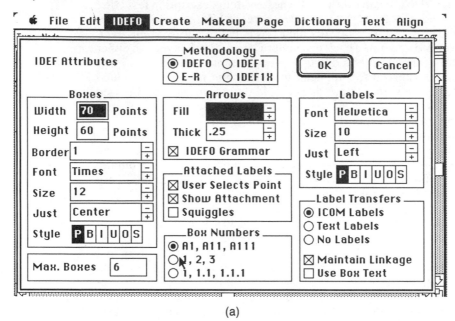

(a)

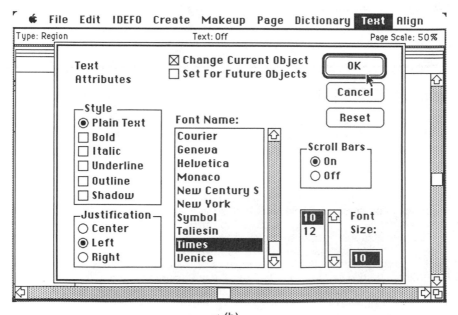

(b)

Figure 5-2. continued

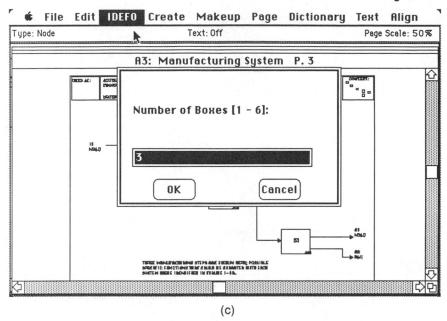

(c)

(d)

Figure 5-2. continued

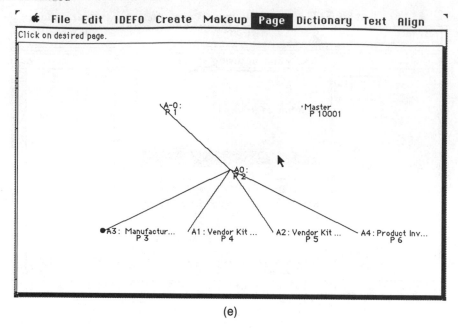

(e)

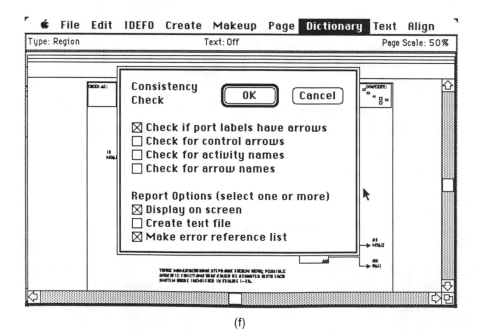

(f)

been performed and ensure that the desired relationships have been established. Figure 5-2f illustrates how the dictionary capability of Design/IDEF can be used for consistency checks to ensure that all aspects of model have been appropriately considered.

Design/IDEF provides many useful tools to develop the desired types of functional drawings and to cross-check to ensure the quality of the final product. The user interface emphasizes the user of windows and cursors to prompt the user and to assist in achieving the desired system description.

Another IDEF-related product, called IDEF/LEVERAGE™, is offered by D. Appleton Company (DACOM). IDEF/LEVERAGE is a family of software tools that supports the $IDEF_0$ and the $IDEF_1$ portions of the IDEF standard methodology and thus can be used to study both the processes and data requirements within an organization. In addition to providing support for creating graphic models of the manufacturing process, IDEF/LEVERAGE translates these graphic models into a machine-readable modeling language that can be transferred to a host computer for model integration and analysis. During host processing, IDEF/LEVERAGE examines the logical associations within each user-generated model and then validates the models. The host-level analysis provides a global picture of the shared data for all aspects of the system, allowing an integrated data base approach. Thus, IDEF/LEVERAGE provides a simple tool for coordinating the needs of many users linked to a common data base and system.

D. Appleton Company (DACOM) offers IDEF-based techniques and tools to support industrial modernization. The IDEF/LEVERAGE product can be used in a variety of ways to support planning groups as they undertake the design of advanced manufacturing systems.

The variety of IDEF-related products and the spread of this methodology indicate the widespread need for a structured methodology for CIM system planning and evaluation. As the hardware and software capabilities increase, it may be expected that additional modeling products will be made available to manufacturing system planners. These products will offer structured approaches to understanding the nature of the opportunities and constraints that face manufacturing operations and allow these operations to engage in effective decision making.

5-3. Simulations

One of the most detailed approaches to modeling a manufacturing system involves the development of a simulation. In a typical computer-based simulation, simulated products are moved step by step through a series of defined processes. Such a simulation is run many times in order to develop a statistical description of the performance of the manufacturing system. Simulations can be used to study both transient and steady-state effects.

Many different simulations are available for use in a manufacturing setting (Wild and Port 1987). Some simulations are not user-friendly, and require the ability to enter detailed data statements and system design statements into the computer program. More recently, simulations with improved user interfaces have become available with visual aids to support data input and system design output and multicolor animation of the "factory" to help explain the significance of program results.

A number of different simulation languages and programs can be used to support manufacturing design studies. Three of the best known are GPSS, SLAM II, and SIMSCRIPT II.5. GPSS (General Purpose Simulation System), a process-oriented simulation for modeling discrete systems, is available in a number of versions and makes use of either written input instructions or graphical symbols for input. SLAM II and SIMSCRIPT II.5 are computer languages that can be used for many types of simulation modeling and are widely employed to model factory systems; these languages are discussed further below.

The system-environment simulation (SES) model, introduced in Chap. 3, provides a simple simulation that is oriented toward explaining key CIM principles. This model is useful for initial insight and understanding. Only a nominal manufacturing case is considered, without statistical variations. This model can be used to study some of the simple interior aspects of manufacturing systems and to study ways in which the system is linked to the environment. As configured, the SES model can be used to obtain preliminary evaluations of manually oriented or highly automated manufacturing configurations and to contrast these two approaches to design.

5-4. SLAM II, XCELL+, and SLAMSYSTEM

This section introduces several simulation products developed by Pritsker Corporation (Pritsker 1986; Standridge and Pritsker 1987; Pritsker Corp. 1988a, 1988b). Three different simulation approaches are available, depending on user requirements.

SLAM II® is a general-purpose simulation language. This Simulation Language for Alternative Modeling is thoroughly documented and has been applied in over 2000 installations worldwide. By using SLAM II, it is possible to develop simulations with a wide range of complexity. SLAM II is most appropriately applied to manufacturing system design where specific detail is needed. Versions of SLAM II are available for mainframe computers, workstations, and personal-class computers. At present, SLAM II is used for educational programs at many universities.

SLAM II is a proven simulation language that can support a wide range of modeling efforts in manufacturing system design. The development of a SLAM II simulation model will typically begin with the definition of a flow diagram that graphically portrays the flow and processing of products through the system. The *nodes* (at which processing is performed) are connected by *branches* (which define the routing that takes place in the system).

A person who wishes to use SLAM II in a simulation can create a system description using graphical symbols to represent the system. The user can then either manually convert this graphical description to a set of data statements or make use of existing computer capabilities that enable direct use of the symbols as the program input.

A SLAM II network model thus consists of a set of interconnected symbols that depict the operation of the system under study. The symbols can be either manually or automatically converted for input into the program that analyzes the model using simulation techniques. The output of the SLAM II simulation produces a statement listing a summary of the initial state of the simulation model as interpreted by SLAM II, a system trace or other intermediate reports if requested, and a SLAM II summary report that displays the statistical results for the simulation.

SLAM II can be a powerful design tool. On the other hand, significant learning time is required in order to understand the way the model operates and to prepare the input and interpret the output. SLAM II functions at a detailed system level and provides a detailed level of information. SLAM II is written in FORTRAN, and various versions can be operated on computers ranging in size from microcomputers to mainframes.

The use of SLAM II can be enhanced by a program called The Extended Simulation System (TESS)℗. TESS provides a data base–oriented simulation environment that can achieve automatic translation from graphical definition of system input to the simulation kernel, and then can produce graphical output, including animations, that will enhance user understanding. TESS is based on an integrated data base and provides a user-friendly environment for SLAM II.

In a number of applications, detailed system performance should be explored only after initial first-cut studies have been completed. The XCELL+® simulation product is an easy-to-learn and use software product for manufacturing system simulation. It is particularly helpful during the original system design when choosing among design alternatives. XCELL+ thus may be regarded as a rough-cut prototyping tool. A predefined set of nodes and symbols is available, and the software can produce only discrete event modeling.

XCELL+ by Pritsker Corporation is an easy-to-learn and use software product for manufacturing system simulation. During initial system design, this product can support the planning group as it considers various design alternatives.

XCELL+ is an effective tool for taking "quick looks" at alternative system configurations. However, this product is limited in terms of the available nodes and symbols when compared with SLAM II, which can incorporate any worldview and a wide range of custom nodes and symbols. The XCELL+ manufacturing simulations can operate on a wide range of personal computers.

SLAMSYSTEM℗ is another simulation that has a user-friendly graphical interface. SLAMSYSTEM is built around an embedded version of SLAM II, but provides a much broader graphics interface and thus is an easier system to apply. SLAMSYSTEM meets the need for user interfaces with a "windows" capability. Typically, this type of presentation makes it easier to learn how to use the simulation. SLAMSYSTEM allows the user to conduct a complete SLAM II modeling project, while taking advantage of a simpler interface. SLAMSYSTEM can be operated on standard microcomputers.

In many applications, it may be reasonable to start a simulation effort with XCELL+ in order to obtain a "quick look" and then use either SLAM II and TESS or SLAMSYSTEM (depending on personal preference and experience) to explore further one or more system concepts that seem to have potential.

EXAMPLE 5-2. SLAM-II SIMULATIONS

Consider the simulation of a simple manufacturing system that consists of products arriving at an inspection station. The rate at which the inspection station can perform its function is limited so that if products arrive too rapidly a queue develops. The simple system being modeled then consists of arriving products, a potential queue of products at the inspection station, and the flow of inspected products out of the inspection station. How can SLAM II be used to represent this manufacturing process and extended to more complex systems?

Results

(A) Simple Process: The three building blocks that are necessary to describe this system using SLAM II symbols are shown in Fig. 5-3. Figure 5-

Figure 5-3.
SLAM II language application as discussed in Example 5-2 (courtesy of Pritsker Corporation). (a) The QUEUE node. (b) The CREATE node. (c) The TERMINATE node. (d) Combination formed by the QUEUE, CREATE, and TERMINATE nodes.

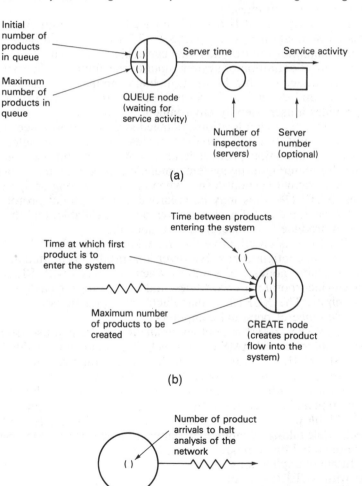

(a)

(b)

(c)

Figure 5-3. continued

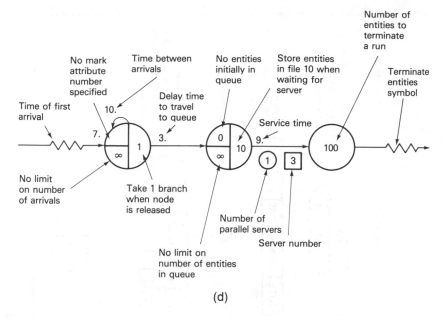

(d)

3a shows the basic QUEUE node that describes the products waiting for service. The node itself is described by a circle with labels that give the initial number of products that are in the queue and the maximum number of products that can build up in the queue. The process is described by the branch or arrow, with data on the time required to perform the activity and the number of servers available. The branch or arrow represents the service activity.

Figure 5-3b shows a CREATE node that can be used to create the product flow into the system. This node may specify the time at which the first product is to enter the system, the time between the products entering the system, and the maximum number of products to be created. Figure 5-3c shows a TERMINATE node that will halt the simulation once a given number of products has arrived and completed the inspection process. It is also possible to simulate for a specified length of time. (Several additional parameters that enter into the actual SLAM II modeling have been omitted for simplicity.)

Figure 5-3d illustrates how a simple manufacturing process can be modeled using SLAM II language. A wide variety of symbols exists in order to support descriptions of complex manufacturing systems. Figure 5-3e lists the network symbols for SLAM II. Note that the ACTIVITY, CREATE, QUEUE, and TERMINATE symbols appear on the list.

(B) Extended Setting: Figure 5-4a illustrates a typical setting in which SLAM II might be applied. As shown, a flexible manufacturing system (FMS) performs machining operations on castings. Castings are initially loaded onto *pallets* and sent to one of two lathes on a conveyor belt, to a wash/load area and then by an automatic guided vehicle (AGV) to the machining center. The machining center consists of ten horizontal milling machines. Various combinations of operations and *fixturing* are available for each machine.

A typical simulation effort would address system balance and productivity, by comparing different combinations of resources that could be applied to achieve system objectives. Multiple simulation runs could be employed to decide on an optimum FMS configuration. Figure 5-4b shows how an animated version of the system could be developed using SLAM II and TESS.

Figure 5-3. continued

SLAM II Network Symbols

Name	Symbol	Statement
ACCUMULATE		Accumulates a set of entities into a single entity
ACTIVITY		Specifies delay (operation) times and entity routing
ALTER		Changes the capacity of a resource
ASSIGN		Assigns values to attributes or global system variables
AWAIT		Holds entities until a resource is available or a gate is open

Name	Symbol	Statement
FREE		Makes resources available for reallocation
GATE		Logical switch definition and initial status
GOON		Continuation node
MATCH		Holds entities in QUEUE nodes until a match on an attribute is made
OPEN		Opens a gate

(e)

178

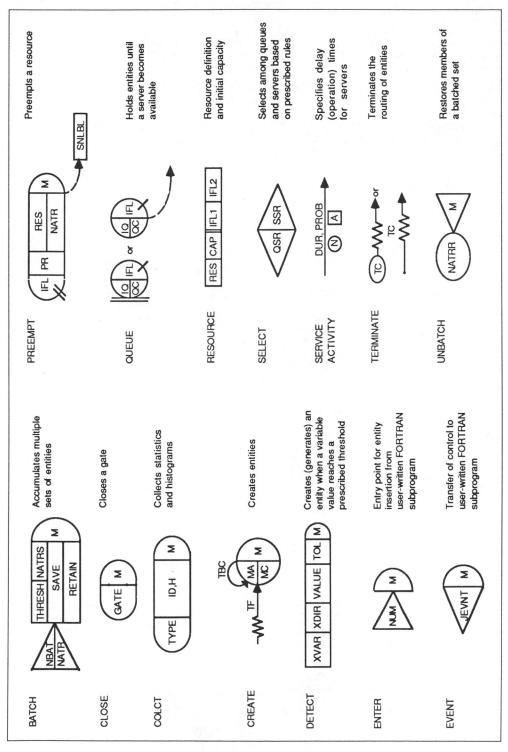

BATCH — Accumulates multiple sets of entities

CLOSE — Closes a gate

COLCT — Collects statistics and histograms

CREATE — Creates entities

DETECT — Creates (generates) an entity when a variable value reaches a prescribed threshold

ENTER — Entry point for entity insertion from user-written FORTRAN subprogram

EVENT — Transfer of control to user-written FORTRAN subprogram

PREEMPT — Preempts a resource

QUEUE — Holds entities until a server becomes available

RESOURCE — Resource definition and initial capacity

SELECT — Selects among queues and servers based on prescribed rules

SERVICE ACTIVITY — Specifies delay (operation) times for servers

TERMINATE — Terminates the routing of entities

UNBATCH — Restores members of a batched set

(f)

179

Figure 5-4.
SLAM II application
(courtesy of Pritsker
Corporation). (a) Phys-
ical description of a
flexible manufacturing
system (FMS). (b)
Functional description
of the FMS, used for
simulation. (c) Pie
chart graphical output
that might be obtained
from the simulation.
(d) Bar chart showing
resource utilization for
various equipment
units. (e) Throughput
comparison bar chart
for various equipment
combinations.

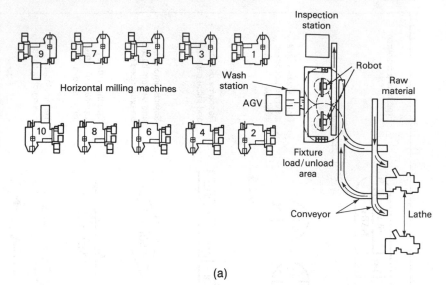

(a)

(b)

Figure 5-4. continued

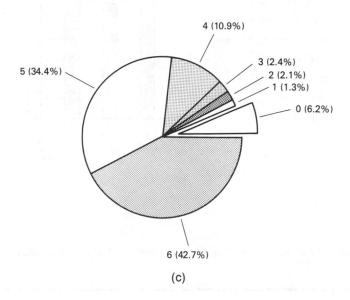

(c)

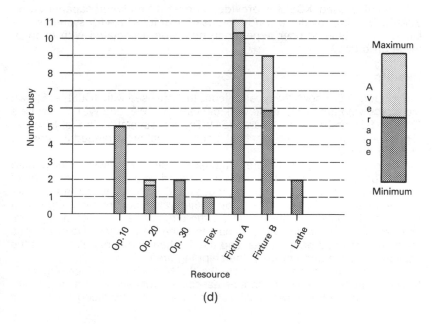

(d)

Figure 5-4. continued

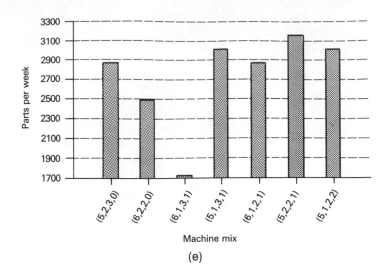

Machine mix

(e)

Figures 5-4c through 5-4e illustrate the types of graphical outputs that might be obtained from the simulation.

EXAMPLE 5-3. XCELL+ AND SLAMSYSTEM

As discussed, XCELL+ provides a first-cut simulation capability, and SLAMSYSTEM provides a windows-based user interface for system simulation. What type of user experience might be associated with each of these products?

Results

Figure 5-5a shows the types of icons that are used with XCELL+. Shown are a receiving area (R1), three work centers (W1, W2, and W3), two buffers (B1 and B2), and a shipping area (S1). The representation in Fig. 5-5a describes a factory with three identical machines that are arranged in series to produce a single product.

Figure 5-5b shows a more complex manufacturing system for printed circuit boards. A number of pallets move around on a conveyor belt. Each of the stations represents a different process and equipment unit in the manufacturing activity. Pallets move through the system continuously. Operation begins at the receiving station, where the circuit boards enter the system. The circuit boards are loaded onto pallets at the load station, five operations are then performed on the parts, and the finished parts are unloaded at the unloading station and moved to the shipping area.

Figures 5-c, d, and e illustrate the SLAMSYSTEM interface. Figure 5-5c shows the hardware that might be associated with such a simulation. Figures 5-5d and 5-5e show actual screens that were produced from a test

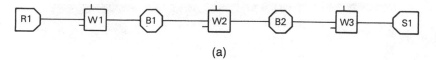

(a)

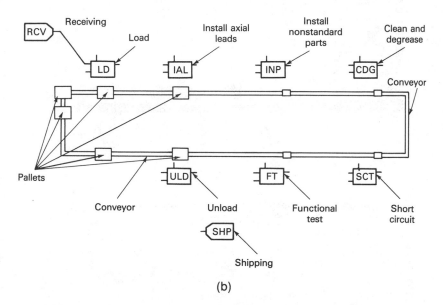

(b)

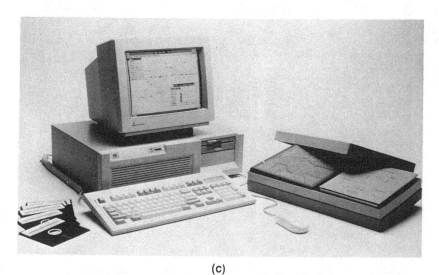

(c)

Figure 5-5.
Illustration of screens associated with XCELL+ (courtesy of Pritsker Corporation). (a) Types of icons used with XCELL+. (b) A more complex manufacturing system for printed circuit boards. (c) Illustration of SLAMSYSTEM operating on a microcomputer frame. (d, e) SLAMSYSTEM screens produced from a test project.

Figure 5-5. continued

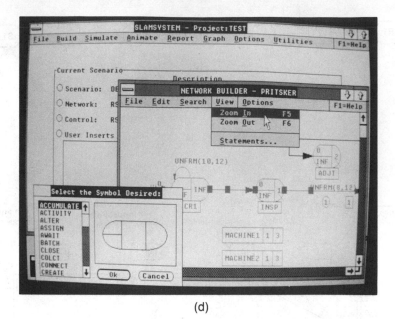

(d)

(e)

project. Figure 5-5d includes SLAM II symbols. Figure 5-5e illustrates some of the output statistics associated with the use of this product.

As illustrated, SLAM II, XCELL+, and SLAMSYSTEM are alternative simulation tools that can be used by the system designer. It is important to assess carefully the nature of any design problem, consider the simulation objectives, and then choose the most appropriate simulation product for use in the given application.

5-5. SIMSCRIPT II.5, SIMFACTORY II.5, NETWORK II.5, and COMNET II.5

This section introduces simulation software produced by CACI Products Company (Law and Larmey 1984; CACI Products Co. 1988; Cheung, Dimitriadis, and Karplus 1988; Mills 1988). SIMSCRIPT II.5® is a general simulation language; SIMFACTORY II.5®, NETWORK II.5®, and COMNET II.5™ are off-the-shelf models for manufacturing systems, computer networks, and telecommunications networks.

SIMSCRIPT II.5 is a well-defined and documented simulation language. In order to develop a SIMSCRIPT simulation, it is necessary to structure the model, provide information, and choose the desired output. As noted by Law and Larmey (1984), each SIMSCRIPT simulation program begins with a preamble section that is used to define the building blocks for the model. As they observe, "the preamble is also used to define the global variables, the basic unit of time for the simulation clock, and the desired measures of system performance (p. 1–8). The execution of a SIMSCRIPT simulation begins with the READ.DATA routine to obtain the required input. Variables are then initialized and the simulation begins. When all scheduled processes and events are complete, the program terminates.

The SIMSCRIPT II.5 simulation language is designed to support model development. Many of the model support functions are built-in and do not have to be addressed by the user. The use of SIMSCRIPT II.5 requires the ability to learn the statements that are included in the programming language and to understand how to combine these statements to produce the desired model.

As a general-purpose simulation language, SIMSCRIPT II.5 automatically provides tools that simplify the model-building process. However, as noted by Cheung, Dimitriadis, and Karplus (1988), "users still need a modest amount of training and experience before they can become efficient users of these simulation languages. Frequently, system designers who need performance data do not necessarily have this type of experience or the time to acquire it." For this reason, it is often desirable to use an already developed model. The SIMFACTORY II.5 product has been developed for manufacturing planning.

The SIMSCRIPT II.5 simulation language has been used to create the SIMFACTORY II.5 model so that the user does not need to write any computer code. The user enters data by means of a simplified interface, and the SIMFACTORY II.5 program then performs the simulation and reports information on factory performance. SIMFACTORY also allows the user to watch an animation of the factory and thus gain insight into the dynamic behavior of the system. In order to implement the SIMFACTORY II.5 program, the user need only learn how to use the interface, which is explained in detail in the instruction manual.

The program includes many different *icons* for use in developing a simulation. The system can be defined in terms of these standard icon representations, or in terms of custom icons created by an editor. While the simulation is operating, a variety of charts can be displayed to present ongoing system statistics in a graphical form. Pie charts, histograms, and trend charts can be developed by using function keys. It is possible to observe animation while the simulation is taking place. Output reports and graphics describe the utilization of equipment in the system, the status of queues, the level of throughput and inventory produced, and overall system performance. Figure 5-6 shows some of the standard icons available with the SIMFACTORY II.5 product. Example 5-4 discusses a typical SIMFACTORY II.5 application.

CACI Products Company also offers off-the-shelf models called NETWORK II.5 and COMNET II.5. NETWORK II.5 is intended for the analysis of computer systems and local area networks (LANs), and COMNET II.5 is intended for the analysis of telecommunications networks. NETWORK II.5 is typically used to describe how a given computer network will perform under specific information flow requirements. In order to build a NETWORK II.5 simulation, the user describes the hardware used in the computer network and the software programs that are to be operated. The simulation then provides a description of the utilization of each computer, the information flow between computers, and the overall time delays associated with having the network perform defined operations. An application of NETWORK II.5 to a computer network is discussed in Example 7-6 of Chap. 7.

Figure 5-6.
Some of the standard icons available with SIMFACTORY II.5 (courtesy of CACI Products Company).

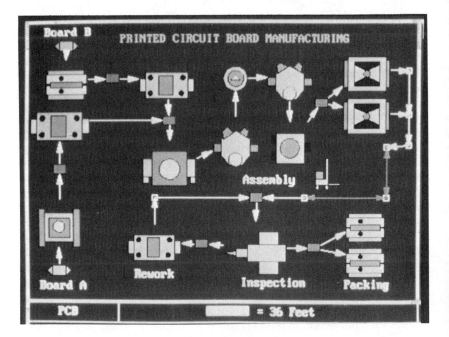

EXAMPLE 5-4. SIMFACTORY II.5 APPLICATION

One of the design objectives of the SIMFACTORY II.5 simulation is to simplify the user interface and achieve user-friendly operation. What type of interaction with this software product might be expected by the user?

Results

(A) Simple Manufacturing System: Consider the simple manufacturing system in Fig. 5-7a. This system takes in raw turbine blades, then performs a grinding operation followed by an inspection to produce finished turbine blades. Queues can develop for both the grind and inspection operations.

In order to apply SIMFACTORY II.5 to this system, the LAYOUT editor is used to develop a functional system flow chart. Figure 5-7b describes this simple factory. Specific icons have been selected to represent the two queues and the two operations. Once the functional flowchart has been defined, the next task is to define the properties associated with each process and with each queue. The STATION editor performs this task, as shown in Fig. 5-7b.

Now the queues must be described. Figure 5-7c shows a typical screen that would result when defining the queue characteristics. Note that the name *Grindq* has been given to the particular queue of interest here. The material being stored in the queue is the *part,* the capacity is 1000 parts, and the raw material input to this queue is the *raw turbine blades.* The RAW MATERIAL editor shows that the mean time between arrivals of the blades is 60 minutes and the mean arrival quantity is 15. The DISTRIBUTION Editor shows that all of the blades will be arriving at one time.

The next task is to describe the processing stations (for grinding and inspection). Figure 5-7d shows the SIMFACTORY II.5 screen that can be used to define each of these processes. The inputs to the two processes are the raw and machined blades. The particular icon that has been selected is called *white part,* and the processing time for the grind operation follows a normal distribution with a mean of 3.0 minutes and a standard deviation of 0.2. The processing time for the inspection station is described for a triangular distribution with a minimum of 2 minutes, a mode of 3 minutes, and a maximum of 5 minutes. The output is a finished blade. The QUEUE and PROCESS PLAN editors may then be combined with the factory layout shown in Fig. 5-7b to completely define the system.

Once all the data have been provided to the simulation, the simulation may be run and display presentation graphics used to describe the performance of the system. A variety of reports and data are produced, as defined by the OUTPUT Editor, to review the results. The stations can be reviewed to see if equipment utilization is within the desired limits and to observe sizes of the queues and the utilization of the equipment units.

(B) Printed Circuit Manufacturing: Figure 5-7e shows a somewhat more complex factory to be used for printed circuit board manufacturing. It is assumed that two types of boards, type A and type B, are processed separately; beginning with the queue at the Touch Up operation, the boards flow through the same stations. Each icon in Fig. 5-7e represents a particular operation, and the circles represent queues. The products arrive at the Touch Up queue, pass through the Touch Up operation, move to the Wash operation,

Figure 5-7.
A simple manufacturing system application (courtesy of CACI Products Company).
(a) Sequence of processes for the manufacturing system. (b) Using the STATION editor to describe the characteristics of each station. (c) Typical screen to be used for defining the queue characteristics. (d) SIMFACTORY screen that can be used to define each process. (e) A more complex factory to be used for printed circuit board manufacture.

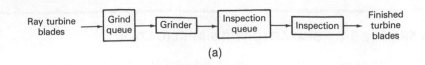

(a)

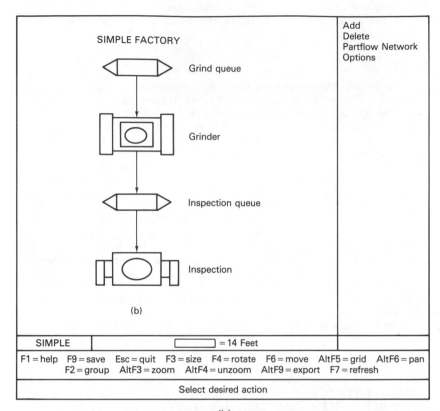

(b)

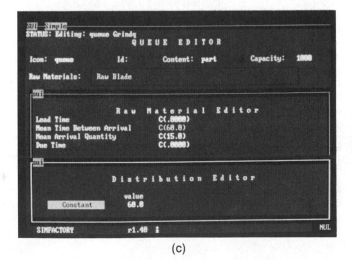

(c)

Figure 5-7. continued

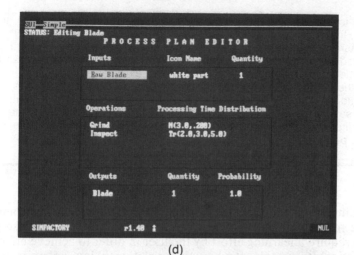

(d)

(e)

and then to a Transfer operation. A second queue develops at Final Assembly. A transporter is utilized to move the boards from final assembly to inspection; after inspection some of the boards will require rework and the nondefective boards are sent directly to the packing area. After being reworked, the defective boards are returned to inspection. Figure 5-7e illustrates how a realistic factory study effort may be performed to consider the utilization of the various portions of the system and to understand how well the system is balanced.

SIMFACTORY II.5 will automatically check the input data for internal consistency and then start the simulation. If problems exist, error messages will be provided on the screen. Once all of the data are confirmed, the animation window will open and the user can watch the animated view of the factory. It is possible to interrupt the simulation and to look at reports describing the state of the factory at an instant, then return back to the run. This aspect of SIMFACTORY II.5 is called a "snapshot" capability. It is also possible to obtain statistics about the stations and queues at any time the ani-

The SIMFACTORY simulation by CACI Products Company offers a simplified user interface and a highly interactive mode of operation. It is possible to develop an animated simulation while periodically interrupting the simulation to observe system performance and to gather statistical information at any time while the animation is running.

mation is running. Once the results of the simulation are observed on the screen, they may be printed to any standard printer.

As noted on this screen the colors of the icons can vary during the animated simulation. A key in the lower left-hand corner of the screen indicates the various status conditions that may exist for each equipment unit. When the simulation is performed, the user can observe the results of the simulation in action and thus find out the status of each equipment unit. Once one or several simulation runs are complete, the user may draw on the OUTPUT Editor to create a wide range of output tables and graphs to summarize system performance.

Simulation products of the type described are an important resource for manufacturing design. At the same time, all of these programs require a learning effort by the user in order to achieve effective application. It is important when contemplating their use of a simulation to ensure that extensive documentation and support are available for installation, training, and problem solving.

5-6. MANUPLAN II

MANUPLAN II by Network Dynamics Incorporated (NDI) provides a simplified model for use in rough-cut or initial design of a manufacturing system. When dealing with large, complex settings, significant insights can be gained by performing a quick scan of a wide range of potential configurations.

Simulations have the advantage of providing detailed insight into the properties of a manufacturing system. On the other hand, they often require extended setup times and long computer run times to explore the cases of interest, with many different products flowing through varied processing steps. Simulation is an excellent tool to use for detailed study of a fairly well defined system.

In some cases, during the *rough-cut* or initial design of a manufacturing system, a preliminary modeling strategy is desirable to allow a quick scan of a wide range of potential system configurations and give reasonable estimates of the strengths and weaknesses of the various alternatives. Once the rough-cut evaluation has been performed, it may then be useful to study the final few choices through a more limited set of detailed simulations.

A modeling philosophy appropriate to this point of view has been developed by Network Dynamics, Inc. (NDI). NDI's software product called MANUPLAN II™ is a *rapid modeling technique* (RMT) for rough-cut evaluations of potential system configurations (NDI n.d.; Suri and Whitney, 1984; Suri and Dille, 1985; Suri, 1988; Suri and Diehl, 1987).

MANUPLAN II relies on mathematical modeling rather than on simulation. Thus, approximate formulas are used to predict the overall statistical performance of a manufacturing system. A set of equations has been developed, and these equations are used to describe the functions of the system under steady-state conditions. This is in contrast to a simulation, where multiple runs must be conducted in order to produce an indication of the statistical system performance.

The MANUPLAN II software can support the functional modeling requirements specified in Fig. 4-5. However, unlike simulations,

MANUPLAN II is an analytic model. The manufacturing system is treated as a number of processing stations that are described by reliability characteristics; alternative routing structures or topologies among these processing stations; and a network of queues that describe the movement of product through the system. MANUPLAN II cannot treat transient or changing conditions in the manufacturing system; rather, this analytic software model can be used to study steady-state conditions of the manufacturing system and to study trade-offs among various combinations of equipment units, characteristics of the manufacturing units, and routing structures among the units.

The input to MANUPLAN II includes equipment types, capacities, and reliabilities; product demands; process routings; and lot sizes. Outputs include whether or not desired production rates are achieved, product lead times and work-in-progress, equipment utilization and downtimes, and queues at each work center. The input and output data are available to the user in a standardized format. The advantages of MANUPLAN are the ease with which modeling can be performed on many different factory configurations and the short processing time needed to produce rough-cut estimates of the performance of each configuration.

Several important problems arise in the application of such software. It must be remembered that this is a steady-state model that provides average system characteristics and does not provide time-varying insights into the system operation. And the queuing formulas used in such models have often been developed with specific restrictions regarding the underlying probability distributions of the service times and of the nature of the queue lengths that can exist.

The formulas have been extended to include approximations for situations that occur in factory settings. It is reasonable, then, to explore the magnitude of the errors that are introduced through the use of formulas with derivation assumptions being violated. Considerable literature has developed around this issue, and it has been shown through theoretical investigations and by comparisons between analytic models and simulations that the queuing network formulas are remarkably robust, so reasonable modeling efforts can often be obtained even when the assumptions implicit in these formulas are not rigorously followed (Suri 1983, Suri 1985, Suri and Diehl, 1986).

These underlying studies of model limitations are important to establish the validity and applicability of MANUPLAN II and similar modeling techniques. Since these modeling techniques use equations that assume steady-state conditions and often assume formal limitations on the types of systems that are being considered, it is important to determine the sensitivity level of the results to violations of the assumptions inherent in these equations.

Fortunately, it has been determined, through theoretical studies and experience, that the MANUPLAN II formulation often provides a reasonable first-cut approach for system design. However, at all times when such models are in use, the restrictions on these models must be remembered, and the models must be used in such a way that they cannot mislead the user.

The basic operation of the MANUPLAN program is illustrated

in Fig. 5-8. Raw materials are provided from dock D and passed through a variety of manufacturing processes. The final product is delivered to stock S. The basic parameters used in MANUPLAN relate to the delay time t_D associated with each manufacturing process and to the percentage of scrap from each manufacturing process. Input from the dock to the first processing station is driven by the processing time t_{D1} and lot size. As illustrated, lot size is taken in by station 1 from the dock, and then in a given time t_{D1} this lot size is processed to become an output from station 1. Another lot is then input from the dock to station 1. Therefore, the rate at which material is removed from the dock to processing station 1 is the same as the rate at which the given lot size is processed through station 1.

For continuous flow manufacturing, a lot size of 1 can be chosen as appropriate. If this selection is made, it is important to note that the delay time t_{D1} is the time associated with the spacing between products being output from processing station 1, not the time it takes for an individual product to proceed from the input to the output. Emerging from station 1, a defined percentage of the product is routed to scrap,

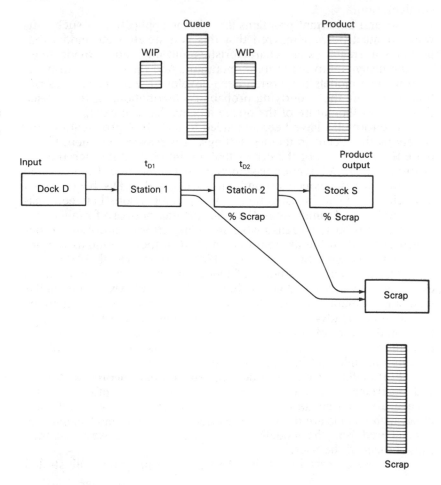

Figure 5-8.
Example of MANU-PLAN II product flow. Raw materials originate on dock D; processes are performed at station 1 and station 2, producing stock (S) and scrap (courtesy of Network Dynamics, Inc.).

and the remainder proceeds to station 2. If the delay time associated with station 2 is greater than the delay time associated with station 1, then a queue will develop at the input to station 2. Product through station 2 passes to stock, with a percentage again being routed to scrap.

MANUPLAN II is designed to run over defined periods of time, and the queue sizes are thus determined. The size of the queues that develop, whether or not the product output achieves a given level, and the scrap formed are determined by the production period, usually taken to be one year. Figure 5-8 illustrates the basic operation of a simple MANUPLAN analysis. Once the lot size is defined, the delay times are defined, the percentage of scrap is defined for each manufacturing station, and a given period of manufacturing is defined, then the product output, the scrap, the queues, and the work-in-progress are defined on a steady-state basis.

Within this basic structure, several important considerations can be factored into MANUPLAN II. First, the equipment stations are, in general, limited to 95 percent utilization, and this will constrain the output. Second, there are a number of factors that can be used to modify the effective delay times. These modifiers reflect the real-world nature of each station. The mean time before failure, mean time to repair, and mean time of repair for each station can essentially increase the effective processing delay.* As the equipment becomes more prone to failure and to repair requirements, the processing time (and effective variability) will increase. The statistical distribution of the repair times can be introduced through a variability factor that defines a standard deviation associated with the processing time.

MANUPLAN II is structured to tell the user whether or not a given desired product output can be obtained within a year. The production goal is used as an input, and the simulation will tell the user whether or not this objective can be obtained. In system studies of the type discussed in Chap. 4, it is necessary to find the maximum production capability of the system. For this case, it is necessary to make several runs of MANUPLAN II and to interpolate, adjusting the desired productivity/year, to find those values just obtainable and those that are just not obtainable. The maximum productivity can then be established between these limits. The percentage of operating time/year is treated as input variable, and the number of hours/day that the system is operating can be adjusted to reflect multiple shifts.

The basic output of MANUPLAN II is thus information as to whether or not the desired productivity level can be met. Through multiple runs, the maximum productivity of the system can be obtained. The analysis will estimate queue sizes at the end of the operation, work-in-progress, the amount of scrap produced, and the utilization rates for all equipment.

Again, as noted above, the lot size is associated with the degree of batch processing. Lot size will decrease as movement is made toward continuous flow manufacturing. MANUPLAN II does not accommodate a pull system analysis, but basically reflects a push system

---

* These parameters directly relate to the MTOI and MTI equipment descriptors introduced in Chap. 4.

with the first manufacturing station pulling input from the dock and creating a product flow through the system. The queues that result are associated with varying kinds of mismatches in the production line capabilities and structure.

EXAMPLE 5-5.   MANUPLAN II APPLICATION 1

Consider a simple manufacturing configuration, as shown in Fig. 5-9a, with a single product being manufactured. A vendor kit (or input material) is taken from the dock, processed through a single equipment unit, and delivered to stock. Assume that 30 minutes per part is required for average processing time. What insights into this configuration can be obtained from a MANUPLAN II analysis?

Results

The input data for MANUPLAN II analysis are shown in Fig. 5-9b. The minutes worked per day by the system is assumed to be (24 hours per day) × (60 minutes per hour) = (1440 minutes per day). The equipment is working 365 days per year, so this particular manufacturing equipment unit will be in continuous operation on a year-round basis.

The utilization limit of 95 percent, which is the maximum that can be entered in MANUPLAN II, places an upper limit on the fraction of time that the equipment can be in use during the year.* It is assumed in MANUPLAN II that the arrival of parts at the dock has a variability around the mean arrival time and that equipment performance has a variability around the mean processing time. As shown here, variabilities of 30 percent are assumed for both arrivals and equipment. The variability in arrival times indicates a distribution (around the scheduled time) for the arrival of material for the system. The variability of equipment indicates a distribution (around the average time) for equipment operation. Thus, it may be expected that a system with variability in arrivals will not perform as well as a system where each part arrives exactly as scheduled (that is, with each interarrival time equal to the mean arrival time).

The equipment unit is designated as FMS1 and there is one equipment unit in FMS1. The mean time to failure of 180,000 minutes corresponds to 125 days. To simplify the situation for this example, the mean time to repair is set at zero, which means that even when the equipment fails there is no repair time required, and therefore no downtime will be associated with failure. The overtime factor can be used if the equipment unit can be run at a rate greater than its scheduled productivity level; this factor allows the equipment to run faster and achieve extra hours of operation within a scheduled time. The value of 1.0 used here means that the equipment will work a standard day. The setup and run speed factors allow the user to change setup and run times without having to change individual routing data for all numbers of items. The value of 1.0 means that the standard times specified in the equipment unit operations section are used. The variability factor adjusts the time basis for a particular unit of equipment.

The part name assigned is HYB1, and for this particular simulation it

* For utilization greater than 95 percent, queue lengths are very sensitive to utilization rates, and MANUPLAN II cannot maintain the necessary level of accuracy.

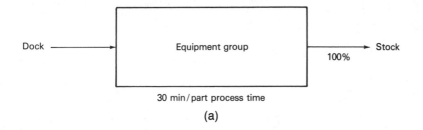

Dock → Equipment group → Stock 100%

30 min/part process time

(a)

Figure 5-9.
MANUPLAN II application for Example 5-5 (courtesy of Network Dynamics, Inc.). (a) System configuration. (b) Input parameter values. (c) Output parameter values.

```
MANUPLAN INPUT TITLE LINE
  EXAMPLE 5-5
*
*
* VERSION
II/1.0
*
* Opern  Flow time  Demand              MINs worked        DAYs worked
* Unit   Unit       Period              per DAY            per YEAR
  MIN    DAY        YEAR                1440               365
*
* Utilization   Variability %    Variability %
* Limit         in Arrivals      in Equipment
  95.0          30.0             30.0
*
* equip    no.in reliability-(MINs) overtime   speed-factors variability
* name     group     mttf      mttr    factor    setup   run      factor
*
  FMS1       1     180000      0        1         1      1         1
  DONE
*
*
* part     demand     lot      demand    speed-factors variability detail
*name/num per YEAR     size     factor    setup     run    factor    flag
*
  HYB1       1000       1        1         1        1      1         1
  DONE
*
*
* OPERATIONS FOR PARTS
*
*
*
* OPERATION ASSIGNMENT FOR ITEM (part)
HYB1
*
* Opern    Equip    Proportn time/lot   time/pc
* name     Name     Assigned (setup)    (run)
*
  PRASM    FMS1        1        0         30
  DONE
* Routing for Item (Part) HYB1
* From     To       Proportion
* Opern    Opern
*
  DOCK     PRASM       1
  PRASM    STOK        1
  DONE
* * * * * *
```

(b)

Figure 5-9. continued

```
MANUPLAN OUTPUT REPORT
EXAMPLE 5-5
version code
II/1.0

Opern    Flow Time Demand   MINs  worked        DAYs  worked
 Unit     Unit     Period    per DAY            per YEAR
 MIN       DAY      YEAR        1440                365

   Utilization      Variability %     Variability %
    Limit           in Arrivals       in Equipment
     95                 30                 30
    EQUIPMENT UTILIZATION SUMMARY
  % of capacity required      TOTAL    WORK-IN-PROCESS (in lots)
   for       for      for     Utili-        at        in
  SETUP      RUN    REPAIR    zation     EQUIP    QUEUE    TOTAL
    0        5.7      0        5.7       0.06       0       0.1


Desired production can be achieved
 Good      Scrap    WORK IN FLOW TIME
 Prodn     Prodn    PROCESS dock-stok
(pieces) (pieces) (pieces) in DAYs
  1000        0       0.1      0.02
   TOTAL PIECES:       0.1




HYB1
 PRODUCTION WAS ACHIEVED
WORK IN    TIME     TOTAL FLOW
PROCESS    spent    TIME spent
(pieces)   /visit   /good piece
   0.1      30.2       30.2


For each piece of GOOD prodn:
no. of pieces that are routed
through this branch
       1
       1




  DETAILS FOR            EQUIPMENT GROUPS

EQUIPMENT GROUP:  FMS1

  Part    Opern      Utilization in      WIP
Name/Num  Name       SETUP      RUN   (pieces)

HYB1      PRASM        0       5.71      0.1

END OF OUTPUT
```

(c)

is assumed that 1000 parts per year are to be produced. The lot size is one, which means the equipment will manufacture one product at a time. The demand factor is a real number that modifies the period of demand, and the value used here is 1.0. The speed factors allow changing of the time base associated with each part and are set at the standard value of 1.0.

The operation being performed here is called PRASM, and the equipment name that performs this process is FMS1. The manufacturing system as designed takes a part from the dock, processes the part through PRASM in

30 minutes, delivers the part to stock, and repeats the process. This equipment is assigned full time to this operation with a mean production time per part of 30 minutes. The routing is from dock to the PRASM operation, and from PRASM to stock. All product (100 percent) goes through this flow.

Figure 5-9c shows the output. As expected, the equipment is working 1440 minutes per day, 365 days per year, with a 95 percent utilization limit and the defined variability in arrival times. The equipment utilization rate is 0 percent for setup, 5.7 percent for run, and 0 percent for repair, for a total utilization of 5.7 percent. This is the expected value since the number of parts that can be produced per day is given by (1440 minutes per day) ÷ (30 minutes/part) = (48 parts); the number of days it will take to produce the desired 1000 parts is (1000 parts) ÷ (48 parts per day) = (20.8 days). Then (20.8 days) ÷ (365 days) = 5.7 percent of the available manufacturing time in a year.

Because of the low utilization rate, there is no significant effect caused by the variability in arrival times and equipment processing times for this particular illustration. The average work-in-progress at the equipment is found by taking the (total number of minutes of production time) ÷ (total minutes in a year); the result is (1000 × 30) ÷ (1440 × 365) = .06 percent, as shown. Since the manufacture of these individual units is spaced out widely over a year, the variabilities in arrival times and processing do not affect the dock-to-stock time. There is essentially no queuing, and the total work-in-progress (in lots) when rounded off equals 0.1.

It is determined that the desired production can be achieved when 1000 pieces are produced with zero scrap. On the average, work-in-progress is 0.1 piece, and the average flow time from dock to stock (in days) is given by 30 minutes per part divided by 1440 minutes per day, or 0.02 day. The desired production was achieved; the work-in-progress on the average is 0.1 piece; the time spent per visit and the total flow time for each good piece are approximately 30.2 minutes. (The slight difference from the nominal value of 30 minutes reflects a small impact due to the variability in arrival and equipment times.) For each piece of good product, all product is routed through this particular branch. The utilization of equipment group FMS1 is 5.7 percent, and on the average the equipment is working on 0.1 piece.

EXAMPLE 5-6. MANUPLAN II APPLICATION 2

Now consider Fig. 5-10a. The same manufacturing configuration is being used; however, the objective now is to achieve the maximum possible production from the equipment. Input data are shown in Fig. 5-10b. The major change is that the demand per year has been increased from 1000 to 16,637. This number has been obtained by estimating operation of (1440 minutes per day) ÷ (30 minutes per part) × (.95 maximum utilization rate) = 16,644 parts is the absolute maximum number that can be produced using mean values. Otherwise, the input data used here are the same as those in Example 5-3.

Results

The output of the MANUPLAN II modeling is shown in Fig. 5-10c. Note that equipment utilization is now 95 percent; which is the maximum possible, and the work-in-process has measured to 0.95 of a part at any time, which

Figure 5-10.
MANUPLAN II application for Example 5-6 (courtesy of Network Dynamics, Inc.). (a) System configuration. (b) Input parameter values. (c) Output parameter values.

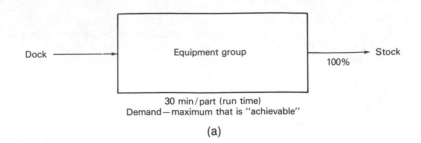

(a)

```
MANUPLAN INPUT TITLE LINE
EXAMPLE 5-6

* VERSION
II/1.0
*
* Opern   Flow time  Demand                MINs worked        DAYs worked
* Unit    Unit       Period                per DAY            per YEAR
   MIN    DAY        YEAR                    1440               365
*
* Utilization       Variability %      Variability %
*   Limit           in Arrivals        in Equipment
    95.0               30.0               30.0
*
* equip       no.in reliability-(MINs) overtime  speed-factors variability detail
*  name       group    mttf     mttr   factor    setup    run    factor      flag
*
   FMS1         1    180000        0      1         1       1        1         1
   DONE
*
*
* part       demand      lot      demand   speed-factors variability detail
*name/num  per YEAR     size      factor   setup    run    factor     flag
*
   HYB1       16637        1         1       1        1       1        1
   DONE
*
*
* OPERATIONS FOR PARTS
*
*
*
* OPERATION ASSIGNMENT FOR ITEM (part)
HYB1
*
* Opern    Equip    Proportn  time/lot   time/pc
* name     Name     Assigned  (setup)    (run)
*
   PRASM    FMS1        1         0         30
   DONE
* Routing for Item (Part)  HYB1
* From     To       Proportion
* Opern    Opern
*
   DOCK     PRASM       1
   PRASM    STOK        1
   DONE
******
 DONE

OUTPUT AREA FOR EQUIP.GRP DETAILED RESULTS (BELOW)
```

(b)

Figure 5-10. continued

```
MANUPLAN OUTPUT
EXAMPLE 5-6

version code
II/1.0

Opern.  Flow Time  Demand   MINs  worked        DAYs  worked
  Unit     Unit    Period    per DAY              per YEAR
  MIN      DAY      YEAR      1440                   365

  Utilization     Variability %      Variability %
    Limit         in Arrivals       in Equipment
     95               30                 30
    EQUIPMENT UTILIZATION SUMMARY
  % of capacity required      TOTAL    WORK-IN-PROCESS (in lots)
    for       for      for    Utili-      at        in
  SETUP       RUN    REPAIR    zation    EQUIP     QUEUE      TOTAL
    0          95       0        95      0.95       1.61        2.6


Desired production can be achieved
  Good       Scrap    WORK IN  FLOW TIME
  Prodn      Prodn    PROCESS  dock-stok
(pieces)   (pieces) (pieces)  in DAYs
  16637        0        2.6       0.06
  TOTAL PIECES:        2.6




HYB1
 PRODUCTION WAS ACHIEVED
WORK IN    TIME      TOTAL FLOW
PROCESS    spent     TIME spent
(pieces)   /visit    /good piece
   2.6      80.9       80.9


For each piece of GOOD prodn:
no. of pieces that are routed
through this branch
      1
      1



  DETAILS FOR              EQUIPMENT GROUPS

EQUIPMENT GROUP:   FMS1

  Part     Opern    Utilization in      WIP
Name/Num   Name     SETUP      RUN    (pieces)

HYB1       PRASM      0       94.96      2.6
```

(c)

is the same maximum. The number of parts in the queue has gone from 0 to 1.61 because the variabilities in arrival times and equipment processing are causing a queue to develop at the dock. The total work-in-progress in both dock and equipment at any time is thus 2.6 parts. The average flow time from dock to stock in days has increased from 0.02 to 0.06 because of the queuing process that has developed owing to the variabilities in arrival times and equipment performance. The production shown was achieved by having 2.6

parts either in a queue or in the manufacturing process at any time. The time spent per visit to this manufacturing system is up from 30.2 to 80.9 minutes because of the queuing problem. All product is routed through this system branch, and the only equipment group is FMS1.

EXAMPLE 5-7.   MANUPLAN II APPLICATION 3

The simple manufacturing configuration is now modified as shown in Fig. 5-11a so that 90 percent of the output goes to stock and 10 percent to scrap. The desired level of production is set at a fairly low level of 2000 parts per year. What insight can MANUPLAN II provide in this situation?

Results

The results of a MANUPLAN II analysis are shown in Figs. 5-11b and 11c. The input is shown in Fig. 5-11b. The demand per year is set at 2000 parts and the routing is 100 percent from dock to PRASM. Routing from PRASM to stock is 90 percent and from PRASM to scrap is 0.1, or 10 percent. The results of the modeling are shown in Figure 5-11c. The equipment utilization is a little more than twice that required to produce the 1000 parts, which is what would be expected. The total work-in-progress at the equipment is again a little more than twice that experienced with a product goal of 1000 parts. (The reason why it is slightly more than twice as large is because of the interaction between the variabilities in arrival times and production manufacturing times.) In order to produce 2000 good pieces, total production of 2222 pieces is required. Then 10 percent of this total goes to scrap and 90 percent of this total goes to good output. In any instant of time, 0.1 piece of work is in process, which is almost the same as the first configuration for producing 1000 parts; this is to be expected because the interaction between the parts is small and the dock-to-stock flow time is still 0.02 day.

The desired production level can be achieved. The time spent per visit is 30.4 minutes, which is up slightly from the 30.2 minutes of Example 5-5.

Figure 5-11.
MANUPLAN II application for Example 5-7 (courtesy of Network Dynamics, Inc.). (a) System configuration. (b) Input parameter values. (c) Output parameter values.

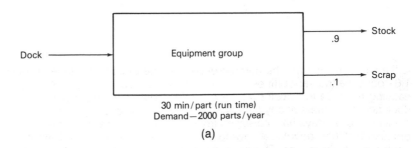

Dock → Equipment group → .9 → Stock
→ .1 → Scrap

30 min/part (run time)
Demand—2000 parts/year

(a)

Figure 5-11. continued

```
*
* MANUPLAN INPUT TITLE LINE
  EXAMPLE 5-7
*
*
* VERSION
II/1.0
*
* Opern  Flow time  Demand          MINs worked     DAYs worked
* Unit     Unit     Period          per DAY         per YEAR
   MIN      DAY      YEAR             1440             365
*
* Utilization    Variability %     Variability %
*  Limit         in Arrivals       in Equipment
   95.0             30.0              30.0
*
* equip    no.in reliability-(MINs) overtime   speed-factors variability
*  name    group      mttf     mttr  factor     setup    run   factor
*
   FMS1      1    180000       0       1          1      1        1
   DONE
*
*
* part      demand      lot     demand    speed-factors variability detail
*name/num  per YEAR    size     factor    setup    run   factor    flag
*
   HYB1      2000        1        1         1      1       1         1
   DONE
*
*
* OPERATIONS FOR PARTS
*
*
*
* OPERATION ASSIGNMENT FOR ITEM (part)
HYB1
*
* Opern     Equip    Proportn time/lot  time/pc
*  name     Name     Assigned (setup)   (run)
*
   PRASM     FMS1        1        0        30
   DONE
* Routing for Item (Part)  HYB1
* From     To         Proportion
* Opern    Opern
*
   DOCK     PRASM        1
   PRASM    STOK        0.9
   PRASM    SCRP        0.1
   DONE
******
```

(b)

Figure 5-11. continued

```
MANUPLAN OUTPUT REPORT
EXAMPLE 5-7

version code
II/1.0

Opern    Flow Time Demand  MINs  worked      DAYs  worked
 Unit      Unit    Period  per DAY            per YEAR
 MIN       DAY      YEAR     1440               365

   Utilization       Variability %     Variability %
    Limit            in Arrivals       in Equipment
     95                  30                 30
    EQUIPMENT UTILIZATION SUMMARY
   % of capacity required    TOTAL    WORK-IN-PROCESS (in lots)
     for      for      for  Utili-      at        in
    SETUP     RUN    REPAIR  zation    EQUIP    QUEUE    TOTAL
      0       12.7      0     12.7     0.13       0       0.1


Desired production can be achieved
  Good      Scrap   WORK IN FLOW TIME
  Prodn     Prodn   PROCESS dock-stok
 (pieces) (pieces) (pieces) in DAYs
    2000    222.2     0.1     0.02
    TOTAL PIECES:     0.1



HYB1
 PRODUCTION WAS ACHIEVED
WORK IN   TIME     TOTAL FLOW
PROCESS   spent    TIME spent
(pieces)  /visit   /good piece
   0.1     30.4      30.4


For each piece of GOOD prodn:
no. of pieces that are routed
through this branch
    1.11
       1
    0.11


  DETAILS FOR              EQUIPMENT GROUPS

EQUIPMENT GROUP:   FMS1

   Part   Opern     Utilization in     WIP
Name/Num  Name      SETUP      RUN  (pieces)

HYB1      PRASM       0      12.68    0.1
```

(c)

EXAMPLE 5-8.   MANUPLAN II APPLICATION 4

Figure 5-12a shows a configuration in which 50 percent of the output goes to stock and 50 percent goes to scrap. The demand is 2000 parts per

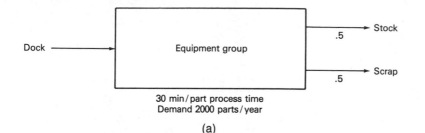

Figure 5-12.
MANUPLAN II application for Example 5-8 (courtesy of Network Dynamics, Inc.). (a) System configuration. (b) Input parameter values. (c) Output parameter values.

(a)

```
*
* MANUPLAN INPUT TITLE LINE
  EXAMPLE 5-8
*
*
* VERSION
II/1.0
*
* Opern   Flow time  Demand              MINs worked        DAYs worked
* Unit    Unit       Period              per DAY            per YEAR
   MIN     DAY        YEAR                 1440               365
*
* Utilization     Variability %     Variability %
*  Limit          in Arrivals       in Equipment
   95.0              30.0              30.0
*
* equip    no.in reliability-(MINs) overtime     speed-factors variability
*  name    group    mttf     mttr   factor       setup    run    factor
*
   FMS1      1    180000        0      1            1      1        1
   DONE
*
*
* part     demand      lot    demand     speed-factors variability detail
*name/num per YEAR    size    factor     setup    run    factor    flag
*
   HYB1      2000       1        1         1        1       1         1
   DONE
*
*
* OPERATIONS FOR PARTS
*
*
*
* OPERATION ASSIGNMENT FOR ITEM (part)
HYB1
*
* Opern     Equip    Proportn time/lot  time/pc
*  name     Name     Assigned (setup)   (run)
*
   PRASM     FMS1        1        0        30
   DONE
* Routing for Item (Part) HYB1
* From     To       Proportion
* Opern    Opern
*
   DOCK      PRASM        1
   PRASM     STOK       0.5
   PRASM     SCRP       0.5
   DONE
* * * * * *
```

(b)

Figure 5-12. continued

```
MANUPLAN OUTPUT REPORT
EXAMPLE 5-8
version code
II/1.0

Opern   Flow Time Demand  MINs  worked        DAYs  worked
 Unit    Unit    Period    per DAY              per YEAR
 MIN      DAY     YEAR      1440                  365

    Utilization       Variability %      Variability %
     Limit             in Arrivals       in Equipment
      95                   30                 30
    EQUIPMENT UTILIZATION SUMMARY
    % of capacity required     TOTAL     WORK-IN-PROCESS (in lots)
     for       for      for    Utili-      at        in
    SETUP      RUN     REPAIR   zation    EQUIP     QUEUE     TOTAL
      0        22.8      0       22.8      0.23      0.01       0.2


Desired production can be achieved
  Good      Scrap    WORK IN FLOW TIME
  Prodn     Prodn    PROCESS dock-stok
(pieces) (pieces) (pieces)  in DAYs
   2000      2000      0.2      0.02
  TOTAL PIECES:        0.2




HYB1
 PRODUCTION WAS ACHIEVED
WORK IN    TIME     TOTAL FLOW
PROCESS    spent    TIME spent
(pieces)  /visit   /good piece
   0.2      30.8       30.8


For each piece of GOOD prodn:
no. of pieces that are routed
through this branch
       2
       1
       1




   DETAILS FOR            EQUIPMENT GROUPS

EQUIPMENT GROUP:   FMS1

   Part    Opern      Utilization in      WIP
 Name/Num  Name       SETUP      RUN    (pieces)

 HYB1      PRASM        0       22.83      0.2

END OF OUTPUT
```

(c)

year, so now the equipment must manufacture 4000 parts per year. The
input in the model is shown in Fig. 5-12b. In the routing, 50 percent of the
output of PRASM goes to stock and 50 percent goes to scrap.

Results

The results of the modeling are shown in Fig. 5-12c. The percentage
of capacity required per run increases substantially because now a total of

4000 parts must be produced through the equipment unit; the work-in-progress increases to 0.23 part in any instant of time and a small effect of queuing is noted. The desired production can be achieved, as expected, with 0.2 piece in process at any given time and the mean flow time from dock to stock is, as expected, 0.02 day. The time spent per visit per each piece is now up to 30.8 minutes, again owing to interaction among the variability factors.

These four examples illustrate simple systems in order to clarify use of the MANUPLAN II model. The advantages of this approach to system analysis can be better appreciated in realistic settings that involve dozens or hundreds of part numbers, rework loops, and numerous, complex equipment units. The interaction between rework and scrap from one part of the system with equipment queues, work-in-progress, and flow times of other parts of the system is not obvious in a typical application, and significant insights can be gained from the model. A further application of MANUPLAN II is provided in Chap. 10, where the software is used to determine the performance of specific manufacturing system configurations as part of a comprehensive system design effort.

It is difficult to model adequately the complexities of real manufacturing systems. At the same time, as discussed in Chap. 1, oversimplified models can produce misleading results. User-friendly simulation and analysis software packages that allow the user to cope with realistic complexity are an important aid to the planning group.

5-7. Costing Models

The software products discussed in this chapter are intended to support the design of manufacturing systems. They enable the system designer to understand the manufacturing capabilities that are required for the system and to evaluate alternative system configurations. Analytic and simulation modeling can be used to select alternative equipment units and configurations on the basis of productivity.

In addition, it is essential to be able to predict the costs associated with alternative configurations. Given both types of data, it is possible to calculate cost per unit and other figures of merit for the manufacturing system in order to select the most appropriate alternative.

A general viewpoint of system costing is illustrated in Fig. 5-13. The costing depends on two data bases, the system data base, which describes the nature of the manufacturing system and can be used to calculate productivity, and the financial data base, which describes the costs of all aspects of operations. These two data bases can be combined by using a variety of different methods to achieve production cost estimates.

The above sections have described how system configurations may be modeled and the productivity of the system determined. This section discusses financial data bases and their use to support cost-effectiveness analysis.

In Chap. 3, alternative systems are compared in terms of fixed and variable costs. This is a common way for developing a data base. Chapter 6 addresses cost estimation for work cells and factories based on the individual equipment and integration costs. As illustrated in

The strategies used to design and optimize a manufacturing system will determine system production levels and total system costs. The costs associated with specific products can be calculated by allocating total costs among the products that are created by the system. Unsold products (due to quality problems or market restrictions) and overhead costs (due to system management) will increase the total costs associated with specific products.

Figure 5-13.
Dependence of system
costing on two data
bases, associated with
the system description
and financial informa-
tion.

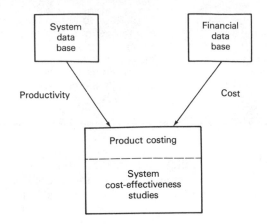

Chap. 6, a variety of methods can be applied to link the financial and performance information to achieve the calculations that are required. Chapter 10 describes the performance of a manufacturing facility in terms of the labor and capital requirements to achieve a given level of production, and describes costs in terms of personnel and capital costs. The cost-benefit analysis applied in Chap. 10 enables selection of the lowest cost per unit production strategy.

The materials described in Chaps. 3, 6, and 10 provide an overview of the costing issue. However, it is important to recognize that in actual facility design, implementation, and operation, much more detailed and sophisticated costing efforts must take place. Once a general system configuration has been selected, it is essential to determine the costs associated with specific products being produced by the factory in order to develop a mechanism for preparing product cost analyses and quotes that will ensure adequate profit for the company.

One of the major difficulties in advanced manufacturing today is that traditional cost accounting methods do not adequately describe the types of systems that are being developed. When the accounting paradigms that are in use do not adequately reflect the objectives and environment of the enterprise, decision making will be distorted. Changes in financial analysis are critically needed.

If accounting practices do not match the reality of the manufacturing setting, then it is difficult or even impossible to justify the types of advanced manufacturing system configurations that are required for competitive operations. A company caught in this situation can spend its time turning down opportunities to create more competitive manufacturing systems because the accounting system does not allow justification of the steps that must be taken.

This situation is made more difficult by the established approach of using direct labor as the base for calculating product costs. As equipment becomes more fully automated, direct labor costs decrease while technology costs increase. A distorted picture of the manufacturing system can result if all costs are expressed in terms of direct labor and combined overhead costs. It is necessary to change accounting practices in order to adequately reflect the opportunities that are available.

As illustrated in Fig. 5-14, increased levels of automation and integration often lead to a basic restructuring of the cost categories of a manufacturing facility. The increased use of capital reduces inventory costs while increasing engineering and technology costs. This kind of shift requires major rethinking of the standard methods of accounting. The discussion in Chap. 10 addresses this costing issue by combining all personnel and capital costs and by calculating a cost per unit without developing ratios that depend on the direct labor base.

In any system design, the final configuration selected is critically dependent on the methods of costing and cost allocation. Costing efforts applied to individual equipment units often do not provide the kind of perspective that is needed. Rather, it is necessary to deal with the productivity of the total system and to consider the implications of capital investment in this setting. As noted in Chaps. 6 and 10, investment in a particular area of automation may not be cost-effective, whereas investment in automation on a factory-wide basis may be cost-effective, if approached appropriately. The accounting profession is under substantial fire to address costing and cost-benefit methods that are appropriate to the objectives of manufacturing companies. Poor costing strategy can result in the selection of the wrong system and the ultimate demise of the enterprise.

In order to predict the costs of potential systems during the design phase, and to operate the system once implemented, costing software of many different types has become available. These software products allow the user to calculate costs for any product that is manufactured in a facility, using standardized data bases and user-friendly interfaces.

Established costing methods will often weigh against desirable system changes. If the accounting methods in use do not match the objectives of the enterprise and are biased toward certain types of manufacturing systems, the benefit-cost analyses of the planning group will be distorted. The determination of system objectives and the application of financial analysis systems congruent with these objectives are necessary for successful decision making.

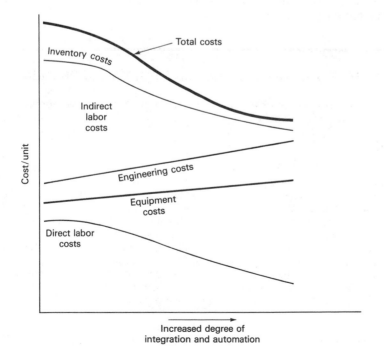

Figure 5-14.
Graph describing how the cost per unit for a product can be related to increased degrees of manufacturing automation and integration. For the product shown, direct labor costs drop, while equipment and engineering costs increase. The indirect labor costs drop, due to highly integrated information flow, and inventory costs are reduced due to application of just-in-time strategies.

The software can be used to support the design activity and the performance of "what-if" studies. The costs associated with all products can be estimated, to allow the company to evaluate whether the new system will represent a more competitive enterprise. Once the system is implemented, the cost estimating software can be used to estimate individual product costs as business opportunities develop.

Many cost estimating software products can be linked to a computer system so that the cost estimating process results in output data that can be downloaded to the factory to drive the manufacturing equipment. The cost estimating function thus links to the detailed design of the manufacturing processes for the factory. The detailed design strategy for a product not only forms the basis for product costing but also forms the basis for managing the production cycle in the factory. Such computer-aided cost estimating (CACE) has developed rapidly over the past several years. The cost estimating software is integrated with the design process.

Design and costing are performed as a single function in order to ensure that the method of manufacturing is the most competitive one available to the factory. As the designer considers alternatives, it is essential that the costs of each design be available so that the lowest-cost design philosophy can be followed. The design for manufacturing (DFM) concept is introduced in Chap. 3 and discussed further in Chap. 9.

Cost estimating is thus an essential aspect of both system and product design. The system must be configured so that the aggregate costs associated with producing a desired product line will allow low-cost, high-quality manufacturing. At the same time, the system must allow the product designer to select the manufacturing process that makes maximum use of this configuration and to isolate the costs associated with each specific product.

EXAMPLE 5-9. COST ESTIMATING SOFTWARE PRODUCTS

Describe some of the software products that may be applied to estimate production costs.*

Results

(A) MiCAPP: Various software products are available to support the cost estimating function. The MiCAPP (*MicroComputer Assisted Process Planning*) detailed cost estimating system, offered by MiCAPP Corporation, is distributed by the Society of Manufacturing Engineers (SME). The MiCAPP product, a series of 18 computer programs tied together with a common data base routine, is expressly designed for the machine shop. The system is menu driven in order to make it as easy to use as possible. The software is machine specific, which means each program has a selection of specific machines typically found on the shop floor. Costing can be performed by making use of the data base included with the system, combined with custom information provided by the user. This cost estimating system has many features that

---

* Several of these products have also been discussed in the literature (Goldberg 1987).

allow the user to adjust to the types of machining operations to be performed and the materials to be used. Times and costs are calculated for both setup and operation. Once the user defines the machining functions and the equipment (applying the standard data base regarding the nature of the product and the material), times can be associated with each step in the manufacturing process. These times are then converted into costs, and the manufacturing cost of the product is determined.

Figure 5-15a shows the cost estimating main menu of the MiCAPP product. This menu allows the user to set up a new part number file, create an estimate using the list of machining functions shown, produce a summary, or use the help resources of the menu. The user selects the particular function and produces reports of the type shown in Figs. 5-15b and 5-15c for a turning estimate and a summary of the estimate.

MiCAPP is a cost estimating system designed for the machine shop, and includes a data base that describes the specific machines typically found on the shop floor. The user can produce a wide range of costing estimates by combining standard and custom time and cost factors.

(B) E-Z QUOTE: Another cost estimating product is available from E-Z Systems. E-Z QUOTE™ is described as one of the most comprehensive cost estimating tools available today. The E-Z QUOTE product helps the user become familiar with the program on a step-by-step basis. First, the use of the specific keyboard and function keys is developed. Then a step-by-step process is used to build customized data bases, which allow definition of setup and production rates for each equipment unit. In addition, the data base includes minimal dollar amounts that are to be charged for running each station and other specialized features. E-Z QUOTE is also designed for machine tool production. An understanding of this particular manufacturing area is designed into the data base and into the methods for user interface and calculating the types of output materials that are required. Once costs are calculated, the markup percentages can be applied, depending on the type of customer.

Figure 5-16a shows the operations data base that might be used with E-Z QUOTE. Note that specific machining operations are described for each

E-Z QUOTE is a comprehensive cost estimating tool that allows the user to link many manufacturing system operations. Operations and materials data bases can be linked to the preparation of estimates, and the customer purchase order can be linked to production routing sheets and shop floor control documents.

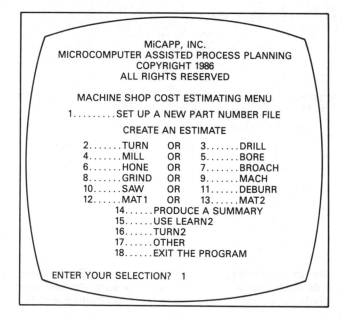

(a)

Figure 5-15.
Sample screens from the MiCAPP cost estimating system (courtesy of MiCAPP, Inc.). (a) Main menu. (b) Turning estimate. (c) Summary of the estimate.

Figure 5-15. continued

```
*********************************************************************
TURNING ESTIMATE

CUSTOMER NAME   B & M MACHINE   DATE   7/21/89

PART NUMBER     123  EST BY N
OPERATION NUMBER   10
FREE MACHINING ALLOY I.E. AISI 4140 Q & T  BRINELL HARDNESS  225


S.S. N/C LATHE
THE MAXIMUM RPM FOR THIS MACHINE IS  1500
---------------------------------------------------------------------
STATION 1
LOAD & UNLOAD
DIAMETER 0 LENGTH OF CUT 0
MACHINING TIME 0 TOTAL TIME .15
RPM 0          FEED 0.0000


---------------------------------------------------------------------
STATION 2
TURN OR FACE MULTIPLE CUTS                      CARBIDE
DIAMETER  2.5  LENGTH OF CUT  5
DIAMETER TOLERANCE = +- .001   AND DEPTH OF CUT   .1150001
MACHINING TIME 2.722696 TOTAL TIME 5.155426
NUMBER OF PASSES  4  AND HORSE POWER REQUIRED  9
LAST PASS TIME  2.25273 FEET RATE  0.0037
RPM 598        FEED 0.0092


---------------------------------------------------------------------
STATION 3
TURN OR FACING OPERATION                        CARBIDE
DIAMETER  2.25  LENGTH OF CUT  3
DIAMETER TOLERANCE = +- .002 AND DEPTH OF CUT   .15
MACHINING TIME .6921674 TOTAL TIME .7521674
RPM 670        FEED 0.0065


---------------------------------------------------------------------
STATION 4
DRILLING OPERATION
DIAMETER .5 LENGTH OF CUT  3
MACHINING TIME 1.07093 TOTAL TIME 1.13093
RPM 420        FEED 0.0070


---------------------------------------------------------------------
STATION 5
FORMING OPERATION                               CARBIDE
DIAMETER 2 LENGTH OF CUT .06
100% WIDTH .75
MACHINING TIME 9.310286E-02 TOTAL TIME .1531029
RPM 429        FEED 0.0015


---------------------------------------------------------------------
NUMBER OF PARTS PER CYCLE 1                              0
TOTAL OPERATION TIME 7.341626      SHOP RATE 0 35  OFF STD. FACTOR .2
TOTAL ADJUSTED TIME PER PC. 7.341626
TOTAL OPERATION TIME WITH OFF STD. 8.809352
TOTAL COST WITHOUT SET UP $ 5.139138
TOTAL SET UP TIME = 35 MIN.
TOTAL SET UP COST = $ 24.5
TOTAL PARTS PER HOUR AT STANDARD 8.172576
TOTAL PARTS PER HOUR AT ACTUAL   6.81048
```

(b)

work center and for each operation to be completed. Setup hourly costs, run hourly costs, and minimum run costs are included in the data base. The user can modify the data base in order to assume conformance to the particular facility.

Figure 5-16b shows the materials data base that can be included, de-

Figure 5-15. continued

```
**************************************************************************
SUMMARY OF THE ESTIMATE

CUSTOMER NAME      B & W MACHINE
PART NAME    SAMPLE  PART NUMBER  123
DATE 7/21/89

        MACHINE          TIME/PC.   PC./ CYCLE   B/U TIME   OFF STD.   W.C. RATE

S.S. N/C LATHE        7.341626         1          35         .2        $ 35

MILL                  1.25             1          60         .25       $ 30

HEAT TR               .85              1          25         .15       $ 30

------------------------------------------------------------------------------
      TOTAL           9.441626                   120

ODD LOT PRODUCTION OPERATION AND SETUP COSTS

LOT NO.       LOT SIZE      TOTAL COST PER PIECE

  1            100          $ 7.172889
  2            200          $ 6.791014
  3            300          $ 6.663722
ESTIMATE WITH MATERIAL COST - PRICES ARE PER PIECE

LOT NO.       LOT SIZE      OPER COST      MAT COST       TOTAL

  1            100          $ 7.172889     $ 3.5          $ 10.67289
  2            200          $ 6.791014     $ 3.45         $ 10.24101
  3            300          $ 6.663722     $ 3.25         $ 9.913721
GRAND TOTAL ESTIMATE WITH MATERIAL AND OSS COST

LOT SIZE      OPER COST      OSS COST      MAT COST      G TOTAL
 100          $ 717.2888     $ 0           $ 350         $ 1067.289
 200          $ 1358.203     $ 0           $ 690         $ 2048.203
 300          $ 1999.117     $ 0           $ 975         $ 2974.117

TOOLING COSTS
MACHINE     S.S. N/C LATHE            TOOLING COSTS $  200
MACHINE     MILL                      TOOLING COSTS $  100
MACHINE     HEAT TR                   TOOLING COSTS $  0
------------------------------------------------------------------------------
OTHER TOOLING COSTS                       $ 0
ENGINEERING COSTS                         $ 0
------------------------------------------------------------------------------
TOTAL TOOLING                             $ 300

GRAND TOTAL UNIT COSTS FOR OPERATIONS, OSS AND MATERIAL

LOT NO.       LOT SIZE      G TOTAL COST / PC.
  1            100          $ 10.67289
  2            200          $ 10.24101
  3            300          $  9.913721
```

(c)

scribing the types of commonly used information and the weight factors to be considered in costing. E-Z QUOTE contains a large data base previously constructed for the user, who can then modify this resource to accommodate particular needs. The final quote produced will typically look like that in Fig. 5-16c. As this quote is produced, the software will also produce a quotation letter to be sent to the customer, as shown in Fig. 5-16d. After the customer's purchase order is received, a routing sheet can be printed for the quantity ordered. This printout, shown in Fig. 5-16e, can be used as a shop travel ticket. The bill of material section will also support the purchasing department. Estimated times are printed in a format that allows for actual times to be noted for later job costing.

Figure 5-16.
Sample screens from E-Z QUOTE (courtesy of E-Z Systems, Inc.). (a) Operations data base. (b) Materials data base. (c) A complete estimate developed directly from the E-Z QUOTE software. (d) A quotation letter developed by the software. (e) A routing sheet developed by the software.

```
                           OPERATIONS DATABASE

OPERATION     Description of                        Set-up    SU      Run      Min $
NUMBER     Work Center or Operation     Formula     Rate $    Hours   Rate $   Run
------     ------------------------**********  -------   ------    ------  ------   ------
  10       BELT MACH DEBURR               HLF.F     35.00     0.00    35.00    0.00
  20       DEBURR BNCH HND RAD            HPC.C     35.00     0.00    35.00    0.00
  30       MIG/TIG WELDER WELD            HPC.C     35.00     0.00    35.00    0.00
  40       SPOT WELDER WELD               HPC.C     35.00     0.00    35.00    0.00
  50       GRIND/CHIP HND GRND            HPC.C     35.00     0.00    35.00    0.00
  60       MTL PREP TANKS                 HSF.F     35.00     0.00    35.00    0.00
  70       PRE-PAINT PREP                 HPC.C     35.00     0.00    35.00    0.00
  80       WET BTH PRIME COAT             HSF.F     35.00     0.00    35.00    0.00
  90       WET BTH PRE&FINIS              HSF.F     35.00     0.00    35.00    0.00
 100       SHEAR SHEAR                    HSF.F     35.00     0.00    35.00    0.00
 110       SHEAR SEC.SHEAR                HPC.C     35.00     0.00    35.00    0.00
 120       CUT-OFF EQ. SAW                HPC.C     35.00     0.00    35.00    0.00
 130       AMADA TURRET PUNCH             HH.HP     35.00     0.00    35.00    0.00
 140       BEHRENS TURRET PUNCH           HH.HP     35.00     0.00    35.00    0.00
---------------------------------------------------------------------------------
F1  = Help          F2  = Clear Screen
Esc = Command       F10 = Exit without update
```

(a)

```
                           MATERIALS DATABASE

              Material Type          Weight
              Description            Factor    Shape      Supplier / Phone No.
-------------------------------------------------------------------------------
A2014    ALUMINUM 2014-T6           0.098000  SHEET      Alco. Inc.
CRS      COLD ROLL STEEL            0.283600  SHEET      Shannon Supply
CRSP     COLD ROLL STEEL PLATE      0.283600  PLATE      Miller Company
A5052    ALUMINUM 5052              0.097000  COIL       Miller Company
CRSC     COLD ROLL STEEL COIL       0.283600  COIL       Carpenter Steel
IROC     52100 CAST IRON            0.290000  PLATE      Shannon Supply
A1.5     1.5" ALUMINUM ROUND 2011   0.102000  BAR,ROUND  Shannon Supply
C2.5     2.5" CRS ROUND             0.283600  BAR,ROUND  Shannon Supply
A2.0F    2.0" FLAT ALUMINUM 3003    0.099000  BAR,FLAT   Miller Company
S330     330 STAINLESS              0.286000  BAR,FLAT   Miller Company
NI200    NICKEL 200                 0.321000  TUBE       Ace Supply
BR36     BRASS                      0.307000  PIPE       Ace Supply
CRST     COLD ROLL STEEL TUBE       0.283600  TUBE       Carpenter Steel
AANG     ALUMINUM ANGLE 5056        0.095000  ANGLE
-------------------------------------------------------------------------------
F1  = Help          F2  = Clear Screen
Esc = Command       F10 = Exit without update
```

(b)

COSTIMATOR 3 is designed for use with machine-tool (metal-working) settings, and provides a data base that contains a wide range of standard elements. An integrated software strategy combines support for cost estimating, process planning, and the development of operations sheets into a single product.

The objective of E-Z QUOTE is to provide a product that enables the cost estimator to access quickly the types of data that are required to complete accurate cost estimates and thereby achieve maximum factory efficiency.

(C) COSTIMATOR 3: Another product is COSTIMATOR® 3, available from Manufacturers Technologies®, Inc. For use with machine-tool (metalworking) settings, the COSTIMATOR 3 system offers a data base that includes a wide range of machine and piece handling standard elements, as well as a flexible low-cost system for storing and creating additional standards. Many types of materials can be handled and numerous report types produced. The COSTIMATOR 3 package includes a sonic digitizer pen that enables an operator to trace cutter tool paths on the print of a part. Cost estimates are made

Figure 5-16. continued

```
        E - Z QUOTE (tm)   THE EXPERT ESTIMATOR (tm)    Release 5.2C
                              E - Z Systems
              ESTIMATE    ITEM    1  OF   1   PART#  : 458762 Rev A
   Estimate Date : Thu Jul 06 20:12:52 1989 Bill of Material Item Ref. # 1 of 1


   1. Customer    : Interior Designs      6. Assembly #  :
   2. Buyer       : Mr. Olson             7. Assy. Desc. :
   3. Estimator   : Dan Eremenchuk        8. Part #      : 458762 Rev A
   4. RFQ #       : 4589-2                9. Part Desc.  : Table Top
   5. Quote #     : 19730-2              10. Comments    : 2nd Quote


        QUANTITIES (Chapter 2)              Quantity Unit Multiplier  = 1.00
                                            Pricing  Multiplier       = EACH
      Pieces of this Part per Assembly = 1
      Quantities Quoted      This Part           Assembly
                                5                    5
                               50                   50
                              100                  100
                              500                  500
                             7500                 7500


        MATERIALS (Chapter 3)
```

```
Material Type ALUMINUM 2014-T6                  Weight/Piece      1.60000
Stock Shape    SHEET         Thickness    0.1250 Weight Factor    0.098000
Blank Length    13.5000      Cut Width    0.0000 Scrap Option     W/SCRAP DO
Blank Width      9.6750      Outside Dia. 0.0000 Do in inches     4.00
Pieces/Blank       1         Inside Dia.  0.0000
             Stock                        Mark   Pcs    Stock Scrap
Quantity     Sizes       Total Weight Cost/Lb Up %  Reqd.  %Each %Last $ PER EACH
--------  --- --- ---   ------------- ------- ------ ------- ----- ----- ----------
       5   36 x  72         31.75     2.1000  10.00       1 74.80 74.80   14.6694
      50   48 x  96        112.90     2.1000  10.00       2 20.63 37.64    5.2158
     100   48 x 120        211.68     2.1000  10.00       3 18.37 36.51    4.8898
     500   48 x 120        987.84     1.7500  10.00      14 18.37 27.44    3.8032
    7500   48 x 120      14747.04     0.9500   5.00     209 18.37 72.79    1.9614
==============================================================================
```

```
        OPERATIONS (Chapter 5)
                              $SU   SU  $Run                            Min/
Description          Op#  Seq# Rate Hours Rate Formula   x    by   x   EACH
Layout               999   1  0.00  0.00 45.00 TIMEH   4.0000              0.00
Shear                100   2 35.00  1.25 35.00 HSF.F   0.0500 0.900        2.70
Amada Turret Punc    130   3 35.00  2.50 65.00 HH.HP3000.0000 54.000       1.08
Belt Mach Deburr      10   4 35.00  0.50 35.00 MP      1.5000              1.50
Spot Welder Weld      40   5 35.00  0.25 35.00 HPC.C   0.0200 4.000        4.80
PAINTING            1110   6 36.00  0.25 38.00 CE      1.2200              1.93
Clean and Pack      1120   7  0.00  0.00 29.00 MP      1.5000              1.50
----------------------------------------------------------------------------
Total: TIMEH:    4.00   SU Hrs:    4.75        4.44 Per/Hr & Min/EACH  13.5063
----------------------------------------------------------------------------
```

```
Labor/Burden factor = 1.000
     AMORTIZED FIXED COSTS (Chapter 6, Page 1)           Fixed
  No.  Description of fixed cost / Vendor                 Cost
       Additional Operation COST  Chgs.                   0.00
   1.  NC Tape Prep                                     125.00
```

(c)

Figure 5-16. continued

```
        OTHER COSTS PER EACH   (Chapter 6, Page 2)
No.   Description of each cost / Vendor                    Cost Per EACH
 1    Silk Screening / Screen Plus                            0.2500

        AMORTIZED LOT COSTS (Chapter 6, Page 3)       Fixed      Cost Per
No.   Description / Vendor             Quantity       Cost       EACH
 1.   Anoodizing                           5         55.00       0.0000
                                          50         55.00       0.0000
                                         100         25.00       0.5000
                                         500          0.00       0.5000
                                        7500          0.00       0.4550

        HARDWARE COSTS (Chapter 6, Page 4)      Each     Each     Cost Per
No.   Part Number / Description of Hardware     Qty      Cost     EACH
 1    Rubber Pads 1/2" Sq.                        4     0.0550     0.2200

      MARKUP PERCENTAGES (Chapter 7)        Quantity    G&A%     Mark-Up%
                                                 5      18.00     20.00
                                                50      18.00     20.00
                                               100      18.00     20.00
                                               500      18.00     20.00
                                              7500      12.00     14.00
```

```
*********************************************************************************
```

ESTIMATE IN $ PER EACH   Table Top   PART # 458762 Rev A

Quantity	Material	Set_Up	Direct Labor	Min. Run & Addon	Fixed Cost Per EACH	Other Cost Per EACH	Per EACH Total Cost
5	14.6694	33.3000	44.3650	30.9600	36.0000	0.4700	159.7644
50	5.2158	3.3300	11.9650	0.0000	3.6000	0.4700	24.5808
100	4.8898	1.6650	10.1650	0.0000	1.5000	0.9700	19.1898
500	3.8032	0.3330	8.7250	0.0000	0.2500	0.9700	14.0812
7500	1.9614	0.0222	8.3890	0.0000	0.0167	0.9250	11.3142

Quantity	G&A	Mark-Up	Credits	% OF COSTS Mat.	Per Labor	EACH Sales Price	$ Sales Totals	Man Hours
5	28.7576	31.9529	0.00	9.18	67.99	220.4749	1102.37	9.88
50	4.4245	4.9162	0.00	21.22	62.22	33.9215	1696.07	20.01
100	3.4542	3.8380	0.00	25.48	61.65	26.4819	2648.19	31.26
500	2.5346	2.8162	0.00	27.01	64.33	19.4320	9716.02	121.30
7500	1.3577	1.5840	0.00	17.34	74.34	14.2559	106919.41	1697.04

```
ADDITIONAL CHARGES (Chapter 8)
   No.   Description of additional charges               $ Amount
   1.    Special Tooling                                   450.00
   2.    Jig Fixture                                       225.00
   3.    Material Costs Checked By _____ on _____        0.00
   4.    OK to send QUOTE Letter per _____ on _____        0.00
```

(c continued)

Figure 5-16. continued

```
----------------------------------------------------------------------
Comments:
                                              Inspected By [        ]
======================================================================
Seq. No.     6  Oper. No.   1110   Descrip.  PAINTING
Special Inst.
             Estim.      Actual       31 Pcs/Hr. PRODUCTION  Estim.       Actual
Set-Up Hrs.  0.25      [      ]    550 Qty Run Time Hrs.   17.658    [        ]
----------------------------------------------------------------------
Comments:
                                              Inspected By [        ]
======================================================================
Seq. No.     7  Oper. No.   1120   Descrip.  Clean and Pack
Special Inst.
             Estim.      Actual       40 Pcs/Hr. PRODUCTION  Estim.       Actual
Set-Up Hrs.  0.00      [      ]    550 Qty Run Time Hrs.   13.750    [        ]
----------------------------------------------------------------------
Comments:
                                              Inspected By [        ]
======================================================================
Totals: Estim. TIMEH      4.00  Actual TIMEH  [        ]
Estim. Set-Up Hrs      4.75   Estim. Run Hrs per     550  pieces =   123.81
Actual Set-Up Hrs [      ]   Actual Run Hrs per [      ] pieces = [      ]

Additional  Operations / Vendor                 Incoming    O.K.
                                                Ins.  Date  (Y/N) Init.
1.  NC Tape Prep                                [      ] [   ] [    ]
2.  Silk Screening / Screen Plus                [      ] [   ] [    ]
======================================================================
                    BILL  OF  MATERIALS
Material Type/Description    ALUMINUM 2014-T6        Stock Shape   SHEET
BLANK Width    9.6750   Length  13.5000   I.D.   0.0000  Piece/Blank   1
STOCK Width    48       Length  120       O.D.   0.0000  Piece Weight  1.60000
Stock Thickness 0.1250 Cut Width 0.0000 Pcs. Req.   15 Total Weight    1086.62


Hardware Part No. / Description           Each Quantity    Total Quantity
1.  Rubber Pads 1/2" Sq.                        4              2200

ADDITIONAL CHARGES (Chapter 8)
        No.  Description of additional charges
        1.   Special Tooling
        2.   Jig Fixture
        3.   Material Costs Checked By _____ on _____
        4.   OK to send QUOTE Letter per _____ on _____
```

(c continued)

Figure 5-16. continued

```
                    E - Z Systems
                    1433 W. Fullerton Ave, Suite M
                    Addison, IL  60101
                    (312) 953-1555

Interior Designs                            7/06/1989
Mr. Olson
9034 N. 6th Street
Kansas City          MO   64104


Dear Mr. Olson

We are pleased to quote the following:

RFQ #          :  4589-2
Part #         :  458762 Rev A
Part Desc.     :  Table Top

QUANTITY            PER EACH       ADDITIONAL CHARGES
--------            ----------     ------------------------------------
       5              220.47       Special Tooling            450.00
      50               33.92       Jig Fixture                225.00
     100               26.48
     500               19.43
    7500               14.26




Terms          : Net 45 Days
Freight Method : Yellow Freight
Delivery       : Within 15 days ARO

This quote is valid for thirty days.  All prices are based upon
the bluprints submitted with your RFQ # 128-AD. If you have any
questions, call me at the office before Wednesday.  Please note
that this is the second time we have quoted these tables for
you.


Thank you for the opportunity to quote your product for you.
Please call me at the above number to place your order.

Very truly yours,
E - Z Systems


Dan Eremenchuk
```

(d)

Figure 5-16. continued

```
          E - Z QUOTE (tm)   THE EXPERT ESTIMATOR (tm)   Release 5.2C
                              E - Z Systems
                      ROUTING  SHEET   Part #  : 458762 Rev A
     Estimate Date : Thu Jul 06 20:16:32 1989 Bill of Material Item Ref. # 1 of 1

      1. Customer    : Interior Designs      6. Assembly #  :
      2. Buyer       : Mr. Olson             7. Assy. Desc. :
      3. Estimator   : Dan Eremenchuk        8. Part #      : 458762 Rev A
      4. RFQ #       : 4589-2                9. Part Desc.  : Table Top
      5. Quote #     : 19730-2              10. Comments    : 2nd Quote


     P.O.  No.          :        3543     JOB #  :     2987

     Scheduled Start Date [          ]   Scheduled Completion Date [          ]
     Actual Start Date    [          ]   Actual Completion Date    [          ]

     Assembly Quantity   :      550   Pieces of this Part per Assembly =  1
     Quantity of this part required  :      550



     Operations
     ================================================================================
     Seq. No.     1 Oper. No.   999   Descrip.  Layout
     Special Inst.
                  Estim.      Actual       138 Pcs/Hr. PRODUCTION   Estim.       Actual
     Set-Up Hrs.  0.00      [       ]       TIMEH :          4.000  [        ]
     --------------------------------------------------------------------------------
     Comments:
                                                          Inspected By [          ]
     ================================================================================
     Seq. No.     2 Oper. No.   100   Descrip.  Shear
     Special Inst.
                  Estim.      Actual        22 Pcs/Hr. PRODUCTION   Estim.       Actual
     Set-Up Hrs.  1.25      [       ]       550 Qty Run Time Hrs.  24.750  [        ]
     --------------------------------------------------------------------------------
     Comments:
                                                          Inspected By [          ]
     ================================================================================
     Seq. No.     3 Oper. No.   130   Descrip.  Amada Turret Punc
     Special Inst.
                  Estim.      Actual        56 Pcs/Hr. PRODUCTION   Estim.       Actual
     Set-Up Hrs.  2.50      [       ]       550 Qty Run Time Hrs.   9.900  [        ]
     --------------------------------------------------------------------------------
     Comments:
                                                          Inspected By [          ]
     ================================================================================
     Seq. No.     4 Oper. No.    10   Descrip.  Belt Mach Deburr
     Special Inst.
                  Estim.      Actual        40 Pcs/Hr. PRODUCTION   Estim.       Actual
     Set-Up Hrs.  0.50      [       ]       550 Qty Run Time Hrs.  13.750  [        ]
     --------------------------------------------------------------------------------
     Comments:
                                                          Inspected By [          ]
     ================================================================================
     Seq. No.     5 Oper. No.    40   Descrip.  Spot Welder Weld
     Special Inst.
                  Estim.      Actual        13 Pcs/Hr. PRODUCTION   Estim.       Actual
     Set-Up Hrs.  0.25      [       ]       550 Qty Run Time Hrs.  44.000  [        ]
```

(e)

Figure 5-17.
Sample process plan developed by COSTIMATOR 3 (courtesy of Manufacturers Technologies, Inc.)

```
                     COSTIMATOR 3
                     Process Plan
-----------------------------------------------------------------
Part Number    : N197345         Date Created  : 6-13-89
Part Name      : TRANSMISSION CASE  Date Modified : 6-21-89
Customer Name  : Tractor Company  Done by       : TRF
Job Number     : 88-185           Quantity      : 180
Material Type  : CAST IRON        Page          : 1
-----------------------------------------------------------------
  10 28 "MAZAK" 60x80x20 VERTICAL CNC

   Milling

    FACEMILL - NEGATIVE INSERT - Coated Carbide

      Tool Diameter                     :        4.000
      Revolutions per Minute    (RPM) :      172.
      Inches per Minute         (IPM) :        2.63
      Surface Feet per Minute   (SFM) :      180.
      Inches per Revolution     (IPR) :        0.0153
      Multiples                         :        4.000

      Distance Traveled                 :       25.000

      Machining Time     (hours/piece) :        0.634
      Machining Time     (minutes/piece) :     38.025

   Drill

    C'SINK - High Speed Steel

      Diameter                          :        0.3750
      Depth                             :        0.1875
      Number of Holes                   :       14.
      Revolutions per Minute    (RPM) :      403.
      Inches per Minute         (IPM) :        1.03
      Surface Feet per Minute   (SFM) :       40.
      Inches per Revolution     (IPR) :        0.0026
      Multiples                         :        1.000

      Machining Time     (hours/piece) :        0.042
      Machining Time     (minutes/piece) :      2.539

    DRILL - High Speed Steel

      Diameter                          :        0.2570
      Depth                             :        1.2778
      Number of Holes                   :       14.
      Revolutions per Minute    (RPM) :      789.
      Inches per Minute         (IPM) :        1.94
      Surface Feet per Minute   (SFM) :       53.
      Inches per Revolution     (IPR) :        0.0025
      Multiples                         :        1.000

      Machining Time     (hours/piece) :        0.154
      Machining Time     (minutes/piece) :      9.240
```

from a series of multiple choice menus. The COSTIMATOR 3 system checks the input data for mistakes in routing, and process plans generated by the software list the operations, tools, and machine settings needed to produce the part at minimum costs.

The optional Op-Sheet™ section of the COSTIMATOR 3 system enables the fast creation of *operations sheets*. A scanner can be used to input a part drawing to the computer, while a "mouse" can be used to change the drawing as required for a specific manufacturing operation. Using this feature, each machine operator can be supplied with complete instructions and a picture to guide operations. Operations drawings created on the software may be printed out as separate reports or electronically incorporated into process plans and printed in an integrated format. Another optional package, CIM Conversion Utilities™, can be used to communicate data from the COSTIMATOR software to other computers in the manufacturing system in order to integrate operations.

Figure 5-17 shows the typical results obtained from a COSTIMATOR 3 process plan. Illustrated are several manufacturing processes that are applied to a particular part (a transmission case) for the customer (a tractor company). The material used is cast iron, and the process plan called for here starts with milling and drilling operations. Each process is described in terms of the specific characteristics of the operation to be performed, based on a built-in program understanding of the parameters that are appropriate to describe each process. The calculated machining time is provided in terms of hours per piece and minutes per piece. Many such operations can be combined to produce the total machining time, materials handling time, and manufacturing time for each piece to be produced in the factory. These data may then be multiplied by the appropriate cost factors to produce a cost estimate for the product.

The three products discussed here have both advantages and disadvantages, depending on the particular application and setting. These particular software products, which are designed for machining operations, are typical of those that can be developed for any manufacturing setting. The design and implementation of any manufacturing setting will require cost estimating software for use in selecting the preferred system design and in utilizing the resulting system to compete most effectively. Cost estimating must be seen as an integral part of the product design and system application procedures. It is essential that products be designed and the system be utilized in such a way as to minimize costs and take advantage of the competitive nature of the facility.

ASSIGNMENTS

5-1. In order to achieve maximum educational insight from Chap. 5, it is helpful to stimulate class discussion of the concepts that have been included, develop critical reviews that probe the advantages and limitations of the concepts, and explore alternative ways for approaching the topic. Select one or several topics from the bold-faced headings that introduce sections or examples in this chapter. Come to class prepared to explain this topic, critically describing the text materials and suggesting ways the topic discussion might be strengthened.

5-2. If you have a copy of Design/IDEF software, use the software to describe a manufacturing system. Explore the different features of the software and the way it can be used as a system design tool. Come to class prepared to show the model you have developed and to discuss your experiences with using this product. What features do you find most useful? What additional features might make the product more useful for your particular application?

5-3. If you have a version of the SLAM II or SIMSCRIPT II.5 software, develop a simple simulation to illustrate an important aspect of a manufacturing system. Operate the simulation and collect performance data. Be prepared to describe in class the simulation you have developed and the meaning of the outputs you have obtained. Based on your experience, what are the major difficulties of making use of these products? In what ways could they be further simplified based on your personal experience? What were the most important insights you obtained from the simulation?

5-4. Repeat Assignment 5-3 using the XCELL+ or SIMFACTORY products.

5-5. Repeat Assignment 5-3 using the MANUPLAN II software.

5-6. Select a textbook on computer simulation and identify an aspect of the book that is of particular interest to you. Prepare an outline to explain the topic and what you believe are its most interesting aspects.
(a) Come to class prepared to present these materials and to participate in a discussion.

(b) After class discussion, prepare a brief paper describing your insights and conclusions based on this review.

5-7. Repeat Assignment 5-6 using an article from a journal that publishes articles on computer simulation.

5-8. Describe the ways in which the types of software modeling discussed in this chapter can most effectively be integrated into the overall manufacturing system design process. In what ways should this connection be made in order to serve the overall design objectives most effectively? What potential dangers exist in using such modeling efforts, and how might these potential dangers be addressed in advance?
(a) Come to class prepared to present your insights and discuss them with the class.
(b) After the class discussion, prepare a brief paper describing your insights in this area. What types of additional software modeling tools might be useful to support the system design process? What would be the characteristics of these models and how would they be applied?

5-9. Identify someone in an industry who has made or makes significant use of software for modeling support in daily job activities. Discuss with this person a description of this software and the industry experiences with this software. If possible, observe the group making use of this software in the operational setting. Discuss the strengths and weaknesses of this product and decide on your own how effectively the product is supporting its intended purposes within the company.
(a) Be prepared to present your materials and to discuss them in class.
(b) Prepare a brief paper describing the results and insights obtained.

MANUFACTURING EQUIPMENT AND AUTOMATION

In order to develop improved manufacturing systems, it is necessary to understand the types of manufacturing equipment and automation that are available for consideration. Only systems that can be implemented in hardware will have practical significance.

This chapter provides a linkage between an understanding of the hardware technologies that can be drawn upon, the functional performance of such equipment, and the types of settings of interest. The technology insights in the chapter provide an input to system modeling efforts; in turn, modeling efforts can indicate the types of equipment and technology performance that might lead to the most effective systems.

The purpose of this chapter is to "map out" the technology limitations and opportunities that must be considered as part of any manufacturing activity. This issue is first addressed in a general treatment, which describes manufacturing equipment and performance limitations that are applicable to many different industry settings. Also considered are the relationships between equipment technology and performance, and the associated costs, to allow cost-benefit trade-offs to be performed.

Manufacturing equipment technologies are then explored in detail for two particular areas, involving machine tools for metalworking and the manufacture of small electronics products. These areas of emphasis are important in their own right and provide a useful focus for equipment discussion. At the same time, these studies can also provide insight into other manufacturing areas of interest.

Central to the development of any manufacturing facility is the selection of appropriate physical equipment to implement the desired product line. The degree to which a factory configuration can be implemented depends on the capabilities and costs of the available equipment.*

Figure 6-1a shows how this chapter relates to the relevant portions of the system design strategy of Fig. 4-5. The following discussion deals with layers 1 and 2 of the multilayer model and with the relationships between these two layers. Of concern here are how the physical aspects of the equipment and technology being utilized (associated with layer 1) help determine the functional performance associated with the equipment units (of layer 2) and how the functional requirements associated with system design (from layer 2) can be used to help determine the appropriate equipment design and technology for use (for layer 1).

Chapter 6 addresses the physical and functional aspects of manufacturing equipment. The physical characteristics of the equipment (describing how it is implemented in hardware) shape the boundaries of possible functional performance; the desired functional performance helps determine the required physical characteristics (the hardware selection that must be made). The hardware and functional performance of the equipment are closely related aspects of manufacturing system design.

Figure 6-1b indicates how Chap. 6 approaches the study of manufacturing equipment and automation. As indicated by the planning procedure introduced in Chap. 4, the manufacturing equipment performance specifications determine the physical constraints associated with layer 1 of the system. (The relationships between manufacturing equipment capabilities and the design of a CIM system are initially discussed in Secs. 4-3 and 4-4.) The physical performance of the manufacturing equipment can be associated with parameter values that link layers 1 and 2. The parameters describe the functional performance of the equipment, and the limiting values determine the extreme performance capabilities that are possible within commercially available or custom state-of-the-art equipment.

When designing a CIM system, the design group must have an appreciation for the functional parameter limits of the equipment of interest, as these parameters will constrain the range of possible design configurations. Obviously, it is not useful to design a CIM factory that cannot be implemented because it exceeds the physical equipment capabilities. In turn, it is desirable to match manufacturing equipment to the application. It is not useful or cost-effective to apply the most sophisticated equipment, with the highest performance parameter limits, to every possible application. Rather, it is important that the design process determine the preferred parameter values. The preferred values must make use of available equipment values and should be those appropriately associated with the design of any particular factory.

This cyclical design process between layer 1 and layer 2 is illustrated in Fig. 4-5, steps 6 and 10, and in Fig. 6-1b. The parameters associated with available equipment drive the functional process modeling of facility design. The parameters determine how the real-world equipment constraints will be incorporated into the functional design process. In turn, as many different functional configurations are considered, the cost-benefit of these alternatives can be evaluated and preferred parameter values determined. So long as these preferred val-

---

* Several resources discuss the issues involved in equipment automation and the current state of the art (National Academy of Sciences 1986; National Institute of Standards and Technology 1988).

ues are within the limits of available equipment, the design group is assured that the manufacturing equipment can fulfill the design requirements. To the degree that the optimum design configurations exceed present equipment capabilities, new equipment design requirements can be developed to provide guidance to original equipment manufacturers.

The purpose of this chapter is to provide an overview of the types of manufacturing equipment that are in common application today and that are anticipated for the near future, to provide a foundation for understanding the parameters and the parameter limits of modern-day technology, and to discuss the preparation of equipment cost estimates that are driven by equipment performance parameters. In order to appreciate the origins of the parameter limits, methods for achieving various degrees of equipment automation are considered. Sensors, actuators/effectors, controllers, and control loops are discussed in order to provide an appreciation of the fundamental limitations associated with manufacturing equipment. These aspects of an equipment unit were initially introduced in Sec. 3-9.

Discussion in Chap. 4 has indicated that many levels of automation can be associated with manufacturing equipment. The objective should be at all times to choose the degree of automation that is appropriate for the facility being designed, as revealed through the cost-benefit studies of the type discussed in Chap. 4.

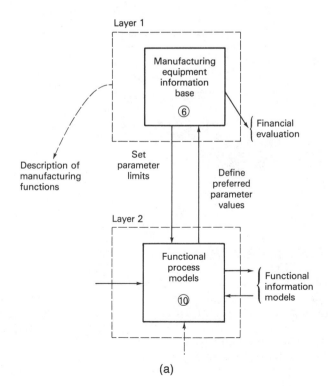

(a)

Figure 6-1.
Overview of Chap. 6.
(a) Discussion emphasis, related to the design model of Fig. 4-5.
(b) Linking steps 6 and 10 of the system design procedure shown in Fig. 4-5. This figure illustrates how the functional parameters may be used as a linkage between the physical layer (layer 1) and the functional layer (layer 2).

Figure 6-1. continued

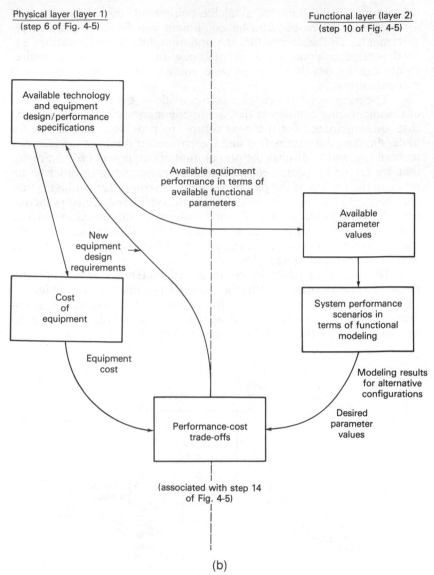

Physical layer (layer 1)
(step 6 of Fig. 4-5)

Functional layer (layer 2)
(step 10 of Fig. 4-5)

Available technology
and equipment
design/performance
specifications

Available equipment
performance in terms of
available functional
parameters

New
equipment
design
requirements

Available
parameter
values

Cost
of
equipment

System performance
scenarios in
terms of functional
modeling

Equipment
cost

Modeling results
for alternative
configurations

Desired
parameter
values

Performance-cost
trade-offs

(associated with step 14
of Fig. 4-5)

(b)

6-1. Equipment Unit Parameters

As introduced in Chap. 4 and summarized in Fig. 6-2, the design of a CIM facility can be achieved by describing each equipment unit to be used through a set of functional parameters. Five parameters have been introduced for use, as follows: the number of product categories for which the equipment can be used (with software download between product types); the mean time between operator interventions (MTOI); the mean time of intervention (MTI); the yield or percentage product output of acceptable quality; and the mean processing time

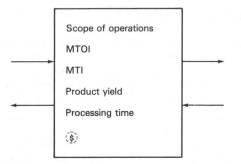

Figure 6-2.
Parameters used to
link layers 1 and 2.

per product unit. As part of the design process, ranges of available values and limiting values for each of these parameters must be associated with each manufacturing equipment unit. An idealized equipment unit would be infinitely flexible so that the equipment could handle any number of categories desired, would require no operator intervention between setup times, would produce 100 percent product of acceptable quality, and would have an unbounded production capability.

The five parameters used here to describe the functional performance of an equipment unit are introduced in Sec. 4-3. These parameters describe the nature of the equipment-operator interaction and the productivity characteristics of equipment-operator combinations.

The degree to which real equipment can evolve toward this reference concept depends on the physical constraints associated with the design and operation of the equipment. The performance of the equipment in each of the five parameter areas will be related to details regarding the way in which the equipment operates. This chapter addresses those factors that will affect equipment performance in these five areas. Relationships are developed between the physical descriptions of equipment operation and the functional parameters that will be associated with this operation. The objective is to link together the physical design of the equipment and its functional performance in the factory setting. This discussion provides insight into areas in which future equipment improvement would be advantageous, and also suggests the magnitude of the cost-benefit payoffs that might be associated with various types of equipment designs.

An understanding of the relationships between the equipment characteristics and the performance parameters can be used to enable an optimum equipment selection based on the parameter requirements associated with a given factory configuration. In this way, the design team can survey alternative types of available equipment and select the units that are most appropriate for each potential configuration.

The first parameter listed above refers to equipment flexibility. The equipment is described in terms of the number of product categories for which the equipment unit can be used for manufacturing, with only a software download to distinguish among product types. A completely fixed piece of equipment that cannot respond to computer control might be able to accommodate only one product category without a manual setup. On the other hand, a very flexible manufacturing system might be able to accommodate a wide range, or even all, of the product categories. This parameter will thus be driven by the breadth of processes that can be performed by an equipment unit and the ability of the equipment unit to respond to external-control data in order to shift among these operations.

The most cost-effective solution will depend on the factory configurations that are of interest. Thus, original equipment manufacturers (OEMs) are always concerned with anticipating future types of factories in order to ensure that their equipment will be an optimum match to the intended configurations. There is a constant trade-off between increases in the flexibility parameter and cost. In general, more flexible and "smarter" manufacturing equipment will also be associated with higher cost. Therefore, the objective in a particular setting will be to achieve just the required amount of flexibility, without any extra capability built into the equipment unit.

The following discussion provides more specific details regarding alternative equipment design strategies to achieve varying degrees of flexibility.

The second parameter is the mean time between operator interventions (MTOI). The related objective is to produce equipment with an MTOI value that is matched to the factory configuration in use. In a highly manual operation, it may be acceptable to have operator intervention required on a constant basis. On the other hand, if manual servicing is to be eliminated (the objective is to achieve operator-independent manufacturing between manual setups), then the equipment must be designed so that the MTOI is longer than the planned duration between manual setups. The equipment manufacturer must try to assess the ways in which factories will be configured and produce equipment that can satisfy manufacturing needs without incurring any extra cost due to unneeded features.

The third parameter is the mean time of intervention. Each time an intervention is required, it is desirable to have the intervention and service time matched to system design. If the intervention time becomes large with respect to the mean time between operator interventions, then the efficiency of the item of equipment drops rapidly in terms of the fraction of the time available to manufacture the desired product.

The fourth parameter is the percentage product of acceptable quality, or yield. In a competitive environment, it is essential that all manufacturing equipment emphasize the production of quality product. If the equipment produces a large quantity of product that either must be discarded or reworked, then the operation of the factory is strongly affected, and costs will increase rapidly. The objective must be to determine the product yields that are required for given configurations and then design equipment that can achieve these levels of yield. Achieving higher yield levels will, in general, require additional sensing and adaptability features for the equipment. These features will enable the equipment to self-adjust and to monitor itself so that if it becomes out of alignment it will not continue to operate.

The fifth parameter is the mean processing time. If more product units can be completed in a given time, equipment cost can be more widely amortized. As the mean processing time is reduced, the equipment can produce more product units in a given time, reducing the manufacturing cost per unit. Again, equipment will generally become more expensive as the processing time is reduced.

Trade-offs will generally exist among improvements in these five parameters and the cost of equipment that is to be used. If very high performance equipment is to be employed, high cost will often be associated with that equipment, and the factory configuration must make effective use of the equipment capabilities. On the other hand, if the factory configuration does not require extreme parameters, then it is far more cost-effective to choose equipment units that are less sophisticated but adequate for the purposes of the facility. This interplay between parameter values and equipment design and cost is an essential aspect of the system design, and is reflected in steps 6, 9, and 10 of Fig. 4-5.

The remainder of this chapter discusses the hardware and software constraints that are experienced in the design of equipment to improve performance in these five parameter areas and consideration of cost estimating in this context. Examples are given of various types of equipment and their physical design and performance. In addition, research trends are noted in order to provide a framework for understanding the types of equipment that will likely be available in the near future.

EXAMPLE 6-1. AVAILABLE AND PREFERRED EQUIPMENT PARAMETERS

Figure 6-3a lists equipment parameters associated with various technologies and equipment versions; Fig. 6-3b lists optimum parameter ranges that are required to achieve a desired manufacturing system. Which technologies are presently available to the system being designed and which will require additional research and development to develop new equipment versions?

Results

As illustrated in Fig. 6-3, none of the existing type A equipment units has an adequate MTOI; available equipment units require more operator intervention than is associated with the desired equipment properties. On the other hand, the MTI is satisfied by all three equipment versions (A1, A2, A3). The yield requirement, of greater than or equal to 90 percent, is satisfied only by unit A1; the processing time requirement, of less than or equal to 8 minutes, is satisfied only by A3.

This type of situation arises when the manufacturing system is pushing the state of the art. The data suggest that the optimum parameter range might be available with additional research and development despite the fact that no present equipment version can achieve this level of operation. The MTI, yield, and processing time requirements can all be satisfied by differing units, and the MTOI requirement can almost be satisfied by the A3 unit. Thus, it might be reasonable to evaluate the R&D costs required to develop a new equipment version that can satisfy the optimum parameter range and produce an enhanced manufacturing capability.

Figure 6-3.
System parameters for Example 6-1 (The same available parameter values are used in Example 4-4, Fig. 4-10). (a) Available parameter values. (b) Optimum parameter values.

Equipment Function	Equipment Version	Available Parameter Values			
		MTOI (min)	*MTI (min)*	*Yield (%)*	*Process Time (min)*
A	A1	0.1	0.1	90	12
	A2	1.0	0.1	85	10
	A3	10	1.0	80	8
B	B1	1.0	0.1	95	10
	B2	10	0.5	90	2
C	C1	0.1	0.1	98	3
	C2	5.0	1.0	98	2
	C3	8.0	2.0	96	1

(a)

Optimum Parameter Range				Notes
MTOI (min)	*MTI (min)*	*Yield (%)*	*Process Time (min)*	
>15	≤2.0	≥90	≤8.0	Requires R&D to develop new equipment version
>5.0	<1.0	≥90	<3	Within state of the art
>15	≤2.0	≥90	≤3.0	Requires R&D to develop new equipment version

(b)

For equipment function B, the optimum parameter range can be satisfied by equipment unit B2. Equipment function B is within the present state of the art and available equipment versions. For equipment unit C, no equipment version has an adequate MTOI. All three have an adequate MTI and an adequate yield. The processing time is also adequate for all three versions. Thus, the problem focuses on increasing the MTOI of one of the versions to 15 minutes. This closeness between the available and optimum parameter ranges again suggests that it would be worthwhile to explore R&D to develop a new equipment version.

EXAMPLE 6-2. EQUIPMENT COST DATA

Figure 6-4 shows the types of data that can be collected to evaluate the costs and benefits associated with the development of new equipment units that will have optimum properties. Interpret the data shown.

Results

A4 is a new equipment version that can be achieved based on research and development. As noted, all available parameter values now fall in the optimum parameter range. The R&D costs for the equipment unit are $280,000; production costs are shown for each of the four units. The system designers can consider the costs associated with all options and decide if the investment and production costs associated with unit A4 are cost-effective. (Analysis of this type is considered in detail in Chap. 10.)

Since equipment version B2 satisfies equipment function B, no additional R&D is required here. C4 is a new equipment version that satisfies all optimum parameter requirements. Again, the R&D and production costs are indicated. Given the type of information shown in Fig. 6-4, the system designers can evaluate whether or not the development of new equipment versions will achieve a sufficient performance advantage to be worth the R&D and production investment.

Figure 6-4.
System parameters for Example 6-2.

Equipment Function	Equipment Version	Available Parameters Values				R&D Costs ($000)	Production Costs ($000)
		MTOI (min)	MTI (min)	Yield (%)	Process Time (min)		
A	A1	0.1	0.1	90	12	—	50
	A2	1.0	0.1	85	8	—	75
	A3	10	1.0	80	10	—	85
	A4	18	1.0	90	8	280	155
B	B1	1.0	0.1	95	10	—	150
	B2	10	0.5	90	2	—	300
C	C1	0.1	0.1	98	3	—	125
	C2	5.0	1.0	98	2	—	250
	C3	8.0	2.0	96	1	—	300
	C4	20	2.0	96	1	540	400

One of the difficulties associated with U.S. manufacturing strategy is that many companies procure manufacturing equipment only from commercial vendors and do not consider producing their own custom equipment. Custom modification can produce a significant manufacturing advantage, but also requires the company to expand both its planning scope and its product development skills. The type of analysis shown here enables a company to determine whether investment in custom manufacturing capability would be worthwhile. Alternatively, for those companies without the necessary resources, the R&D costs may be associated with contracting out the development of the optimum equipment in such a way that the sponsor retains proprietary rights for a period of time.

The types of analysis discussed in Examples 6-1 and 6-2 illustrate how equipment unit parameters can be used to compare alternative equipment versions that are available for given functions. The analysis can support decision making as to whether or not to invest R&D funds for equipment improvement and upgrading.

6-2. Range of Equipment Technologies and Automation Available

Figure 6-5a illustrates how a modified IDEF$_0$ model (based on the material introduced in Secs. 1-5, 3-3, and 5-2) can be used to de-

The five equipment unit parameters introduced here can be used to compare alternative equipment versions, assess which versions can satisfy given functional requirements, and identify needed areas of research and development to improve equipment performance.

Figure 6-5.
Application of a modified form of IDEF$_0$ to describe a typical process workstation. (a) Illustration of a complete workstation, including material input, process performance, and material output. (b) Workstation implementing manually oriented functions. (c) Workstation making use of semiautomated equipment with operator support. (d) Workstation making use of fully automated equipment. (e) Alternative workstation strategies.

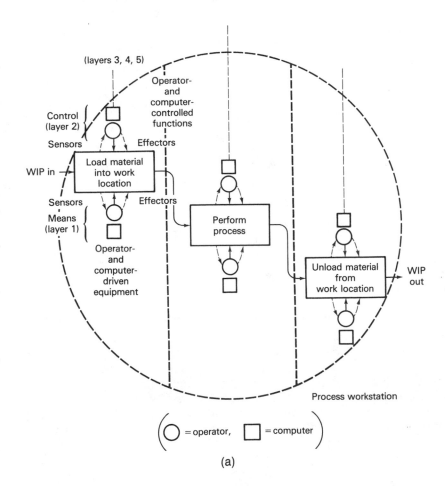

(a)

Figure 6-5. continued

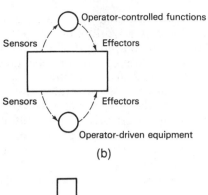

Operator-controlled functions

Sensors

Effectors

Sensors

Effectors

Operator-driven equipment

(b)

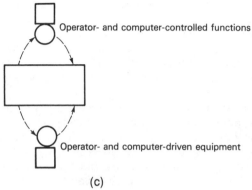

Operator- and computer-controlled functions

Operator- and computer-driven equipment

(c)

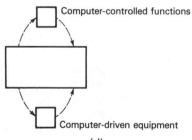

Computer-controlled functions

Computer-driven equipment

(d)

Type of Workstation	Loading	Processing	Unloading
Manual	Manually oriented equipment	Manually oriented equipment	Manually oriented equipment
Semiautomated process	Manually oriented equipment	Semiautomated equipment	Manually oriented equipment
Fully automated process	Manually oriented equipment	Fully automated equipment	Manually oriented equipment
Fully automated workstation	Fully automated equipment	Fully automated equipment	Fully automated equipment

(e)

scribe a typical process workstation. The objective of the process is a material transformation. *Input* work-in-progress is loaded into the workstation, transformed into *output* WIP, and then unloaded from the station. The *control* of each step in the process takes place through both manual and automated means, according to equipment and product specifications, through combined operator and computer actions. The *means* for each step in the process are achieved through the manufacturing equipment itself, in turn driven by operators and computers. As can be seen in the model, the various aspects of manufacturing can be associated with the layers of the five-layer model. Layer 1 is associated with process means, layer 2 with the process control, and layers 3–5 with operator and computer control of workstation performance.

There are many different ways to describe manufacturing equipment. In order to enhance the comparison among equipment types and to develop a structured analysis, it is useful to define common elements of all manufacturing equipment and use these elements as a means of description.

As discussed here, equipment units consist of sensors, actuators and effectors, and controllers and control loops. These subsystems can be viewed as fundamental building-block units for manufacturing equipment. By developing an understanding of these system elements, the designer can appreciate the differences among various equipment hardware options. These subsystem elements provide a general approach to the physical (hardware) system aspects analogous to the way in which the five operating parameters provide a general means for describing functional performance.

One strategy is to describe each equipment unit in terms of the sensors, actuators, effectors, controllers, and control loops associated with the unit. *Sensors* provide a means for gathering information on equipment operations or the process being performed. In many (typical) cases, a sensor is used to transform a physical stimulus into an electrical signal that may be analyzed by the equipment unit and used for decision making regarding the operations being conducted. An *actuator* can be used to convert an electrical signal into a mechanical motion. The actuator acts on the product and equipment through an *effector*. The effector serves as the "hand" that achieves the desired mechanical action. A *controller* is a computer of some type that receives information from sensors and its internal programming, and uses this information to operate the equipment (to the extent available, depending on the degree of automation). The controller produces mechanical motion by making use of an actuator, which converts an electrical signal to a mechanical action. The sensor, actuator, effector, and controller are linked through a *control loop*. In limited-capability control loops, little information is gathered, little decision making can take place, and limited action results. In other settings, "smart" equipment with a wide range of sensor types can apply numerous actuators and effectors to achieve a wide range of automated actions.

The purpose of a sensor can be to inspect work-in-progress, to monitor the work-in-progress interface with the equipment, or to allow self-monitoring of the equipment by its own computers. The purpose of the actuators and effectors is to transform the work-in-progress according to the defined processes of the equipment unit. The function of the controller is to allow for varying degrees of manual, semiautomated, or fully automated control over the processes. In the most fully automated case, the controller is completely adaptive and functions in a closed-loop manner to produce automated-system operations. In other cases, human activity is involved in the control loop.

In order to understand the ways in which the physical (layer 1) equipment properties affect the functional parameters (of layer 2) associated with the equipment, and in order to determine the types of physical equipment properties that are necessary to implement the various desired functional parameters, it is necessary to understand the technologies that are available for manufacturing systems that use automation and integration to varying degrees. Figures 6-5b–d show one way of describing such systems.

The least automated equipment makes use of detailed operator control over all equipment functions. Further, each action performed by the equipment unit is individually directed by the operator. Manual equipment thus makes the maximum use of human capability and adaptability. Visual observations can be enhanced by the use of microscopes and cameras, and the actions that are undertaken can be enhanced by the use of simple effectors. The linkages between the sensory information (from microscopes or through cameras) and the resulting actions are obtained by placing the operator in the loop as shown in Fig. 6-5b.

This type of equipment is clearly limited by the types of sensors used and their relationship to the human operator, the types of effectors that can be used in conjunction with the human operator, and the capabilities of the operator.* An equipment unit that is designed using a manual strategy must be matched to human capabilities. The human-equipment interface is extremely important in many manufacturing applications. Unfortunately, equipment design is often not optimized as a sensor-operator-actuator/effector control loop from this point of view.

For effective application, manufacturing equipment must be described by considering how the equipment-operator combination will function in the work environment. The features of the subsystems will in general depend on combinations of equipment and operator efforts, and functional performance will depend on integrated activity. The equipment and operator together define the manufacturing capability of the system.

A second type of manufacturing equipment may be associated with semiautomated performance in which some portion of the control loop is replaced by a computer, as shown in Fig. 6-5c. This strategy places a number of new requirements on equipment design. Specifically, sensors now must provide input data for both the operator and the computer, and appropriate types of data must be provided in a timely manner to each of these control loops. Semiautomated equipment must have the capability for a limited degree of self-monitoring and control associated with the computer portion of the decision-making loop. An obvious difficulty in designing such equipment is to merge the computer- and operator-controlled activities in an optimum manner. The computer must be able to recognize when it needs operator support, and the operator must be able to recognize which functions may be appropriately left to computer control. A continuing machine-operator interaction is part of normal operations.

The next manufacturing concept (Fig. 6-5d) involves fully automated equipment operations. The processing within the unit itself is fully computer controlled. Closed-looped operations must exist between the sensors and actuators/effectors in the unit. The equipment must be able to monitor its own performance and perform decision making for all required operations. For effective fully automated operation, the mean time between operator interventions must be large when compared with the times between equipment setups. The processes in use must rarely fail, and the operator will intervene only when

---

* Blache (1988) describes the limitations on operator functions in a manufacturing setting.

these failures occur. In such a setting, the operator's function is to ensure the adequate flow of WIP and respond to machine failures.

Several types of workstations are noted in Fig. 6-5e. The most sophisticated involves fully automated processing and materials handling. Computers control the feeding in of work-in-progress, the performing of the manufacturing process, and the removal of the work-in-progress. Equipment of this type provides the opportunity for the most advanced automated and integrated operations. The equipment must be modified to achieve closed-loop operations for all of these functions.

Figure 6-6 illustrates how the various functional performance pa-

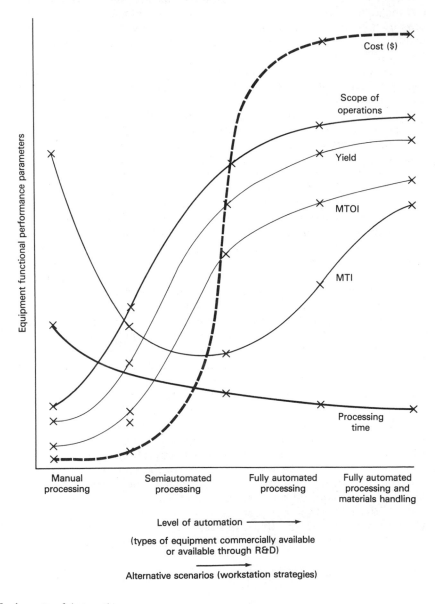

Figure 6-6.
Illustration of how functional performance and equipment unit costs can be related to the degree of equipment and system automation, as reflected in a range of scenarios.

Equipment functional performance parameters

Cost ($)

Scope of operations

Yield

MTOI

MTI

Processing time

| Manual processing | Semiautomated processing | Fully automated processing | Fully automated processing and materials handling |

Level of automation ⟶

(types of equipment commercially available or available through R&D)

⟶

Alternative scenarios (workstation strategies)

rameters and equipment unit costs can be related to the degree of automation associated with each item of equipment in order to develop the scenarios that are necessary for step 10 functional analysis (refer to Fig. 4-5). As may be noted, for the particular family of equipment units being described here, the MTOI and yield of the equipment increase with higher levels of automation. The mean time of intervention (MTI) drops and then rises again as the equipment becomes more complex. The average processing time, which is taken to be the time between work-in-progress product units emerging from the equipment, decreases with increasing levels of automation. Finally, the costs increase rapidly with the degree of automation.

A trade-off exists between equipment performance and cost as a function of the degree of automation. These various degrees of automation can be used to define scenarios, and then each scenario can be explored in terms of its cost-benefit relationships. An example of such a study is described in detail in Chap. 10.

When many different equipment-operator combinations can be used to satisfy a given set of manufacturing requirements, a family of equipment units may be said to exist. In general, a family of equipment units can include manually oriented, semiautomated, and automated equipment capabilities. The equipment designer must select the level of capability that results in the preferred system performance and minimum cost. However, costing must be performed on a system level, not on an individual equipment basis, if an optimum design is to result.

EXAMPLE 6-3. SELECTING AN OPTIMUM LEVEL OF EQUIPMENT AUTOMATION (CASE I)

Figure 6-7 describes three equipment units in terms of the performance parameters considered above (see also Fig. 6-6). Discuss the significance of the data provided.

Results

Figure 6-7 can be used as a data base to evaluate which member of an equipment family (associated with a specific scenario) will be the most appropriate for a given application. By performing a cost-benefit analysis (as discussed in Chap. 10), it is possible to determine whether the more highly automated equipment types, with higher costs, are justified by the improved functional performance parameters.

Such cost-benefit studies must be performed in a system context to ensure that a balanced level of performance is maintained throughout the factory. If the cost-benefit analysis is performed on a single factory function or equipment type without consideration of performance of the total integrated facility, decision making is likely to be flawed.* The analysis must consider the combined effort of all equipment units and the total cost performance of the entire operation.

Figure 6-7 indicates a case that might arise in facility design, with three different equipment units considered individually and then in combination. The types of data in Fig. 6-6 have been drawn on to define the individual benefits and cost of each equipment type.

As shown, a cost-benefit analysis of equipment unit A alone justifies a high level of automated processing and materials handling. On the other hand, stand-alone cost benefit analyses performed of equipment units B and C justify only manual operations. As is often the case, if equipment units are

---

* The need for system-level costing studies is also discussed in Sect. 5-7.

Figure 6-7.
Description of three equipment units in terms of performance parameters. For the cases shown here, it is assumed that values for scope of operations, yield, and MTOI increase with level of automation; values for MTI and processing time decrease with level of automation. (a) Equipment unit A evaluated separately justifies a high level of automated processing and materials handling. (b) Equipment unit B evaluated separately, with a low level of automation justified. (c) Equipment unit C, also with a low level of automation justified. (d) The level of automation justified by combined units A, B, and C.

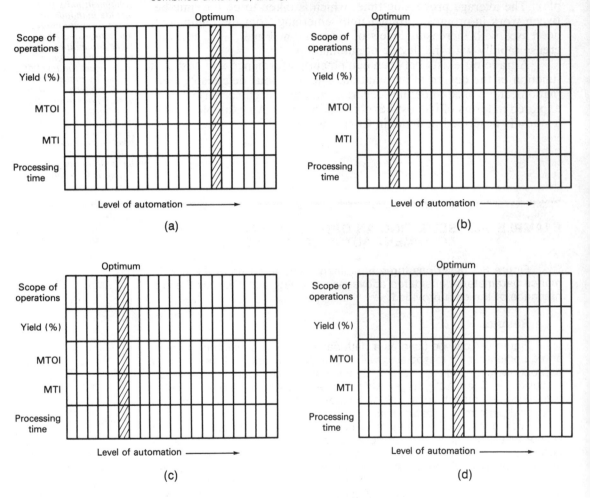

evaluated in terms of their individual cost-benefit, then this manufacturing company would choose to buy one "island of automation" and to link this automated unit to manual equipment units. This type of individual-unit analysis, which does not consider integrated system effects, can produce unsatisfactory results for an enterprise.

Figure 6-7d shows a result that might be obtained by considering the combination of equipment units, their composite performance and composite costs. The overall system justification is now at a level between semiautomated and fully automated processing. Thus, a very different type of system will be configured by considering the cost-benefit of the entire system rather than the separate cost-benefit factors of the individual equipment units.

As discussed in previous chapters, there is no substitute for a system overview of the manufacturing enterprise. Individual equipment studies that are performed without a system context often produce inadequate results.

EXAMPLE 6-4. SELECTING AN OPTIMUM LEVEL OF EQUIPMENT AUTOMATION (CASE II)

Figure 6-8 illustrates another set of data with both individual and composite cost-benefit analyses. Discuss the significance of these results.

Figure 6-8.
Description of three equipment units in terms of performance parameters. For the cases shown here, it is assumed that values for scope of operations, yield, and MTOI increase with level of automation; values for MTI and processing time decrease with level of automation. (a) Equipment unit A, for which an optimum implementation applies a low level of automation. (b) Equipment unit B, for which a low optimum is also obtained. (c) Equipment unit C, with an intermediate optimum level of automation. (d) A high level of optimum automation is determined for combined units A, B, and C.

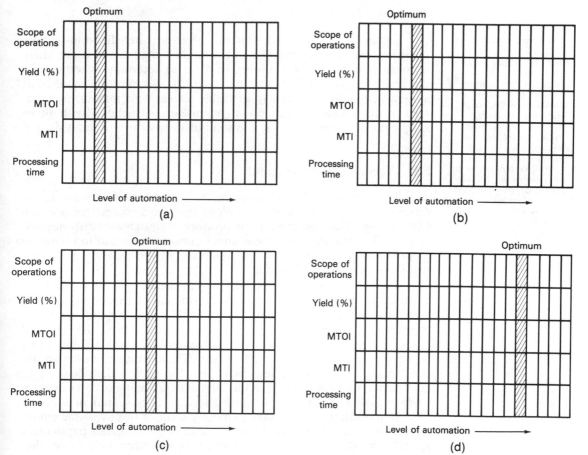

Results

Figure 6-8 illustrates a study in which a company applies classical cost-benefit analysis to each equipment unit, and concludes that only low levels of automation are justified for each unit. Based on this type of finding, no company will be willing to make the transition from manual operations to a highly integrated CIM system. This type of individual-unit analysis often has such an outcome. However, as is illustrated in Fig. 6-8d, a system study would justify fully automated processes.

The cost-effectiveness and competitive advantage of a highly automated system can often be realized only when all of the equipment units achieve a balanced level of automation and integration. If the enterprise continues to perform equipment justification on an individual-unit basis, then the enterprise may never realize that the justification for a total CIM system exists despite the fact that individual equipment units cannot be justified. This is because if a single unit is upgraded, then the upgraded unit is bound in performance by the other units that cannot reach a matching level of performance. As discussed in Chap. 10, there is often a critical level of automation and integration at which all equipment units can work together effectively to achieve the optimum desired performance level; less balanced systems often cannot be justified.

The performance of cost-benefit analyses without careful trade-off studies of the larger system context is likely to result in an unbalanced system design, with individual equipment units and workstations that are not matched to one another, and with an overall system performance far below the optimum desired. There is no substitute for a system overview of the manufacturing enterprise.

Examples 6-3 and 6-4 illustrate that cost-benefit studies performed without an examination of the larger system context are likely to produce unsatisfactory results. In order to determine whether system upgrading is justified, it is necessary to describe the families of equipment that are potentially available for each manufacturing function, to determine the functional parameters associated with these equipment units and perform a system-level cost-benefit analysis.

6-3. Technology Assessment

This section addresses a means that can be used to evaluate the various types of equipment technology that are available for use in a CIM setting. The objective is to produce a straightforward methodology to describe the equipment under consideration and to relate the properties of the equipment to the functional parameters that determine system design and performance.

The methodology begins with preparation of a list of the types of manufacturing equipment that are to be studied, by function (including both materials processing and materials handling functions). A description of the types of equipment currently available can be prepared based on equipment manuals and catalogs that describe equipment by type, operation, specification, and cost. The starting point for an understanding of any technology is to develop such a data base. This documentation should provide the best understanding of the commercially available equipment that can be applied to the system of interest.

It may often be necessary to modify some of this available equipment in order to achieve the desired functional performance parameters specified by the system design. Thus, it is also necessary to develop a data base to understand the types of equipment modifications and upgrades that are available, based on present state-of-the-art research and development technology, and to have a sufficient understanding as to the types of custom equipment that can be implemented should it prove to be cost-effective.

Once an adequate data base has been prepared, the second task is to define the technical parameters associated with each type of equipment. It is useful to divide these technical parameters into those associated with: (1) the sensors in use in the equipment, (2) the actuators and effectors in use in the equipment, and (3) the controllers and control loops associated with the equipment. Dividing the technical parameters into these three categories provides an additional means for comparing equipment units and for achieving a better understanding of performance options.

Figure 6-9 illustrates one way in which sensor, actuator/effector, and controller and control loop equipment characteristics can be described, depending on whether the equipment is controlled by an operator, by a computer, or a combination of the two. The three equipment categories can be combined in various workstation configurations as shown in Fig. 6-5. Each square in the table of Fig. 6-9 provides a description of the technical parameters associated with a particular aspect of the equipment unit. For example, an entry in the upper left-hand corner of the table would describe the technical parameters of those sensors that relate to manually oriented equipment. A table such as the one in Fig. 6-9 can be prepared for each equipment function (or *family* of equipment units) included in the data base.

Figure 6-10 illustrates how the *technical* parameters of various families of equipment can be related to the *functional* parameters of equipment performance, by combining many data summaries of the form shown in Fig. 6-9. The functional parameters of the equipment are related to the technical parameters, with different curves drawn to represent manually oriented equipment, semiautomated equipment, and fully automated equipment. Data describing families of equipment units, targeted to a given manufacturing function, can be combined as in Fig. 6-10 to show how the functional parameters for all members of

An assessment of available manufacturing technologies involves a study of both the equipment currently available and the modifications and upgrades that are possible based on current technology. In many cases, the preferred system design will depend on a degree of equipment customization to meet system requirements. Changes of this nature can result in a substantial competitive edge for the enterprise.

Figure 6-9.
A method for describing equipment units in terms of sensor, actuator/effector, and controller and control loop characteristics.

Type of Equipment	Sensors	Activators/ Effectors	Controller and Control Loop
Manually oriented equipment (operator control)			
Semiautomated equipment (combined operator and computer control)	Technical parameters		
Fully automated equipment (computer control)			

Figure 6-10.
Relating equipment technical parameters to functional parameters.

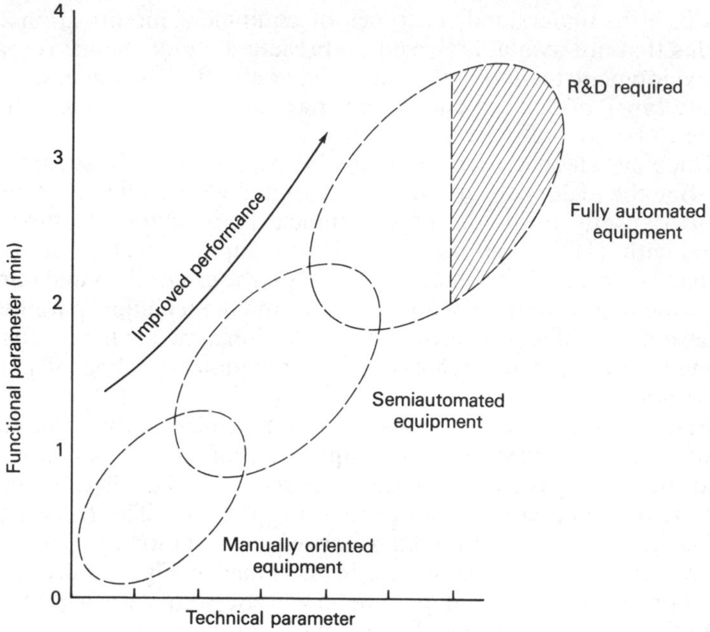

R&D required
Fully automated equipment
Improved performance
Semiautomated equipment
Manually oriented equipment
Functional parameter (min)
4
3
2
1
0
Technical parameter

the equipment family are related to the associated technical parameters.

It is useful to relate the technology parameters of the equipment to the functional equipment performance through various types of "maps." The objective is to understand the types of technologies required for a given level of functional performance and to appreciate the technology implications associated with setting performance objectives for the system.

The type of representation in Fig. 6-10 might be described as a "map" that relates the capabilities and limitations associated with the technical performance of the equipment to the way in which the equipment can function in a manufacturing environment. Figure 6-11 extends this mapping concept to show how the cost of equipment also varies as a function of the technical parameters being evaluated. Thus, by combining Figs. 6-10 and 6-11, it is possible to draw conclusions about the choices of functional parameters that are possible for the families of equipment to be used in the manufacturing system and for the costs that relate to the functional parameters and the technical parameters.

The procedure described above provides a means for linking the technical and cost descriptions of the families of equipment that are available for the manufacturing setting to the functional parameters and costs that will be associated with these equipment units. This method of technology assessment forms a bridge between the technological nature of the equipment and the functional system performance of the equipment.

The method of discussion provided here may be viewed as a means for relating the types of system studies described in Chap. 4 (see particularly steps 6 and 10 of Fig. 4-5) with the detailed equipment data base from which the planning and implementation team must work.

The above discussion introduces various types of work cells that are available for the creation of manufacturing systems. The more automated work cells require the development of operator-independent

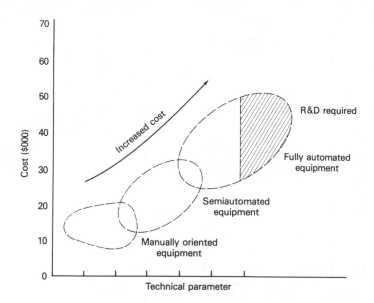

Figure 6-11.
Relating equipment technical parameters to equipment costs. Note that the same technical parameter boundaries are used in Figs. 6-10 and 6-11.

Figure axis labels: Cost ($000); Technical parameter; Increased cost; R&D required; Fully automated equipment; Semiautomated equipment; Manually oriented equipment

processes and materials handling systems (associated with robot functions).

As shown in Fig. 3-27, robot performance is a function of the ways in which sensors, actuators/effectors, and controllers and control loops work together to achieve the desired operating characteristics. The following discussion provides further detail about some of the technologies that are available to support such automated operations. A major challenge is to decide how to draw upon these technologies to achieve most appropriately the enterprise objectives. The emphasis should be on deciding which technologies make sense for the enterprise and will help achieve cost-effective systems.

There are many ways in which to describe automated equipment. The term *robot* can be used in a general sense to describe any type of automated equipment. On the other hand, "robot" is often restricted to a configuration using a manipulator* (arm) connected with an effector (hand). The approach used here is a more generic one; as defined here, a robot is any automated equipment that is driven by a computer controller.†

EXAMPLE 6-5. RELATING TECHNICAL AND FINANCIAL PARAMETERS

Figure 6-12 describes the MTOI and costs associated with a family of equipment units, building on the general methods illustrated in Figs. 6-10 and 6-11. Discuss the significance of the data in Fig. 6-12.

---

* A detailed survey of robotic manipulator arms, including methods of design, has been prepared by Andeen (1988).

† The following discussion describes various types of robotic operation that can be obtained from automated equipment; references on the state of the art in robotics are generally available (Ayres et al. 1985; Lane 1986; Fuchs 1988; Stauffer 1988a).

Figure 6-12.
Selection of equipment units based on functional and cost parameters. (a) Relating equipment technical parameters to functional parameters for specific equipment units. (b) Relating equipment technical parameters to equipment costs for specific equipment units. (c) Calculation of the figure of merit (MTOI/cost) for the specific equipment units selected.

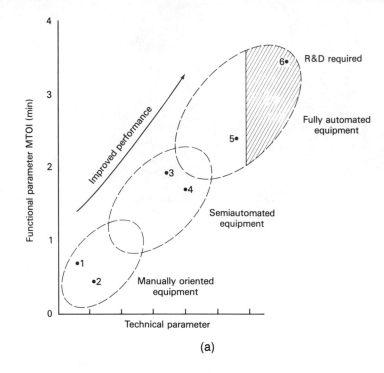

(a)

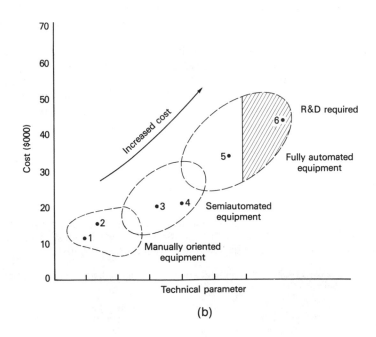

(b)

Figure 6-12. continued

| | Equipment Unit | | | | | |
	1	2	3	4	5	6
MTOI (min)	0.75	0.50	1.85	1.75	2.50	3.50
Cost ($000)	10	14	22	23	35	43
MTOI/Cost (min/$000)	0.075	0.036	0.084	0.076	0.071	0.081
Value ranking based on figure of merit	4	6 (worst value)	1 (best value)	3	5	2

(c)

Results

Figure 6-12 shows the MTOI and costs associated with six members of an equipment family. The technical parameter is assumed (in this case) to provide a measure of the pattern-recognition capability of a machine vision operation. Assume for the moment that only the MTOI and costs are relevant to an important decision to be made. (In realistic cases, a wide range of functional parameters would have to be reviewed in such a situation.)

As shown in Fig. 6-12, the MTOI increases in moving from manual to semiautomated equipment and then to fully automated equipment. At the same time, costs increase in moving from manual to higher levels of automation. The MTOI performance drops in moving from equipment unit 1 to unit 2, despite an increase in cost. If no other considerations enter in, unit 1 is preferable for development or purchase over unit 2. (Again, in realistic settings, many other functional parameters will enter into this decision making; the purpose here is to provide a simple example of how the method can be applied.) The same experience is noted with moving from unit 3 to 4; the MTOI drops despite the increase in cost, again allowing a simple selection between these two units. For units 5 and 6, the higher MTOI is associated with higher costs.

Many different figures of merit can be used to compare data of this nature. The figure of merit may involve individual equipment performance combined with system-level performance or may depend completely on system-level predictions. For the existing case, a desirable equipment unit will have a higher MTOI and a lower cost. Thus, a possible figure of merit is the ratio MTOI per cost. The larger the value of this figure of merit, the more attractive the product will be.

Figure 6-12c illustrates the MTOI per cost for the six equipment units described. Based only on this criterion, the preferred equipment selection would be unit 3. On the other hand, the highly automated R&D equipment version is close to the same figure of merit and should be viewed carefully to determine other potential advantages associated with such a choice.

As discussed above, customized manufacturing equipment often provides a substantial competitive edge to a company. If a highly automated capability can be developed in a cost-effective fashion, the simplistic figure of merit (MTOI/cost) that has been selected may be inadequate for making a final decision. The equipment unit is being evaluated by itself, without a viewpoint of the entire system performance, which can be misleading. The type of selection shown in Fig. 6-12, which considers a single family of equipment using a single figure of merit, is common and tempting. At the same time, the solutions obtained may be inappropriate to the long-term competitiveness of the enterprise.

Sections 6-4 through 6-6 discuss equipment and technology in terms of sensors, actuators/effectors, controllers, and control loops for various manufacturing purposes. Section 6-7 then illustrates in detail how equipment performance and cost can be linked. (Sections 6-8 and 6-9 address machining centers, cells and systems, and system costs. The remainder of the chapter discusses the manufacture of electronics assemblies.)*

6-4. Sensors

Most of the robots in use today are not very smart. They do not make use of external sensors that enable them to monitor their own performance. Rather, they depend on internal positioning sensors to feed back (to the control system) information regarding manipulator positions and actions. To be effective, this type of robot must have a rigid structure and be able to determine its own position based on internal data (largely independent of the load that is applied). This leads to large, heavy, and rigid structures.

The more intelligent robots use sensors that enable them to observe work-in-progress and a control loop that allows corrective action to be taken. Thus, such robots do not have to be as rigid because they can adapt. The evolution toward more intelligent, adaptive robots has been slow, partly because the required technologies have evolved only in recent years and partly because it is difficult to design work cells that effectively use the adaptive capabilities. Enterprises are not sure whether such features are cost-effective or how to integrate smart robots into the system.

The emphasis in this text is on the building-block elements that are necessary for many types of processing. If the most advanced sensors are combined with the most advanced actuator/effector concepts

Many types of robots exist, with wide variations in capability dependent on the design strategy in use. However, industrial evolution toward the use of intelligent, adaptable robots has been slow because of uncertainties over the benefits associated with the increased capabilities. Both the original equipment manufacturers and the manufacturing companies that purchase equipment are interested only in features that will increase system performance in a cost-effective way. Evaluation of high-capability robotics for a manufacturing system is often a difficult task because of the system perspective that must be taken.

---

* The product and equipment descriptions included in this chapter are largely based on catalogs and brochures from original equipment manufacturers. The equipment and companies were selected by the author from available sources. The equipment units described here are intended to illustrate how manufacturing systems may be designed.

The author has developed interpretive discussions of the equipment based on available information from equipment vendors and on past personal experience where relevant. The characteristics of the products and equipment are those presented by the manufacturer and in most cases do not involve firsthand experience by the author. In general, the manufacturers have reviewed the discussions for technical accuracy.

There are several items of equipment with which the author has had firsthand experience, particularly the Teledyne TAC Assembly System and Hughes Aircraft Company's 2460-II wire bonder, discussed in Examples 6-18 and 6-19. The author has worked with these equipment units in the laboratory and has received funding from these companies for related R&D projects. The author has observed many other types of equipment units in industry-wide application. Given the objective of the discussion here and the sources of material, the author cannot guarantee the particular performance features of a given equipment unit and is not recommending any of these products for purchase.

and state-of-the-art controllers and control loops, very sophisticated equipment can result. On the other hand, much more rudimentary sensors, effectors, and controllers can produce simple types of actions.

Figure 6-13 illustrates alternative ways in which sensors may be viewed. In Fig. 6-13a a physical stimulus acts on the sensor; the change in the sensor is pre-processed, and a signal is passed along to the processing loop. In many cases today, sensors are analog (they involve a continuously changing output property) and control loops make use of digital computers. Therefore, an analog-to-digital (A/D) converter between the pre-processor and the digital control loop is often required, as shown (Fig. 6-13b).

The sensor may operate either passively or actively. In the passive case, the physical stimulus is available in the environment and does not have to be provided. For an active case, the particular physical stimulus must be provided. One of the most common types of sensors involves machine vision (Edson 1986; Larin 1986; Gould 1987; Hall and Hall 1988; Schrelber 1988; Stauffer 1988b). Machine vision is an active means of sensing, because visible light must be used to illuminate the object before a physical stimulus can be received by the sensor. Laser detectors are also active sensors. Passive sensors include infrared devices (with the physical stimulus coming from infrared radiation that is associated with the temperature of the body) and sensors to measure pressure, flow, temperature, displacement or proximity, humidity, and other physical parameters.

Figure 6-14 shows how sensors can be used to create a vision system. The composite sensor consists of an array of devices that are sensitive to visible light, forming a television camera. The output from the camera is pre-processed and then applied to identify the desired features of the product in view. Because of the large number of elements (called *pixels*) in such a television picture it is necessary to preprocess information as much as possible. Special integrated circuit IC "chips" have been designed to combine the output of the sensor with pre-processing in order to simplify this problem.

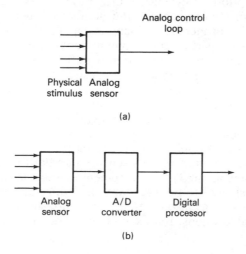

(a)

(b)

Figure 6-13.
Two alternative ways in which sensors can interact with control loops. Higher performance systems often lead to more complex processing strategies.

Figure 6-14.
Illustration of a way in
which sensors may be
used to create a vision
system.

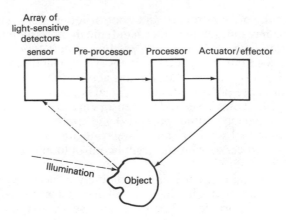

For a machine vision system, design efforts should focus on obtaining the needed information as easily as possible, while minimizing the processing complexity. Therefore, it is desirable to build as much information as possible into the object and into the control system in advance. Bar coding and other methods of labeling provide one simple solution to assisting machine vision systems in object identification. Labeling keys can assist the vision system and make the identification job much easier. Labeling is probably underutilized as a way of building a priori knowledge into a vision system.

If the vision system is intended to perform more general functions, the task can become quite difficult. The system must be able to identify a range of shapes in different orientations and under different lighting conditions. It is most effective to decide on the range of objects to be viewed and the functions to be performed by the machine vision system and attempt to pre-process information as much as possible to achieve the desired types of decisions that are needed. Efforts may be made to pre-process the identification of lines, edges, or specific shapes in order to provide a simplified output for decision making.

Machine vision is more widely used today in settings for which labeling or a priori knowledge is drawn upon. As the use of vision systems becomes more general, the number of such aids will be reduced. There is an increasing focus on application-specific machine vision systems for particular equipment needs. On the other hand, as the system becomes more application specific, the equipment becomes less flexible and may not achieve the desired level of adaptability.

As noted, sensors are often of an analog nature, providing a continuous output based on the physical stimulus. Typically, sensors will produce an electrical output signal that can then be processed. In some cases, the stimulus response is pre-processed and then converted to electrical signals. Multiple sensors can be used in a *sensor synthesis* to combine the effects of various sensors in order to produce a higher level of discrimination. The development of such sensor *fusion* capabilities is considered to be a research and development topic today. Arrays of sensors may also be used to measure simultaneously parameters at several locations. For example, an array of optical sensors can be used for machine vision and an array of pressure sensors can allow for a tactile (touch) capability.

Signal pre-processing can involve various types of discrimination, filtering, and recognition functions in order to support the performance of the type of control system that is being designed. It is important to do as much pre-processing as possible because of the complexity of the sensor signals. The number of different signals received and the rate at which they vary can overload a digital system if "brute force" processing is employed. Pre-processing of the signal as much as possible may be needed in order to produce a workable volume of data for the processing and control loop.

The current trend is toward sensors that are more rugged, reliable, and compatible with digital computer control. A major emphasis is on smaller size. The same processes that are used to make integrated circuits are also being used to make very small sensors. The creation of sensors "in silicon" using these processes has many distinct advantages in terms of device size and weight. It is also possible to combine the sensor itself with ICs that perform much of the required pre-processing. An intelligent sensor can be much more effective in many applications.

Packaging must be designed so that the sensor can function as desired in the equipment setting, and sensor flexibility is desired to achieve a range of tasks.

An inadequate linkage exists between sensor research and those who are designing equipment for manufacturing systems. There is need for an integrated effort to develop sensors that are most appropriate for the equipment needs that are being experienced. It is often difficult to achieve integration between the sensor technology in use and the equipment technology being employed in order to achieve a cost-effective manufacturing system. There is always a trade-off between sensor flexibility and the specific sensor application.

EXAMPLE 6-6. SENSORS

A wide range of sensor types can be used to create fully automated equipment. Indicate some of the sensor products that are available to the equipment design engineer.

Results

A wide range of sensor types is available to industry today, providing the equipment design engineer with many alternative strategies for creating automated equipment. This example considers position and velocity, machine vision and bar-code scanning, and pressure and temperature sensors to indicate some of the capabilities that are available.

(A) Position and Velocity Sensors: Industrial robots and automation equipment involving mechanical operations often require sensors that can be used to determine the mechanical position or rotational speed of the equipment. These capabilities are essential for open-loop robot systems; in such

settings, the robot arm must determine its location and direct its operations based on internal position and speed signals rather than on external signals from machine vision. The ability to monitor internally the position and speed of mechanical operations is essential to robots that do not have external sensors.

Encoders are often used for this purpose. An encoder is a sensor that sends out a signal that gives the position or rotational speed of a shaft or a mechanical moving part. Because of the importance of digital controllers and digital control loops, direct digital measurement has lead to wide use of optical encoders (Fig. 6-15a). An absolute optical encoder is designed so that the position of a shaft can be determined by shining a light through a coded disk; the light that passes through the disk pattern falls on a line of photodetectors that directly produce a digital output signal. A computer can read the output of the photodetector and obtain a digital word that tells it exactly the position of the shaft. Figure 6-15a shows the operation of an absolute encoder that provides a digital word to describe the absolute position of a shaft.

An incremental encoder creates a series of pulses that correspond to the mechanical increments; a computer can be used to count the number of pulses to determine the location of the shaft or measure the time between pulses to determine velocity. Generally, incremental encoders provide more resolution at lower cost than similar absolute encoder versions; the former typically will have fewer interface problems because they have fewer output lines. Figure 6-15b shows the design of an incremental encoder.

A typical optical encoder involves the bearing housing assembly, a light source, code disk, photodetector assembly, and necessary printed circuit electronics. (The manufacture of such electronics circuits is discussed below.)

Figure 6-15c shows the BEI Motion Systems Company Model H25 incremental optical encoder. This product was designed for the industrial machine-tool marketplace. The H25 offers such features as electromagnetic shielding, heavy-duty bearings, matched thermal coefficients on critical components, and custom high-efficiency optics. Typical applications include machine control, process control, and robotics. The model H25 is available in a range of configurations. The type of application will determine the housing configuration, the shaft seal configuration, the number of cycles per turn, the number of output channels, the type of illumination to be used, and characteristics of the output terminal.

Figures 6-15d through 6-15f illustrate a number of optical encoder applications. Figure 6-15d shows the encoder driven by moving wire to encode the motion of a lifting mechanism. Alternatively, shown in Fig. 6-15e is a lathe numerical control display; the purpose is to encode the motion of the lead screw and thus to enable automatic control. Figure 6-15f shows a linear actuator, which encodes the motion of the rack and pinion.

These optical encoders represent an important type of sensor for use in the development of automated manufacturing equipment. Any time a position or rotational speed signal is to be applied as input to a control loop, the encoders described here can be effective. Such products are widely available commercially and in many configurations, with properties that can be adapted to specific industrial applications.

(B) Machine Vision and Bar-Code Scanning: One of the most rapid growth areas for sensors involves the development of sophisticated machine vision systems. Such systems require stable high-resolution cameras as sensors.

Figure 6-15g shows a Pulnix America, Inc., Model TM-840 high-reso-

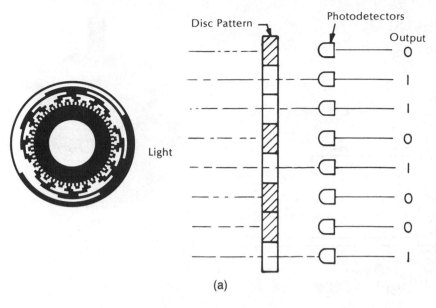

Disc Pattern

Photodetectors

Output

Light

	Output
	0
	1
	1
	0
	1
	0
	0
	1

(a)

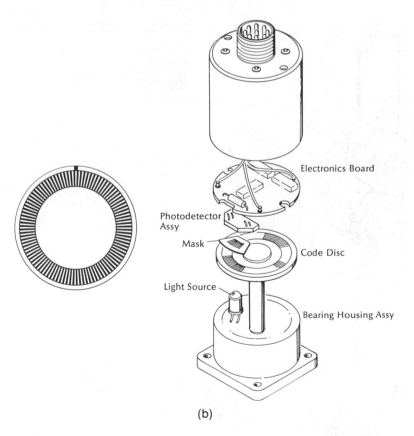

Electronics Board

Photodetector Assy

Mask

Code Disc

Light Source

Bearing Housing Assy

(b)

Figure 6-15.
Alternative sensors. (a) Design of an 8-bit absolute optical encoder (courtesy of BEI Motion Systems Company). (b) Design of an incremental optical encoder (courtesy of BEI Motion Systems Company). (c) Model H-25 incremental encoder (courtesy of BEI Motion Systems Company). (d–f) Encoder applications (courtesy of BEI Motion Systems Company). (g) Pulnix Model TM-840 CCD camera (courtesy of Pulnix America, Inc.). (h) Pulnix Model TM-540R camera with remote optic assembly (courtesy of Pulnix America, Inc.). (i) LazerData Model 310 bar-code scanner (courtesy of LaserData Corporation). (j) Viatran Model 122 pressure transducer (courtesy of Viatran Corporation). (k) Mikron Instrument Company Series 500 infrared temperature sensor (courtesy of Mikron Instrument Company, Inc.). (l) Mikron Instrument Corporation Model M78 infrared temperature transmitter (courtesy of Mikron Instrument Company, Inc.)

Figure 6-15. continued

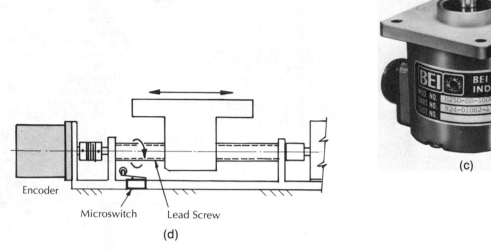

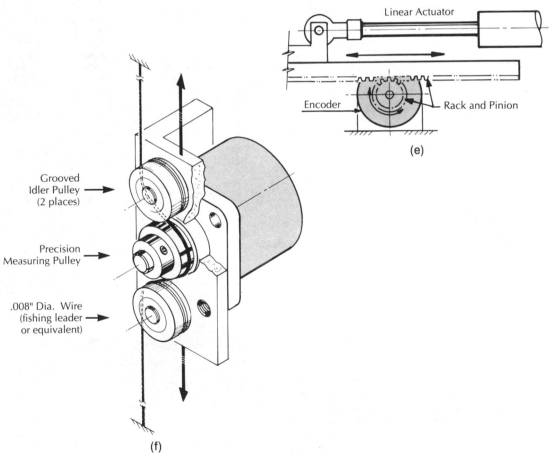

Encoder

Microswitch   Lead Screw

(d)

(c)

Linear Actuator

Encoder   Rack and Pinion

(e)

Grooved
Idler Pulley →
(2 places)

Precision
Measuring Pulley →

.008" Dia. Wire
(fishing leader →
or equivalent)

(f)

Figure 6-15. continued

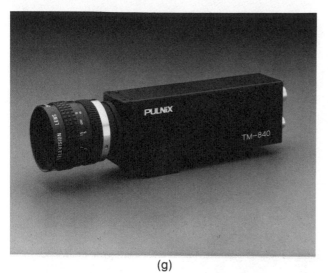

(g)

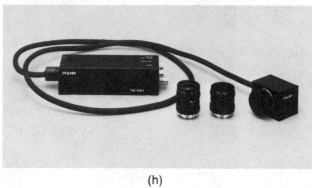

(h)

(i)

Figure 6-15. continued

(j)

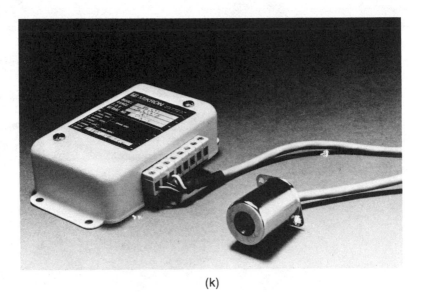

(k)

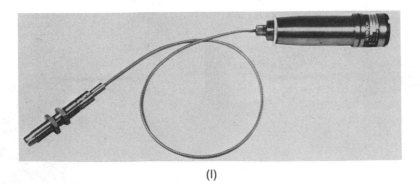

(l)

lution charge control device (CCD) camera with an imager that can form a picture that is 800 × 490 pixels. This camera is designed for high-resolution vision systems and can be used in a wide range of manufacturing operations. The camera measures about 4.5 × 3.2 × 13.6 cm (about 1.8 × 1.3 × 5.4 in.) and weighs 250 g (about 8.8 oz). This type of camera has become a standard for use in machine vision systems.

In a number of applications, it is not feasible to mount the entire CCD camera close to the object that is to be inspected. To meet this need, the model TM-540R shown in Fig. 6-15h has a remote imaging capability. This remote imager CCD camera permits the mounting of the lens/imager optics assembly within a confined area or within a structure requiring a low mass. As shown in Fig. 6-15h, the imager is attached to the main camera model by a flexible cable that carries all video and power. Various cable lengths are available by custom order. To facilitate mounting, several different versions of the imager module are available to provide alternative optical orientations. This type of remote imaging allows machine vision to be applied in many settings in which the complete camera interferes with the mechanical operations of the manufacturing system.

A different type of optical sensor is used to create bar-code scanning systems. To be effective, bar-code scanners must allow the rapid and automated identification of components and work-in-progress for automated manufacturing systems.

One of the most common methods of scanning a bar code uses a laser scanner. Figure 6-15i shows a LazerData™ Model 310 data scanner. For this product, the hardware is kept simple; only a single moving part is required to perform the desired laser scan. A single helium-neon laser provides the illumination source, with beam redirection using mirrors to create a pattern that can capture any bar code passing through the scan area. The optics are packaged separately from the electronics in order to provide simple maintenance, and real-time decoding is provided for identification.

Model 310 provides stationary bar-code scanning for general applications. It allows reprogramming through data terminals, and can thus achieve integration with computer communications networks. It achieves a high scanning rate for rapid and accurate reading of labels. The system is designed with completely modular subassemblies so that units can be quickly interchanged. The result is a standard high-performance scanner that can be used as a sensor for product identification.

Figures 6-15g, 6-15h, and 6-15i represent two quite different types of optical systems. The cameras in Fig. 6-15g and 6-15h use ambient or controlled external light, and scan a large number of pixels in order to assemble a simultaneous picture of the view being seen by the camera. This camera view is used as the input to the control loop. On the other hand, the scanner in Fig. 6-15i uses laser-generated illumination and sequential scanning to search for specific patterns that can be recognized by the scanner. These equipment subsystems are only two indications of the many optical sensors that are available for automated equipment.

(C) Pressure and Temperature Sensors: Figure 6-15j shows a Viatran Corporation pressure transducer that can be used to measure gas pressures associated with a process and, therefore, to drive a control loop. Pressure transducers are widely used in manufacturing settings where gas pressure is an important part of the input to the control process. The Viatran transducers are based on the strain gauge principle, which has been available for many years and is a basically simple operation. Many such gauges are available, with design for use in a wide range of environments. The pressure sensed by the strain gauge is converted to an electrical signal that can then be used as the input to the control system. The electrical output is linear over a wide range and can be directly input to many computer controlled devices.

Stainless steel is used in the manufacture of the complete assemblies. The assemblies are designed to be as small as possible, and the gauges incorporate various types of calibration and compensation. The interior of the unit contains sophisticated electronics signal conditioning, adjustment, and compensation circuits (that can make use of printed circuit and hybrid manufacturing technology, as discussed below in detail). These pressure sensors are available with a wide range of electrical connections and operational specifications.

Figure 6-15k shows a Mikron Instrument Company, Inc., Series 500 infrared temperature sensor that can be used to remotely determine temperatures of 0 to 1000°F in a manufacturing setting. The rugged and compact sensing head converts heat-generated infrared energy into an electrical signal. The sensor output is translated into a calibrated temperature signal that can be transmitted to the electronics package, which can be mounted 5 to 100 feet away. The output from the electronics module can be used as input to a control loop. Temperature measurements in restricted areas are simplified by the small remote-sensing head. An optional cast aluminum jacket can be used around the sensor to protect it from physical damage in demanding industry environments. Air or water cooling can be used, depending on the surrounding environment.

Shown in Fig. 6-15l is a Mikron® M78 two-color infrared temperature transmitter for measuring temperatures above 1300°F accurately, independent of emissivity and target size. This capability is important for applications where obstructions or severe environmental conditions make direct sighting impossible. The unit consists of the sensor itself, a fiber-optic cable, and a lens assembly. The two-color principle allows temperature measurements to be made independent of target characteristics. The M78 transmitter can be provided with fiber-optic cable links of up to 40 feet and with various lens assemblies and extension tips that can be used for sighting the target. For extreme environments, a cooling jacket and air purge assembly are available to protect the lens. This model illustrates noncontact temperature sensors that can be used in numerous manufacturing settings.

Sensors are available today to measure almost any physical stimulus. However, major difficulty is experienced in deciding on the combination of sensors and sensor-system interactions that will result in equipment performance and cost matched to manufacturing needs. Too often, sensor research is not strongly linked to equipment design, and the result is underutilization of sensors.

The above discussion of sensors provides a brief introduction to the potential opportunities for the design of automated equipment based on the ability to self-monitor processes and conditions. Sensors of almost any type are now commercially available to measure those equipment parameters necessary to achieve fully automated operation. Design issues often relate to selecting those sensors that are most appropriate for the use at hand and performing the system integration that is necessary to achieve the desired operations.

6-5. Actuators/Effectors

Figure 6-16 shows a possible configuration for an automated equipment unit. The product is held in place by *fixturing* (Talavage and Hannam 1988). In some types of equipment, fixturing can be automatically changed to hold different types of products. The product is

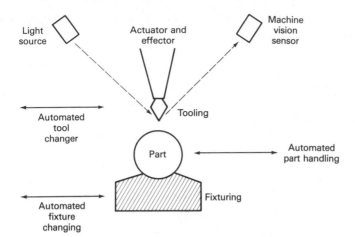

Figure 6-16.
One possible strategy
for combining sensors,
actuators, and effec-
tors for automated
equipment.

acted on by a *tool;* an automated tool changer may be available to allow the incorporation of a range of tools into the given equipment unit. The tool can be held in place by a robot arm, or manipulator. The effector (or hand) at the end of the manipulator is used to hold the tool, moving in response to signals sent to the actuator. An external sensor may or may not be present to guide the robot arm; alternatively, internal sensors may control the arm.

The effector can modify the product through welding, drilling, or other process, or it may be used for assembly (associated with picking up a component and inserting the component into a product). An automated materials handling system can provide a flow of parts into the system on an automated basis. Many other actuator/effector designs may also be employed. The problems encountered often involve the need for flexibility and adaptability. If a limited number of fixtures and tools is available, then only a limited number of processes can be performed. The more product-dedicated the elements of the system are, the more constraints will be experienced in enabling the enterprise to change over time. If there is an open-loop control system, the robot arm and tooling fixture must be rigid; with a closed loop, flexibility in the system may be obtained but integration difficulties will increase.

In assembly operations, it is often necessary to combine the type of equipment configuration shown in Fig. 6-16 with an effector capable of picking up the components that are being assembled. One way to do this is to arrange the components in a carefully ordered fashion so that the effector can grasp them. Essentially, each component is fixtured into place.

Another approach to developing an automated system is to eliminate the fixturing to hold the product and thus have fixtureless operations (Hoska 1988). The robot is designed to be smart enough to identify the components in an arbitrary arrangement, and is able to orient the components as desired and pick them up. This type of function requires a sophisticated sensor system combined with tactile sensors (obtained by arrays of pressure-sensitive devices) and complex processing.

Actuators and effectors provide a means for manufacturing equipment to act on the work-in-progress. Responding to sensory and preprogrammed input, an actuator can serve as the source of the act, and the effector as the means of action. Design skill is required to choose sensors, actuators, and effectors that will work together in an effective way with both equipment and operator to achieve the desired manufacturing operations.

Actuators and effectors provide the means by which manufacturing equipment acts on work-in-progress. The actuator serves as the source of the act, and the effector as the means of action. Together, the actuator and effector provide the equipment with an ability to act on the environment.

Many different types of actuators and effectors are available (Malone 1987). Typically, an actuator will produce a mechanical motion in response to an electrical input. An effector (which serves as the hand of the equipment unit) is driven by the actuator and can interact with the work-in-progress through tools, grippers, or other special-purpose devices. The manipulator (which serves as the arm of the automated equipment and provides motion capability for the effector) can consist of a robot arm, an X-Y motion grid, or other mechanical device.

The characteristics of the actuators and effectors, combined with the load, determine system performance. A massive actuator/effector combination is difficult to move and will require powerful actuators. The nature of the task to be performed by the equipment will determine the preferred design for the effector, which in turn will determine the preferred design for the actuator. The type of action to be performed by the equipment, and the interaction of the effector with the environment, will determine the preferred design for this portion of the equipment system.

The development of equipment automation is a demanding task. A wide range of sensor types is available, along with many different strategies for handling the work-in-progress. Difficulties are experienced in choosing the best design approach. It is quite possible to design a complex system and not have it work as expected. For example, systems often incorporate machine vision, but the machine vision may operate in an inadequate way, so that an operator must always be present to correct mistakes. Even with sensors and materials handling subsystems, it is still difficult to produce an integrated system to achieve the desired functional parameters for the equipment.

Even if such a system can be satisfactorily designed and implemented, issues involving flexibility and adaptability for the future are of critical concern. It may be that by the time a particular automation problem has been solved, this problem no longer exists; then a new problem appears that is not readily addressed by the existing equipment. Thus, original equipment manufacturers are cautious in moving toward the sophisticated use of robots, because they want to make sure they are meeting the perceived needs of the system integrators. It is often necessary to gain significant experience with any new aspect of automation before it can be considered ready for manufacturing application.

Difficulty may be experienced in transferring from research and development and theoretical equipment designs to manufacturing equipment that is sufficiently reliable to achieve the desired operating parameters on the manufacturing floor. Such efforts must follow the overall system design procedure introduced in Chap. 4. The equipment manufacturers must combine a top-down overview of the types of automated equipment parameters that would be desirable, with a bottom-up understanding of available technology and the evolutionary steps that are necessary to achieve the desired parameter levels. The equipment industry tends to progress incrementally from one capability to the next in order to avoid producing a transition product line that is not commercially feasible.

EXAMPLE 6-7. ACTUATORS AND EFFECTORS

Many types of actuators are used to achieve mechanical motion for automated manufacturing equipment. Discuss the relative features of motors and moving coil and moving magnet actuators. Discuss some of the ways in which sensors, actuators, and effectors can be combined to produce robotic performance.

Results

Many types of motors can be used to achieve mechanical motion in automated systems. Stepper motors turn by a precisely determined angle with each energized pulse. The motor is designed to move between a defined set of rotary positions. By controlling the rotary motion one step at a time, a stepper motor provides control over shaft positions and movement. Stepper motors can be used under open-loop conditions or with feedback; performance is improved in the latter case. Because of the digital stepping feature, such motors are widely used as an interface between computers and mechanical operations (Acarnley 1982; Kenjo 1984).

As noted by Acarnley (1982):

Stepping motors have been available for many years, but commercial exploitation only began in the 1960's when improved transistor fabrication techniques made available devices capable of switching large dc currents in the motor windings. This feature of switched winding currents gives the stepping motor its unique "digital machine" properties, which are a considerable asset when interfacing to other digital systems. The rapid growth of digital electronics through the 1970's (and continuing today) (has) assured the stepping motor's future, and today there is worldwide interest in its manufacture and application (Preface).

The essential property of the stepping motor is its ability to translate changes in [electrical current] into precisely defined increments of rotor position ("steps"). Accurate positioning of the rotor is generally achieved by magnetic alignment of the iron teeth on the stationary and rotating parts of the motor (p. 1).

Other types of housed and unhoused motors are also widely used for automated manufacturing. Figure 6-17a shows brushless motors from BEI Motion Systems Company that are specifically designed for machine tools and robotics applications. These motors feature high torque-to-size and torque-to-weight ratios and can include encoders (as discussed in Example 6-6) to control exact motor positioning.

Figure 6-17b shows various linear and rotary moving coil and moving magnet actuators produced by BEI. These actuators provide a highly linearized force over a range of stroke distances, and are widely used to meet positioning requirements of automation. Linear actuators typically achieve motions of a few thousandths to several inches and a force ranging from ounces to pounds. Rotary actuators typically range between 5 and 30 degrees.

A diversity of actuator types and configurations is available to achieve

(a)

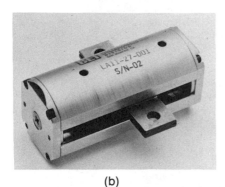

LA11-27-001
S/N-02

(b)

Figure 6-17.
Sensors and actuators. (a) Brushless motors (courtesy of BEI Motion Systems Company). (b) Linear and rotary actuators (courtesy of BEI Motion Systems Company). (c) Instrumented gripper (courtesy of Lord Corporation). (d) Instrumented gripper used for printed circuit board assembly (courtesy of Lord Corporation). (e) Force/torque wrist sensor (courtesy of Lord Corporation). (f) Model QC50 automatic tool changer (courtesy of Lord Corporation).

(c)

Figure 6-17. continued

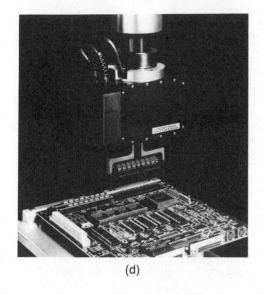

(d)

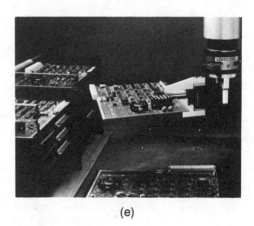

(e)

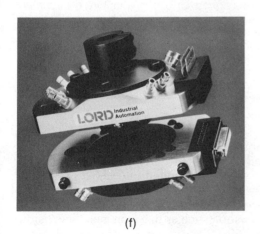

(f)

the type of motion that is desired for an automated mechanical system. These motion products are often integrated with sensors and an effector in order to provide a combined sensor/actuator/effector assembly.

Figure 6-17c shows a Lord Corporation Instrumented Gripper that is designed to provide a flexible end effector for electronic component insertion or light mechanical assembly tasks. This unit was developed to provide a sensor-based programmable gripper as an alternative to single-function grippers. It allows the user to handle a wide number of different components and makes use of interchangeable fingers; the unit also provides a verification of whether the part was properly inserted.

This electrically actuated gripper has independent control of its two parallel fingers. Each finger has both position sensing and force/torque sensing to enable one finger to stop when a programmed threshold force is detected, even though the other finger continues to move. This permits the fingers to grip off-center parts without damaging or losing the part. The fingers can be programmed to move using either force or position control.

The assembly in Fig. 6-17c provides an integrated sensor/actuator/effector capability. Figure 6-17d shows how this assembly may be used to attach components to a printed circuit board. The positions of the fingers used for pickup are sensed by incremental encoders that are mounted on the finger positioning shafts (refer to Example 6-6 for a discussion of encoders). A pre-processor is attached to the robot arm near the gripper, providing digitized signals from the force/torque strain gauges for transmission to the system controller. By communicating with the robot controller, the system can provide a wide range of assembly operations.

Figure 6-17e shows a Lord Corporation Force/Torque Wrist sensor. This device features wrist-mounted tactile sensors that allow feedback for adaptive control. The sensor delivers real-time feedback on the magnitude, location, and direction of applied forces and torques. Each system features the force/torque transducer designed to measure a wide range of force levels. The type of wrist sensing system shown in Fig. 6-17e is designed to be able to provide force feedback information for a wide range of applications in automated manufacturing.

Figure 6-17f shows another Lord Corporation product, an end effector exchange system for use in assembly operations. This automatic tool changer allows automated equipment to change end effectors or other tooling without operator intervention. A wide range of end-of-arm tooling can be applied, regardless of individual end effector requirements.

The unit is light (about 2 pounds) and can handle a payload of up to 50 pounds. This tool changer is designed for use in a factory environment and can be used in conjunction with grippers, vacuum pickups, or other end-of-arm tooling. With this type of automatic tool changer, a single robot can perform many different tasks and operations. No lubrication of the unit is required, and all seals, springs, and electrical contacts are user-replaceable for easy maintenance.

Actuators may be combined with a wide range of effectors to achieve essentially an unlimited range of automated operations. In many cases, internal sensors will be integrated into actuator/effector assemblies in order to allow for feedback with internal positioning data. In other cases, the feedback from the internal sensors will be combined with feedback from external sensors (such as that from machine vision) to produce a manufacturing operation that can monitor its own activities from both internal and external prospectives. A major task in designing automated equipment is to decide on the types of sensors, actuators, and effectors that are needed to achieve the desired operations. Many technologies and capabilities are available for the design of manufacturing equipment. Often the most difficult task is to decide on those technologies that are most appropriate and to integrate them to achieve the desired operation.

6-6.   Controllers/Control Loops

Figure 6-18a shows how a sensor, actuator, effector (with load), and a controller can be combined in a control loop for automated equipment operation. The sensor observes the object being controlled. Data from the sensor are input to the controller, which compares the sensor input with the desired status input. Based on the observed and desired conditions, the controller produces a signal that drives the actuator, which in turn results in motion of the effector and load. The controller, sensor, and actuator/effector are thus embedded in a control loop. The objective of the control loop is to produce the type of motion that is desired.

It might seem that such a system would be simple to develop; however, this is not the case. Performance of the control loop depends on the sensitivity of the sensors and actuators, the inertia associated with the effector and the load, the finite capability of the actuator to achieve rapid motion, and the response time and delays associated with the controller.

The design of control loops that will produce the desired dynamics for complex systems is a demanding task. It is difficult to obtain the "performance envelope" desired when equipment must provide a wide range of manufacturing functions in many different situations, with constraints on the sensors, actuators, and effectors in use.

Consider the type of motion shown in Fig. 6-18b, which involves having a tool move from one location to another and then back to the original position. The ideal performance of such a system involves having the tool move from location A to location B as quickly as possible, remain at location B, and then return back to location A.

Figures 6-18c and 6-18d show various types of actual motions that can result from control loops. In Fig. 6-18c, the system vibrates around the desired final position before "settling down." This type of overshoot can produce disastrous collisions with work-in-progress; under any conditions, a long wait is required before operations can proceed. On the other hand, Fig. 6-18d shows a situation in which the motion is so slow that a significant delay occurs in achieving the desired action. It turns out to be quite difficult to achieve idealized performance for combined effector/load actions when a wide range of effector types and loads must be considered. The same type of problem is experienced whether the effector is holding a tool that is acting on a part or whether the effector is integrated into a "pick and place" unit. Care must be taken to produce a control loop that will have the dynamic properties that are desired for the equipment.

Figure 6-18a describes a control loop that is associated with fully automated equipment. For semiautomated equipment, for which the automated equipment and the human must work together to achieve the desired operations, it can be extremely difficult to design the type

Figure 6-18.
Control loops. (a) Il-
lustration of how a
sensor, actuator, effec-
tor (with load), and
controller can be com-
bined in a control loop
for automated equip-
ment operation. (b–d)
Alternative descrip-
tions of control loop
performance based on
design characteristics.

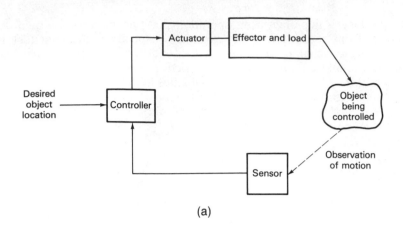

(a)

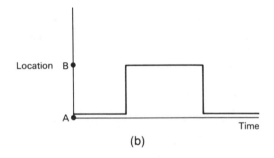

(b)

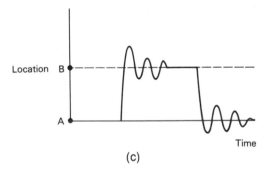

(c)

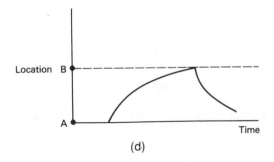

(d)

of control loop that is required. If the control loop depends on inter-action between the human and the computer controller, the dynamics of the system that results may be quite unexpected.

Figure 6-18a can be associated with an internal or external control loop. For an internal-sensor control loop, the sensor does not receive feedback from the environment of the effector, and absolute position measurements from the sensors must be used to guide the manipulator, effector, and a load to the desired positions. For an external-sensor control loop, the sensor receives environmental information. Absolute position information is not required, since an adaptive response can be produced.

EXAMPLE 6-8. CONTROL LOOPS

In order to integrate internal and external sensors and actuators and effectors into a working automated system, it is necessary to create a control loop. Discuss some of the features of motion control loops.

Results

Control systems are available in component form or as integrated systems. BEI Motion Systems Company offers an Ultra-Loc™ control system that makes use of a motor-encoder assembly coupled to a shaft whose rotational position or velocity is to be controlled. The encoder continuously monitors the position of the shaft as the motor drives the shaft to the commanded position or at a desired rate. Electronic circuitry is used to compare the output of the encoder with the input command to produce a signal to the motor. The encoder thus functions as the sensor as the motor produces the desired motion. A computer is used to link the sensor (encoder) and the actuator (motor).

The control loop can make use of analog or digital control processes; however, because of the wide use of digital computers today and because of the higher levels of performance that can be obtained with such systems, digital control loops are presently in the forefront of control system design. Typically, the sensor signal either will start out in digital form (obtained from an encoder) or will be passed through signal processing to convert an analog signal to a digital signal. The digital signal is input to the computer, which has been programmed to achieve specific system operations. The computer interprets the incoming signal and then produces an output digital signal that can drive the actuator. Typically, the actuator requires a pre-processing unit that receives the digital computer control signal and converts this signal to the appropriate electrical output to achieve the type of motion desired. Figures 6-19a and 6-19b show a mirror positioning system and a laser beam positioner system that have been developed for use with the Ultra-Loc control system.

Figure 6-19c shows another sophisticated control strategy that is available with machine vision systems from Cognex. The Cognex Model 2000 and 3000 series of machine vision systems provides advanced object location and recognition technology. These systems integrate the camera sensors (discussed in Sec. 6-4 and Example 6-6) with computer processing and decision-making circuitry that can be used for operator decision making or for automated action. Each of these vision systems consists of a hardware and software element; the developer chooses the vision processor and hardware

(a)

(b)

Figure 6-19.
Control loop applications. (a) Dual-axis mirror positioning system (courtesy of BEI Motion Systems Company). (b) Dual-axis laser beam positioner (courtesy of BEI Motion Systems Company). (c) Cognex machine vision system (courtesy of Cognex Corporation).

(c)

that best fit the application needs and applies the software to create custom application programs. A variety of software tools are provided to support the specific application program.

Typical functions performed by such a machine vision assembly might include alignment, guidance, inspection, and gauging. Alignment applications include parts positioning. The vision system determines the location of the part and transfers the information to the controller; the controller then drives the actuator to produce the desired operations. The guidance function can be used for guiding robots to pick up unfixtured parts. Inspection can locate features for position-based inspections and perform measurements of how closely an object matches a desired model. Gauging can be used to locate two or more points and calculate the distance or angle between these points to perform remote measurement.

All of these applications involve the integration of the sensor and a mechanical action through a control loop to achieve the desired performance. The Cognex products provide the hardware and software that is necessary to produce this control loop. These machine vision systems have a wide variety of tools to support automated operation. Included in these is an auto-focus tool that enables the developer to develop automatic camera focusing into the machine vision applications. The system is also designed to make use of a wide range of image-processing capabilities to enhance system performance.

The Cognex vision systems combine grey-scale image processing with artificial intelligence techniques to interpret images quickly and precisely. Thus, the controller used in the control loop is a sophisticated one that can make maximum use of the input from the camera sensor. The products are based on a custom integrated circuit chip that helps interpret images. This technology accommodates the design of faster and more compact vision systems.

Many other machine vision products are on the market and available for manufacturing applications.

The development of automated systems for manufacturing requires a careful determination of the technical features of the application and the selection of a preferred solution after evaluation of the range of available hardware. The software is used to integrate the system components. The necessary hardware is assembled, and the software is written to achieve the desired function. This is a sophisticated design activity that requires a team with both hardware and software skills, and a significant level of research and development activity to produce the desired product.

The limited use of automation in manufacturing is often a reflection of the cost associated with producing customized control loops and application features, rather than the result of limitations of component technologies. Automated control loops for robotics typically must be optimized for specific operations. The more broadly the operation is defined in order to achieve more generic manufacturing operations, the more sophisticated and complex the necessary controllers and control loops must become.

Original equipment manufacturers (OEMs) find that it is desirable to produce optimized equipment units for specific industry applications

The properties of digital control loops are determined by the sensor, actuator and effector hardware, the characteristics of the computer controller being used, and the software design. The integration of hardware and software requirements to produce the desired system equipment operating characteristics is a demanding task.

Given system complexities, technical uncertainties, and cost pressures, original equipment manufacturers often find it desirable to produce application-specific equipment units instead of more generic units. The constraints and opportunities in the field have encouraged the development of "islands of semiautomation and automation" rather than equipment optimized to systemwide application.

rather than to work toward more generic manufacturing units. This is one reason why the "islands of automation" approach to manufacturing has become so prevalent. It is sufficiently complex to achieve a highly automated single equipment unit; it is often not cost-effective to add a wide range of adaptability and application features that are required for highly integrated operation. Rather, the OEM often focuses on the specific task on hand, the available hardware, and the requirements of the specific controllers and control loops to produce a highly automated stand-alone unit. An additional level of design and investment is required to produce modular manufacturing equipment units that can be integrated in terms of information and materials handling flow.

6-7. Linking Equipment Performance and Costs

Figure 6-1 summarizes the interactions that take place between the physical equipment constraints of layer 1 and the functional equipment requirements that are developed in order to study layer 2. From one point of view, the equipment design task requires application of the methods established in Chap. 4 to define the system cost-benefit payoff in terms of equipment performance parameters. In this way, the equipment designer can define the most cost-effective types of parameter improvements. From a reciprocal point of view, the discussions of Chap. 6 provide an understanding of state-of-the-art equipment technology and the capability of the equipment manufacturers to produce equipment with desired parameter values. Thus, the equipment constraints and the equipment requirements interact closely. The design effort must be to match manufacturing equipment capabilities to system needs in an ongoing dynamic process.

The functional activity descriptions of layer 2 (discussed in Chap. 4) consider the equipment performance in terms of functional parameters. Limiting values of these parameters (from layer 1) are used to define constraints on the functional design. In turn, preferred values of these parameters are used to establish equipment design requirements (for layer 1).

An understanding of layer 1 requirements involves a review of related technologies and equipment design strategies. An understanding of equipment constraints and how to best accommodate the preferred parameter values is necessary. As discussed, this activity can be supported by various methods for mapping out technologies to indicate the ranges of performance parameters that are associated with each type of technology. In this way, an appreciation can be obtained for associations between the technology in use and the functional parameters that describe equipment performance. Automation typically involves sensors, actuators and effectors, controllers, and control loops, so mapping-out efforts must be particularly concerned with these technical areas.

The layer 1 studies must also provide a means for estimating the costs associated with various types of equipment. Each time a type of technology is selected, cost estimates must be prepared to enable cost-benefit studies to be performed. A foundation must be prepared for this cost-estimating activity.

As illustrated in Fig. 6-6, costs can generally be considered a function of the degree of automation at a given time. However, the curves change with time and technology. Therefore, system design must be associated with a baseline (study) year and a target (application) year. It is necessary that an organization continually review the costs of varying degrees of automation in order to determine whether decreased costs have shifted the optimum manufacturing configuration. Given today's rate of technology growth, there will be a continuing shift toward higher degrees of automation.

One of the major challenges facing a CIM planning and implementation team is to gain an understanding of the costs associated with obtaining each type of equipment that is under consideration. So long as the equipment to be used is available from the market, and is not to be modified in any way by the purchaser, the costing can be done directly by obtaining quotations from the various vendors. The cost versus performance levels of the equipment can then be evaluated.

However, during the design of the manufacturing system it may be desirable to consider design options based not only on existing equipment but also on more advanced types of equipment that might enhance overall system performance. If an enterprise is interested in selecting an optimum degree of automation and integration, it will likely not want to restrict its choices to equipment on the market today. Consideration will be given to equipment that might be available in the future (during the time period of interest) and the type of equipment that could be produced to customize operation.

A significant competitive advantage can accrue to a manufacturing system that makes use of equipment that is not commercially available to competitors. The manufacturing capability of the particular enterprise will have unique features that can be used to best address and compete in the marketplace. Thus, the design of a CIM system will often require an ability not only to gather information on existing equipment and its performance and costs but also to develop estimates of the types of upgraded equipment that could be manufactured for use in the enterprise. Performance parameters and costs must then be determined for this customized equipment.

One method for developing equipment cost estimates in this environment is indicated in Fig. 6-20. The costing exercise begins with a composite description of the equipment unit design and with a description of the design's functional parameters. The equipment unit can then be described in terms of subsystems, where often these subsystems can be conveniently described in terms of sensors, actuators/effectors, and controllers/control loops. A decision can then be made as to whether these various subsystems exist or must be customized. For the custom case, it must then be determined whether the company should arrange with a vendor to produce these subsystems or attempt to manufacture them internally. The "buy or make" choice must always be evaluated whenever customized equipment is being considered.

This process of disaggregation continues to the point where the

Equipment cost estimates change with time and technology. System design must be a continuous, learning-oriented process in order to assess ongoing changes in equipment that may lead to new system transition stages, operational improvement, and competitive advantage.

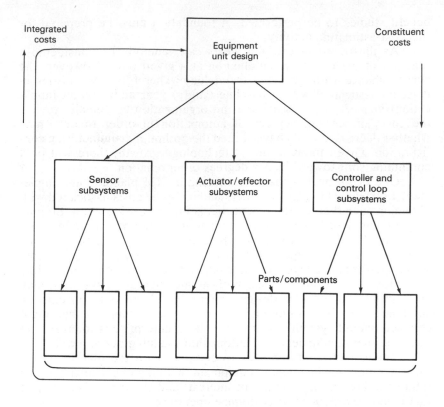

Figure 6-20.
Bottom-up method for developing equipment cost estimates.

designer has decided on the means of obtaining each subsystem or component for the desired unit of equipment, understands how the components will perform, and has estimated what the related costs will be. It is then necessary to estimate the costs of integrating all of the components and subsystems to produce an operating equipment unit. If a new subsystem is going to be incorporated into an upgraded equipment unit, it is essential to estimate the integration and assembly costs of the upgraded unit.

An alternative approach for costing is illustrated in Fig. 6-21. To perform an initial first-cut estimate of cost, it is sometimes possible to use *proxy*, or substitute, variables. These variables can provide an initial indication of the likely costs to be associated with the equipment. This approach is not as accurate as the method indicated above, but sometimes it is used for a first-cut effort.

Historical experience is used to show how equipment costs vary for a particular equipment function in terms of selected variables. The selected variables might usefully be those introduced in Chap. 4 and summarized earlier in this chapter (MTOI, MTI). Another approach is to develop sets of equations relating such parameters to equipment costs.

If such shortcuts are taken during initial planning, it is necessary to assess carefully whether or not qualitative changes in equipment prevent such curves or equations from being appropriate. At the same

It is sometimes helpful to perform first cut approximate costing studies before beginning detailed cost studies (that are based on component subsystem costs combined with integration costs). Accuracy and timing are ongoing issues for all costing strategies. Costing is particularly difficult when considering the advantages of custom equipment modifications. Despite these problems, effective costing methods are an essential aspect of system design.

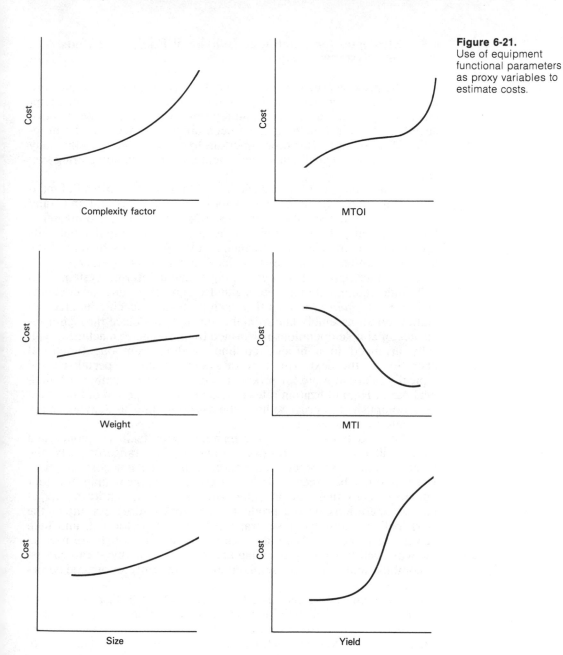

Figure 6-21.
Use of equipment
functional parameters
as proxy variables to
estimate costs.

time, as the detailed cost-benefit comparisons are made among different system designs, eventually it will be necessary to perform the type of detailed, top-to-bottom and integrated bottom-to-top analysis discussed above, in which the equipment is considered in terms of subsystems and components and then the integration and assembly issues are considered.

6-8. Strategies for Combining Individual Equipment Units into Systems

The previous sections of this chapter have provided a framework for the study of individual equipment units. The various types of equipment that are available for manufacturing systems have been introduced, and several methods have been applied to describe the properties of these units. It is now important to consider methods that may be used to combine individual equipment units into manufacturing systems.

As introduced in Chap. 3 (Sec. 3-7), an important aspect of manufacturing system design is how various types of equipment units (and process functions) should be grouped together to optimize manufacturing capability. In some applications, the location of individual units is determined solely by convenience and the physical structure of the factory. Another alternative is to place similar units together.

Another approach to developing a manufacturing system from individual equipment units is to make use of mass production concepts and *transfer equipment* units that each perform a specific, limited operation on an assembly line. Manufacturing takes place through a sequence of simple operations performed by the transfer machines, typically arranged in a production line, leading from one functional operation to the next. This type of system is highly specialized and typically is appropriate for production of very large numbers of identical parts. High utilization rates over an extended period of time must be accomplished in order to have the assembly-line strategy as an appropriate system design.

For situations in which the manufacturing facility is concerned with smaller volumes of production and a wider range of parts, the assembly-line/mass production approach is often not cost-effective. Here, it often becomes desirable to group equipment units together into *work cells* that can, in composite, perform an entire family of related operations on the product. The work-in-progress enters the work cell, remains while several functions are performed, and then leaves the work cell. The individual equipment units that are used in the work cell (for both processing and materials handling) can consist of combinations of manual, semiautomated, and fully automated equipment.

System design can often make effective use of work cells that achieve combinations of manufacturing functions. Selecting the preferred work cell organization for a specific manufacturing system is an essential aspect of system design in order to achieve optimum performance.

As discussed in Sec. 3-7, group technology (GT) provides one approach to work cell definition. The manufacturing units are grouped by function into clusters or work cells. The equipment units within each cell have a logical functional relationship to one another that enhances the overall manufacturing system. If a decision is made to seek ways to group equipment units for maximum effectiveness and efficiency, issues are raised regarding the most appropriate functions to group together and the most effective means for creating the desired relationships among the individual units.

Figure 6-22 shows one way in which the formation of work cells may be viewed. The horizontal axis describes the degree of automation associated with each equipment unit or function being considered, ranging from the simplest manual equipment, to semiautomated equip-

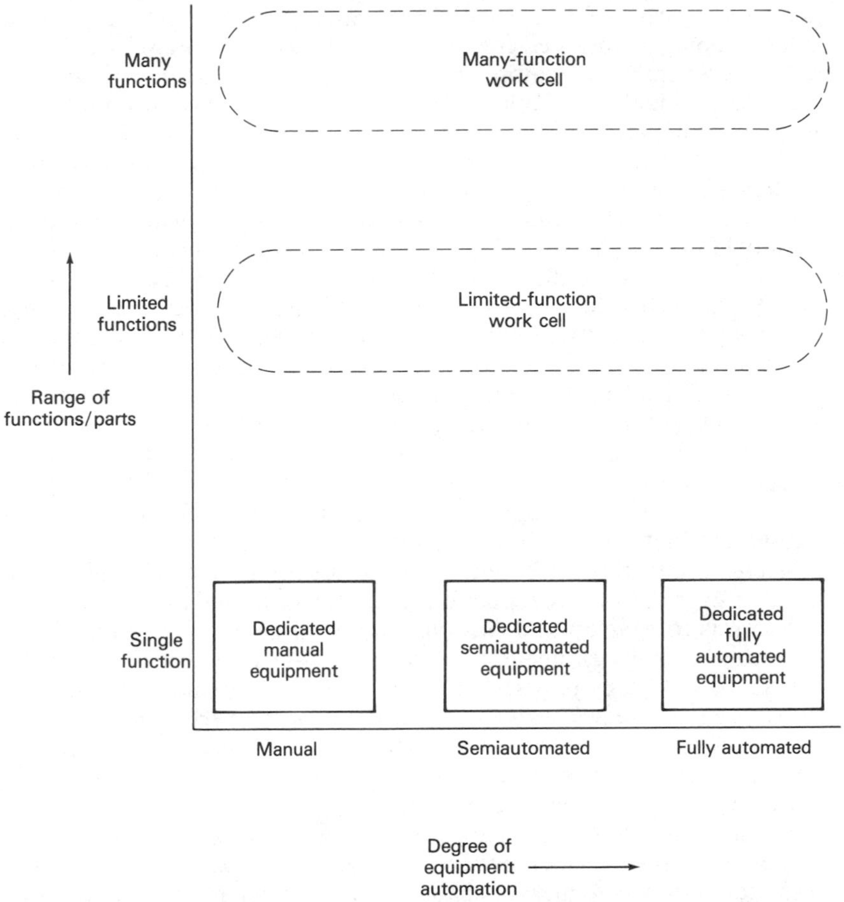

Figure 6-22.
Alternative manufac-
turing cell concepts.

Figure 6-22. Alternative manufacturing cell concepts.

ment which still requires operator support, to fully automated equipment with an MTOI sufficiently large to limit operator involvement.
The vertical axis shows the range of functions associated with an equipment cluster or work cell. The simplest level deals with single-function
equipment, while more complex levels deal with a limited set of functions and many-function work cells.

The concept of a flexible manufacturing system (FMS) is introduced in Sec. 3-7. A flexible manufacturing system is a type of work
cell in which the materials handling and processes involved have been
fully automated. Thus, an FMS consists of equipment and functions
that are linked together to achieve automatic and integrated operation.
In the more general sense, however, it is possible to develop a work
cell that is partly integrated and still requires some degree of operator
support. Such a system may sometimes be the most effective approach
to system integration. The progression shown in Fig. 6-22 provides an
overview of a wide range of manufacturing system strategies that may
be employed to group equipment units into work cells, with an FMS
as the most automated and integrated approach.

As shown in Fig. 6-23, different philosophies may be employed in developing a work cell for a given application. Strategy A calls for applying a number of equipment units, from manual to fully automated, and integrating these standard units. This typically involves modifying standard equipment units as necessary, adding in a materials handling system with the desired level of automation, and developing an integration strategy to tie the composite equipment units together. An alternative approach, strategy B, involves starting out with desired individual functions and developing an FMS that is optimized to perform this specific range of functions. For this case, it is necessary that the range of parts and the range of functions be precisely defined, and that it be found most cost-effective not to integrate separate equipment units but to produce a specialized product that has all the needed FMS properties. As might be expected, such an FMS would generally take longer to produce and would result in more specialized production than would strategy A, combining standardized equipment and functions into an integrated work cell.

Strategy C is a situation for which there are so many types of equipment units and such a wide range of product types that it is not feasible to integrate all of the units (or functions) into a single work cell, or even into an acceptable number of different work cells. One solution is to develop an FMS integration strategy that allows for the exchange of units as needed in order to modify the functions of the FMS. This strategy is perhaps at an opposite extreme from strategy B; whereas strategy B optimizes the FMS for the performance of specific functions, strategy C maximizes the flexibility of the FMS by allowing various units to be added to or removed from the system in order to achieve the desired operations.

In many settings, industry is evolving toward situations in which more customization is required for products to allow for a wider ability to match products to market changes and for much more rapid product evolution. Therefore, there is constant pressure to achieve the widest range of manufacturing capabilities at the lowest cost. If the volume of operations is sufficiently high, a large number of work cells can be developed to produce the full product range for the factory, or a transfer/assembly line might be appropriate. On the other hand, it may be more cost-effective to develop a work cell structure that can accommodate modular, exchangeable equipment units and thereby respond to market changes not by reconfiguring the entire factory but by adapting individual work cells (*Robotics World* 1987).

In order to follow strategy C, it is necessary to consider ways in which different types of manufacturing equipment can be used in a flexible, adaptable approach and to find integration strategies that will allow the use of exchangeable units. This particular approach to producing more flexible work cell assemblies is discussed further below.

The designer of a manufacturing system is thus faced with a range of alternatives to use in achieving equipment integration. Solutions can include the three strategies shown in Fig. 6-23, as well as other approaches. It is important that the designer give best consideration to all of the integration concepts in order to select the most cost-effective one.

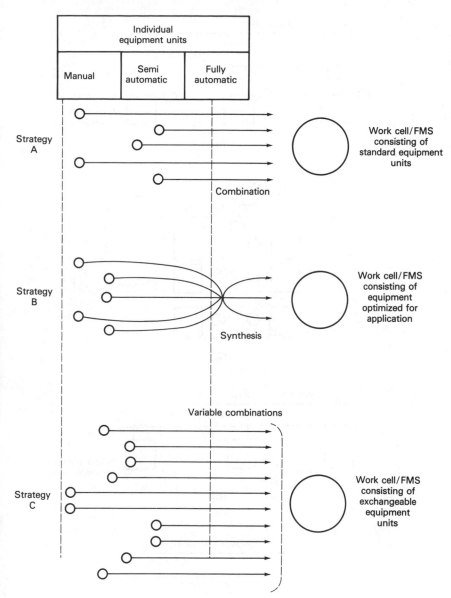

Figure 6-23.
Alternative work cell
configurations.

Figures 6-24a–e show some of the ways in which work cells can be formed. Figs. 6-24a and b show how an assembly line might be transformed into a work cell. Figures 6-24c and d show how an automated materials handling system may be applied. The operator is replaced by an automated materials handling system that emulates the operator. Figure 6-24e shows alternative FMS configurations. Figure 6-24f shows an FMS structure that makes use of exchangeable units around a core unit that provides the integration mechanism. (This strategy is discussed further in Chap. 10.)

Figure 6-24.
Alternative equipment configurations. (a) Assembly line. (b) Work cell with manually operated equipment units. (c) Extension of assembly-line concept for full automation. (d) FMS concept as work cell extension.
(e) Alternative FMS configurations. (f) Configuration with exchangeable equipment units.

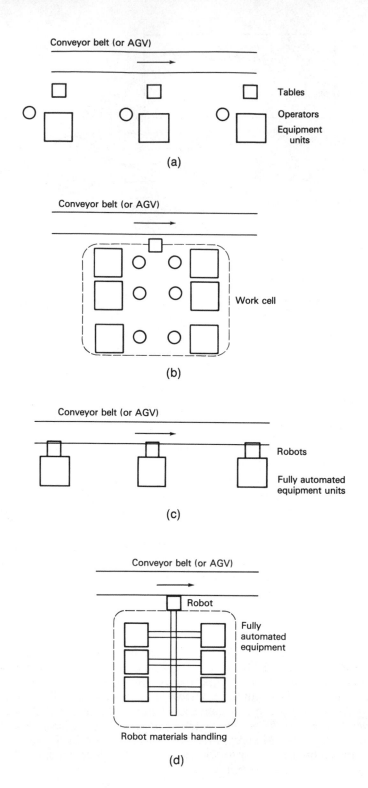

Conveyor belt (or AGV)

Tables

Operators

Equipment units

(a)

Conveyor belt (or AGV)

Work cell

(b)

Conveyor belt (or AGV)

Robots

Fully automated equipment units

(c)

Conveyor belt (or AGV)

Robot

Fully automated equipment

Robot materials handling

(d)

Figure 6-24. continued

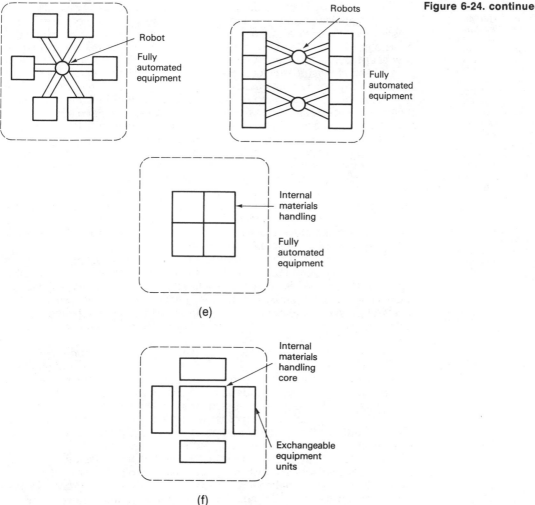

Robot

Fully automated equipment

Robots

Fully automated equipment

Internal materials handling

Fully automated equipment

(e)

Internal materials handling core

Exchangeable equipment units

(f)

EXAMPLE 6-9. MACHINING CENTERS, CELLS AND SYSTEMS

What are some of the opportunities and constraints associated with the application of advanced CIM concepts in the machining industry?

Results

(A) General Concepts: In this text, the term *work cell* is used to describe a grouping of equipment units. Work cells can make use of manually oriented equipment, semiautomated equipment, or fully automated equipment, and various levels of integration can be developed among the units. A work cell is thus a generic term associated with the clustering of equipment units to achieve the group technology objectives.

Machining centers, cells, and systems are essential strategies for the application of CIM concepts to metal-working manufacturing systems. Example 6-9 introduces general concepts associated with these manufacturing approaches, then discusses several specific products. The example links building-block machining centers and cells with a number of system configurations that can be implemented. A useful strategy is often to use a modular approach, expanding the system configuration as manufacturing and market conditions permit.

As quoted by Martin (1989), Wyznowski has developed a similar definition: "A manufacturing cell is the grouping of people and processes into a specific area dedicated to the production of a family of parts or products." Or, as amended by Scott (in the same reference), "A cell is the grouping of dissimilar types of machinery needed to manufacture a complete family of parts." In all of these definitions, the nature of the specific equipment or the integration strategy remains quite general.

A *flexible manufacturing cell* (FMC) is a small work cell that makes use of high degrees of automation and integration. A *flexible manufacturing system* (FMS) is a large or complex flexible manufacturing cell. In general, no "hard and fast" line exists between the FMC and FMS concepts.

The equipment to be employed in a work cell can be of many different kinds, ranging from manually oriented to fully automated. In machining applications, a wide range of equipment types are in use. A manually oriented machining work cell can be developed by co-locating a number of manufacturing equipment units in such a way that the operators can coordinate work among the units. At the more automated level, *machining centers* have become widely employed constituent equipment units in machining work cells (Gayman 1987; Wick 1987; Larin 1989; Martin 1989).

Typically, machining centers can range from small to large, from micromachining to macromachining applications, and can be treated as a building-block element for flexible manufacturing cells and systems. A typical machining center can undertake multiple machining functions, working on a variety of products to produce many different parts. The machining can be performed with many different geometric arrangements, depending on the exact nature of the machining center.

It is often difficult to achieve untended, fully automated operation for a machining center with a wide range of center flexibility. All too often these are competing characteristics. More fully automated equipment becomes fairly rigid and not sufficiently adaptable to changes in products and markets. On the other hand, a very flexible machining center may require more tended, less automated operation. A major issue in such centers is the potential for upward growth of the center (in terms of sensors, actuators/effectors, controllers/control loops, and communication/control options).

As noted in Chap. 4, a typical CIM system is continually evolving through a series of robust growth stages. If a large capital investment is made in a machining center that cannot be easily adapted to future growth stages, then major constraints are introduced to the manufacturing system. Such machining centers typically introduce trade-offs between automation and adaptability, and thus cannot be viewed as being able to perform all desired functions in the future. As the enterprise environment changes, the requirements placed on the machining center will also change. A number of solutions to this dilemma have been introduced into industry, resulting in machining centers of many different types that are matched to current enterprise needs and those needs that are anticipated for the reasonable future.

A machining center can form a building block for a machining work cell in which a number of different equipment units are integrated. Typically, an FMC or FMS will require automated materials handling and an integrated information flow among the various components of the machining work cell. In the same way that hardware restrictions will affect cell or system adaptability, integration will also affect center performance. Restrictions on cell-level design are often associated with an inability to conduct the needed design procedure, hardware restrictions, and (particularly) software and integration requirements.

Often the computer controller in a cell will actually be as rigid in its own way as the hardware. The development of software and integration capabilities for the cell may often be associated with the dominant portion of the total cost of the cell. It is difficult to produce the software that is necessary to "knit together" the equipment units into a machining work cell. At the same time, solutions are often quite rigid in terms of customization of the existing equipment and requirements of the cell at the time of design. Upward mobility of the cell level controller may be limited.

Thus, the objective of achieving flexible machining work cells that can grow with the enterprise is often frustrated by an inability to produce system designs and software that allow for a graceful integration growth. This is one of the major issues facing such cells today.

The ability to achieve flexible machining work cells that can evolve with the enterprise is often limited by the orientation of original equipment manufacturers toward more application-specific products. Hardware and software constraints incorporated into the building-block elements of the system often make it difficult for the system to evolve as new opportunities and insights develop.

(B) Kearney and Trecker Machining Centers: Figure 6-25a shows a Kearney and Trecker Corporation (KT)® Model 2300 ORION® horizontal machining center, designed for the production of medium to large-sized work pieces. The ORION family of machining centers handles a wide range of workpiece size and materials requirements. In Fig. 6-25a, the workstation is at the center; to the left is the automatic tool storage area, and in front are two load/unload stations. The work piece is mounted on a pallet and fixtured in place. The work-in-progress moves on the pallet from a load station to the worktable, where the machining operations are performed automatically using tools that are automatically accessed. When the machining operations are complete, the product is transferred to an unload station. This KT machining center is designed to be a flexible component for a wide range of CIM configurations; the unit may be used in a stand-alone mode or in integrated manufacturing system configurations.

This machining center emphasizes flexibility, with emphasis on ensuring that a wide range of products can be handled with high quality. The ORION series is designed with maximum rigidity to maintain high spatial consistency and high product repeatability. At the same time, the design must be for maximum flexibility to accommodate a variety of products and for rapid materials handling for throughput.

The expandable tool storage and automatic tool changing capability provide a high range of operational flexibility. Once preset tooling is selected, the modular design allows a choice of tool storage modules that hold 50, 100, or 150 tools, up to 150 mm (4.53 in.) in diameter, weighing up to 27 kg (60 lb), and with a length limit of 500 mm (19.6 in.). The automatic tool changer features an electronically controlled tool changing cycle to minimize nonproductive time. Tool search and partial tool change to the ready position are done during the machining cycle. Tool search is bidirectional, dependent on the shortest path, and tool change to the spindle takes only 6 seconds.

The tool gauging system detects worn or broken tooling and measures tools for offsets. This allows the machining center to monitor its own tooling. KT has developed a proprietary computerized numerical control (CNC) system; however, its ORION products are compatible with control systems from other vendors. The control system (to the right in Fig. 6-25a) provides a series of menus and prompts that lead the operator step by step through the desired procedures. The control module makes use of state-of-the-art electronics, with simple, replaceable printed circuit boards to simplify servicing, maintenance, and tailoring to specific needs.

The KT ORION machining centers are typical of the machining capabilities that are available today. Many different types of machining centers have been developed to allow for automatic product and tool handling with flexible computer control. Figure 6-25b indicates how such machining centers can be combined for application in realistic factory environments in order to achieve cost-effective manufacturing that is matched to enterprise requirements.

Figure 6-25.
Machining centers and cells. (a) Kearney and Trecker Corporation (KT) Model 2300 ORION horizontal machining center (courtesy of Kearney and Trecker Corporation). (b) Alternative factory configurations (courtesy of Kearney and Trecker Corporation). (c–f) Giddings & Lewis machining centers and installations as they have been implemented in factory settings (courtesy of Giddings & Lewis Corporation).

(a)

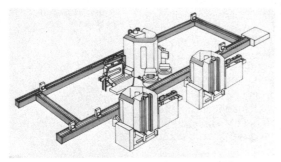

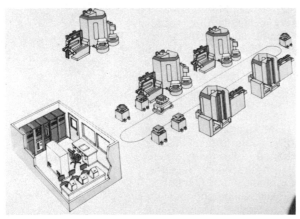

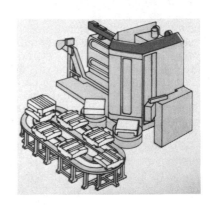

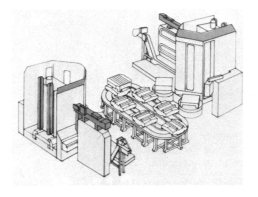

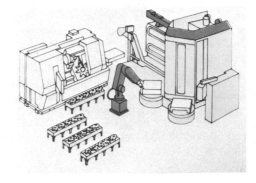

(b)

Figure 6-25. continued

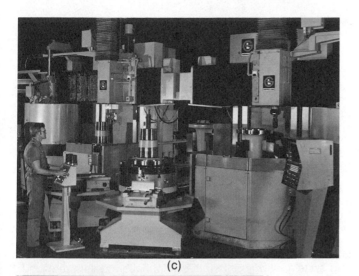

(c)

(d)

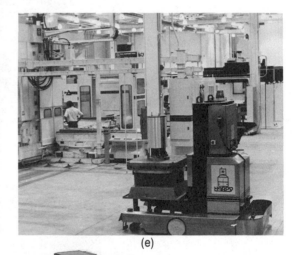

(e)

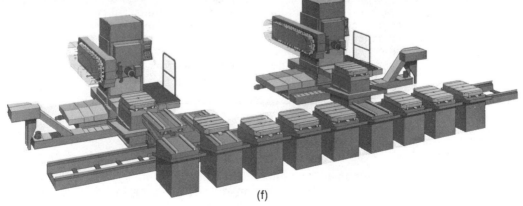

(f)

Manufacturing Equipment and Automation   **279**

KT has developed software that can be used to enhance control of machining centers by allowing many operator functions to be replaced with the decision-making capability resident in the control system. The software allows the use of self-monitoring features (linked to sensors) that can detect variations in stock and monitor a wide range of functions. The software also allows for computer networking so that high-level information exchange can take place between a specific machining center and other computers. The software is designed to adapt to a wide range of situations, including partial pallet loads and multiple fixtures, first and second operations not required on progressive fixtures, and other capabilities. Predetermined rules that establish allowable variances and tolerance limits are accomplished without operator intervention.

(C) Giddings & Lewis Flexible Manufacturing Systems: Figure 6-25c shows a Giddings & Lewis® (G&L) flexible manufacturing system that produces a family of six traction motor frames for locomotives. The work-in-progress is being moved among manufacturing stations by the tracked materials handling system. Several of the motor frames are shown in process at various workstations. In this system, nine equipment stations, an automated materials handling system, and an integrated information system replaced 29 stand-alone equipment units. Part completion was reduced from 16 days to 16 hours; employee productivity increased 240 percent with a 38 percent increase in production capacity, and required floor space was reduced by 24 percent. Improvements in parts quality and uniformity were also noted. This flexible manufacturing system illustrates how CIM concepts can be applied to the manufacturing of large, heavy products to achieve much more efficient and effective manufacturing systems.

Figure 6-25d shows a G&L vertical turning cellular system that features two side-by-side vertical turning centers with 12-station tool changers, 3 park stands, and a rotary work changer. This cell is used to produce a family of seven parts for tractors. A G&L computer controls and manages all operations. Figure 6-25e shows a cellular system that machines missile parts and features several different equipment operations, including horizontal machining, coordinate measuring, and vertical turning. Shown here is a wire guided vehicle for delivery of parts. The cellular system includes 12 pallets and 7 load/unload stations and is controlled by a G&L CM9000 cell manager. Figure 6-25f illustrates a G&L cellular system that consists of two palletized horizontal machining centers served by a computerized materials transporter on a runway. This type of illustration can be an important aid in considering alternative system designs and in understanding the physical nature of the configuration that will result.

G&L considers a wide range of cell configurations for any system under design. A phased program is often used for transition from simple to more complex system configurations. For example, initial phases may make use of a minimal number of operations and a simple materials transport system. The next phase may involve additional pallets to automate parts flow and optimize machining time. In phase three, other machining centers may be added to create cell-level efficiencies.

6-9. System Costs

In considering alternative manufacturing system designs, one of the most important issues involves estimating the cost of the capital equipment that will be required for full factory development. As shown below in Fig. 6-26a, capital costing must begin with an overview of the entire factory. Costs may then be divided into those associated with specific work cells and from there linked to individual equipment units.

For a manually oriented manufacturing system, it may be possible to combine the individual equipment unit costs (and the associated operator training) to gain a baseline idea of systemwide costs. For more automated manufacturing systems, additional major cost items are incurred with computer integration and related operator training. In a move toward more CIM-oriented manufacturing systems, it is necessary to achieve computer communication among work cells and throughout the factory and to develop integrated decision making and data collection. These tasks may become the driving consideration in system costs. (Chapter 7 provides a further discussion of computer network considerations, and Chap. 8 addresses some of the issues associated with operator performance.)

The method described (in Fig. 6-20) for costing a particular equipment unit can thus be extended to costing out work cells or a complete factory. In essence, the strategy is to divide the total system into smaller and smaller elements and to calculate the costs associated with each of the elements. Once the elements are defined, the integration process must begin. This effort can be both expensive and time-consuming. In many manufacturing system settings today, the costs of software development and system integration are greater than the hardware costs. At the same time, the integration and software requirements may drive system performance much more than do the specific technologies applied in the equipment units.

One approach to such a costing effort is to estimate the number of people, the skill levels, and the hours of activity that will be required to implement the system. In many settings, because there is a lack of familiarity with integration or software costs, these estimates may tend to be too low. The use of scaling models (introduced in Sec. 6-7) can be helpful as a means to check against the experiences of other settings and past personal experiences to ensure that cost estimates are reasonable. Such scaling models can also be used to perform first-cut estimates based on experience at the rapid trade-off stage.

Figures 6-20 and 6-26a illustrate some of the cost considerations that must enter into the design trade-off. At the equipment level, subsystems must be combined to achieve the desired manufacturing processes and the desired level of functional performance, and integration must be obtained among the subsystems. Equipment design for highly automated settings will require computer controllers and computer communications systems that will enable the equipment to be integrated into work cells.

A work cell can be composed of a number of individual equipment units or can be a custom-designed and integrated unit that does not build on commercially available component equipment units. The choice between whether to assemble several existing equipment units (perhaps with minor modification) into a work cell or develop new work cell equipment that integrates the functions of several commercial equipment units is an option to be evaluated during system design. At

Similar methods can be used to develop cost estimates for both equipment and systems. Detailed approaches require subdividing the equipment or system into subsystems and components, developing costs for these elements, and estimating integration costs. In many advanced manufacturing settings, integration costs may drive the final equipment or system cost estimates. Approximate, heuristic costing efforts based on proxy variables can sometimes be a helpful cross-check in such settings.

the cell level, costs will be experienced with respect to integration hardware and software. In general, the hardware will be technology limited, the system design will be limited by system understanding and the design methodologies available, and system integration will be limited by the ability to develop the needed software. In combination, the restrictions experienced at the equipment and cell levels can be a major cost driver.

At the factory level, additional cost consideration must be given to computer aided design and product design linkages to the manufacturing system (as discussed in Chap. 9). In addition, system management and command and control functions must be evaluated carefully.

The costing of the capital equipment for an advanced manufacturing system can be a difficult task. It is necessary not only to consider each element of the system but also to evaluate ways in which the elements can be combined to produce the desired operating whole. The costs associated with system design and integration, particularly the software design and development, can be the driving cost item in such a setting. There must be an active concern not only with the technologies that are used but also the ways in which these technologies will be combined to produce the desired system function.

The equipment and work cells that are available for factory design can cover a wide range of configurations, as shown in Fig. 6-26b. At one extreme is equipment with rigid, manually driven effectors that operate on a fixed product. The operator's sensor and control capabilities are used to monitor the equipment. In such a typical manual setting, the human operator becomes the means for system integration, and no extensive software requirements exist. At another extreme are highly automated work cells with on-line flexible equipment that is able to produce a wide range of product types without hardware change. Such systems have the potential for the development of entirely new factory concepts. If a work cell can produce a wide range of operations without hardware change, then a fully automated building-block unit can be combined for the factory. However, as noted above, the integration and software issues associated with linking several such cells together in a factory are major issues that must be considered in addition to the internal system design issues associated with the cell.

EXAMPLE 6-10.   CAPITAL COSTING

Figure 6-26a illustrates one way of viewing a capital costing cycle. Based on this cycle, Figs. 6-26c and 6-26d provide data that might be associated with two different settings. Discuss the significance of these results.

Results

Figure 6-26c illustrates a manually oriented factory for which the capital costing cycle is dominated by the individual equipment unit costs. The factory is divided into work cells and then into equipment units; the factory makes use of stand-alone units with manual operations. The constituent costs can

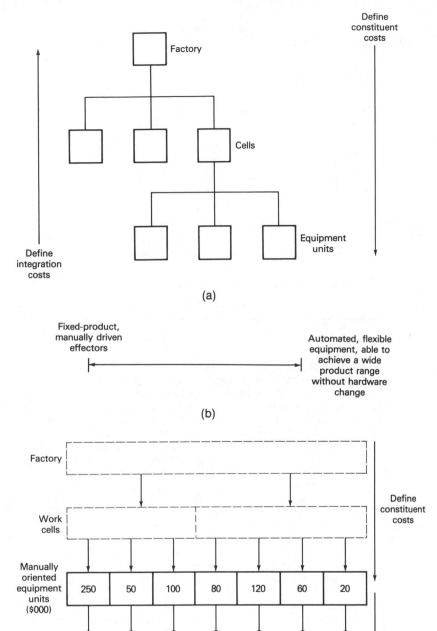

Define
constituent
costs

Factory

Cells

Equipment
units

Define
integration
costs

(a)

Figure 6-26.
System costing. (a)
Capital costing cycle
involving the definition
of constituent and in-
tegration costs. (b)
Equipment and work
cells that are available
for factory design can
cover a wide range of
configurations. (c)
Cost analysis for a
manually oriented fac-
tory. (d) Cost analysis
for a highly automated
factory.

Fixed-product,
manually driven
effectors

Automated, flexible
equipment, able to
achieve a wide
product range
without hardware
change

(b)

Factory

Work
cells

Define
constituent
costs

Manually
oriented
equipment
units
($000)

| 250 | 50 | 100 | 80 | 120 | 60 | 20 |

Work
cells
($000)

| 400 | 280 |

Define
integration
costs

Factory
($000)

| 680 |

(c)

Figure 6-26. continued

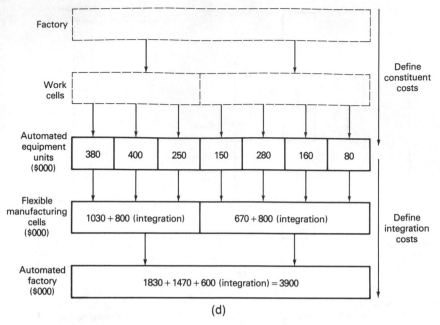

(d)

be determined by deciding on the nature of the work cells and equipment units that are required for factory operation and combining these costs as shown. Integration costs are assumed to be small, so the total cost can be estimated by combining the component costs. This is a classical costing approach that is familiar to many factory designers. Equipment units are viewed as associated with stand-alone costs, and the cost-effectiveness of each equipment unit is determined by a figure of merit associated with that individual equipment unit rather than by systemwide impact. Not surprisingly, this approach often results in a noncompetitive manufacturing system.

Figure 6-26d shows a capital costing cycle associated with highly automated equipment and with constituent costs that are higher than those in Fig. 6-26c. For the fully automated system of Fig. 6-26d, the costs are substantially higher for each equipment unit and the cost of integrating the cells increases significantly. Driving costs are associated with the integration of the equipment into cells and the integration of the cells into the total factory. The software costs of such a highly automated, integrated system can well exceed the equipment hardware costs. This is particularly true when some of the equipment units have to be modified for integration into the system.

The analyses shown in Fig. 6-26c and 6-26d produce different costing structures. The system in Fig. 6-26d is much more expensive than that in Fig. 6-26c. At issue is whether or not the effectiveness of the system is sufficiently increased for the additional cost to be warranted. The type of integrated system in Fig. 6-26d may be necessary in order to take advantage of highly automated equipment units. An effort to introduce "islands of automation" into the manual system of Fig. 6-26c may not produce cost-effective performance because of the misbalances that exist in the system.

It may be difficult to develop evolutionary phases to lead from the manual to the more automated system versions, unless realistic expectations are developed with respect to the transition stages. It may be that in moving from manual to automated systems, only a limited improvement in effectiveness, or even a drop in effectiveness, will be observed until a certain critical level of automation is obtained.

6-10. Manufacture of Electronics Assemblies

In studying manufacturing equipment and automation, it is helpful both to consider the characteristics of manufacturing equipment in general and to focus on particular types of manufacturing in order to provide a more targeted focus for understanding. In this text, the manufacturing sectors selected for emphasis are associated with machine tools and the production of small electronics assemblies (refer to the discussion in Sec. 4-9). Example 6-9 employs machine tools and machining cells that provide advanced manufacturing capabilities. The following discussion relates to a number of processes associated with electronics manufacturing.

The discussion so far has emphasized machine tools for metal-working applications. The remainder of this chapter addresses the manufacture of small electronics assemblies. Many different processes are experienced in this manufacturing environment, leading to an opportunity to address a wide range of equipment types and families.

Electronics-related products give rise to one of the largest industries in the United States (National Academy of Sciences 1984; Institute of Electrical and Electronics Engineers 1989). It is of essential importance to the United States to maintain a strong manufacturing capability in the electronics sector as it relates to computers, commercial products, state-of-the-art aircraft, space satellites, and numerous other applications.

No specific knowledge of electronics or electrical engineering is required to understand the basic nature of the related manufacturing processes. The following discussion introduces some of these processes and indicates the limiting factors in their application to the selected product area. The focus on the generic characteristics of these manufacturing processes illustrates many aspects of manufacturing equipment for more general consideration.

One of the most important manufacturing issues related to electronics assembly involves the techniques to be used for the production and assembly of building-block subsystems for electronics products. Typically, the development of these products involves the manufacture of a circuit board or mounting surface of some type, the development of the necessary interconnections between components, the mounting of components into place, and the final packaging.

Figure 6-27a shows the dominant electronics subsystem options in use today. The building-block components for modern electronics are typically the monolithic integrated circuits (or IC "chips") with hundreds of thousands of individual devices manufactured on the surface of a silicon wafer. The specific manufacturing processes associated with these monolithic integrated circuits are specialized and beyond the scope of this text.

The method of circuit fabrication will often determine the size, weight, performance, and cost of the circuit. The most widely used, standardized method involves the use of "through-hole" printed circuit boards (Fig. 6-27b). This type of printed circuit board consists of one or several layers of plastic (epoxy), with copper conducting patterns

Figure 6-27.
Electronics manufacturing. (a) Illustration of the dominant electronics manufacturing processes in use today. (b) Development of a through-hole printed circuit board as a method of electronic circuit integration. An individual IC component is inserted into a plastic package and the plastic package is then attached to the printed circuit board. (c) For hybrid microcircuits, the individual components are assembled on a ceramic surface or substrate; the substrate is then enclosed in a metal, ceramic, or plastic package.

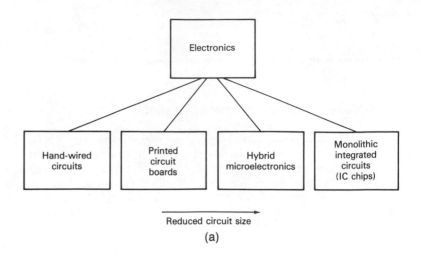

(a)

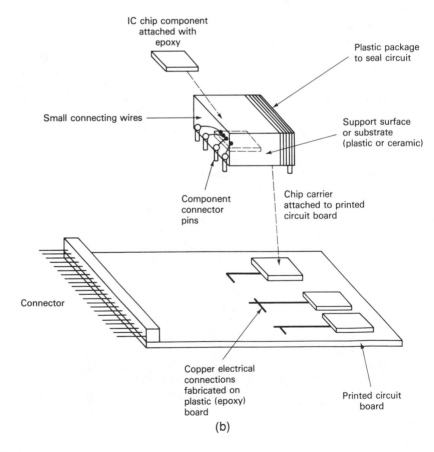

(b)

Figure 6-27. continued

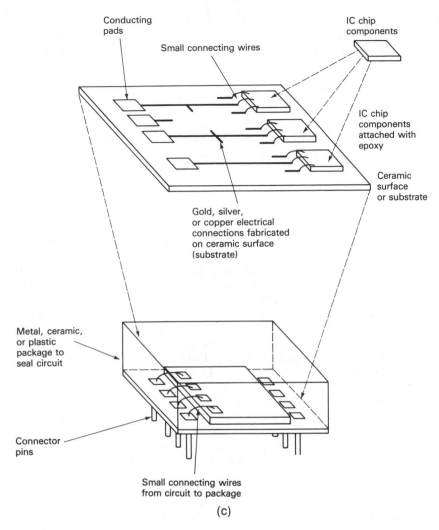

Conducting pads

Small connecting wires

IC chip components

IC chip components attached with epoxy

Ceramic surface or substrate

Gold, silver, or copper electrical connections fabricated on ceramic surface (substrate)

Metal, ceramic, or plastic package to seal circuit

Connector pins

Small connecting wires from circuit to package

(c)

produced on each layer to provide for interconnection between components, and holes drilled in the boards for the insertion of components. The components are individually packaged in metal or plastic containers with wire pins extending from the container for electrical connection. The pins on these packages are inserted into the holes prepared in the printed circuit board. The printed circuit board is then passed over a hot liquid solder bath, which attaches the component pins to the copper circuit patterns on the board. The result is a standard electronics product that is low in cost and can achieve high performance. Almost all components used today are available in a form that can be used for through-hole printed circuit board manufacture. The printed circuit board in turn becomes a building block for a larger circuit.

Surface-mount printed circuit boards represent an evolutionary step from the through-hole printed circuit board. Each individual component is still in its own package, but the metal pins on these packages are replaced by small metal tabs. The printed circuit board no longer has holes drilled through it, but has matching copper tabs prepared on its surface. A "sticky" liquid solder is applied to these pads, and the components are placed so that the package and board tabs are held together by the solder. The circuit is then heated; solder reflow bonds the surface-mount components to the board.

The move toward surface mount is significant, and commercial manufacturing capabilities are available. Typically, a shift from through-hole to surface-mount printed circuit boards can achieve a reduction of 50 percent in circuit size and weight for the designer. Many of the competitive leading-edge products are making use of surface-mount capability. Fewer components are available for surface mount, and surface-mount board costs tend to be higher than those for through-hole.

Another fabrication option makes use of *hybrid microelectronics* (discussed in detail in Sec. 6-11). Instead of individually placing each component in its own package and then combining these packages on a printed circuit board, the strategy is to work with the bare components and to attach these bare components directly to a material layer (often called a *substrate*) on which the necessary conducting patterns have been prepared (Fig. 6-27c). A common method for doing this is to use thick and thin film technology to create conducting patterns on a ceramic substrate and then to attach (with epoxy) a variety of bare components directly on the surface. The connections between the components and the conducting patterns on the ceramic are made by using wire bonders to connect very small wires (typically, $\frac{1}{1000}$ inch or 1 mil in diameter) from the components to the conducting layers.

It is possible to use the hybrid process to create portions of a system or to modify an entire system to make use only of hybrid technology. Typically, this approach achieves an additional factor of 5 reduction in circuit size and weight below those values associated with surface-mount technology.

6-11.   Hybrid Microelectronics Emphasis

Because of the diversity of processes involved, the many families of equipment required, and the diversity of the equipment families now available, hybrid microelectronics is a useful area of focus for the study of manufacturing equipment, automation, and integration.

In this text, particular focus is placed on the manufacture of hybrid microelectronics products. As will be illustrated, this selection provides many insights into the state-of-the-art manufacturing issues facing companies today. Because of the small size of the products being produced, the manufacture of hybrid microelectronics products (or simply hybrids) makes state-of-the-art demands on equipment. The handling and assembly of very small components results in high-value-added products that are essential to competitiveness. The manufacture of hybrids involves a wide variety of processes and materials and a state-of-the-art understanding as to how to transfer materials process research to the manufacturing setting.*

* Further background on hybrids is provided in Lizari and Enlow (1988), Tummala and Rymaszewski (1988), and *Computer Design* (1989).

It has been estimated that in the early 1990s, about $14 billion worth of hybrid circuits will be produced worldwide annually, with about $10 billion of these produced in the United States (*Hybrid Circuit Technology* 1986; Bader 1989). Hybrid microcircuits can be the building-block components for many types of electronic circuits and systems, so leadership in this area can carry over into other related electronics markets.

The manufacture of hybrids involves machine vision systems, robots for assembly, the joining together of many different materials and processes, industrial uses of lasers, bar coding for products, sophisticated performance testing, and many other representative areas that are reflected throughout industry at large. Significant effort is being devoted to increased levels of automation for these processes and applications (Pirocanac 1986; Bader 1989). For example, one of the initial sites selected by International Business Machines for application of its corporate factory computer network strategy involves the manufacture of hybrid microelectronic products. The application of advanced manufacturing concepts to this product area is serving as a learning site for many other industries. This IBM-supported CIM development program, conducted by Teledyne Microelectronics, is discussed in detail in Example 7-8.

New applications in the hybrid manufacturing area are being implemented in a variety of settings. Integrated circuit technology has reached the state where packaging is a major determinant of performance. The ability to combine creatively many different technologies to produce electronic systems that are able to utilize the potential of ICs is another example of this emphasis on hybrid technology.

The semiconductor industry has focused attention on the intense nature of international competition involved in the manufacture of IC chips; much less attention has been paid to the manufacture of microelectronic circuits and systems using these building-block chip components. Yet competitiveness and product performance are related to the combining of components into systems and producing a complete product. Hybrid fabrication can result in a smaller, denser, more attractive product, but the use of hybrids is limited today by the manufacturing methods available.

6-12.   Printed Circuit Manufacturing

Because of the similarities in manufacturing equipment used for various types of electronics production, it is useful to provide a brief introduction to the equipment used to fabricate printed circuits. Emphasis can then shift to hybrid manufacturing.

Until recently, most of the printed circuit boards produced in the United States have used chip carriers and other components with pins sticking out of the case. In order to create a printed circuit board, components are inserted into the board so that the pins stick through the board itself and make contact with conducting pads formed on the board. The pins are then attached to the pads through a solder process;

the pins and conducting pads on the board are passed over a solder wave so that the solder bonds together each pin and pad.

As discussed above, an alternative printed circuit board strategy has been developed during more recent years. This approach applies surface-mount technology. Instead of using pins that extend through the printed circuit boards, small conducting flaps or tabs are produced on the bottom of each component and attached to the surface of printed circuit board. Solder is printed onto the surface of a printed circuit board. While the solder is still in a liquid/adhesive state, the surface-mount components are placed on the desired pads. The entire unit then passes through a solder reflow or heating process; the solder flows and bonds the conducting flaps to the conducting pads on the printed circuit board. The primary advantage of the surface-mount strategy is to increase circuit density and therefore reduce the size of the overall system. Surface-mount technology represents a transition stage between the traditional pin-through printed circuit board technology and the multichip hybrids that use bare IC chips.

A variety of methods thus can be used to attach individual electronics components to a substrate and then integrate the substrate into an entire electronic system. Many similarities exist among these techniques, and the equipment used for one process can often be modified for use in another. Equipment to achieve pin-through attachment of components to printed circuit boards has evolved to equipment that provides surface-mount capability. In turn, the surface-mount printed circuit board capability in many ways is closely related to hybrid manufacturing (discussed in Sec. 6-13).

EXAMPLE 6-11. PRINTED CIRCUIT BOARDS

Describe the types of equipment that are available to manufacture printed circuit boards.

Results

(A) Universal Instruments Corp. There is a growing need in today's electronics industry for flexible, retoolable automated assembly equipment. In response to this need, Universal Instruments Corporation has developed a flexible assembly work cell. Figure 6-28a provides a view of the Vari-Cell™ II Model 6515 flexible assembly work cell, and Fig. 6-28b shows an assembly operation in progress. The Vari-Cell II Model 6515 can handle a wide range of printed circuit assembly operations. It can be configured in a stand-alone mode, or as part of an in-line or on-line system to handle both standard and nonstandard components.

The system includes a sophisticated machine vision system and advanced robotic technology. The work cell is configured with one standard arm-mounted camera, with provision for up to seven cameras to be added. The computer controller directs all of the Vari-Cell II's functions, including the fully integrated vision system, all component feeders, and an active lead clinch system for use in through-hole component applications.

In order to enhance product quality, the Vari-Cell II computer can detect missing components at the pick-up nozzle and can also use the vision system

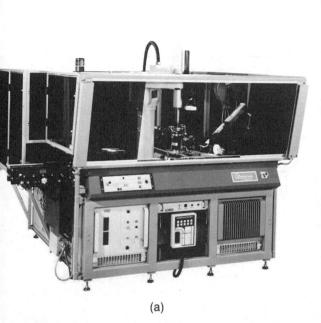

(a)

(b)

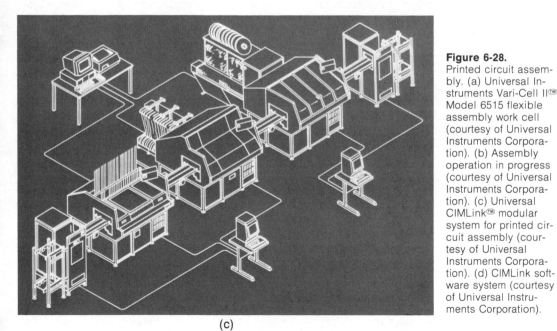

(c)

Figure 6-28.
Printed circuit assembly. (a) Universal Instruments Vari-Cell II™ Model 6515 flexible assembly work cell (courtesy of Universal Instruments Corporation). (b) Assembly operation in progress (courtesy of Universal Instruments Corporation). (c) Universal CIMLink™ modular system for printed circuit assembly (courtesy of Universal Instruments Corporation). (d) CIMLink software system (courtesy of Universal Instruments Corporation).

Figure 6-28. continued

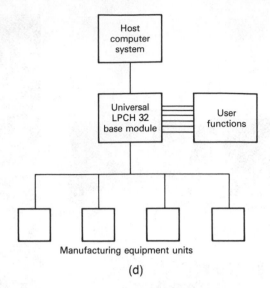

Manufacturing equipment units

(d)

to inspect and orient components. A tactile sensor is used to measure the force/placement pressure. The system can also be programmed to retry an assembly step when the first attempt is not successful.

Universal® is concerned with producing printed circuit manufacturing systems that are highly automated and integrated, and is an advocate of CIM strategies. Universal offers CIMLink™ modular system networking software that can be used to integrate a number of equipment units, as illustrated in Fig. 6-28c. The CIMLink software can be used to implement a CIM system in steps or phases and to achieve a learning and testing experience. The CIM-Link software product provides a well-rounded solution to smaller production needs, while at the same time providing a growth path for larger facilities.

The elements of the CIMLink software system are shown in Fig. 6-28d. As noted here, the LPCH 32 CIMLink base module is used to coordinate a number of stations. The CIMLink software includes a management information reporting utility (MIRU) module, bar-code module for tracking work-in-process, a repair cell controller module used to prepare error reports, and a CAD module that enables the manufacturing system to achieve direct download of data from the CAD system to the factory. The LPCH 32 base module can be linked directly to a wide range of manufacturing equipment, including the flexible assembly work cell.

The LPCH 32 base module provides the software architecture that allows the user to transfer data between a host computer and the various assembly processes. This module allows an operator to monitor and control either single machines or up to 20 machines. Insertion or placement programs can be downloaded and uploaded from the host to the assembly units.

The CIMLink system networking software is one of the products now on the market that enables a manufacturer to combine equipment units into highly integrated operations. Another software solution is discussed in Example 7-8.

Universal offers to analyze totally the manufacturing processes required by a particular enterprise and to help the enterprise develop a manufacturing system to meet production requirements. The system teams offer to supply both the software and hardware that will be necessary to achieve a high level of integrated operation. Printed circuit boards are tracked from their entry into the system, and throughout the processes, by making use of bar-code labeling and scanning.

Assembly stations are available in modular form to allow boards to be populated in the most efficient sequence. Fully populated boards exit the system by way of magazines or cassettes. The system is designed to allow expansion as production volume increases or component mix varies. Any equipment units that are not in operation due to maintenance can be automatically bypassed.

(B) Amistar Corp. Amistar Corporation also provides highly automated flexible assembly of printed circuit boards. A wide range of modular equipment units may be combined for printed circuit assembly. Figure 6-29a shows Amistar's automated flexible assembler. This equipment unit can be used to populate circuit boards with through-hole, surface-mount, and odd-shaped components. The system can be used with a variety of feeders for different components, and the vision system allows for self-taught programming and pattern recognition.

The flexible assembler consists of two independent work cells working in parallel. Boards are transported through the flexible assembler by an adjustable-width, in-line board mover that connects to other in-line machines either directly or through conveyor belts. Up to four flexible assemblers (eight work cells) may be connected and managed from one controller to achieve high throughput. A work cell is committed to either surface-mount or through-hole operations by selecting the appropriate tooling kit.

The Amistar flexible assembler shown in Fig. 6-29a represents a manufacturing strategy that is oriented toward full information integration and materials handling automation. The emphasis on developing flexibility in the system is necessary in order to achieve cost-effective operations for a wide range of product types.

Figure 6-29.
Printed circuit assembly. (a) Amistar Corporation automated flexible assembler for printed circuit boards (courtesy of Amistar Corporation). (b) Amistar Model AI-6448 automatic axial lead inserter (courtesy of Amistar Corporation).

(a)

(b)

Figure 6-29b shows the Model AI-6448 automatic axial lead inserter that can be used as a stand-alone system to sequence and insert components in one operation. As may be seen here, the result is a complex equipment unit that can automatically select components from a wide range of feed sources, correctly populate the board with these components, and then automatically transfer the board to a new storage area in preparation for solder reflow. The entire unit is microprocessor controlled. The microprocessor both controls operation and maintains surveillance over machine operation. Communications are also permitted with an external computer to both transmit and receive insertion programs.

Printed circuit board manufacturing capability has extended to the point where very highly integrated and automated manufacturing solutions are available. When such a wide range of capabilities are made available to manufacturers, it may be concluded that such systems are cost-effective in a wide range of applications.

An important equipment development strategy in many manufacturing settings is to modify available equipment to conduct a process that is related to but distinct from the initial equipment performance. An overview of the problems and opportunities experienced in efforts to modify printed circuit equipment for hybrid manufacturing can provide insight into such efforts.

As discussed, hybrid manufacturing is in a transition phase. The equipment discussed below ranges from more manually oriented to completely automated equipment. The equipment described here for printed circuit board manufacturing in many cases can be modified for use in hybrid manufacturing. On the other hand, unique requirements for hybrid manufacturing often result in the most cost-effective solutions being those that make use of equipment that is only partly automated and still requires some degree of operator support. Thus, where the printed circuit board manufacturing equipment often represents extensive use of automation and integration, the hybrid manufacturing area represents a transition in which both operator- and computer-controlled equipment are often combined to achieve the most cost-effective manufacturing capability.

6-13. Hybrid Manufacturing

Figure 6-30 describes a set of manufacturing processes that can be used to create hybrids. A wide variety of processes can be used in such cases to create different types of hybrids; the particular operations shown here have been selected as representative and suitable to illustrate the CIM concepts being developed.

Also illustrated in Fig. 6-30 are the vendor-supplied materials required for each hybrid, including the substrate or mounting surface to be used, the thick film pastes to be applied to the substrate, the components to be assembled onto the substrate, and the package that will contain the final substrate. After completion of the manufacturing processes, it is typical to conduct leak tests to determine that a hermetic seal has been obtained during packaging, electrical tests to ensure performance, and various environmental tests to ensure the stability of circuit performance under different conditions.

Once the hybrid product is complete, it can be integrated into an electronic system. This will require another assembly operation, with the hybrid now treated as a supercomponent consisting of many sub-components. The hybrid is often assembled on a printed circuit board, as discussed earlier.

A variety of means are used to package the final hybrid circuit. As shown in Fig. 6-31a, conducting pads can be produced on the surface of the substrate; when manufacturing is complete, the substrate may be inserted directly into a connector to achieve a pressure contact. Figures 6-31b and 6-31c show alternative methods by which connecting pins can be attached to the substrate through use of a solder reflow operation. Solder is deposited on the surface of the substrate, on top of conducting pads, and connecting pins are placed in contact with the solder. The pins are held to the substrate through a pressure fit (Fig. 6-31b) or by inserting each pin through a hole manufactured in the substrate (Fig. 6-31c). The solder is then heated and the reflow creates a bond between the pin and the substrate.

Figures 6-31d–e show an alternative method of packaging, in which the substrate is attached by epoxy to the interior of a package that has already been manufactured with connecting pins. A wire bonder (discussed below) can then be used to form connections between the conducting pads on the substrate and the pins that extend through the package.

These and other types of packaging place different requirements

Figure 6-30.
Typical manufacturing processes for thick film hybrid microelectronics.

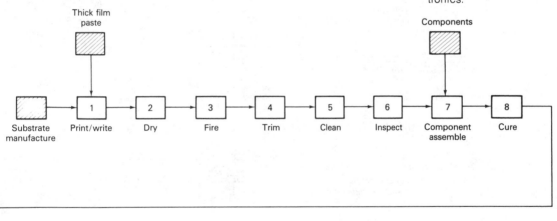

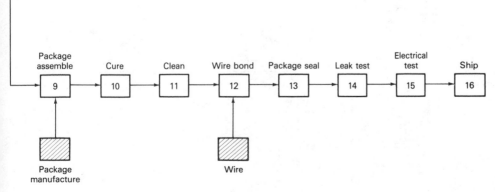

Figure 6-31.
Alternative methods for attaching a hybrid to the surrounding circuit. (a) Pressure-fit connector. (b) Pin attached by solder reflow. (c) Alternative method for pin attached by solder reflow. (d–e) Placement of the substrate into a preformed package, wire bonding from the substrate to the connecting pins, sealing the hybrid package with a lid, and then soldering the pins to the printed circuit board. (f) An alternative method for attaching the hybrid package to a surface-mount printed circuit board.

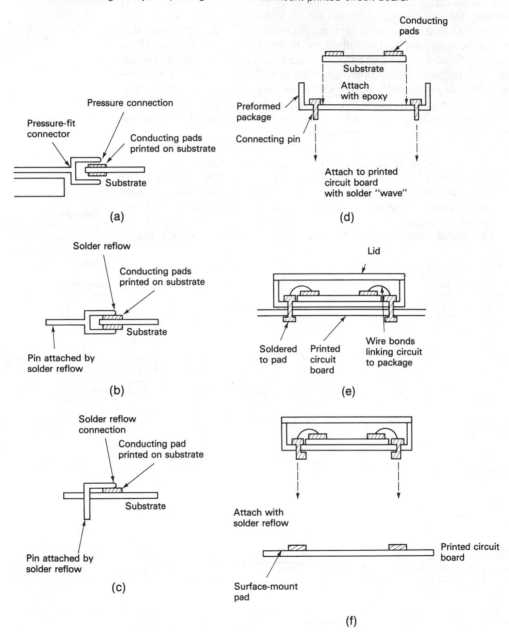

on the manufacturing process. Depending on the integration method, packaging operations may occur at different steps during the manufacturing process and require different types of equipment. For the hybrid products discussed below, the connecting mechanism shown in Fig. 6-31d to 6-31f is used. It is assumed that all hybrids will be packaged into vendor-supplied packaging units (made of ceramic or metal) and that the wire bonding will be performed to attach the hybrid to the package pins. The resulting unit can be treated as a single electronic component and attached to a printed circuit board or inserted in a socket.

The hybrid manufacturing strategies discussed below may be viewed from two points of view. One, the hybrid manufacturing process provides a means for combining many different semiconductor components and creating a complete microelectronic circuit. The hybrid thus provides a method of integration and packaging. Alternatively, the resulting hybrid can then be viewed as a supercomponent, which becomes a building block to create the types of printed circuit boards desired. The advantage of the hybrid is that by combining many components in an unpackaged stage, a much higher average density can be achieved for the circuits.

Attachment of the hybrid to a printed circuit board is only one method of application. Numerous other methods exist that do not involve the attachment of the hybrid component to a printed circuit board. The following discussion emphasizes the manufacture of hybrids to be used as supercomponents for printed circuit boards. This defines the particular packaging needs that are useful and appropriate. Emphasis throughout the following discussion is on how equipment technology may be combined in overview during system design. The manufacturing processes and equipment units also support the system design techniques presented in Chap. 10.

The particular processes discussed below are typical ones that can be used to create what are called *thick film hybrids*. The concepts introduced here apply to all types of hybrid manufacturing and in many ways apply to all efforts to manufacture products that require assembly and packaging of components that must relate to one another. Many different types of processes and equipment can be employed in the manufacture of thick film hybrid microelectronic products. The processes that have been selected in Fig. 6-30 are typical ones that may be associated with a wide range of equipment functions and degrees of automation and integration. Product flow generally leads sequentially from one function to the next. In some applications, repeat loops must be made through several of the processing steps.

Equipment for each process function is discussed. For each manufacturing function, several types of equipment, associated with varying degrees of automation and integration, can be used to describe an equipment *family* (Fig. 6-32). A complete equipment family can begin with operator-intensive procedures that make use of manual tools, proceed to automated processes under local control (islands of automation), advance to automated processes and equipment that can make use of computer communications capabilities to relate to a work cell or the larger factory environment, and finally conclude with fully au-

Figure 6-32 illustrates families of equipment for a range of manufacturing processes. It is often helpful to describe the manufacturing system under design in terms of the families of equipment that can be considered for use. Entries into a chart (such as the one in Fig. 6-32) can be backed up by detailed catalogs, quotations, and technical materials from vendors.

tomated equipment for which the processes in use do not require operator support. Fully automated equipment allows computer communications for two-way information flow, and the materials handling associated with the equipment is fully automated (through the use of appropriate robot capability).

6-14.   Printing and Writing Operations

The first step in producing the hybrid is to create a pattern of interconnections on the substrate (mounting surface) to which IC chips and other components will be attached. Interconnecting "wires" are manufactured on the substrate by making a liquid paste with the desired conducting properties and printing this paste on the substrate in the

Figure 6-32.
Equipment families for thick film hybrid manufacturing.

Degree of Automation and Integration	*1* Screen Printer (SP)/Direct Write Unit (DWU)	*2* Drying Oven	*3* Firing Furnace	*4* Laser Trimmer
	Equipment Function			
Manual process (with use of manual tools)	SP: Manual control over print process, manual product insert and removal	Manual control over temperature and timing, manual product insert and removal		Manual control over laser, manual product insert and removal
Automated process	SP/DWU: Automated print/ dispense, under local control, manual product insert and removal	Automated local control over temperature and conveyor belt	Automated zone furnace, local control over temperature and conveyor belt	Automated laser functions under local control, manual product insert and removal
Automated process plus computer communications capability	DWU: Remote computer control over dispense pattern, data upload	Remote computer control over temperatures and conveyor belt, data upload	Remote computer control over temperature and conveyor belt, data upload	Remote computer control over laser functions, data upload
Fully automated (process, communications, and materials handling)	DWU: Remote computer control over dispense pattern, automated product insert and removal	Remote computer control over temperature and conveyor belt, automated product insert and removal	Remote computer control over temperature and conveyor belt, automated product insert and removal	Remote computer control over laser functions, automated product insert and removal

desired pattern. The same process can also be used to "print" selected components directly on the substrate (see process 1 in Fig. 6-32).

Several methods may be used to apply the paste to each substrate in the desired pattern. For high-volume production runs, a *screen printer* may be most efficient. In the silk-screen process, layers of material are deposited on each substrate through a fine mesh stainless steel screen. To create the desired master pattern on the screen, a layer of photographic emulsion is bonded to the screen, then exposed to ultraviolet light through a photoreduction of the circuit pattern (or *topography*) that is desired. The emulsion hardens when exposed to ultraviolet light and remains soft elsewhere. The unexposed areas of the emulsion are then washed away, leaving the desired pattern.

This *photolithographic* process produces a screen master that can be used to print the desired material pattern on the substrate. Each

5	6	7	8
Equipment Function			
Cleaning Unit	Inspection Unit	Component Assembly Unit	Curing Oven
Manual cleaning	Manual Review through microscope	Manual placement of epoxy and components on substrate	Similar to drying oven (process 2)
Automated local control of cleaning operations and conveyor belt	Machine vision and pattern recognition under local control	Automated placement of epoxy and components under local control, manual substrate insert and removal	
Remote computer control over temperature and conveyor belt, data upload	Remote computer access for reference patterns, data upload	Remote computer control over placement of epoxy and components, data upload	
Remote computer control over cleaning operations and conveyor belt, automated product insert and removal	Remote computer access for inspection, automated product insert and removal	Remote computer control over pick and place operations, automated product and component feed, automated product removal	

Figure 6-32. continued

Degree of Automation and Integration	9	10	11	12
	Equipment Function			
	Package Assembly Unit	Curing Oven	Cleaning Unit	Wire Bonder
Manual process (with use of manual tools)	Manual placement of epoxy and substrate on floor of package	Similar to process 8	Similar to process 5	Manual location of bonds
Automated process	Automated placement of epoxy and substrate under local control, manual insert and removal of substrate and package			Automated location of bonds under local control, manual product insert and removal
Automated process plus computer communications capability	Remote control over placement of epoxy and substrate, data upload			Remote computer control over location of bonds, data upload
Fully automated (process, communications, and materials handling)	Remote computer control over assembly, automated substrate and package feed, automated product removal			Remote computer control over location of bonds, automated product insert and removal

Variations on the screen printing process are used in many manufacturing settings to paint special materials onto surfaces with high accuracy. The equipment available for this purpose ranges from manual to sophisticated automated and integrated types.

screening cycle uses a paste that is specifically designed to achieve a desired circuit function and a screen master that can apply the paste to selected areas of the substrate. The silk-screening process (which is actually a very old manufacturing process, widely used in many other fields from the printing of T-shirts to the printing of fine painting detail on mechanical products) is applied to "paint" special pastes onto the substrate and create interconnections between the components that are going to be assembled. Other pastes can be used to create selected components directly on the substrate, using alternative screen patterns.

Another strategy is to use what is called a *direct write unit* (DWU); the paste is loaded into a hypodermic-like device and applied through a very small nozzle to create the desired patterns. As might be expected, the DWU is much slower than the screen printer and is most

13	14	15	16
Equipment Function			
Package Sealing Unit	*Package Test Unit*	*Electrical Test Unit*	*Package for Shipping Unit*
Manual control over sealing process	Manual leak test station	Manual probe station, manual test equipment	Manual wrap and box seal
Automated local control over sealing process	Automated leak test station under local control	Automated probe station under local control, automated test equipment cycling	Automated wrap and box seal under local control
Remote computer control over sealing process, data upload	Remote computer control over leak test, data upload	Remote computer control over probe station and test equipment cycling, data upload	Remote computer control over wrap and box seal
Remote computer control over sealing, automated product insert and removal	Remote computer control over leak test, automated product insert and removal	Remote computer control over location of probes, automated product insert and removal	Remote computer control over wrap and box seal, automated product and box handling

appropriate for reduced production levels. On the other hand, the DWU is highly adaptable and allows for software control to achieve alternative products.

Column 1 in Fig. 6-32 summarizes an equipment family associated with this function. A manual screen printer provides operator control over the printing of pastes and manual product insert and removal. An automated process can be achieved with either a screen printer or direct write unit, with automated print or dispense under local control and with manual product insert and removal.

The next level of automation and integration involves the use of a direct write unit that allows remote computer control over the dispense pattern. (This option is not available to the screen printer because the photolithographic process does not allow remote computer control over the pattern in use.) Finally, a fully automated system (shown as a direct write unit) allows remote computer control over the dispense pattern combined with automated product insert and removal. Those equipment versions with computer communications ability will allow both downloading and uploading of information through computer networks (as discussed in Chap. 10).

In many cases, the fully automated versions of equipment shown in Fig. 6-32 are not commercially available (as of the writing of this text). These fully automated versions would require additional research and development.

EXAMPLE 6-12. SCREEN PRINTERS AND DIRECT WRITE UNITS

Discuss some of the technologies for depositing thick film materials on hybrid substrates using screen printers and direct write units.

Results

Figures 6-33a and 6-33b show two very different screen printer technologies that are available from Affiliated Manufacturers Inc. (AMI) for hybrid manufacturing. The model 462 thick film screen printer shown in Fig. 6-33a is a low-cost, operator-controlled system. Printers of this type have been available for the last 20 years, and the particular version shown in the figure reflects much of the learning that has taken place regarding how to make such operator-oriented machines. The mechanical operations have been simplified over the years to achieve maximum convenience for operator use and operational effectiveness. This particular model is intended for small-scale production in the laboratory or for quality control testing. This simple bench model includes many features that have not been available in earlier years for the most complex screen printers, and thus illustrates the continuing learning process that can take place for basic manufacturing equipment.

Figure 6-33b illustrates an AMI® Model 885 screen printer design that makes much increased use of automation. This model has also been available for many years, and reflects continuous upgrading with state-of-the-art improvements. Designed and constructed to achieve a high-quality, reproducible product, thousands of these units are in the field today. This unit can be combined with a variety of options to tailor production to individual facility requirements. The printer can achieve fully automated substrate handling by adding the appropriate modules and tooling. The system is also available with an automatic machine vision capability to check the printing screen alignment during pattern changeovers. Using the pattern recognition capability, the screen may be automatically positioned for printing. The machine vision capability can also be used as a quality control tool for checking the pattern-to-substrate alignment.

Figure 6-33c shows how this model can be integrated with automated drying and materials handling hardware to develop an automated flow of substrates through the printing and drying processes. The integrated systems offered by AMI are based on a modular concept in which various components can be combined to satisfy specific needs (Fig. 6-33d).

A variety of screen printers and configurations is available. The principal design requirement is to decide which version will be most cost-effec-

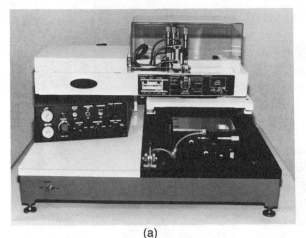

(a)

Figure 6-33.
Thick film printing and writing. (a) AMI/Presco® Model 462 thick film screen printer (courtesy of Affiliated Manufacturers, Inc.). (b) AMI/Presco Model 885 thick film screen printer (courtesy of Affiliated Manufacturers, Inc.). (c) Integrated screen printing and drying system with fully automated materials handling (courtesy of Affiliated Manufacturers, Inc.). (d) Typical system process flow (courtesy of Affiliated Manufacturers, Inc.). (e) Micropen Model 400 direct writing and dispensing system to apply thick film materials to hybrid substrates (courtesy of Micropen Corporation). (f) Operations flowchart for the direct writing system (courtesy of MicroPen Corporation).

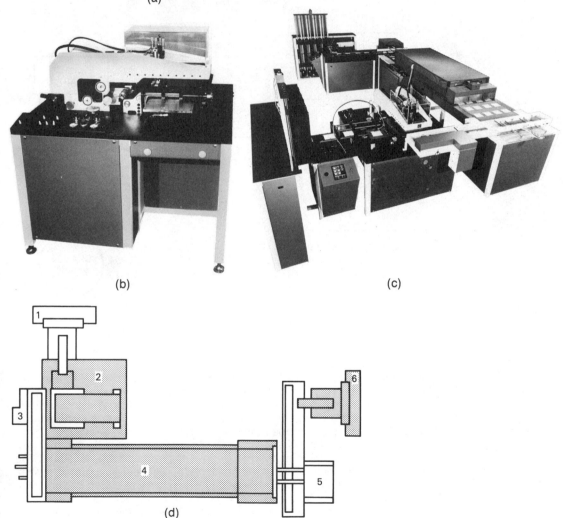

(b) (c)

(d)

Figure 6-33. continued

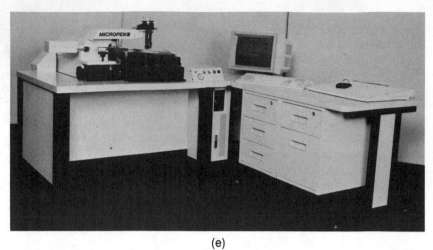

(e)

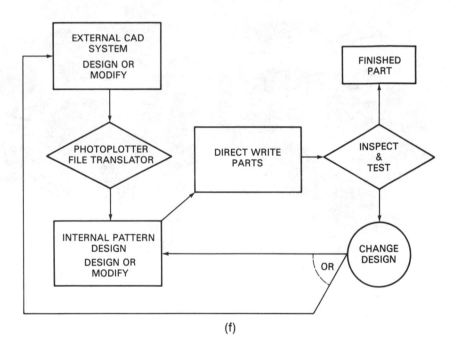

(f)

tive. For high-volume production runs of a single substrate design, screen printers can be very efficient. On the other hand, owing to the use of photolithographic processes, screen printers are not able to accommodate software control of the pattern being printed. Thus, a printer cannot achieve automated data file download to shift from one product to another.

Figure 6-33e shows a Micropen® Model 400 direct write unit that provides a fully adaptable method to deposit thick films on a substrate. The Micropen Unit functions as a self-contained, completely integrated system for the design, redesign, and manufacture of thick film hybrid circuits. The operations flowchart for the system is shown in Fig. 6-33f.

This direct write unit can achieve precision deposition of any fluid material. The system shown in Fig. 6-33e contains an internal design system that can be used to develop the desired substrate pattern; the system also can be linked to an external CAD system, which allows data to be transferred from a computer-aided design center to the direct write unit. Data exchange between this unit and external CAD systems can be made through a variety of communications networks, as discussed in Chap. 7. The mechanical aspects of the Micropen unit are under computer control at all times. Each of the writing and dispensing variables may be individually programmed.

This direct write unit may be contrasted with the automated screen printing system in a number of ways. Where the automated screen system is optimized for high-volume production of a specific pattern, using automated material handling, the Micropen model is designed for hand-loaded and unloaded substrates, with complete computer control of the direct write unit to allow rapid pattern changeover. The direct write unit can thus be integrated into a computer-based information handling system to change the printed pattern. The screen printer shown in Fig. 6-33a to 6-33c does not allow for this type of integration. Thus, a trade-off exists between the two systems.

The direct write process has the potential for integration with an automated materials handling system to create a product that can achieve both functions. At issue is how to combine the relatively low speed of the direct write unit with a high-speed production setting in order to maximize both throughput and automation. Such a high-speed, fully automated direct write unit represents an extension of today's state of the art.

If it is desired to individually record a serial number on each substrate during thick film application, then a means must be available through the printing or writing system for providing such a control number. In the case of the direct write unit, this labeling can be achieved through software that changes the code number from substrate to substrate. For the screen printer, an additional mechanism must be added in order to provide a serial code on each substrate.

6-15. Drying Operations

Once the liquid paste has been applied to the substrate by using either a screen printer or direct write unit, the paste must be dried in order to remove the solvents that had been added (to achieve a printable liquid state). Refer to process 2 in Fig. 6-32. Typically, the substrate is allowed to dry in a controlled-temperature environment that ensures removal of all of the resins in the paste. The residue will be a dry printed layer that consists of finely ground-up metal and glass particles.

One common approach to achieving the drying process is to use a simple drying oven with manual control over the temperature and the timing of the drying process, combined with manual product insert and removal. A more automated process involves local automation control over the temperature of the oven, combined with a conveyor belt to move the substrate through the heated area at a controlled rate. The next member of the equipment family might allow remote computer control over the oven temperature and the conveyor belt speed, and would allow computer download of control data combined with upload (or data logging) of information regarding the processes performed. A fully automated drying oven would allow remote computer control over

Heat is required for many different industry processes. The ovens used here for hybrid manufacturing can also be used for baking, dehydration, annealing, conditioning, evaporating, curing, preheating, and other applications.

the temperature and the conveyor belt, combined with automated product insert and removal.

EXAMPLE 6-13. DRYING OVENS

Discuss the technologies that can be used to dry the thick film once it is applied to a substrate.

Results

Figure 6-34 shows a manually loaded and unloaded, low-cost Grieve Corporation oven that can be used for drying the thick film. This oven has digital temperature controls and is intended to achieve a uniform temperature for closely controlled production and laboratory work. This model is available with a forced exhaust, to ensure that the flammable solvents and vapors that are baked out of the paste will be removed (to satisfy safety and quality requirements). This type of oven is available in models that achieve temperatures up to 500°F, 650°F, or 850°F. The type of unit shown here can be effectively utilized by an operator who inserts substrates (perhaps mounted in a substrate holder or rack) into the oven and then sets the timer for the desired drying time. This is a manual operation that can require extensive written logs to document when each substrate enters the oven and when it is removed.

Figure 6-33c shows an AMI conveyor dryer system that can be used in conjunction with automated materials handling. This product can be inte-

Figure 6-34.
Grieve Corporation gas-fired Model HA-500 cabinet oven (courtesy of the Grieve Corporation).

grated with a thick film screen printer, as shown. AMI's dryer unit is designed for short drying times, uniform penetration of heat, and repeatable drying properties. Both free-standing or integrated units are available. The dryer makes use of infrared heating units to provide the desired performance. Controls are set by an operator; a closed-loop control system maintains the desired temperature and the speed at which the product moves through the oven.

The two oven configurations discussed here provide different methods for drying. The former is operator-oriented, requiring manual documentation and control, whereas the latter is designed for automated product handling and data logging. The preferred choice will depend on the system design.

6-16. Firing Operations

Once the paste has been applied to the substrate and dried, the next manufacturing step is to heat the paste to a high temperature so it will have the desired electrical characteristics. This is typically done by using a *firing furnace*. The substrates are placed on a conveyor belt that slowly moves them through the desired temperature profile. They are slowly heated to a high temperature and then allowed to cool. This process melts and bonds the paste to the substrate so that a secure interconnection pattern is developed. Many different types of paste can be sequentially applied to the substrate. Some of these pastes must have very precise electrical characteristics.

Figure 6-32 (process 3) describes several versions of firing furnaces. No manual processes are indicated, as the simplest firing furnaces in use today involve a significant degree of process automation. An automated zone furnace (with sequences of heated zones used to control the temperature profile) allows local control over the temperature and the speed of the conveyor belt.

Conveyor-belt zone furnaces provide a carefully controlled temperature profile. These furnaces are representative of a larger range of equipment products that move work-in-progress through controlled heating zones to achieve the desired temperature profiles.

If computer communications capability is added, then remote computer control can be achieved over the temperature and the conveyor belt, and data logging can take place to confirm equipment operation. A fully automated firing furnace allows remote computer control over the temperature and the conveyor belt combined with automated product insert into and removal from the furnace.

EXAMPLE 6-14. FIRING FURNACES

Discuss the types of firing furnaces that can be used to process the thick film after it has been dried on the substrate.

Results

Figures 6-35a and 6-35b show two state-of-the-art thick film firing furnaces that are available today. Shown in Fig. 6-35a is a Watkins-Johnson unit that is designed to be a production-oriented, cost-effective conveyor furnace. The zoned furnace is designed for precise thermal processing of thick film

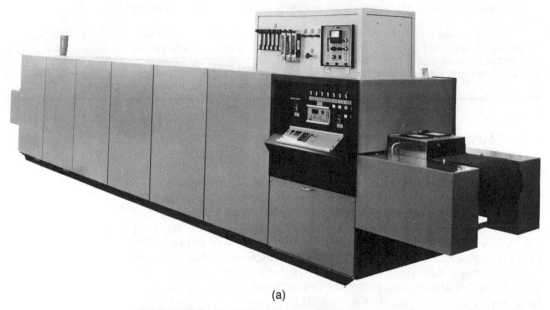

(a)

Figure 6-35.
Thick film furnaces. (a)
Watkins-Johnson thick
film firing furnace
(courtesy of Watkins-
Johnson Company).
(b) Lindberg thick film
firing furnace (cour-
tesy of Lindberg, a
unit of General Sig-
nal).

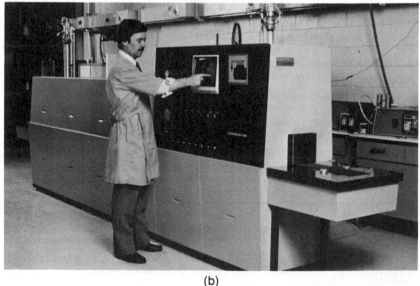

(b)

materials. The multiple zones are used to control the temperature profile
through which the substrates pass, and each zone is controlled by a microprocessor to ensure the stability of the temperature profile. The furnace also
provides control over airflow to ensure exhaust of any by-products. Such furnaces are available in a variety of belt widths (from 6 to 24 inches and up),

with a variety of control zones (typically five to eight) and heated lengths (73 to over 120 inches). The furnace also includes communications capabilities. The system can be monitored or controlled by an operator at the furnace, or the furnace can be monitored from a remote location by using the computer communications capabilities. Extensive data logging takes place to monitor the continuing operation of the furnace.

The Lindberg® furnaces, such as model 816 in Fig. 6-35b, provide another state-of-the-art product. The furnace includes a control system that makes use of a touch-sensitive monitor screen (shown in use). The system is designed to enhance the interactions between an operator and the furnace. System commands are entered by means of the system's touch-sensitive screen, and programming is achieved through self-prompting steps. The Lindberg furnace also includes a data software package that allows a personal computer to communicate with the furnace control system. The furnace controller is designed to interface through a communications link to a personal computer, which may then control the system and provide data-logging capabilities. Detailed information on the user interface with the software is provided by Lindberg.

Both of these products are moving toward the ability to achieve information integration into a CIM system. However, neither furnace is specifically designed to achieve automated materials handling for product flow. Thus, fully automated material integration will require the application of interface units to move the substrate from the drying area into the furnace, and then remove the substrate from the furnace on completion of the firing process. Current design calls for an operator to perform these functions.

6-17. Laser Trimming

Following the firing process, it is often necessary to use a *laser trimmer* to etch away very small parts of the fired pastes so that they will have the desired electrical characteristics.

As shown in Fig. 6-32 (process 4), a manually oriented laser trimmer involves step-by-step operator control over laser processes, combined with manual product insert into and removal from the trimmer unit. An automated process provides for local automated control over the laser functions, combined with manual product insert and removal. Again, an island of automation is created.

The next family member involves the addition of remote computer control over the laser functions and data upload for monitoring. A fully automated laser trimmer would involve remote computer control over the laser functions combined with automated product insert and removal.

EXAMPLE 6-15. LASER TRIMMERS

Lasers are widely used today on the factory floor. Discuss the types of laser trimming equipment units that are available for use in the manufacture of hybrid products.

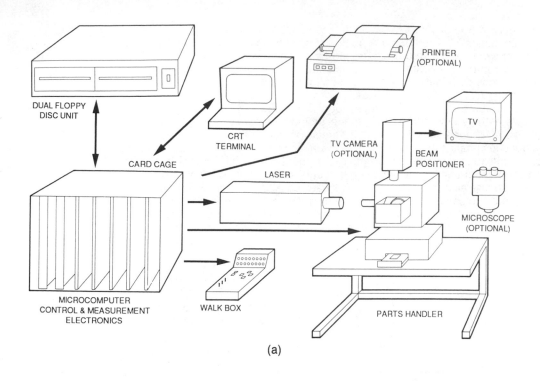

DUAL FLOPPY
DISC UNIT

CRT
TERMINAL

CARD CAGE

LASER

PRINTER
(OPTIONAL)

TV CAMERA
(OPTIONAL)

BEAM
POSITIONER

TV

MICROSCOPE
(OPTIONAL)

MICROCOMPUTER
CONTROL & MEASUREMENT
ELECTRONICS

WALK BOX

PARTS HANDLER

(a)

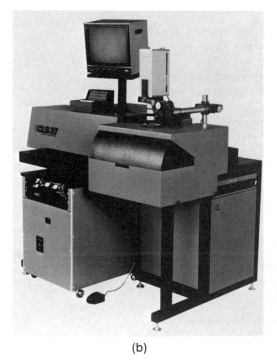

(b)

Figure 6-36.
Laser trimmer systems. (a) Laser trimmer system flowchart
(courtesy of Chicago Laser Systems, Inc.). (b) Chicago Laser
Systems Model CLS-37S laser trim system (courtesy of Chicago
Laser Systems, Inc.). (c) ESI Model 4000A laser trimming sys-
tem (courtesy of Electro Scientific Industries, Inc.).

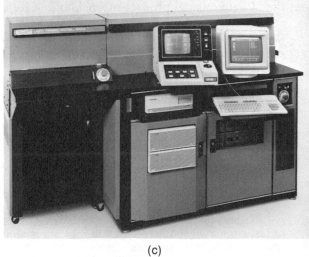

(c)

Results

Figures 6-36a and 6-36b show a laser trim system provided by Chicago Laser Systems (CLS) Inc. Figure 6-36a illustrates the functions of the system, and Fig. 6-36b is a photograph of model CLS-37S. As seen in Fig. 6-36a, the laser trim system consists of a laser, a mechanism to move the laser beam, a control system, and a monitoring system to determine when the trimming is to be terminated. Viewing and parts handling systems may also be included. The type of trim system shown in Fig. 6-36b is computer controlled. The system is designed to trim electronic circuitry to precise parameters. CLS emphasizes its design strategy and customer support. On-site training is available worldwide, and extensive in-depth documentation is provided with all systems. The CLS-37S model shown in Fig. 6-36b emphasizes the high level of throughput that can be obtained and the variety of possible applications. Software control is provided over many previously hardware-only functions. All circuitry is fully modularized for ease of servicing. The system utilizes a custom computer that was designed and built by CLS for the particular application. The CLS laser trimmer thus represents a product for which the hardware and software have been optimized for maximum performance for the specific task at hand. A computer interface allows data logging and remote control by special commands integrated into the CLS operating system.

Figure 6-36c shows an Electro Scientific Industries (ESI), Inc., Model 4000A laser trimming system designed specifically for use with thick film hybrids. Laser beam positioning and measurement subsystems have been designed to achieve the desired product operations. A local area (computer) network (LAN) option allows the system to share operating software, control, and data files. With the LAN, links can be made to CAD stations, and data logging can take place on a continuing basis. The trimmer makes use of a YAG laser with peak-to-peak stability that is better than 5 percent. Automated beam alignment is an optional part of the system. The entire laser and optics assembly is in a single package that slides easily out of the way for free access to the probing area. The work surface is monitored through two closed-circuit television cameras during operations. One camera can be manually zoomed to view the entire process area; the other has a fixed field of view and is used to control the trim position.

The ESI Model 4000A shown in Fig. 6-36c can be ordered with a cassette autoloader option. This allows automated substrate loading and unloading from special cassettes that can be ordered for specific substrate sizes. The software package that is used to control the system is provided by ESI, with easy-to-learn, menu-driven languages. The ESI laser trimming system thus provides highly automated operations with the capability of being combined into an integrated information system. A degree of automated material handling has been provided through the introduction of cassettes. However, in order to achieve continuous, fully automated materials handling throughput, modifications would have to be made to the unit.

Laser trimming systems have become highly automated and more accessible for information integration. The typical equipment unit of today makes use of an on-site operator, but future evolutions might easily emphasize remote operation. The materials handling capabilities allow a degree of automatic feed, but for true automated materials handling, significant developmental changes will have to be implemented.

The laser, which is a very "high-tech" product itself, has thus become a reliable manufacturing tool. The introduction of computer-controlled systems has allowed reliable, high-throughput laser operations. The type of capability demonstrated here for hybrid manufacturing has also been applied to a variety of other manufacturing settings.

6-18. Cleaning

A cleaning process is required at this point to ensure that the substrate and interconnections are sufficiently clean to begin circuit assembly. Manual, ultrasonic, or plasma cleaning can be used for this purpose. Manual cleaning can involve the use of a liquid solvent bath, with insert and removal of the hybrid from the cleaning bath performed by the operator. Typically, manual cleaning involves the use of a *wet bench,* with environmental controls over the solvents used during the cleaning to ensure the operator's safety.

A more automated process allows for automated local control over the cleaning operations, with a conveyor belt to move the hybrid through the cleaning area. The operator no longer has to handle the hybrid during all phases of the cleaning. If computer communications capabilities are added to the cleaning unit, remote computer control is obtained over the temperature and conveyor belt. Fully automated equipment allows remote computer control over the cleaning operations, including the conveyor belt, combined with automatic product placement on and removal from the conveyor belt.

EXAMPLE 6-16. CLEANING UNITS

Discuss some of the processes used to clean electronic circuits and systems.

Results

Cobehn® equipment uses a cleaning process that sprays ultrapure solvents onto the substrate while a continuous flow of filtered warm air dries the device. The solvent removes both organic and inorganic contaminants and provides a suitable surface to continue production of the hybrid. The Cobehn System Model 3000 shown in Fig. 6-37a provides a computer-programmable, operator-free cleaning process. A set of four adjustable spray nozzles applies the solvent to the hybrid (Fig. 6-37b). The hybrid is then dried with warm nitrogen while the next component automatically indexes for processing. The menu-driven software permits selection of cleaning and drying time cycles for each sprayer selected. In order to achieve fully automated operation, ad-

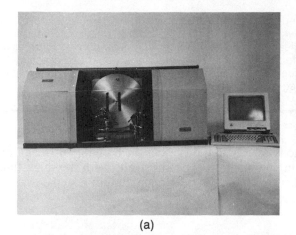

(a)

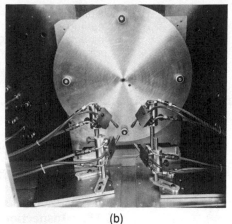

(b)

(c)

ditional materials handling equipment would have to be developed to interface with this cleaning system. An extended version of this system could thus be used in a highly automated CIM environment.

Another method for cleaning makes use of a *plasma* cleaning process. Processing hybrid circuits with argon plasma removes organic contaminants from the wire bonding areas. In the process, the circuits are loaded into carriers and placed in the plasma cleaning chamber. The chamber is evacuated and then filled with low-pressure argon gas. The chamber is then energized with radio frequency energy, which excites the argon gas. The ionized argon mechanically dislodges organic materials on the surface of the hybrid. The March Instruments Model PM-600 system shown in Fig. 6-37c has the appearance of an oven. The instrument is modular in design, making setup and maintenance simple. The process control module uses automatic sequencing to control the cleaning process. Substrate cleaning with argon gas can be used to remove residues of solvents, photoresists, and other organic compounds. Besides being a safe and dry process, argon plasma cleaning will not oxidize the exposed metal components or silver-filled epoxy. The plasma cleaning system shown in Fig. 6-33b requires operator materials handling, although the actual cleaning process is under computer control. In order to upgrade this system for more integrated operation, it would be necessary to develop automated feed and removal of work-in-progress and to achieve computer network integration for control and data logging.

The two methods of cleaning described in Example 6-16 can be used separately or in combination. The substrate, with the thick film on its surface, is then ready for the next manufacturing steps.

6-19. Inspection

The need to inspect work-in-progress continuously is a requirement in almost all manufacturing settings. Human observers are limited in their ability to develop and maintain a high level of inspection skill, particularly with respect to repetitive observation of complex structures. Many types of automated inspection systems that make use of machine vision and pattern recognition are now available. The most cost-effective systems often depend on hardware and software that are customized for the task at hand and make maximum use of a priori knowledge about the work being inspected.

The processes discussed so far provide an interconnection layer and selected thick film components on the substrate and prepare the surface for the attachment of IC chips and other components. At this point, an inspection process is required to make sure that the interconnections and components have been formed correctly. For very detailed, high-density circuits, such an inspection is difficult to conduct manually, and machine vision systems are finding steadily increasing applications.

As shown in Fig. 6-32 (process 6), the simplest inspection is manual review by an operator (making use of a microscope), a tedious and rather ineffective method of inspecting highly complex circuits. The "island of automation" approach involves using machine vision and pattern recognition under local control; the role of the operator is limited to feeding in and removing the product and helping the equipment establish its reference locations with respect to the product before the inspection begins. Computer communications capabilities will allow the equipment to obtain (through remote download of information) the reference patterns that are to be compared with those observed by the machine vision and allow for data upload of the results of the inspection. For a fully automated system, the remote computer access for inspection data is combined with automated product insert and removal.

EXAMPLE 6-17. INSPECTION UNITS

Describe the types of inspection systems that are available to evaluate substrate quality.

Results

Figures 6-38a and 6-38b show automated inspection systems designed for substrate inspection. The Vision Systems Model 830 is designed for operator handling of the substrates, whereas Model 840 is an in-line inspection system with completely automated handling. These inspection systems are

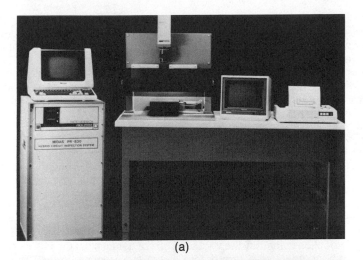

(a)

Figure 6-38.
Inspection methods.
(a) Model 830 auto-
mated optical inspec-
tion unit. (courtesy of
Vision Systems/Divi-
sion of Vanzetti Sys-
tems, Inc.). (b) Model
840 (courtesy of Vision
Systems/Division of
Vanzetti Systems,
Inc.). (c) CerProbe
Model H3500 hybrid
substrate prober can
be used to test the
electrical performance
of the thick film con-
ducting patterns on
the hybrid substrates
(courtesy of CerProbe
Corporation).

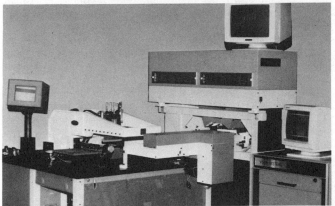

(b)

(c)

designed for high-volume production and make use of state-of-the-art capabilities in machine vision. The systems obtain high inspection throughout and provide an effective means for substrate evaluation. They eliminate operator fatigue, and are more reliable and repeatable than human visual inspection. The systems can accurately inspect up to 80 square inches of circuitry with a $\frac{1}{1000}$ of an inch (1 mil) resolution. Standard off-the-shelf products that are available to the hybrid industry, their setup and operation are designed to be as user-friendly as possible. Required operator training has been held to a minimum.

For each hybrid to be inspected, the system is taught the desired circuit pattern from a known good or "golden" sample. The system thus develops criteria for inspection by viewing a good product. Two reference images are created to establish inner or outer tolerances. The template associated with the "golden" example is then applied to inspect all the hybrids.

A "windowing" technique is available, so that arbitrary rectangles on the circuit can be defined with different template tolerances (to identify extremely sensitive areas or areas of less concern). Postprocessing is used during inspection to identify the various types of defects that may be possible. The system incorporates a proprietary illumination technique that produces a high contrast between the various thick films on the surface. A binary or black-and-white image has been found sufficient to meet all inspection requirements. The system can operate at production speeds in a pass/fail mode or in an operator-interactive mode for defect verification and classification. The system is designed for computer communications, which allows for remote control and data logging.

The systems shown here provide a highly automated means for hybrid inspection and can be integrated into CIM systems in terms of information flow and materials handling. These systems illustrate how microcomputers, vision processors, and improved software can be combined to produce reliable automated inspection units for industry.

In addition to optical inspection, substrate quality may also be tested by several other means. A hybrid substrate prober such as the one shown in Fig. 6-38c can be used to determine whether the desired electrical connections have been produced. The CerProbe Model H3500 has been designed to handle large numbers of substrates in a production setting. It allows a high density of probe points and ongoing visual monitoring during probing. The probe operates by using a dense array of probe pins mounted on a probe plate. The plate is lowered until the pins make contact with the desired circuit points on the hybrid. This model has been used in production line settings, and can be used to confirm that there are no electrical performance flaws in the substrate.

Another test that can be performed at this step makes use of capacitance testing, as provided by the Model SCT-1000 substrate continuity tester produced by Teledyne TAC. A single probe makes contact with all points on a hybrid circuit, and the capacitance values between each point and a ground plane are determined. The test instrument is initially programmed with historical statistical data that define the measured capacitance values at each point for a number of circuits. Subsequent substrates can then be tested to confirm that the capacitance measurements are within the acceptable range. This system has the advantage of single-point probing, eliminating the need for dedicated fixturing. Several different probe configurations are available for different applications, and economic justification is often readily obtained (Crowley 1988).

6-20. Component Assembly

Once the inspection has been completed, the IC chips and other components can be mounted onto the substrate (process 7 in Fig. 6-32). Typically, these components are attached with epoxy and are located on the substrate so they can be interconnected into the circuit. The assembly of these small components onto the hybrid surface is another difficult task. Manual assembly is essentially not feasible, and various types of automated processes must be used to assemble and attach the components.

The most manually oriented component assembly process depends on an operator using a syringe to place epoxy onto the substrate and then a simple vacuum needle to pick up the components and place them on the epoxy for attachment (using a microscope). As might be appreciated, when handling very small components that have specified orientations, this is a difficult method for assembly. A more automated process makes use of equipment that can automatically place the epoxy and components, under local control of the operator, with manual loading and removal of the product. Coordinates are fed into the equipment for each placement, and the equipment then achieves the desired coordinates.

If computer communications capabilities are added, remote computer control may be obtained over placement of the epoxy and the components (based on a data download of the desired locations) and data upload may take place to document equipment operations. For a fully automated system, the computer download and upload capabilities ensure complete automated control over the pick and place operations. The capability can be combined with automated product and component feed and automated product removal.

A requirement for component assembly is characteristic of manufacturing in many settings. The combining of adequate machine vision and other sensor capabilities with the needed handling dexterity and the selection of production controllers and control loops that are sufficiently flexible and adaptable for the task place difficult requirements on assembly units. Many types of units are available, often reflecting custom features dependent on the specific work-in-progress. Because of the difficulty of the assembly process, design for manufacturing strategies often attempt to simplify product shapes and connectors and thus simplify the use of automated assembly. The cost of assembly equipment can be sharply reduced even while automation is increased if improved strategies are incorporated into product design.

EXAMPLE 6-18. COMPONENT ASSEMBLY UNITS

Many techniques are available for component assembly. Discuss the features of typical systems.

Results

Figures 6-39a through 6-39c show an automated hybrid assembly system (HAS) manufactured by Teledyne TAC. This system is a computer-controlled "pick and place" system for selecting microelectronic components from chip trays, vibratory feeders, and tape feeders and depositing these components onto the appropriate location on a substrate. The system includes a closed-circuit television (CCTV) camera for improved programming accuracy.

Shown in Fig. 6-39a are a number of small "waffle packs," which hold bare IC chips. To the left are hybrid substrates that are ready to be populated. The HAS assembly system first applies epoxy to each substrate location

Figure 6-39.
Teledyne TAC Model
HAS-750/PC hybrid
assembly system
(courtesy of Teledyne
TAC). (a) Overview of
the system showing
hybrid substrates and
waffle packs contain-
ing IC chips to be
placed on the sub-
strates. (b) Close-up
view of the IC chips
being transferred from
the waffle packs to the
substrates. (c) Close-
up view of the assem-
bly operation.

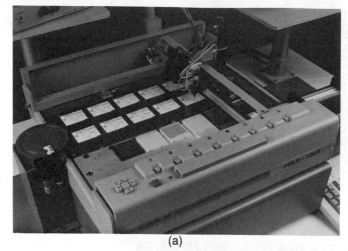

(a)

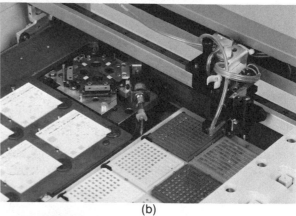

(b)

(c)

where an IC die is to be placed. The system then picks up the IC die, one at a time, locates the correct orientation with respect to the substrate, and places the die onto the substrate. The operator of the system is provided with a CCTV camera to observe the placement and to achieve quality performance. The assembly system is taught the substrate locations to be used by either camera positioning or an x,y digitizing tablet with a dimensioned drawing of the substrate. The input components must be oriented within ±30 degrees and contained in a grid pattern suitable for blind pickup. Components are lifted by vacuum and delivered to the programmable orientation nest. The component is aligned and then transferred to the proper deposit location.

The Teledyne TAC assembly system is operated by a microcomputer with a user-friendly screen that applies windowing software. The operator interacts with the computer screen to achieve all phases of circuit assembly.

Several enhancements to the assembly system will be required for fully integrated CIM operation. More versatile feeding of components and hybrid substrates must be included in order to obtain fully automated materials handling. It would be desirable to apply the existing camera system to achieve a pattern recognition capability (of the type discussed for inspection units) in order to enable the assembly unit to confirm the correct choice and orientation of each component as it is being applied.

At present, the HAS system is being modified to allow for computer network communications. The computer communications capabilities will allow the software files associated with each hybrid to be stored at a remote location and then downloaded when needed. The communications will also allow for continual data logging to document all assembly operations, including production counts and system utilization efficiency. As the information flow, materials handling, and inspection capabilities become more fully automated, the system will be able to achieve operator-free assembly.

As discussed in Example 6-11, Universal Instruments produces a wide range of modular equipment for use in developing computer-integrated manufacturing facilities for printed circuit assembly. The concepts illustrated by the Universal printed circuit assembly unit in Fig. 6-28 can also be applied to hybrid microelectronics. The Teledyne TAC product shown in Fig. 6-39 and the Universal Instruments product in Fig. 6-28 represent two different approaches to providing the assembly function. Based on product volume and other characteristics of the hybrid assembly operations that are to be performed, the preferred strategy can be selected. The choice will depend on the particular application and on a cost-benefit trade-off analysis.

Once the assembly process has been completed, a curing oven must be used to achieve the desired epoxy bond between the components and the substrate (process 8). This operation is performed by using the types of drying ovens discussed in Sec. 6-15.

6-21. Package Assembly

The package assembly operation (process 9) involves dispensing epoxy on the inner mounting surface of the package assembly and then placing the substrate on the epoxy. This type of assembly operation can be performed using equipment similar to that discussed in Sec. 6-20, or by applying the types of printed circuit component assembly operations discussed in Example 6-11.

Once the substrate is placed on the mounting surface, the assembly again must pass through a curing oven (process 10) in order to bond the substrate to the package. The next operation involves cleaning the assembly (process 11) so that the hybrid is ready for wire bonding. The types of cleaning units described in Example 6-16 and others can be used for this process.

6-22. Wire Bonding

The wire bonding manufacturing technology is typical of electronic product development and represents a highly customized approach to interwiring very small electronic devices and circuits. Because of the complexity of the task (in terms of product complexity, small dimensions, required accuracy, and need for large quantities of controlling data), top-of-the-line wire bonders often push the state of the art in automation and integration.

At this point, it is necessary to attach small wires that connect the IC chip and other components to the conducting electrical pattern that has been printed on the substrate and to interconnect the substrate to the pins that are part of the package assembly. The equipment unit used for this process is called a wire bonder (see Figs. 6-40a–h below). Very small wires, typically $1/1000$ of an inch (1 mil) in diameter, are attached first to the IC chip and then to the conductive patterns on the substrate to complete the necessary component interconnections and, similarly, to link the circuit to the package.

The wire bonding operation is typically performed with a small tool called a *capillary*. Gold wire is threaded through the capillary (see Figs. 6-40b and 6-40f). Bonding starts with the use of a flame or spark to create a small "ball" at the end of the bonding wire. The capillary is lowered to place the ball in contact with a metal pad on the substrate or IC surface. A combination of pressure, temperature, and ultrasonic vibration creates a bond between the wire and the metal pad. The capillary then lifts up, continuing to feed wire according to a programmed profile, and guides the wire loop to a second bond location. The second bond is created in a similar way, using a sideways pressure on the wire. The capillary then lifts up, clamps and breaks the wire, and is ready to create another bond.

As shown in Fig. 6-32 (process 12), a manual wire bonder can be used to create the desired bonds. For this case, the operator must manually control the locations for the beginning and end of each bond, a tedious and difficult process. An automated process involves bond placement under local automated control. The desired coordinates for the beginning and end of each bond are fed into the wire bonder, and the bonder itself then produces the desired interconnection. Manual product insert and removal is required. Typically, for such a wire bonder, the operator must place the substrate in the desired location, assist a machine vision system in orienting itself to the substrate, enable the machine to begin the bonding process, and monitor the bonding process to obtain the desired locations.

Computer communications capabilities, when added to the wire bonder, will allow remote computer control over the locations of the bonds (through software download) and data upload to record the exact bonding locations. A fully automated system combines remote computer control and computer communications capabilities with the automated product insert and removal.

EXAMPLE 6-19. WIRE BONDERS

Describe the types of equipment available to electrically interconnect the components with the conducting pattern on the substrate and
with the interconnection pins mounted in the package.

Results

The Hughes Aircraft Company 2460-II wire bonder shown in Fig. 6-40a
has been designed to satisfy hybrid manufacturing requirements. This equipment unit requires operator insertion and removal of the hybrid. The bonder
uses a sophisticated pattern recognition technique to guide the bonding process. For each hybrid to be manufactured, a bonding program is developed
by an operator who leads the bonder manually, step by step, through each
wire bond. The operator identifies the endpoints of each bond that is to be
formed. This information is stored in a computer inside the wire bonder. Then,
when operation begins, the pattern recognition system is able to locate the
bonding locations on the substrate and on the die and to perform automatically the actual bonding operation. A wide range of parameters is control-
lable for each bond. At the same time, bonder software minimizes the required
programming.

Once the bonder has been provided with the necessary data about the
bonds to be made, the operator inserts the hybrid into the bonder, confirms
that the pattern recognition system has correctly located its reference points
on the substrate, die, and components, monitors performance of the bonder
throughout the bonding process, and then, after bonding is complete, removes the hybrid from the bonder.

The Hughes 2460-II wire bonder contains sophisticated computer capabilities both for storing of information and managing of the bonding process. The bonder also has provision for communication with remote stations.
The bonder can be remotely operated; software is available for remote bonder
operation and for extensive data-logging information for each bond created.

The bonder represents another compromise in technology between the
operator-oriented and fully automated modes of operation. The bonding cycle
itself is fully automated but the input and removal of hybrids is achieved
manually. In the future, as systems become more integrated, it may be expected that the unit will be modified to allow for automated materials handling
in addition to automation of the bonding process itself. The wire bonder would
then be able to function in a completely operator-free environment. The
present bonder can be used as a stand-alone unit or integrated into an automated production line controlled by the host computer. Figures 6-40b and
6-40c show the wire bonding process and the nature of the bonds that are
created. In Fig. 6-40b, note the white capillary that controls bond placement
and formation.

Hughes also offers a wire bonder pull tester that can nondestructively
test each bond. Information regarding each bond is downloaded from the
wire bonder computer to the pull tester computer. The pull tester then uses
a small hook to pull gently on each wire bond to confirm that it has achieved
a satisfactory bond with the components and substrate. The pull tester also
can check to make sure each bond is in the desired location, thus performing
a quality control function.

Kulicke and Soffa Industries, Inc., manufactures a full family of wire

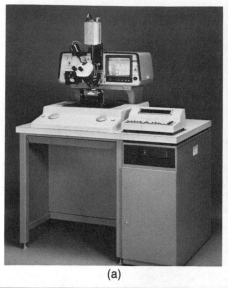

(a)

(b)

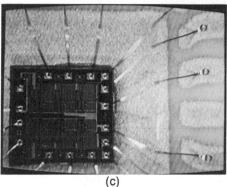

(c)

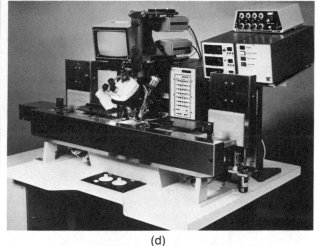

(d)

Figure 6-40.
Wire bonders. (a) Hughes Aircraft
Company Model 2460-II automatic
wire bonder (courtesy of Hughes
Aircraft Company). (b) Close-up
view of the wire bonding process
(courtesy of Hughes Aircraft Company). (c) The completed wire
bonded product (courtesy of
Hughes Aircraft Company). (d) Kulicke and Soffa Industries, Inc.,
Model 1419/DAWN automatic wire
bonder. Photograph by Leon Oboler
(courtesy of Kulicke and Soffa Industries, Inc.). (e) Close-up view of
the automated feed system. Photograph by Leon Oboler (courtesy of
Kulicke and Soffa Industries, Inc.).

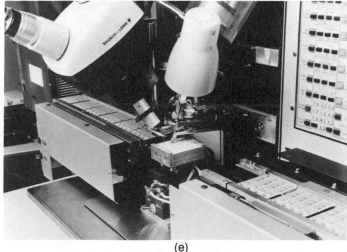

(e)

(f)

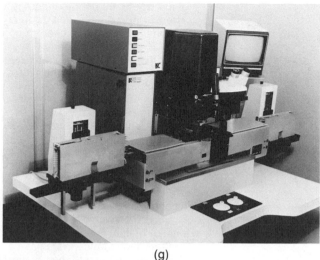

(g)

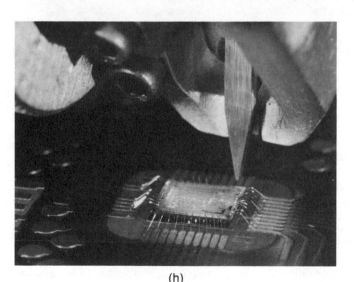

(h)

Figure 6-40. continued
(f) Wire bond in process. Photograph by Leon Oboler (courtesy of Kulicke and Soffa Industries, Inc.). (g) Kulicke and Soffa Industries, Inc., Model 1470 automatic wire bonder. Photograph by Leon Oboler (courtesy of Kulicke and Soffa Industries, Inc.). (h) Close-up view of aluminum wedge bonding. Photograph by Leon Oboler (courtesy of Kulicke and Soffa Industries, Inc.).

bonders, from simple manual models to versions designed to achieve fully automated operation. The Model 1419/DAWN™ automatic wire bonder (Fig. 6-40d) is designed specifically for the special requirements of the hybrid industry, and is intended to achieve fully automated operation. This bonder operates with several different automatic materials handling systems, which can feed in the hybrid substrates and components, automatically aligns the substrates using machine vision recognition, completes the wire bonding, and automatically removes the work-in-progress.

The operator bonds the first hybrid manually in order to "teach" the bonding program and the location of the necessary reference points. The bonder can then automatically wire bond all identical circuits.

Figure 6-40e is a close-up view of the automated feed system for model 1419, with wire bonding taking place on the heated work holder in the center of the figure. Figure 6-40f shows a wire bond in process. Note the fine gold wire extending below the capillary. The bond has just been completed and the wire will now be broken in preparation for performing the next bond.

The pattern recognition system makes use of advanced optics and software in order to allow operator-free operation. The system includes automatic focus for multiple bond heights and grey scale discrimination. The pattern recognition system is able to inspect a new substrate, locate the needed reference points, and direct the wire bonding operation.

Figures 6-40g and 6-40h show the model 1470 automatic wire bonder for aluminum and gold wedge bonding of ICs or hybrids. Figure 6-40g shows the wire bonder with an automated materials handling system in place, and Fig. 6-40h shows a close-up of aluminum wedge bonding in progress. The model 1470 combines state-of-the-art microprocessor controls and innovative design features to achieve the desired degree of accuracy and flexibility.

These wire bonders provide a fully automated capability that can significantly reduce operator involvement in the manufacturing cycle. When interconnected with other robotic materials handling systems, they can operate without human intervention, except when it is necessary to teach new circuit configurations to the system.

Many different types of wire bonders are commercially available. The system designer has access to a family of equipment units that range from manually oriented equipment (with each bond individually placed and directed by an operator) to fully automated and highly integrated systems (that can function without operator intervention). With this spectrum of equipment options, the challenge is to determine the level of automation and integration that is most appropriate for a given facility, in order to obtain the most cost-effective and competitive operations.

6-23. Package Sealing

Hermetic sealing for many types of products is required to prevent performance degradation caused by environmental effects. The type of sealing process that is cost-effective will depend on the nature of the product and the range of environments to be encountered. Some extreme environments can produce such demanding requirements that product design will be driven by the need to achieve product survival.

The package sealing function (process 13 in Fig. 6-32) can be performed by using manual equipment to attach and seal the lid. However manual control is slow and tedious. A more automated strategy involves automated local control over the sealing process. Computer communications capability allows remote computer control over the sealing process and data upload to record process functions. A fully automated sealing unit will allow remote control over the full operation, including automated product insert and removal.

EXAMPLE 6-20. PACKAGE SEALING UNITS

Describe an equipment unit for package sealing.

Results

Once the hybrid has been inserted into a package, typically the package is hermetically sealed to protect the circuit from environmental degradation (Dickinson 1989). The Scientific Sealing Technology DAP-2200 hermetic sealing unit shown in Fig. 6-41a can be used for this purpose. The sealing takes place in an environmental chamber, using tooling that is made from high-quality graphite (Fig. 6-41b). The hybrid package and lid are

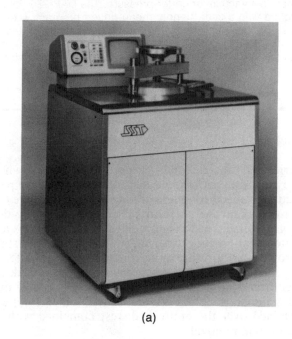

(a)

Figure 6-41.
Scientific Sealing Technology Model DAP-2200 hermetic sealing unit. Used to evacuate and hermetically seal the hybrid package (courtesy of Scientific Sealing Technology).

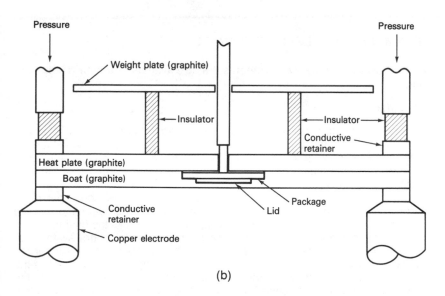

(b)

placed in a "nest" formed by the graphite "boat" and "heat plate," with a free-floating weight to control the pressure applied to the package-lid interface. An electrical current heats the graphite to the desired temperature, thereby heating the outside of the package and lid and producing the desired seal. The temperature is controlled by a control loop that links a thermocouple input to the power supply output providing the electrical current to the graphite.

This unit performs glass sealing, soldering, and brazing that can be accomplished in a vacuum or under pressure. The DAP-2200 system is computer controlled with a user-friendly menu-driven program. A computer port can link to a host computer for data logging and storage of the operational programs. The cycling operations of the system are highly automatic. However, fully automated material handling would require substantial equipment modification to allow the hybrid packages to be inserted into and removed from the vacuum chamber.

6-24. Package Test

A variety of strategies may be used to determine whether leaks exist in the packaging. A manual method is to place the package in a gas or liquid for a specified length of time and then remove the package from the environment and perform a chemical analysis to determine if any of the gas or liquid has penetrated the package and is leaking out. A more automated process involves an automated leak station under local control to perform the same functions. If computer communications capabilities are added, the parameters of the leak test can be downloaded from a host computer and the results can be uploaded. If automated materials handling is added, then remote computer control can be obtained over the entire leak test combined with automated product insert and removal.

EXAMPLE 6-21. LEAK TEST UNITS

Describe alternative strategies that are available for hybrid leak testing.

Results

Sophisticated leak testing can be performed in a variety of ways to confirm hermetic integrity. (Anthony 1985). Gross leak detection (which relates to leaks in the 1×10^{-1} to 1×10^{-5} atm. cc/sec range) may be performed using the InterTest 1014-CII leak pressurization and 1014-CBL leak detection systems.

The pressurization system in Fig. 6-42a includes two chamber pressures that are independently programmable. The operator loads the parts to be pressurized into the chamber, then closes the lid, locks the chamber, and begins the automatic sequencing. The microprocessor control system automatically evacuates the chamber, then fills the chamber with a fluorocarbon under pressure. The fluorocarbon must fully submerge all components in the chamber. The pressure will force the liquid through any small leaks into the hybrid package. After a defined period of time, the system enters a cycle that

(a)

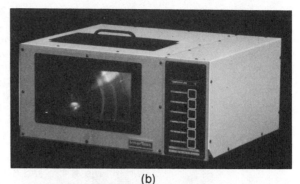

(b)

(d)

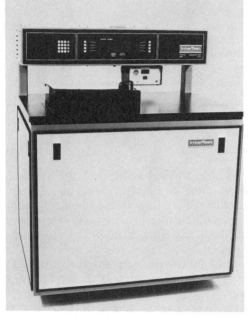

(c)

Figure 6-42.
InterTest Corporation leak test equipment. (a) Model
1014-CII gross/fine leak pressurization system [InterTest
product information (courtesy of ITERE)]. (b) Model 1014-
CBL gross/leak detection system [InterTest product information (courtesy of ITERE)]. (c) Model 1014-BII/S radioisotope pressurization system [InterTest product information (courtesy of ITERE)]. (d) Model 1014-BXL
scintillation crystal ratemeter workstation [InterTest product information (courtesy of ITERE)].

automatically removes the fluorocarbon. The hermetic system indicates to the
operator when the pressurization process is complete and the hybrids are
ready for removal. Once the pressure is removed and the hybrids are taken
from the pressure system, small quantities of the fluorocarbon will remain
inside each hybrid package that leaks.

The hybrid is then placed in the leak detection system (Fig. 6-42b). The part is immersed in another fluorocarbon that has a higher boiling temperature than the liquid used in the pressure cycle. Any of the test material that leaked into the package during the pressurization stage will now leak out and appear on the outside of the package as a flow of bubbles. Using high-intensity lighting, the operator inspects each package through a magnifying window in the chamber for bubble formation.

This combination of a pressurization cycle followed by leak test detection applies integrated microprocessor controls and operator involvement. The system requires ongoing operator participation and decision making, while using a microprocessor to sequence portions of the process.

Equipment used for the radioisotope method of fine leak testing is shown in Figs. 6-42c and 6-42d. An InterTest Model 1014-BII/S radioisotope pressurization system exposes the hybrid to a dry nitrogen gas that includes approximately 1 percent krypton-85, a radioactive gas. If there are any leaks in the package, the gas will penetrate the package. Once the part has been removed from the pressurization system, it can be placed in an InterTest 1014-BXL scintillation ratemeter workstation. Scintillation crystals are used to measure the amount of gas that has penetrated the package. (Krypton-85 fine leak testing detects leaks in the 1×10^{-3} to 1×10^{-10} atm cc/sec range.)

The fluorocarbon pressurization and leak detection systems require an operator to handle the hybrid and to observe any bubbles that indicate leaks. The microprocessor controls the processing sequence and handles the various gasses involved. The radioisotope pressurization and scintillation ratemeter systems present a more automated approach to leak tests, since the operator does not have to visually confirm and evaluate the presence of a leak. More fully automated leak test equipment might be developed by combining the radioisotope pressurization and detection functions with automated materials handling.

6-25. Electrical Test

In general, product testing is an undesirable requirement for a manufacturing system. Costs will be reduced if testing can be avoided. Testing thus reflects failures in the manufacturing system to produce "good" products. Instruments to improve product quality can have a significant impact if testing can be reduced. Statistical quality control (SQC) concepts (introduced in Sec. 3-14) can be applied to make the best, most cost-effective use of testing for a given setting.

Once the package seal has been formed and tested, the hybrid is ready for use. However, a wide range of electrical tests may be performed to confirm product quality (depending on the level of quality control and reliability of the manufacturing processes in use). Such tests can include environmental tests to determine hybrid performance under varying conditions.

It is essential to employ a manufacturing strategy for which high yield will be obtained during final electrical test. If the tests reveal many problems that require rework or scrapping the hybrid, then the final product is going to be excessively expensive and not competitive in the market. Thus all of the previous steps must be completed with a high degree of reliability so that almost all of the final products will pass the functional test.

Manual electrical tests can be performed by using a manual probe station, with small metal probes extended to touch selected areas of the circuit. Manual test equipment can then perform measurements through these probes. This is a time-consuming and tedious process. An automated probe station operates by having the operator place the hybrid in a given location and by using machine vision to orient the equipment. The probes are automatically placed in contact with the desired locations on the circuit, and automated test equipment used to cycle through the desired electrical tests. Such an automated test system can perform thousands of tests in a short time.

Computer-controlled electrical tests can be performed by inserting the completed package into a socket or test fixture and then using computer-programmed automatic test equipment (ATE) to determine whether or not the electrical performance is as designed and desired. This type of automatic test equipment can be difficult to develop and apply because of the sophistication of the hybrid and the wide number of tests that must be performed. Large computer capability is often required to be able to store information on all of the tests to be performed and then automatically to conduct these tests on the hybrid.

If computer communications capability is added to the test equipment, then the appropriate requirements for each test can be downloaded from a host computer and the results of each test uploaded. As might be appreciated, this capability can have a significant impact on the reliability and utility of the electrical tests being performed. If automated materials handling is added to this remote computer capability, the electrical tests can be performed rapidly and with a high level of efficiency.

EXAMPLE 6-22. ELECTRICAL TEST EQUIPMENT

Discuss electrical test equipment for hybrid manufacturing.

Results

Figure 6-43a shows a Micromanipulator Company Model 100PM manual probing station for hybrid circuits. This test station can probe geometries as small as 8 microns, and can be used for both hybrid and integrated circuit probing. The circuit is mounted on the probe station; using a microscope for viewing, the operator moves the probes to make electrical contact with desired locations. The probes are then interconnected with test equipment to achieve the desired test operations. The system uses a vacuum to hold the circuit in place and can accommodate up to four manipulators. Alternative microscopes can be used with different magnification features. Multiple stage assemblies permit probing of flat hybrid circuits that measure from $\frac{1}{2}$ inch $\times \frac{1}{2}$ inch to 4 × 4 inches in size and IC wafers from 4 to 6 inches in diameter. This probing unit also is available with a wide range of accessories.

Figure 6-43b shows a Teradyne Model J953 very large scale integration (VLSI) test system. The J953 series is available in several different models, depending on the device feature size and test complexity. The system consists of a probing station to which the circuit is attached, computer-controlled test equipment, and an operator console for control of the test procedure and data recording. Every model of the J953 series offers up to 256 independent input-output channels. For each device pin, the system provides up to 4 million bytes of pattern memory. The J953 series is designed to allow the user to start with a minimum capability and cost system and to upgrade to a more complex high-performance system. The system thus lets the user select the

(a)

(b)

performance and price combination that best suits the requirements today, with compatibility to upgrade the system later.

The particular series shown here is designed for the testing of complex integrated circuits. The system can also be used to test hybrids that make use of a variety of integrated circuits. The system shown in Fig. 6-43b represents a sophisticated test unit. The system does not have automated materials handling and thus, if it were to be operated in a fully automated assembly-line setting, would require modification. Teradyne offers complete mechanical, electrical, and software interface packages to assist users in applying the test system with handling and probing equipment in an automated environment.

Both the simple and complex types of electrical test equipment in Fig. 6-43 are in wide application. These units (and many other types) are available to test for product quality. In many settings, statistical quality control is being applied (as discussed in Sec. 3-14) so that only sample products have to be tested. On the other hand, in settings that have very demanding applications, it may be necessary to perform a positive test of every hybrid. The degree of testing performed depends on the quality of the manufacturing processes, the type of quality control system that has been implemented, and the ultimate application by the end user.

6-26. Overview

Various strategies may be employed to design the equipment used for the manufacture of hybrids. Each process may give rise to its own manufacturing equipment unit (which is a typical case), or multi-function equipment units can be developed for selected process combinations. A few original equipment manufacturers design their equipment for integrated materials handling and information flow throughout the manufacturing process; many have developed their equipment units as islands of automation with the potential for information integration but without the potential for fully automated materials handling throughout the process line.

Figure 6-32 illustrates how a system design group can examine the equipment families that are available for a manufacturing setting. It is important to be able to describe the functions that must take place in the product flow and to gain an understanding of the various degrees of automation and integration that are available for the equipment. In the above examples, specific equipment units have been described and related to the general outline of Fig. 6-32. (Functional parameters associated with these equipment units are discussed in Chap. 10; see Fig. 10-16 and the related discussion.)

In every CIM design effort, it is necessary for the design group not only to understand generic functions and degrees of automation but also how the specific equipment products on the market relate to this generic overview. In addition it may be useful to look at the available equipment and determine whether additional research and development should be performed to create more integrated equipment functions. It may be that in many cases the most effective CIM system cannot be formed by using standard equipment on the market, but that available equipment must be customized to meet the needs of the manufacturing system. The upgrading of commercially available equipment to more automated and integrated strategies and the integrating of various equipment functions into more cost-effective units (through additional research and development) are options that should be considered by the design group.

6-27. Application of Work Cell Concepts to Hybrid Microelectronics

As noted in the methodology introduced in Chap. 4 (Fig. 4-5), the design of any manufacturing system must reflect an understanding of the business environment in which the system will function. This includes knowledge of both the industry sector itself and of the extended market setting in which the industry exists. The nature of the hybrid microelectronics industry will determine the type of manufacturing strategies that should be considered for evaluation in system design. This industry can be characterized in several important ways:

1. The industry is highly decentralized and includes hundreds of hybrid manufacturing companies.
2. The industry includes a wide range of original equipment manufacturers (OEMs) providing manufacturing equipment to the industry.
3. The industry makes use of a wide range of processes and product types; there are hundreds of processes used in the hybrid manufacturing area, and different combinations of processes can be used to produce thousands of product types.

Hybrid manufacturing provides many insights that are representative of industry-wide issues and problems. A brief overview of some of the equipment families in this area shows how similar studies can be performed for different products and systems.

Thus, the hybrid product area is defined by an extremely high degree of flexibility and potential for customization. As might be expected for such a wide-ranging and flexible product area, production quantities can range from 1 to 10,000 and more for a given product type.

The hybrid industry is thus quite distinct from an industry in which it is expected that the product output will be a large number of similar products using similar manufacturing processes. Rather, the hybrid industry may be characterized as requiring a flexible, adaptive manufacturing setting in which many different processes are used to produce a wide range of product types. In many ways, manufacturing issues such as those related to hybrids are among the most difficult to address. These issues are particularly appropriate for study because there are indications that the evolution of manufacturing systems is away from those that are rigid toward those that are very flexible with respect to the market.

In considering advanced manufacturing strategies for hybrids, the modular work cell approach to developing manufacturing systems (introduced in Sec. 6-8) seems to have significant potential for success. As noted, the hybrid microelectronics product area is associated with hundreds of processes, thousands of product types, and low-to-midrange product volumes. The modular work cell approach may often be most appropriate under these circumstances. Figure 6-24f provides the basis for the detailed system design effort that is presented in Chap. 10.

EXAMPLE 6-23. INTEGRATED HYBRID MANUFACTURING SYSTEMS

The previous sections provide an introduction to some of the types of equipment that can be used to manufacture thick film hybrids. A range of alternative strategies is available to create integrated hybrid manufacturing systems from these and other types of building blocks. How can these strategies be related to the more general discussion of work cells and systems of Section 6-8?

Results

Several companies offer integrated manufacturing systems for hybrids and related products. CIM-oriented system strategies for the manufacture of printed circuit boards (based on products by Universal Instruments and Amistar Corporation) are introduced in Example 6-11, and could possibly be modified for hybrid application.

Other companies are also actively developing such systems. Panasonic® Factory Automation, a division of Matsushita Electric Corporation of America, has devoted major effort to the application of CIM concepts to electronics manufacturing. As stated, the goal of the company is "to be the best and broadest supplier of factory automation hardware and software." Figure 6-44 shows a flexible manufacturing system (FMS) concept described by Panasonic. This FMS concept is intended to apply to a large range of electronics products, including printed circuit boards and hybrid microelectronics.

The manufacturing system is arranged in a series of cells that provide for linear work flow. Much of the materials handling within each cell is done by automated equipment. In addition, a high level of information integration is obtained for each cell and for the factory. As shown in Fig. 6-44, a computer network links the work cells with the CAD facility, the CAM activity, and the SQC area. In addition, plant-level computers and sales offices are linked into the manufacturing system.

Panasonic presently offers to provide such integrated systems to potential customers. Thus, the type of system illustrated in Fig. 6-44 is considered available within today's state of the art. Specific equipment units in use and the details of the integration will vary from location to location, depending on the particular enterprise objectives. Panasonic/Matsushita has developed its hybrid technology (HiT) system based on experience in a wide range of manufacturing sectors. The application of these methods to thick film hybrid microelectronics has become sufficiently important for the company to offer manufacturing systems in this area.

Another company that offers highly automated and integrated manufacturing systems for hybrid microelectronics is DEK Printing Machines Limited. DEK and Universal Instruments Corporation are divisions of Dover Technologies, Inc. DEK has been designing and building automated hybrid thick film printing and circuit production systems since the mid-1970s, and developed one of the first fully automated hybrid manufacturing lines (installed in the United Kingdom). DEK specializes in printing and substrate handling equipment, and has integrated thick film printers, infrared dryers, and cassette-based automated handling systems. Figure 6-45a shows the DEK AutoFeed System 3 configuration, which combines these various functions into a work cell. This work cell (described functionally in Fig. 6-45b) makes use of a modular approach and provides a wide range of operations for automated substrate printing and handling.

DEK also offers customers complete advanced manufacturing systems for hybrids. Figure 6-45c shows how a flexible work cell concept can make use of products from several vendors to achieve a wide range of manufacturing functions. DEK offers to customize work cell design and make use of multivendor equipment in order to meet customer needs.

DEK also cooperates with Universal Instruments in the development of automated and integrated manufacturing facilities for surface-mount printed

Figure 6-44.
Flexible Manufacturing System (FMS) for electronics products (courtesy of Panasonic Factory Automation Company).

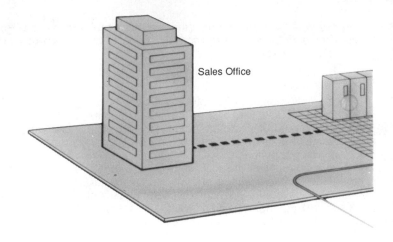

Sales Office

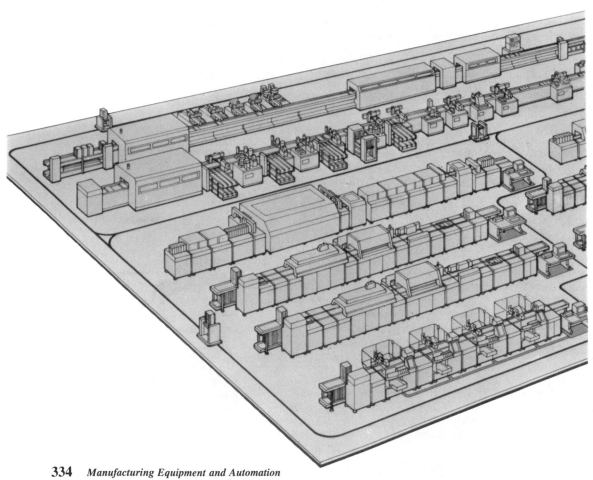

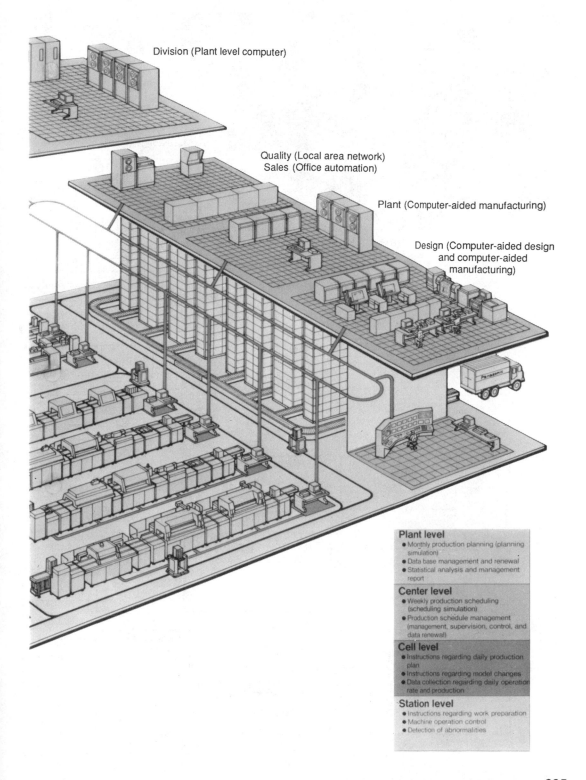

Division (Plant level computer)

Quality (Local area network)
Sales (Office automation)

Plant (Computer-aided manufacturing)

Design (Computer-aided design
and computer-aided
manufacturing)

Plant level
- Monthly production planning (planning simulation)
- Data base management and renewal
- Statistical analysis and management report

Center level
- Weekly production scheduling (scheduling simulation)
- Production schedule management (management, supervision, control, and data renewal)

Cell level
- Instructions regarding daily production plan
- Instructions regarding model changes
- Data collection regarding daily operation rate and production

Station level
- Instructions regarding work preparation
- Machine operation control
- Detection of abnormalities

(a)

Figure 6-45.
Flexible work cell concepts (courtesy of DEK USA, Inc.). (a) Combined printer-dryer using automated materials handling based on cassettes. (b) Definition of equipment units and product flow through the workstation. (c) Alternative flexible work cell concepts.

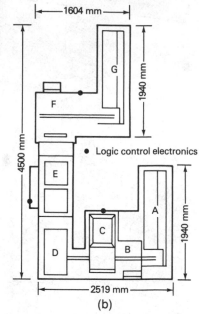

A, G = Multicassette unloader, reloader

B, D = Feed system

C   = Thick film printer

E   = Infrared dryer

F   = Reloader feed

(b)

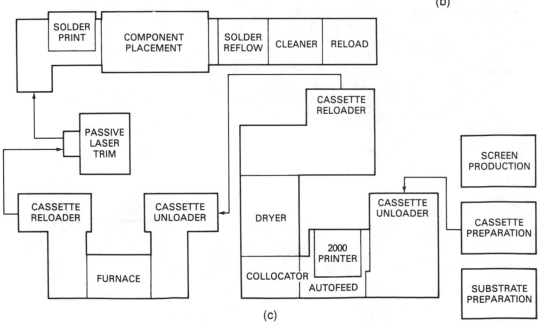

(c)

circuits. These two companies have defined specific areas in which they produce their own manufacturing products, with an emphasis toward modularity combined with high degrees of automation and integration. At the same time, they offer a system integration capability; the companies will assist customers in designing advanced manufacturing systems that draw on their own and other equipment to achieve highly automated and integrated manufacturing capabilities.

Many other types of manufacturing cells and systems can be developed for thick film hybrids. The solutions can involve various degrees of automation and standard or highly customized configurations. Chapter 10 describes an alternative work cell configuration that can be used to apply the system design techniques of Chap. 4 to the hybrid microelectronics industry. The commercially available integration strategies discussed here may be used for comparison with FMS concepts presented in Chap. 10.

ASSIGNMENTS

6-1. In order to achieve maximum educational insight from Chap. 6, it is helpful to stimulate class discussion of the concepts that have been included, develop critical reviews that probe the advantages and limitations of the concepts, and explore alternative ways for approaching the topic. Select one or several topics from the boldfaced headings that introduce sections or examples in this chapter. Come to class prepared to explain this topic, critically describing the text materials and suggesting ways the topic discussion might be strengthened.

6-2. Select a particular equipment unit and describe the unit in terms of the available equipment parameters that are introduced in this chapter. Consider the technical limitations that give rise to these parameter limits and how these limits might be modified through research and development.
(a) Come to class prepared to present your materials and to participate in class discussion.
(b) After class discussion, prepare a brief paper describing your insights and conclusions.

6-3. Join with several other class members to review a number of equipment units that provide alternative strategies for performing the same manufacturing operation. The objective is to evaluate a family of equipment units. Each class member can select one or several units for study. Each member of the group should develop enough information to identify the equipment technical parameters that will determine functional operation. Then, by working as a team, and following the methods discussed in this text and other methods that you find useful, compare these different equipment units as solutions for a manufacturing setting.
(a) Be prepared to present these materials as a group activity and to participate in class discussion.
(b) After class discussion, prepare a brief paper describing your insights and conclusions.

6-4. Select an equipment unit and identify the sensor, controller, actuator, effector, and control loop characteristics of the unit. Describe the performance of each equipment element and how the elements are linked together through the control loop. Discuss the advantages and limitations of this design. Based on a critical review of the processes performed with this machine, discuss ways in which elements of the system could be modified to improve performance. Consider the costs that might be associated with the enhanced performance in each case.
(a) Come to class prepared to present your insights and discuss them with the class. Describe a series of R&D projects that might be conducted

in order to produce a family of products from this single equipment unit.

(b) After the class discussion, and making use of feedback from the class, prepare a brief paper describing your insights.

6-5. Collect information on several types of sensors. Describe the capabilities of the sensors and then consider the types of equipment units in which the sensors might be of particular use.

(a) Come to class prepared to present these materials and participate in a class discussion.

(b) After class discussion, prepare a brief paper describing your insights and conclusions.

6-6. Repeat Assignment 6-5 for actuators and effectors.

6-7. Repeat Assignment 6-5 for controllers and control loops.

6-8. Identify a particular setting in which individual equipment units have been integrated into a work cell structure. Describe the equipment units that are in the work cell, the nature of work cell operations, and the advantages and disadvantages of the cell structure.

(a) Come to class prepared to describe the cell structure with drawings and to indicate the functional performance.

(b) After the class discussion, prepare a brief paper describing not only your work cell but alternative work cell concepts that have been suggested in the class.

6-9. Discuss how the types of equipment performance and cost analyses considered in the chapter can be effectively integrated into an overall manufacturing system design procedure. How can these methods most effectively serve the system design objectives? How can the methods described in this chapter and other methods serve as a linkage between the equipment units available and the operational requirements of the manufacturing system? What limitations are associated with the methods, and how might these limitations be addressed? What alternative strategies might be used for describing the opportunities and constraints associated with the physical equipment units to be used in system implementation?

(a) Be prepared to present your insights and discuss them with the class.

(b) After the class discussion prepare a brief paper describing your insights.

6-10. Find someone in an industry setting who will allow you to watch production operations and observe equipment units in use. Discuss with this individual the reasons for the selection of the equipment units that you observe. If possible, observe workers using this equipment in the operational setting. Discuss the strengths and weaknesses of the equipment units based on your own observations. If you were going to modify the equipment in use, what would be some of the most significant areas in which you would seek change? If changes are only made in selected production areas and not throughout the system, will misbalancing between system elements likely to limit the impact of upgrading efforts? What new equipment family members might be useful to produce a more automated and integrated system? What limitations do you perceive in terms of moving this manufacturing setting toward a broader application of CIM concepts? If the setting already is highly automated and integrated, discuss the reasons for the particular configurations that are in use and how effectively the equipment matches the overall operational system.

(a) Be prepared to present your materials and discuss them with the class.

(b) Prepare a brief paper describing the results and insights obtained from your effort.

7

INFORMATION FLOW AND COMPUTER NETWORKS

Once alternative system configurations have been developed and modeled and various decisions have been made regarding the equipment technology to be applied, it is important to consider the information flow that must take place in the system to produce the desired operations. In general, this will include both the information flow necessary to support human operators and the information flow that takes place as part of any computer network. One of the limiting aspects of system design is the need to apply information concepts that will be accepted by the organization and can be achieved within reasonable cost and schedule.

This chapter provides additional resources for use in the general planning processes that are discussed in Chap. 4. Included are methods for describing system information flow requirements and some of the commercial approaches that are used for computer networking. This chapter provides a baseline that can be drawn upon for more detailed studies of information and computer issues. The intent is to provide a starting point for study of this area and to indicate how the issues associated with information flow link to other aspects of CIM design. A bridge is provided between the overall CIM design issues and information concepts and computer networks.

The information flow for a manufacturing system must enable both operators and equipment units to perform the required tasks. An integrated information strategy is necessary to produce the desired system operation.

Chapter 7 addresses
the types of informa-
tion that must be made
available to operators
and equipment and the
alternative methods
that can be used to
implement the desired
information flow. The
information functional
requirements (of layer
4) and the physical im-
plementation of these
functions (of layer 5)
are critical system as-
pects that are treated
here.

Chapter 4 (Fig. 4-5) describes a method that may be used for the design of CIM systems, and Chap. 6 describes how the equipment and the level of automation for the CIM system can be most effectively defined. The purpose of this chapter is to link the equipment concepts of Chap. 6 with the system-level concepts of Chap. 4. The evaluation of information system properties associated with steps 11 and 12 of the system design strategy shown in Fig. 4-5 is considered in further detail. Figure 7-1 shows how the results of previous chapters lead to the discussion developed here.

The first task is to address the relationships between the functional process descriptions associated with layer 2 and the functional *information requirements* associated with layer 4. The second task is then to consider relationships between the information function requirements of layer 4 and the physical implementation of these functions as associated with layer 5. The initial analysis associated with layer 4 considers methods for mapping *process* functional requirements into *information* functional requirements. The discussion associated with layer 5 emphasizes methods for physically implementing the needed information networks.

This chapter begins by reviewing some of the equipment concepts introduced in Chap. 6 and considering the types of information flow that must take place when such equipment units are integrated to produce an operating manufacturing system. Section 7-1 is intended to provide an understanding of some of the general issues that are involved in evaluating the required level of information flow for a CIM system. Section 7-2 introduces computer networks by making use of the open

Figure 7-1.
Discussion emphasis of Chap. 7 related to the design model in Fig. 4-5.

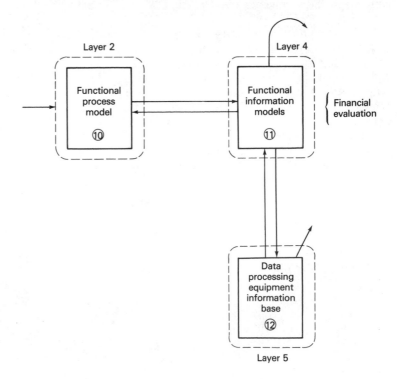

system interconnect (OSI) model. Section 7-3 then applies the OSI model to contrast a number of strategies that can be used to develop computer networks. Sections 7-4 through 7-7 describe several ways in which protocols or standards can be used to implement computer networks. As discussed, there are many issues involved in the application of a protocol approach to linking equipment units together in a network. Section 7-8 introduces an alternative approach that uses microcomputers as "function bridges" between the equipment units and a network to produce a different type of networking concept. Section 7-9 then discusses how information requirements and computer network performance concepts can be integrated into information system design.

7-1. Information Flow for Manufacturing Systems

Figure 7-2 illustrates how the equipment and automation concepts discussed in Chap. 6 may be used as building blocks for the study of information flow in a manufacturing system. As illustrated here, information flow must address both the information required and pro-

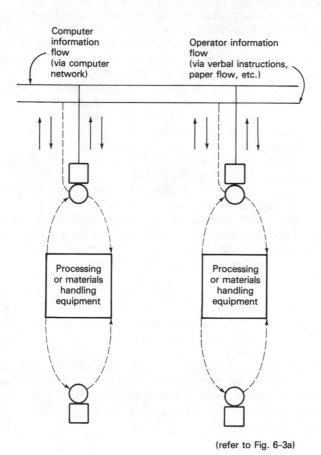

Computer information flow (via computer network)

Operator information flow (via verbal instructions, paper flow, etc.)

Processing or materials handling equipment

Processing or materials handling equipment

(refer to Fig. 6-3a)

Figure 7-2.
Application of equipment and automation concepts discussed in Chap. 6 to the study of information flow.

duced by human operators and the information exchange associated with computer requirements and computer-generated information. The information system must provide not only the information flow necessary to operate the equipment, resulting in the desired manufacturing processes and flow of work-in-progress and materials handling, but must also deal specifically with management functions. To the degree that each equipment unit can manage its own activities, information flow is maintained internal to the equipment unit through closed-loop operations. To the degree that the equipment units must be integrated and managed on a cellwide or factory-wide basis, information flow must take place between the human operators and among the various computers in order to maintain technological operations and to produce the desired management operations.

As may be determined from Fig. 7-2, modeling of the information flow in a CIM system and developing an understanding of the required decision-making processes are difficult tasks. Every level of information flow and decision making must be considered carefully in order to produce a system that will function as desired. In the typical CIM system, the information required by humans and computer-based information must be integrated. Decision making will take place through a variety of combined operator and computer interactions.

The information flow and decision-making strategies for a CIM system must be planned in detail. An ad hoc approach to dealing with the information requirements will likely result in unsatisfactory system operations, reflected in lower productivity, higher costs, and difficulties with quality control.

One of the most difficult requirements associated with implementation of a CIM system is to develop an adequate grasp of the decision-making strategies that will be implemented, the methods that can best be used to take advantage of these decision-making methods, and the performance limitations of these methods. The manufacturing system must be studied as a self-operating mechanism, using a systemwide modeling of the information system. If an effort is made simply to assemble a number of equipment units, connect them together, add on host computers, and then perform information integration on an *ad hoc* basis, the resulting system performance will be far removed from initial anticipation. It is necessary to plan in detail how all information flow will be handled as part of the system. This aspect of system design is equally important as the careful definition of the equipment and automation to be used at each processing stage.

In general, each equipment unit can include both operator and computer interfaces. In order to address both the operator and computer information requirements, it is necessary to consider the ways in which the equipment performance (discussed in Chap. 6) may be associated with the resulting information needs. In doing this, it is useful to consider typical information flow requirements that can be used to illustrate system information needs. Figure 7-3 lists three types of operator information that can be required for a manufacturing facility. Functional equipment performance may require local command and control, which is based on the experience of the operator and on rules provided to the operator. The requirement for local control of operations is based on operator experience and available operation manuals. Local performance monitoring is often associated with operator observation, perhaps enhanced by various methods to improve the human senses for the given purpose.

Figure 7-3.
Three types of opera-
tions information.

Functional Equipment Performance	Functional Information Requirements
Local command and control	Experience and operating rules
Local control of operations	Experience and operating rules
Local performance monitoring	Observation

Figure 7-4 shows how a similar list can be prepared for the computer information associated with a manufacturing system. In order to achieve CIM operations, it is often necessary to implement remote command and control functions for equipment performance. This requires command and control message transfer from a remote controlling device to the equipment unit and an acknowledgment from the equipment unit back to the controller. Another essential requirement for CIM operations is automated equipment operations, based on software instructions that can be transferred over the computer network. This can require both the uploading and downloading of files. File upload from the equipment to a host computer may be required as a result of file preparation at the equipment unit; many types of equipment provide for software development at the equipment, with uploading for storage. Once the programs are stored by the host computer, it will be necessary to download them to the equipment when a particular process is to result.

Another functional equipment requirement is for the remote monitoring of equipment to ensure correct operations and to perform such functions as statistical quality control (discussed in Sec. 3-14). The equipment must be able to perform *data logging;* a flow of data takes place from the equipment to a monitoring station, with acknowledgment from the station back to the equipment. In addition, sometimes shared functions are necessary, with two equipment units achieving direct information transfer from one to the other without going through a hierarchical information management structure. This is particularly true where limitations or linked processes require that the information be transferred quickly from one unit to the other.

Figure 7-5a shows one way in which a computer network can be

Functional Equipment Performance	Functional Information Requirements
Remote command and control	Command and control message transfer
Remote (software) control of equipment operations	File upload/download
Remote performance monitoring	Data logging
Shared functions	Direct information transfer

Figure 7-4.
Two types of computer network information, related to equipment performance and information requirements.

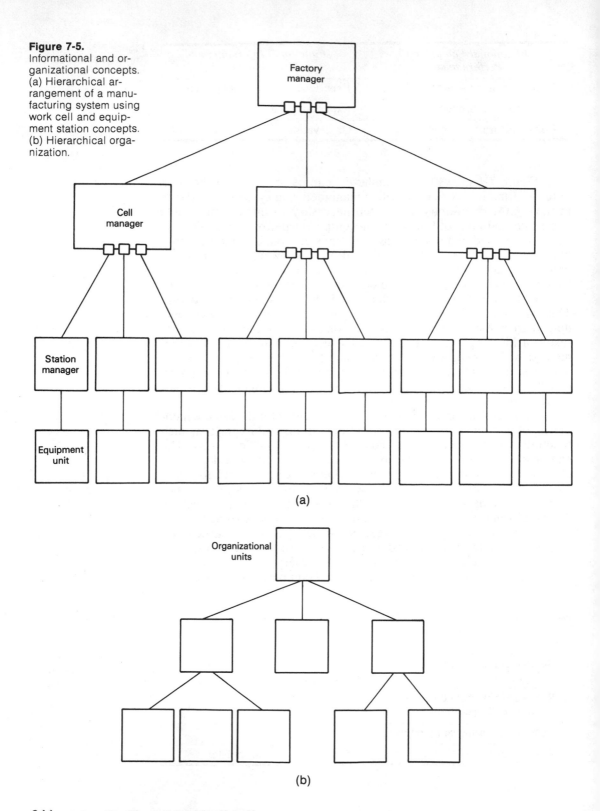

Figure 7-5.
Informational and organizational concepts.
(a) Hierarchical arrangement of a manufacturing system using work cell and equipment station concepts.
(b) Hierarchical organization.

Factory manager

Cell manager

Station manager

Equipment unit

(a)

Organizational units

(b)

configured by employing one-to-one communication linkages to create a hierarchical structure. As shown in Fig. 7-5b, this information structure is in many ways similar to a bureaucratic organizational structure (organizational issues are discussed in Chap. 8). Some of the issues associated with the computer network shown in Fig. 7-5a relate to command and control, response time, the time required for file transfer, the time required for data logging, and the sensitivity of the system to failure in any one of the intermediate nodes of the network. Some of the issues in regard to the corresponding organizational concept introduced in Fig. 7-5b are associated with the operator capabilities in such an organization framework and the efficiency that can be obtained in this type of setting. Other, improved architectures for information flow are examined below.

Major questions exist as to how information should be transferred both in the organizational setting and over the computer network in order to achieve the maximum compatibility between these two aspects of information transfer and maximum system performance. The resulting issues are quite complex, in that the organization and the computer network are complementary approaches to achieving complete information flow. The discussion presented here can help define how much total data will be flowing in each portion of a hierarchical network, where data should be stored, how much memory is required, and many other issues.

As introduced in Chap. 3, computers can be used in a variety of ways to support design and manufacturing and the integration of these functions. As discussed in Sec. 3-8, CAD, CAE, CAPP, and CAM are all methods for using computers to support decision making, information generation, and information flow for the design and manufacturing functions. MRP, MRP-II, and related systems create computer linkages between management and manufacturing activities.

Difficult problems can ensue when trying to take advantage of all of these CIM design principles in a specific setting. Problems can arise in attempting to produce consistent data bases and integrated information flow among all of the component software products. In many cases, the basic structures and definitions of parameters in these products will be inconsistent with one another, and it will not be simple to achieve an interchange of the information as desired. Major questions must be addressed in terms of the types of data bases to be developed, where these data bases should be located, how data integrity will be maintained, how inconsistencies will be resolved, and, generally, how data base management will take place.

If a decision is made to move toward a system in which common data bases are employed for all aspects of system operation, developmental efforts will be required in order to link the flow of information from all necessary data sources to the data base and from the data base to all users. The maintenance of a high-quality data base that will contain the needed information and serve all users is a challenging task in this environment.

It is also essential to develop means to protect the data base from intervention. In general, it should be impossible for any user of the system to erase data from the data base. Access to achieve data modifications must be controlled in a highly protective manner. (Further discussion about the problems of system information flow is presented in Example 9-7.)

Figure 7-6 shows a matrix that may be formed to describe all of the communication requirements among the elements of a manufacturing system. As illustrated, each element of the system might be required to communicate with any other element, and the appropriate types of requirements (such as those shown in Fig. 7-3 and 7-4) can be used to fill out the matrix. This matrix thus provides a summary of the functional information flow requirements for the entire system.

As discussed in Chap. 6, an alternative strategy for linking equipment units together is to form work cells, as shown in Fig. 7-7. Information handling *within* the work cell can be either similar to or quite different from information handling *between* cells. Such cells may or may not make use of direct information transfer from one equipment unit to the other. An alternative may be to develop a cell controller that links all of the equipment units in a hierarchical arrangement. Information storage can take place at the equipment level or at the cell level. It may be possible to achieve reduced intercell information flow with this strategy and to achieve a more distributed decision-making data base. The most effective information structure will depend on the functional relationships between the equipment units and cells.

Alternatively, instead of creating the types of structures shown in Figs. 7-5 and 7-7, it is possible to develop local area networks (LANs) to link equipment units, within either a cell or an entire facility, as shown in Fig. 7-8. Instead of providing hierarchical communications built around point-to-point linkages, a LAN provides a means for many different equipment units to become nodes on a single communications

The information matrix of Fig. 7-6 forces the planning group to consider explicitly all of the communication requirements among the elements of a manufacturing system. The same types of matrices can be used to describe human information exchange and computer network exchange. The completion of all entries required by such a matrix will involve a detailed analysis of the planned system operations. Although difficult and time consuming to perform, such analyses are critical to obtaining the desired system operations.

Figure 7-6.
Matrix description of system communication requirements.

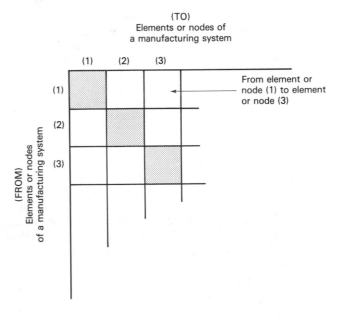

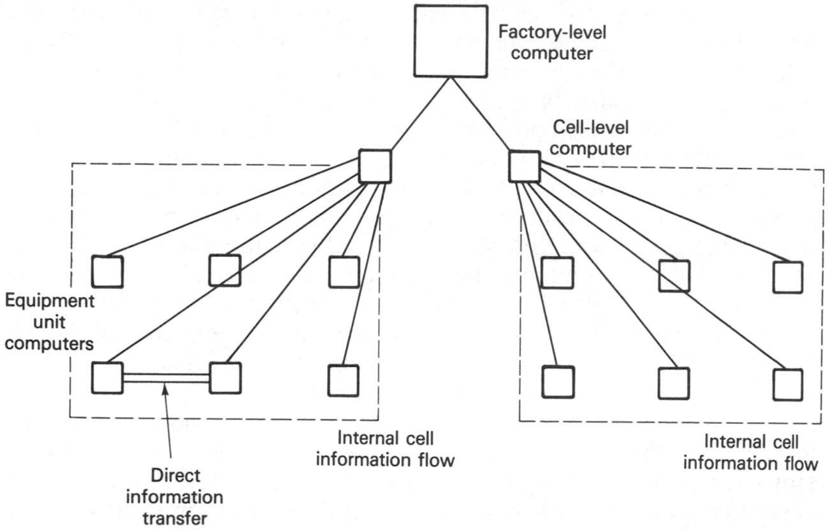

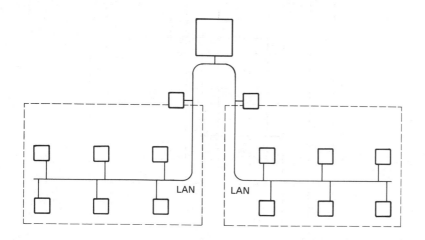

Figure 7-7.
Information flow using work cells and a hierarchical network.

Factory-level computer

Cell-level computer

Equipment unit computers

Direct information transfer

Internal cell information flow

Internal cell information flow

bus and pass messages along this bus from one equipment unit to another. Each message has an address associated with it, so that the correct recipient of the message can be identified by all system elements. An appropriate organizational/management structure can be defined depending on the functional relationships of the system.

A local area network can have many attractive features. Instead of running many individual cables (as required in Fig. 7-7), it is possible to run only one cable with many local attachments or nodes (as in Fig. 7-8). A LAN also has another attractive feature; it is possible to design the information network so that if any equipment node or unit becomes inactive, the network can continue to operate with the remaining elements of the system. Thus, the system is able to achieve a more gradual reduction in operations rather than the sudden failure that can result in the system in Fig. 7-7. Various types of methods for implementing LANs are discussed below.

Figure 7-8.
Information flow using a local area network (LAN).

LAN

LAN

Figure 7-9 shows an alternative matrix for describing the computer information requirements for a system based on the use of work cells to organize the information flow and storage requirements. Many other types of matrices similar to those in Figs. 7-6 and 7-9 may be prepared in the most convenient format for defining the information requirements. It may be necessary to try alternative information configurations for each scenario in order to find the one that is most appropriate. Once several communication matrices have been formed, it is possible to explore the physical implementation requirements associated with each one and therefore obtain a cost-effectiveness comparison. These information matrices thus provide a convenient way to relate the functional aspects of the equipment associated with layer 2 (as developed in Chap. 6) with the functional informational requirements associated with layer 4 (discussed here).

A similar approach may be used to address the manual information flow that must take place among the people in the system. However, it is interesting to note that while much attention is given to computer flow requirements, often much less attention is given to the equally important issue of the manual flow of information. If either aspect is pursued alone, without adequate attention to the other, the total information system will likely fail, or certainly will not achieve its intended level of operations.

The same types of matrices may be used to develop descriptions of the human information exchange or the computer network exchange. In system design, these two aspects should be combined to achieve maximum understanding of the system that is planned. Once these

Figure 7-9.
Matrix description of communication requirements using a work cell structure.

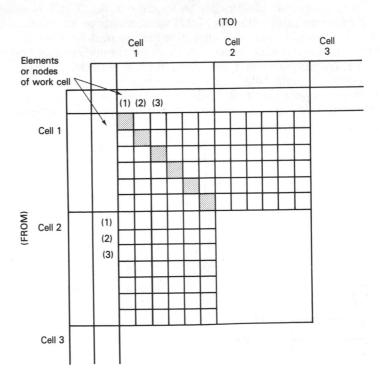

information matrices have been prepared, it is possible to consider alternative implementation strategies (associated with level 5). Each of the information functional requirements that has been developed can then be explored for performance and cost implications.

EXAMPLE 7-1. WIRE BONDER COMMUNICATION REQUIREMENTS

Figure 7-10 describes the performance properties of four different wire bonders. (Wire bonders are discussed in Chap. 6.) Type A is a manual system that makes maximum use of operator control. All information flow associated with this equipment is provided by the operator and the supporting documentation. Wire bonder B allows for computer control of specific bonding steps, but the operator has to set up each bond before it can be made. This unit does not allow for the exchange of data or network communications. Bonder C provides computer control of all bonding operations, with localized loading of computer programs. On-site operator support is required because the MTOI is only of the order of 1 to 2 minutes, and operator intervention is continually required in order to support the pattern recognition system and to resolve processing errors as they occur. Wire bonder D is a fully automated system; the equipment unit is linked to a computer network and the MTOI is greater than 10 minutes. Discuss the information flow requirements that might be associated with each of these wire bonders.

Results

Wire bonder A requires manual documentation and supporting operator actions. Every bond that is to be completed by this manual wire bonder must be based on paper documentation, and the operator must translate the data recorded on paper to the required bonds. The nature of the individual bonds is not recorded, so no data will be recorded directly from the bonder. Any data recorded are provided by the operator after each step or several steps of the bonding activity. As might be expected, this type of operation can produce concerns about the quality of the final process, the intense concentration required of the operator, and the use of paper documentation to define each bond. The lack of automatic data-logging capabilities can severely restrict the ability to monitor this process. In using this type of wire bonder, the major information concerns will be with providing the maximum effective training to the individual operators, making a maximum effort to

Wire Bonder Type	Performance Characteristics
A	Manual system, maximum operator control
B	Computer control of specific bonding steps, with operator setup for each bond
C	Computer control of all bonding operations, with local loading of computer programs; MTOI = 1–2 min
D	Computer control of all bonding operations, with networking for data exchange; MTOI >10 min

Figure 7-10. Performance characteristics of four types of wire bonders.

ensure that the paper documentation is correct, and making a maximum effort to determine that the operator has performed all of the desired functions. This will lead to an extensive flow of paper and to the need for a large document storage area to save the results. In addition, the information system must allow for access to this data base, which will require significant organizational and management support.

Wire bonder B provides the potential for substantial improvement in quality, as the individual steps in the wire bond operation are now performed under equipment control. It is possible for the equipment to print out its step-by-step activities so that a record is maintained. The organization will still need to provide the operator with the necessary training and with paper documentation to define the bond that should be produced. Additional data may now be provided on paper, possibly from a printer, to define the parameters used for each bond. These also must be stored. This type of system may result in expanded requirements associated with the storage and access of information.

Wire bonder C allows the potential for loading computer disks into the unit, in order to tell the unit which bond should be performed, and to allow the results of such bonding to be documented on a disk. The role of the operator becomes more to support the equipment performance and to ensure that the equipment is operating according to the desired performance. This equipment unit has more potential for standardization and quality control, since it is not necessary for the operator to enter individually the parameters associated with each bond, and performance of the unit can be documented on a printout. The requirements for the information flow now shift to making sure that the correct disks are inserted and removed from the equipment unit and that these disks are stored appropriately so that access can be later obtained. If the operator is taking paper notes and these must be linked to the computer disk on demand, care must be taken to produce an appropriate information storage and access system.

Wire bonder D allows complete information flow through the computer network to control the wire bonder and perform data logging. The operator intervenes only when the equipment does not function as desired. This type of system provides the potential for removing all of the paper associated with the wire bonding function and replacing it with computer files. On the other hand, since major system functions are being transferred to the computer system, new problems arise in design of this network and in producing the software that is necessary to obtain the required data flow. In addition, it is necessary to provide adequate documentation control and access so that the quality of the data base can be protected at all times.

In summary, four wire bonder configurations can lead to four different sets of information requirements. In each case, the design of the operator-related information flow and the computer network-related information flow will change, and these two aspects must always be integrated to obtain the desired system operation.

EXAMPLE 7-2. INFORMATION MATRICES

Figures 7-11a and 7-11b show information matrices associated with two work cell configurations, one involving manually based operations and the other involving computer-based operations. Discuss the information systems that might result from these alternative viewpoints and the issues that should be considered as part of each information system design.

Results

Figure 7-11a illustrates operator-based information flow. The work cell consists of four equipment units. (These equipment functions and units are introduced in Chap. 6; a work cell implementation using similar units is discussed further in Chap. 10.) Information flow takes place among the equipment units that are included in the work cell being described and between this specific cell and other work cells and the factory-level information base. It is assumed for this work cell that work-in-progress proceeds sequentially from station to station without reversing the product flow. Information flow can thus take place from (1) unit 1 to units 2, 3, or 4 or outside the work cell, (2) from unit 2 to units 3 or 4 or outside the work cell, and (3) from unit 3 to unit 4 or outside the work cell. In addition, each station receives information from outside the work cell.

As shown in Fig. 7-11a, the inspection unit (station 1) operates on the basis of information received from outside the work cell. A document is received describing the inspection criteria that are to be applied by product type. The operator must become sufficiently familiar with this document to conduct a manual inspection. The output of the inspection is information to the next station (assembly) that a product unit (defined by an identifying code number) has passed inspection. The result of the inspection is a pass or fail ranking; limited reasons for the ranking may be documented in either formal reports or in a notebook that is maintained at the inspection station.

The assembly unit, station 2, receives information about components that are to be added to the product and component placement by product type. A document is provided to the assembly station defining these operations. Once assembly has been performed, the operator passes the product with its code number to the curing oven, and may also enter into a notebook an entry confirming that the components have been placed as desired.

The curing oven operator receives a time-temperature profile by product type, provided in a document that must be used to set and monitor the oven. Once the curing operation has been completed, the operator passes the product and product identification to the solder reflow unit. The operator then documents in a report or notebook that the appropriate time-temperature profile has been observed. Finally, the solder reflow unit operator receives a document that describes additional components to be added and their desired placement as a result of the solder reflow operation. Following this operation, the operator documents that the components have been placed as desired.

These simple operations indicate several important considerations that enter into the design of an information matrix based on operator control. It can be immediately noted that the information input to the work cell consists of many different documents that must be available to each operator at the appropriate time. This type of operation can require a complex document distribution and control system in order to make sure that the information is available to each operator when needed. The operator-based product identification and code documentation must be managed carefully so that the individual products do not become intermixed, or operations become exchanged. Further, the operators must be sufficiently well trained to perform all of the operations that are defined in the input documentation in the standard manner.

The information output from these operations is largely a set of hand entries in reports or notebooks, indicating that each operation was performed

Example 7-2 describes two information matrices that might be associated with manually based and computer-based operations. The former case involves many types of documents that must be available to the operator at the desired time, and the maintenance of hand reports and notebooks. The latter case makes use of bar coding for product identification and automatic record keeping. A successful shift from paper-based to computer-based operations will depend on extensive planning and educational efforts.

Nodes of Work Cell	(1) Inspection Unit	(2) Assembly Unit	(3) Curing Oven	(4) Solder Reflow Unit	To Other Work Cells or Factory Information Base
(1) Inspection Unit		Product type/ code number; passed			Results of inspection (pass/fail) and reasons (document)
(2) Assembly Unit			Product type/ code number; passed		Confirm components and placement (document)
(3) Curing Oven				Product type/ code number; passed	Time-temperature profile (document)
(4) Solder Reflow Unit					Confirm components and placement (document)
From Other Work Cells or Factory Information Base	Inspection criteria by product type (document)	Components and placement by product type (document)	Time-temperature profile by product type (document)	Components and placement by product type (document)	

FROM (row label, left side)

(a)

Figure 7-11.
Information flow. (a) Operator-based information flow. (b) Computer-based information flow.

successfully or unsuccessfully and providing brief reasons for the result. As might be expected, major problems can develop in operating a document control system so that all of the recorded information can be accessed by other work cells on demand. This type of matrix also implies a large amount of informal information flow between the operators in a work cell, and between work cells, in order to achieve the degree of information-based coordination that is required.

Figure 7-11b describes a quite different information flow, based on automated equipment and computer networks. The information to each equipment station is now provided by the download of software and data files through the network. The inspection unit (station 1) receives a download of detailed inspection criteria by product type. The product identifying code is often bar-coded onto the product. When a product arrives at inspection, the

Nodes of Work Cell	(1) Inspection Unit	(2) Assembly Unit	(3) Curing Oven	(4) Solder Reflow Unit	To Other Work Cells or Factory Information Base
(1) Inspection Unit		Product ID number; passed; variances revealed during inspection	Product ID number; passed	Product ID number; passed; variances revealed during inspection	Detailed computer record of each inspection task, by part and equipment ID number (software data log/upload)
(2) Assembly Unit			Product ID number; passed	Product ID number; exact placement of component	Detailed computer record of each assembly operation performed by part and equipment ID number (software data log/upload)
(3) Curing Oven				Product ID number; passed	Detailed computer record of time-temperature profiles by part and equipment ID number (software data log/upload)
(4) Solder Reflow Unit					Detailed computer record of each placement and reflow operation by part and equipment ID number (software data log/upload)
From Other Work Cells or Factory Information Base	Inspection criteria by product type (software download)	Components and placement by product type (software download)	Time-temperature profile by product type (software download)	Components and placement by product type (software download)	

FROM

(b)

operator either can enter directly the serial number into the computer terminal or use a bar-coding "wand" to enter the number into the computer; alternatively, automated equipment can scan for such information. Once product identification is achieved, the equipment unit can request the needed inspection criteria for this particular product type, and the computer network can download the appropriate inspection criteria. The inspection unit performs the needed operations and passes data output to the assembly unit, describing any product variances that have been revealed during inspection. In this way, the assembly operation can adapt to any changes that exist among specific products of a given type. Similar information describing product variances can be passed from the inspection unit to the solder reflow unit, again modifying the processes as to achieve the best operations. A detailed record of each inspection task for each product unit can be uploaded through data logging into a permanent data base. This data base is maintained for each specific product that is passing through the work cell and for each equipment unit that is included in the work cell.

Once the assembly unit identifies a product that has been received, it receives component placement data for this particular type of product through a data download. Once the assembly unit has completed its operation, it passes the product to the curing oven for the next step. At the same time, the assembly unit passes information to the solder reflow unit, to the other work cells, and to the data base regarding the exact placement of components that was achieved during the placement operation. This enables all subsequent operations to adapt to the specific results of the assembly operations. The specific components used, and their exact placement, are integrated into the data base. The curing oven receives the time-temperature profile from the host computer data base. Upon completion of the curing process, the oven confirms the actual time-temperature profile that was used for this particular product unit. Finally, on identification of the specific product, the solder reflow unit receives a data download of component and placement information. Upon completion of the operation, the solder reflow unit can perform a data upload, providing a detailed record of each placement and reflow operation.

As can be readily observed by comparing Fig. 7-11a and 7-11b, the operator-based system develops large quantities of paper and written documentation. The system has a limited ability to pass information about the previous processes from station to station, a limited ability to perform ongoing process control, and a limited capability to maintain detailed records regarding the results of manufacturing processes applied to each product and the long-term operational characteristics of each equipment unit. On the positive side, this type of operation can be quite flexible because the operators can adapt from one type of product to another quite quickly (assuming the operators have received appropriate training) and can move from one type of manual equipment to another as required. This type of operation is much less formal, is performed with much less specific input data, and results in much less documentation of output of the product processes.

In contrast, the computer network information flow makes maximum use of the opportunities to control and monitor processes. Bar coding applied at each specific equipment station enables a high level of detailed control to be maintained for each product. By the time each product is complete, a record will have been developed of every operation provided, the results of each operation, and the specific equipment unit that performed each operation. The ability to perform data download, to monitor the equipment operations performed, and to upload or data-log the actual results of the operations provides the potential for a detailed data base about the processes in use. In addition, the ability to pass information from one equipment unit to another enables each unit to adapt to each specific product unit and, therefore, achieve a higher level of quality.

A problem with this strategy is that such automated activity requires equipment units that can accept the software download of information, produce the operations desired, and then achieve the desired data logging and upload. Such operations may run into technological and cost limitations. Extensive software development is required to achieve the operations described here; programming errors in the software can prevent satisfactory system operation until such errors are located and corrected. The system is completely dependent on computer networks; the system will function effectively within the defined range of parameters but will not be as adaptable as the operator-based system. And software maintenance costs may be significant.

This example contrasts operator-based and computer-based manufacturing systems. The preferred system design will often depend on the specific application. Operator-based systems often require lower investment cost and provide more flexible use of available manufacturing equipment. On the negative side, information flow and record keeping is a continuing difficulty. For the computer-based case, difficulties are associated with cost and technology limitations.

EXAMPLE 7-3. ALTERNATIVE INFORMATION FLOW OPTIONS

Figures 7-12a and 7-12b show alternative information flow architectures that may be associated with the information matrices of Example 7-2. Discuss the relative features of each option.

Results

Figure 7-12a shows an operator-based information flow that would be appropriate for the information matrix in Fig. 7-11a. Two types of information flow take place. The first flow is from the organization to the work cell and to each operator of an equipment unit. This information flow is associated with providing the correct means for operating each equipment unit and for providing acknowledgments back to the system that the equipment units are in appropriate operation. The second type of information flow is achieved by physically associating the product documentation with the product itself. As each product moves through the system, it is accompanied by a paper file that describes every operation that has been performed. As the product passes through each processing station, information regarding its identification and previous operations is read by the operator, and the operator makes new entries into the data packet describing the operations performed at the station. The record keeping associated with each product thus moves with the product.

Several difficulties may be noted with this system. The information flow associated with the organization-to-operator linkage may be inefficient; the operator may have to search for the appropriate operating manual of an equipment unit and for general operating rules. The product-oriented data packet is associated with each specific product. No documentation exists to develop profiles of equipment performance across many products or to examine the profiles associated with different product lines. The system leads to an extensive library of paper documents, which again requires ongoing maintenance. Numerous errors may creep into this system because of routine operator mistakes in reading the documents and in recording data. It is difficult to sufficiently understand the processes that are underway in order to develop the most effective system improvements.

Figure 7-12b shows a computer-based, hierarchical computer network that might be appropriate for the computer system described in Fig. 7-11b. Files are stored at both the factory computer and cell computer levels, and a constant uploading and downloading of information takes place as product bar coding stimulates equipment requests. This strategy removes the "paper bag" associated with the operator system of Fig. 7-12a. Detailed information is now available in the computer data base about all operations performed on every product, and profiles can be readily developed by product line or by equipment unit.

The difficulties associated with this configuration have been discussed

Example 7-3 describes alternative information flow architectures that can be associated with the information flow matrices of Example 7-2. A substantial payoff may be possible in evolving from paper-oriented to computer-based operations. However, formidable hardware and software difficulties may also be encountered.

Figure 7-12.
Alternative information
flow architectures. (a)
Operator-based infor-
mation flow. (b) Com-
puter-based, hierarchi-
cal computer network.
(c) Categories of com-
puter hardware.

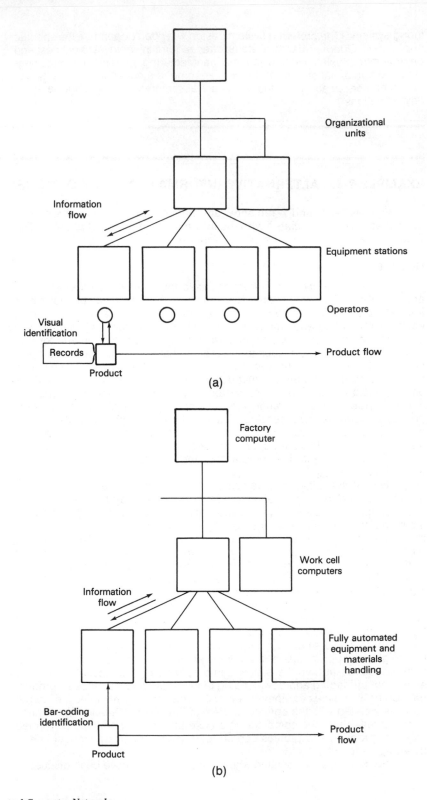

Organizational
units

Information
flow

Equipment stations

Operators

Visual
identification

Records

Product

Product flow

(a)

Factory
computer

Work cell
computers

Information
flow

Fully automated
equipment and
materials
handling

Bar-coding
identification

Product

Product
flow

(b)

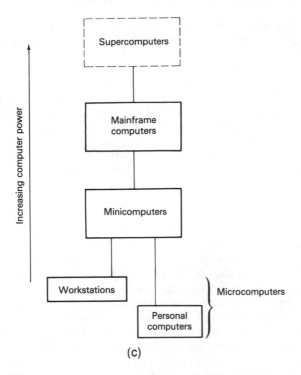

(c)

before. A significant capital investment and extensive software development are required. The system becomes totally dependent on the computer network for its operation, and the function of the organization then becomes supporting the computer network. When this system operates as desired, it will have a high level of efficiency and will allow a high level of control over all aspects of the process. On the other hand, when this system does not operate as desired, formidable difficulties may develop in dealing with hardware and software problems.

The two information flow architectures in Figs. 7-12a and 7-12b are limiting cases. Intermediate cases can make some use of a computer network but also continue to make use of operators. Based on industry experience, it is often cost-effective to attempt to eliminate the paper flow from the floor of the factory. At the same time, the degree to which automation should be applied to the equipment will depend on the particular setting in which the system will function.

7-2. Evolution of Computer Hardware

Computer hardware has evolved rapidly during the past three decades (Fig. 7-12c). During the 1960s, the principal focus was on developing large *mainframe* computers for stand-alone operation. The user would come to the computer location, and provide program and data input through manually prepared punched-cards or tape. The computer would then perform batch operations to read in the data, perform the indicated analysis, and produce paper printouts of the program results. The first mainframe computers were very large by today's standards, had limited memory and speed capabilities, and required a number of specialists to maintain and oversee operations.

This mainframe, or host, approach to computer facilities remained dominant in industry through the 1960s and 1970s. However, during this time, computer capabilities increased as size decreased. It also became possible to attach a number of remote terminals to the mainframe, feed data through these remote terminals, and link several printers to produce the output. Thus, mainframes developed a higher level of performance and were easier to access. Nonetheless, the emphasis continued to be on a central computer with satellite terminals.

The 1970s also saw the introduction of *minicomputers,* which were smaller than the mainframes and reflected improvements in hardware capability. The number of minicomputers and their applications proliferated throughout industry and other settings. The push for minicomputers represented continuation of a focus on increased levels of computing power in smaller, more flexible settings. The minicomputer was still characterized by processing power focused in one location, with links to a number of terminals.

As hardware capabilities evolved during the 1970s, it became possible to assemble smaller and smaller *microcomputers* that could take on some of the functions of the larger mainframes and minicomputers. These microcomputers were developed for specialized applications by those with particular interest in the field, but were limited in usefulness because the hardware was not sufficiently reliable and little commercial software was produced.

The situation changed substantially in 1981, when International Business Machines (IBM®) introduced the *personal computer (PC).* In developing the PC, IBM applied existing technology to create a product that was in keeping with its performance standards. Because of the dominant market position held by IBM, and recognition of IBM's leadership role in the field, the PC had a broad impact. Suddenly a de facto standard was created for microcomputers. Given this standard, a number of companies began to produce hardware enhancements and a wide range of software products for the PC. The market for this type of microcomputer developed rapidly, and other companies began to produce their own versions of the PC through a process called "cloning." During this same period of time, small but powerful computers called *workstations* became widely available on the market. The workstation was an enhanced, improved microcomputer designed to support engineering design and other demanding operations.

The hardware and software available for microcomputers and workstations grew rapidly during the 1980s. Significant increases in hardware capabilities led to a continuing shift from mainframes and minicomputers to microcomputers and workstations. It became possible to purchase a small, relatively inexpensive computer for individual use that would exceed the capabilities of the mainframes of a decade earlier.

This shift in capability encouraged users to consider new ways of using computers to meet industry needs. Instead of applying the mainframe with a number of satellite terminals, it became possible to

Competitive market pressures have resulted in a rapid increase in computer capabilities combined with a rapid decrease in prices. Consequently, the result has been a splintering of the market and a shift from centrally controlled, large, high-cost computers to distributed, small, low-cost computers. This ongoing shift has had a major impact on the feasibility of CIM concepts.

use a number of microcomputers linked together through computer networks to achieve the same purpose. A major shift began to take place to *distributed* systems, which consist of microcomputers linked through a computer network, without the use of a mainframe as the central computing resource. This evolution had a broad impact on computer system operations and on the organizations that use computer resources. Major organizational changes occurred, with a shift away from a large computer run by specialists to decentralized computer facilities available to everyone. This type of network configuration brought the computer to users throughout the enterprise.

The original IBM PC evolved toward the IBM XT® and AT® microcomputer versions, with increasing power and capabilities. The IBM AT, which was the most powerful of these three products, rapidly became the workhorse for industry. The availability of a standard AT class microcomputer and its inevitable clones made it possible to develop a new generation of equipment controllers. Many companies discarded their early controllers to use this high-power, widely recognized product. The growth of equipment unit capability toward automation and integration was enhanced by the availability of such a powerful, standard microcomputer. At the same time, new hardware and software was developed to allow for the networking among computers, providing the means to link these upgraded equipment units.

During this same period, workstation capability also improved dramatically. The distinction in capability began to blur between the minicomputer and workstation, and the engineering designer began to have increased resources in a desktop computer. This evolution stimulated the growth of computer-aided design as an essential aspect of engineering. The workstation became a necessary design tool for use by a wide range of engineers to achieve the desired levels of productivity and product quality.

The distinctions between computer categories are becoming progressively more blurred. Microcomputer and workstation capabilities now overlap with those of the minicomputer. In many cases, the mainframe is linked into a network that includes these other computer facilities. A significant effort is being made today to decide on the most cost-effective combination and configuration of computers for each manufacturing setting. The trend is toward a broader use of distributed processing, with many small computers linked through networks. The eventual role of the mainframe or host computer, and trade-offs in minicomputers versus microcomputers and workstations, are issues that have yet to be resolved, and findings will no doubt depend on the capabilities of each type of computer and on the learning experiences of companies as they attempt to make the most effective use of computer capabilities.

An evolution in computer software and hardware thus has had a major impact on the feasibility of CIM concepts. The development of smart manufacturing equipment and of methods to network computers (through methods discussed below) has made it feasible to consider cost-effective computer applications that enhance manufacturing. In addition, this growth has changed approaches to design (discussed in Chap. 9).

7-3. The Open System Interconnect Model for Computer Communications

The open system inter-connect (OSI) model provides a structured way to think about communications. The complexity experienced in application of the model illustrates the level of detail encountered in information system design. Decision making regarding these aspects of system design must involve a continuing active involvement by the planning group. If this information decision making is delegated to support groups that do not understand systemwide objectives and requirements, the resulting system implementation may have many unexpected and undesirable features.

In order to understand the basic elements of computer communications and to provide guidance for implementation of systems based on computer linkages, it is helpful to describe a layered model of the individual activities that are involved in message exchange among computers. The open system interconnect (OSI) model developed by the International Standards Organization (ISO) provides such a framework (Voelcker 1986).

Figure 7-13 illustrates how two users might use a computer network. As illustrated, the transfer of information takes place from user 1 to user 2. Each message passes through a series of layers that are associated with required message processing. A message is sent from user 1 to user 2, with message acknowledgment sent back from user 2 to user 1. The objective of this communications system is to transfer a message from user 1 to user 2 and to confirm message receipt. The message is developed at layer 7, and passes from there to layer 6, from there to layer 5, and so forth, until the message is actually transmitted over the communications path. The message arrives at layer 1 for user 2, and then proceeds from layer 1 to layer 2, to layer 3, and so forth, until user 2 has received the message. In order for users 1 and 2 to communicate with one another, every message must pass through all the layers.

The reason for developing such a layered approach is to provide

Figure 7-13.
A computer network based on the open system interconnect (OSI) model.

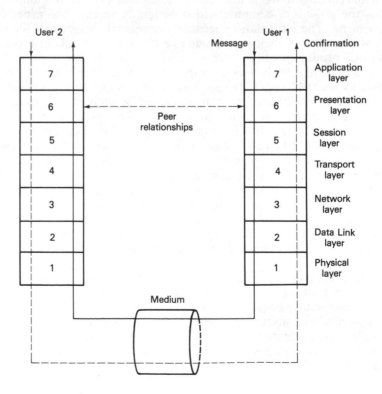

a structure for the messaging procedure. Each time a message moves down from user 1 (from layer 7 to layer 1), additional processing and addressing information is added to the beginning or end of the message. As the original message moves down, new information is being added to obtain the correct communication. Then, as the message moves up to user 2 (from layer 1 to layer 7), this additional information is removed (Fig. 7-14).

The layers work together to achieve "peer" communication. Any information or operating instruction that is added at layer 5 for user 1 will be addressing layer 5 for user 2. The layers thus work as peers; each layer has a given operational or addressing task to make sure the message is correctly communicated from user 1 to user 2. Each layer associates only with the layers above and below itself. The layer receives messages from one direction, processes the messages, and then passes them onto the next layer.

Communication begins when user 1 requests that a message be transferred from location 1 to location 2. In developing this message,

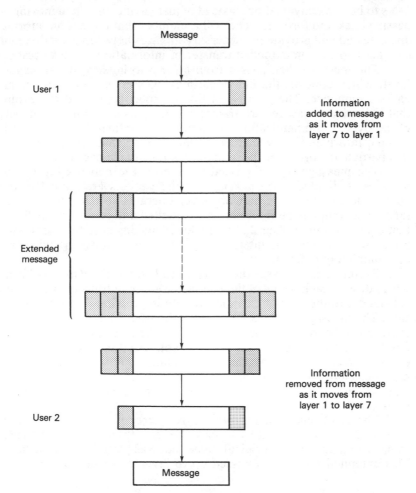

Figure 7-14.
Use of extended messages to apply the OSI model.

it may be necessary to have application software that will provide supporting services to the user. This type of service is provided by layer 7, the *application* layer. Common software application tools enable, for example, the transfer of files or the arranging of messages in standard formats.

The next step involves the *presentation* layer, or layer 6. The initial message is passed down from layer 7 to layer 6, where any necessary translation is performed to develop a common message syntax that user 2 will understand. If the two different users apply different computer or equipment "languages," it will be necessary to define the differences in such a way that the users can communicate with one another. The basic message that began with user 1 is translated to a common syntax that will result in understanding by user 2. Additional information is added to the message at layer 6 to explain to user 2 the nature of the communications exchange that is taking place. An extended message begins to form.

The message now passes from layer 6 to layer 5, the *session* layer. The objective of the session layer is to "set up" the ability for the two users to have a conversation, instead of just having unrelated messages passing back and forth. Layer 5 will remember that there is an ongoing dialogue and will provide the necessary linkages between the individual messages so that an extended transfer of information can take place.

The message then passes from layer 5 to layer 4, the *transport* level, which controls the individual messages as part of a communications sequence. The purpose of the transport layer is to make sure that individual messages are transferred from user 1 to user 2 as part of the overall communications session that is defined by layer 5. Additional information is added to the message so that the transport of this particular portion of the communications exchange results.

The message is then passed to layer 3, the *network* layer. The message is divided into *packets,* and the packets are guided to the correct destination. The network layer operates so that the message (and all accompanying information) is tracked and routed correctly so that it will end up at user 2. This includes making sure that addresses are correct and that any intermediate stops between user 1 and user 2 are completely defined.

System 1 now passes the message to layer 2, the *data link* layer, which directs each *frame* of the message routing in transit. Each frame is loaded into the communications system in preparation for transmission. The message is prepared to leave the user 1 environment and to move onto the communications medium. The next step is from layer 2 to layer 1. At this point, the frame is converted to a series of digital or analog electronic signals that can be placed on the communications medium itself—a wire or coaxial cable, a fiber-optic cable or some other means—to achieve the transmission of the message from user 1 to user 2.

The electronic signal arrives at the correctly addressed location for user 2 and is received by layer 1. Layer 1 then converts the electronic signal back to the original frame that was placed on the medium. This extended message is passed up to layer 2, which confirms that error-free communication has taken place and that the frame has been received at the correct location. Layer 2 directs the frame routing in transit. When the full frame is received at layer 2, the data link layer, the routing information is removed and the remaining information is transferred to layer 3. Layer 3, the network layer, confirms the appropriate routing and assembles the packets. Then the routing information is stripped off the message. The message is then passed to layer 4, the transport layer, which controls the transport of individual messages. Layer 4 confirms that the correct connection has been achieved between user 1 and user 2 and then receives the information necessary to achieve the appropriate connection for this particular exchange.

The remaining information is then passed to layer 5, the session layer, which interprets whether or not this message is part of a continuing communication and, if so, identifies it as part of an ongoing conversation. The information is then passed to layer 6, the presentation layer, which performs any necessary translations to make sure that user 2 can understand the message as it has been presented. The message is then passed to level 7, the application layer, which identifies the necessary support programs and software that are necessary for interpretation of the message by user 2. Finally, user 2 receives the message and understands its meaning.

This step-by-step passing of the message from user 1 "down" to the actual communications medium and "back up" to user 2 involves adding information to the original message prior to the transfer of information and removing this extra information on arrival (Fig. 7-14). At user 1, additional information is added step by step until the communications medium is reached, forming an extended message. When this information arrives at user 2, the additional information is removed step by step until user 2 receives the original message. As noted, a peer relationship exists between the various levels. Each level communicates only with the level above or below it, and the levels perform a matching service. Whenever information that is added to a message by a given level at user 1, the information is removed from the message by the matching level associated with user 2. Communication takes place by having layer 1 communicate with layer 1, layer 2 communicate with layer 2, layer 3 communicate with layer 3, and so forth to achieve an exchange between user 1 and user 2.

This rather complex process involves having an original message modified by additions to the message at the node associated with user 1 and then having these additional materials removed from the message once it is received by user 2. The two users see only the message that is originated and delivered; they do not see all of the intermediate steps. This is analogous to the steps involved in making a telephone call or sending a letter, in which the two users know only that they have managed to start communicating with one another, but do not have any specific knowledge of the details involved in passing the message from one location to another.

This orderly and structured approach to a communications model is useful because it separates out the various tasks that must take place. The model provides a means for assuring that the methods for processing and addressing messages are always the same at every node. Whether a message is being sent or is being received, a sequential processing activity always takes place.

7-4. A Strategy for Comparing Alternative Approaches to Computer Communications

Many advanced manufacturing systems require computer networks to link the various equipment units in the system and to allow communication between the equipment and the work cell and factory controllers. As illustrated in Fig. 7-15, a typical computer network will involve communications capability among the various equipment units and between the equipment units and the higher levels of the system control and management. This requires a communications strategy. This section discusses how the OSI model can be used to compare different computer communications strategies.

Note: The following discussion treats a range of computer network strategies as if they all conform precisely to the OSI model. However, in a number of important cases, these strategies do not exactly relate to the definitions associated with each layer, but must be regarded as only approximately following the definition given above. For purposes of the discussion, these differences are assumed to be secondary to the general emphasis. Of course, if detailed technical issues are being considered, a thorough understanding of such differences is essential. The network strategies are somewhat simplified to aid in initial understanding. The references provide a ready source of further details for those who wish to obtain more specific insights. Tanenbaum (1988) is a useful general reference for computer communications.

Figure 7-16 illustrates the range of solutions that can exist to this problem. The left-hand end of the line (shown at position A) corresponds to a situation in which each equipment manufacturer works independently to develop specifications for manufacturing equipment, without any concern for computer communications requirements. If a computer network for a CIM-oriented system is developed under these circumstances, the system integrator must learn about each equipment

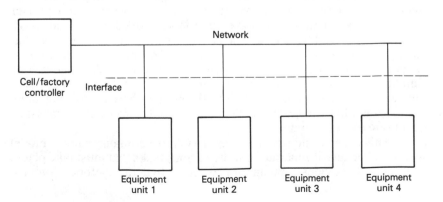

Figure 7-15.
Elements of a typical computer network for a manufacturing system.

Cell/factory controller

Interface

Network

Equipment unit 1

Equipment unit 2

Equipment unit 3

Equipment unit 4

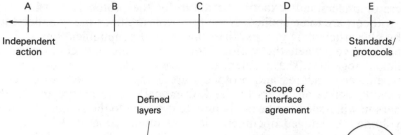

Figure 7-16.
Range of possible solution strategies for network implementation.

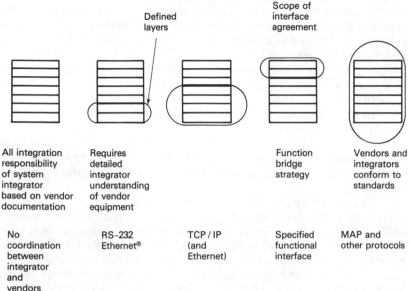

A	B	C	D	E
Independent action				Standards/protocols

Defined layers

Scope of interface agreement

| All integration responsibility of system integrator based on vendor documentation | Requires detailed integrator understanding of vendor equipment | | Function bridge strategy | Vendors and integrators conform to standards |

| No coordination between integrator and vendors | RS-232 Ethernet® | TCP/IP (and Ethernet) | Specified functional interface | MAP and other protocols |

unit in detail and modify the equipment as necessary to produce a functioning network. This is a difficult task that requires the system integrator to make an in-depth study of each equipment unit. Needed information is often not available, or is provided in a form that will not allow the system integrator to obtain network communications easily. This situation is sufficiently difficult that such independent action will often prevent the development of a CIM system. This approach will lead to islands of manual, semiautomated, or automated equipment units that are unable to talk to one another because no effort has been made to build in a communications capability. The system integrator must bear all the costs associated with integration.

At the other extreme (location E) are *standards* and *protocols*. Standards and protocols allow communication among equipment units by defining exactly how each layer of the OSI model is to be implemented. If every layer of activity is precisely defined, then a full-fledged standard or protocol results. So long as everyone conforms exactly to the definitions, then any two units of equipment that are interconnected will be able to communicate with one another. All equipment vendors and system integrators must conform to the exact standards. A number of specific protocols of this nature have been developed (Bux 1981; Borggraaf 1985; Ames 1988; Tanenbaum 1988).

One of the major difficulties in the development of protocols is providing a means so that each participating vendor and system inte-

Figure 7-16 indicates a way in which the planning group can compare a range of network strategies to gain further insight into the relative advantages and disadvantages of the choices. The evaluation of computer network options and selection of a preferred strategy require the ongoing support of highly capable experts in the field.

grator understands exactly the nature of the protocols and how the protocols are to be applied. If the protocols are not exactly understood by all participating groups, then the desired communications will not be achieved when the various equipment units and computers are linked together. The problem of *interoperability* is a significant one whenever standards and protocols are being developed. In order to exactly define all layers of the ISO model and to assure that every person will understand exactly how to conform to the standards, many volumes of detailed documentation are often required. When implementing all of these details, it is easily possible for slight differences to occur between the different groups. If this happens, then when the equipment units are brought together, they will not achieve the desired level of communication.

In order to address this problem, testing facilities are set up to evaluate interoperability for each of the standards being applied. In this way, the various providers of the communications equipment can work together and make sure that their different hardware and software implementations achieve equivalent functions for full information exchange.

Interoperability is not difficult to solve if only one vendor is involved; however, with numerous vendors, the problem can become a difficult one to solve. The two extremes in Fig. 7-16 range from completely independent action by all vendors, which often will result in equipment that cannot achieve any integrated system, to the application of standards and protocols that define every level of the system in such sufficient detail that all vendor equipment will work together in the same environment. In an ideal sense, it may seem that the development of full standards and protocols is always the best approach. However, the problems of having all the vendors and integrators decide on the standards to which they will conform is in itself a difficult task. Also, development of the necessary hardware and software to implement the standards is at best a several-year activity. The time and resource requirements for developing industry-wide standards and protocols are the inhibiting factors. Nonetheless, several well-recognized protocols have been developed and are commercially available.

Between the two extremes other approaches have been taken to achieve computer communication. Option B is one in which the lower layers of the communications system are defined through a "partial protocol" in order to achieve common approaches to physical interconnections and message processing. The middle and upper layers associated with defining messages, addresses, and applications are left undefined for implementation in individual cases. This method requires a detailed understanding of the vendor equipment by the system integrator in order to develop message and application software at the upper layers, but it is a definite improvement over option A in which no agreement is reached. At present, RS-232 and Ethernet®* communications links are commonly used to achieve communications in this way. Many equipment units are provided by vendors with RS-232 or Ethernet connections to the equipment unit to provide for computer-

* Ethernet® is a registered trademark of the Xerox Corporation.

integrated operation. Documentation is provided by the vendor to tell the system integrator how upper layer communication can be implemented. Extensive effort is required by the system integrator to understand the nature of each equipment unit, which makes this approach more difficult to implement than the full standard or protocol, from the integrator's point of view.

For option C, the lower and middle layers of the communications system are defined by protocol, but again the application layers at the top are left to individual decision making. A communications system called TCP/IP, which can exist on top of the Ethernet connection, is available in this category. Further constraints are placed on each equipment vendor, and less individual equipment familiarity is required by the system integrator.

Option D shows a different approach, in which the lower and middle communications layers are left undefined but agreement is reached between vendors and the system integrators as to the functions that will take place across the application layer. This *function bridge* strategy requires that vendors and integrators work together to define the functions that the equipment must perform to meet the needs of the system integrators and then to specify a functional interface. The implementation of the lower levels of this interface is then left to the individual parties. The function bridge strategy is useful in many settings. This approach, significantly different from the others in Fig. 7-16, is discussed in Sec. 7-8.

All of the approaches in Fig. 7-16 are sometimes used by industry. Approach A, which involves independent vendor action, places the full burden of system development on the integrator and is so restrictive that effective CIM operation often cannot be developed. An industry in which original equipment manufacturers take this basic approach may encounter much difficulty in achieving CIM operations.

The approach at point E is at the opposite extreme. For this case, every aspect of communications must be defined through standards or protocols. It turns out to be quite difficult for complex protocols to ensure that the communications interfaces will always follow the exact specifications. Approach E may be effective in large, more stable settings but may not be as widely applicable in industries that are in a rapid state of change and that have a wide range of equipment vendors who have limited capabilities for developing the communications aspects of their products.

Approaches B and C involve developing lower-level standards or protocols. The particular message and application software can be constructed on top of the physical-level capability. These approaches are often used by vendors as a first step toward communications, because they allow the vendors to focus completely on their equipment design and not be concerned with larger system requirements. On the other hand, these approaches place requirements on the system integrator to explore each equipment unit in detail and to learn how to communicate with this individual unit.

The function bridge approach (option D) uses functional definitions to relate the system integrator to the individual equipment vendors. The communications strategy is defined by the system integrator above the functional interface and by the equipment vendors below the equipment interface.

7-5. RS-232–Based Networks

A seemingly simple problem in the development of computer networks is to establish the ability to interconnect any two typical computer system elements. This might involve a computer terminal, a modem,* a printer, or other system elements that must achieve an information exchange. It might seem reasonable that such one-to-one interconnections would follow a very well-defined strategy and take maximum use of a standard. Unfortunately, for historical reasons and because of the wide diversity of the equipment units that are available today, this situation does not hold. In fact, achieving interconnection between typical system elements can be a frustrating experience.

The common RS-232 approach to computer communications makes use of a strategy that was never intended for the purpose. The result is a "guideline" that can often be confusing and difficult to apply, rather than a comprehensive standard. The historical evolution of RS-232 illustrates how a rapidly changing technology can lead to gaps between the available hardware and software and the resources that are desired by user groups.

One of the most common approaches used to interconnect typical computer system elements is associated with a strategy that was never really intended for this purpose. As noted by Campbell (1984), "In 1969, the EIA (Electronic Industries Association), Bell Laboratories, and manufacturers of communications equipment cooperatively formulated and issued EIA RS-232, which almost immediately underwent minor revisions to become RS-232-C." The RS-232 interface was developed to allow data equipment terminals to be connected to modems, to be able to transmit data over the telephone networks. The entire purpose of this standard was to assure that the use of telephone lines to achieve computer communications would be handled in a way that would be acceptable to the telephone company. Unfortunately, this particular problem (connecting a data terminal to a modem) did not address many of the general issues involved in communicating between the elements of a computer network. The documentation only addresses the specific issues that are raised by the terminal-modem interface for which it was intended. Thus, when trying to apply the RS-232 solution to a variety of other settings, many questions are raised that are not resolved by the available documentation.

Thus, in its general application today, RS-232 is not a standard. It is more a guideline used to address some of the issues involved in interfacing equipment. Many issues must be resolved on an individual basis, which leads to the wide potential for difficulty. Essentially, a vendor's statement that a computer system element is RS-232 compatible provides a starting point to consider how two equipment units might be interconnected. However, the detailed aspects of the interconnection will require further understanding of the ways in which the two equipment units are intended to communicate. Campbell (1984) is a helpful introduction to dealing with the issues that result in trying to apply RS-232 concepts.

In a sense, the history of the RS-232 interface illustrates the difficulties associated with creating a well-defined means for allowing the

---

* The word *modem* is an abbreviation for modulator-demodulator, referring to a device that links layer 1 of the OSI model to the physical medium.

various elements of a computer network to interact together. Past experience also indicates how difficult it is to develop standards that will apply in the future to all of the different situations that will be encountered. As it has evolved, the RS-232 approach to a communications interface is an improvement over the "total anarchy" position indicated in location A in Fig. 7-16, but it still leads to a wide range of problems.

An entire computer network can be configured by using combinations of point-to-point RS-232 connections. In fact, a number of networks of this type are in common use. Such networks require that multiple RS-232 interfaces be present on each equipment unit, as is often the case. Each particular interconnection must be customized for the two equipment units being considered. Thus, the system integrator not only must decide on the elements of the system and how they should perform in a functional sense, but also must develop a detailed understanding of the ways in which the RS-232 concepts have been applied to the particular equipment units that are used for the network. The system integrator incurs a substantial expense in achieving the required interconnections.

It is essential to realize that the RS-232 pseudo-standard only addresses the ability to transfer serially information bit by bit from one system element to another. The higher levels of communications protocol shown in Fig. 7-16 are not considered. The RS-232 solution allows the means for running wires from one element to another, in order to allow digital voltages to be conveyed between system elements. The meaning associated with these bits of information is completely dependent on the hardware and software that is implemented in the system elements. The RS-232 solution has resulted in a widely used approach to allowing two computer elements to transfer information from one to another. The situation certainly is much improved over a circumstance in which no guideline is introduced. However, because RS-232 does not completely define all of the relationships that must exist in communications linkages, it falls far short of being a true standard or protocol.

7-6. Ethernet

As illustrated in Fig. 7-16, one approach to local area network (LAN) development is to define a protocol for the first two layers of a communications strategy and then allow individual users to define the upper layers. This approach has been widely applied using a method referred to as Ethernet (Metcalfe and Boggs 1976; Shock and Hupp 1980; Tanenbaum 1988). (There is actually a family of related standards that function in the general method described here.)

In every computer communications system, there must be a means for scheduling when each node will transmit onto the network and will listen to receive messages. One approach to this issue is to use a statistical process. When a unit needs to transmit over the network, it makes an effort to transmit. If another node tries to transmit at the same time, both nodes become aware of the conflict, wait for a

Figure 7-17.
Performance of an Ethernet® computer network, showing delay and throughput as a function of the network load (number of users). Based on L. Kleinrock and S. S. Lam, "Packet Switching in a Multiaccess Broadcast Channel: Performance Evaluation," *IEEE Transactions on Communications*, COM-23, 4 (April 1975), 410. (Ethernet is a registered trademark of XEROX Corporation.)

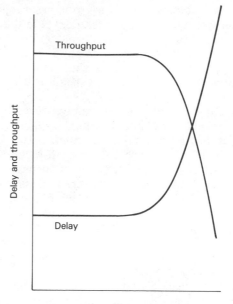

random length of time, and try again. It might seem that this would be an inefficient means for obtaining network control, since the various nodes are essentially trying to claim the network for their own use on a random basis and many collisions may occur. However, as it turns out, for lower communications volume requirements, this method works very well. As illustrated in Fig. 7-17, as the number of nodes on the system and the number of messages being exchanged increases, the number of message collisions taking place between these active nodes goes up and reduces the effectiveness of the system.

This type of access control for a computer network is referred to as CSMA/CD (carrier-sense multiple-access with collision detection). Ethernet and similar solutions are widely applied to create networks, particularly in settings in which a maximum access waiting time to the network does not have to be guaranteed. This type of network is simple to install, and a wide range of hardware and software products are available for support. On the other hand, as indicated in Fig. 7-17, network performance can degrade significantly under high load; therefore, the utility of an Ethernet-oriented network will depend on the particular configuration and loads that are expected for the network.

7-7. TCP/IP

TCP/IP, or transmission control protocol/internet protocol, applies to the transport and network layers shown in Fig. 7-13. TCP/IP thus provides a means for addressing intermediate protocol levels, and in fact is often combined with Ethernet in a communications approach that defines both the lower and middle aspects of the system (Tanenbaum 1988). TCP/IP functions by dividing any message provided to these middle layers into *packets,* and then sending packets of 64K bytes at a time through the communications network. TCP/IP must also reassemble the packets in the correct order at the receiving user. (The TCP/IP protocol does not directly correspond to the detailed aspects of the formal OSI model definitions for the transport and network layers, but is sufficiently similar to be understood in the present discussion.)

TCP/IP provides a common strategy to use for networking. It allows extension of the Ethernet lower layers to a midlayer protocol on which the final application and presentation layers may be constructed.

EXAMPLE 7-4. COMPARING NETWORK STRATEGIES

Discuss the trade-offs that might be associated with RS-232, Ethernet, and TCP/IP as strategies for implementing a computer network.

Results

The RS-232 approach provides a means for creating point-to-point connections between individual equipment units and computers. RS-232 defines only the physical layer of communications requirements and does not provide any information about the upper layers. Many different variations in RS-232 exist, so that even at the physical layer customizing may be required to achieve the desired communications capability. Achieving satisfactory hardware and software interfaces between the units by using RS-232 can be a substantial and time-consuming problem. In addition, such linkages require the development of the necessary software interfaces and application programs to enable the various network nodes to talk to one another. It may often be useful to apply RS-232 as a link between specific equipment units and their controllers, but to use a different communications strategy to develop cell- and factory-level networks.

Ethernet provides a means of creating a local area network by defining the lower two levels of a communications system. Ethernet provides the opportunities associated with a LAN and a potential for more rapid information exchange. The move from point-to-point communication to LAN communication can be effective in many settings. Since Ethernet depends on statistical control of network access, this approach can be taken only if the performance of the system does not require a guaranteed maximum waiting time before network access is achieved. Ethernet still requires a full definition of the upper software layers for a communications system, and will increase system expense over that associated with RS-232. Ethernet is widely used in both office and factory settings as the basis for computer networks.

The decision to use TCP/IP can simplify the software development problems. If an Ethernet LAN is in place, TCP/IP can be added to create middle-layer message control; the customizing of the system can focus on the presentation and application layers of the communications system and on the software drivers and interfaces associated with equipment units.

These different aspects of network development have varying strengths and weaknesses, and each can contribute to meeting computer communications requirements.

Many other alternatives exist for developing computer networks. Application trade-offs must be considered in selecting from among the available choices.

7-8. Manufacturing Automation Protocol (MAP)

The manufacturing automation protocol (MAP) is one of the protocols that has been developed for computer communications systems (Kaminski 1986). MAP was developed specifically for use in a factory environment. General Motors (GM) Corporation has been the leading advocate of this particular protocol. When faced with a need for networking many types of equipment in its factory environment, General Motors decided that a new type of protocol was required. Beginning in 1980, GM began to develop a protocol approach that could accommodate the high data rates expected in its future factories and provide the necessary noise immunity expected for this environment. In addition, the intent was to work within a mature communications technology and to develop a protocol that could be used for all types of equipment in GM factories. MAP was developed to meet these needs. The GM effort has drawn on a combination of IEEE and ISO standards, and is based on the open system interconnect (OSI) layered model (see Fig. 7-13).

MAP reflects an effort to standardize computer communications in a manufacturing setting. The objective has been to "solve" the networking problem. Unfortunately, the resulting strategy is so complex and costly that the application of MAP hardware and software has been limited. Thus, although it has become one of the possible strategies to be applied by a planning group, MAP is not the obvious "solution."

Several versions of the MAP standards have been developed: versions 2.1, 2.2, and, most recently, 3.0. One difficulty among a number of problems has been in obtaining agreement among many different countries and vendor groups on specific standards. Another problem is that the resulting standards are so broad that they have become very complex, making it difficult to develop the hardware and software to implement the system and consequently driving up related costs. Early versions (2.1 and 2.2) of MAP addressed some of the OSI layers to a limited degree, and made provision for users to individualize the application layer for a particular use. Version 3.0 makes an effort to define more completely all the application layer software support as well as the other layers. This has led to continuing disagreements and struggles to produce a protocol that can be adopted by every vendor group in every country to achieve MAP goals.

Because of its complexity, MAP compatibility among equipment units, or interoperability, has been a continuing difficulty. MAP has not been applied as rapidly as was initially hoped for by its proponents because of the complexity, the costs, and disagreements on the way in which MAP should be implemented. Assembling a complete set of documentation for MAP is a difficult activity that requires compiling a file of standards organization reports, a number of industry organization reports, and documentation from all the working committees associated with ISO.

The MAP protocol was developed with several alternatives for the physical layer (layer 1). MAP can be implemented through what is called a *broadband* system (Fig. 7-18). In order to allow the manufacturing units to talk to one another, transmitted messages are placed on the cable; a *headend remodulator* retransmits these messages and di-

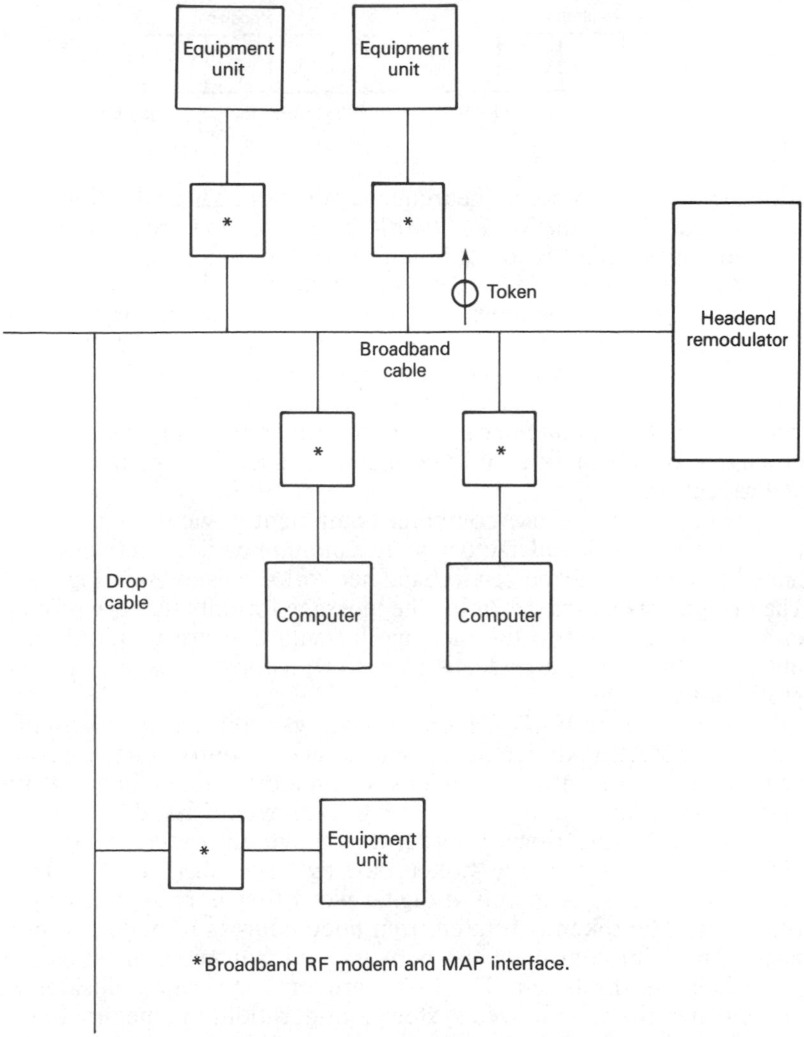

Figure 7-18.
Broadband MAP net-
work.

Equipment unit

Equipment unit

*

*

Token

Headend remodulator

Broadband cable

*

*

Drop cable

Computer

Computer

*

Equipment unit

*Broadband RF modem and MAP interface.

rects them to the receiving stations. The broadband version of MAP has the highest capabilities because it allows several types of communications to take place on the same cabling at the same time. On the other hand, because of this increased flexibility, broadband is more complex and more expensive to install. It requires modems and MAP interface equipment for each item of equipment and a headend remodulator to serve the entire network. The main cable used for broadband is unwieldy (approximately 1 inch in diameter) and is appropriate only for wiring very large factories. Multiple drop cables can be branched off the main cable for the different MAP nodes. Broadband communication can achieve a high data rate of 10 megabits per second and can "split" the frequency spectrum to allow several different communications to take place simultaneously. As shown in Fig. 7-19, the three transmit frequencies and the three receive frequencies are sep-

Figure 7-19.
Broadband frequency spectrum, showing three transmit and three receive channels.

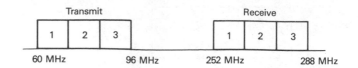

arated from one another in the frequency domain. Three different channels can coexist on the MAP network. The headend remodulator transfers messages from the low frequencies to the high frequencies.

Another type of MAP network makes use of a *carrierband* approach, which uses somewhat less expensive modems and interface units and does not require the same heavy-duty cable (Fig. 7-20). For a small factory, a carrierband version of MAP can be much more cost-effective. The carrierband communication can also achieve a high data rate of 5 to 10 megabits per second, but only one channel can operate on a cable at a given time. A single channel is used for both transmission and reception.

It is possible to use computer equipment devices called *bridges* to enable a broadband factory-wide communications network to be linked to more localized carrierband networks, as shown in Fig. 7-20a. The bridge acts by transforming the message formats that are provided on one side of itself to the message formats that are required on the other. In this sense, a bridge is used to transform from one operating protocol to another.

In developing MAP, General Motors was concerned with assuring that every MAP node would be able to claim control of the network and communicate with other nodes within a certain maximum waiting time. Within this waiting time, every node would be able to initiate required communications. In order to achieve this outcome, MAP implements what is called a "token passing" system (Fig. 7-20b). The token in this case is merely a digital word that is recognized by the computer. The token is rotated from node address to node address; a node can claim control of the network and transmit a message only when it holds the token. The token prevents message collisions and also ensures that, for a given system configuration, the maximum waiting time is completely defined (so long as no failures occur). The token is passed around a logical ring defined by the sequence of node addresses, not necessarily by the physical relationships.

The token is a control word, and each MAP node can initiate communications only when it possesses the token. It is interesting to note that MAP nodes that are not a part of the logical ring will not ever possess the token, but they may still respond to a token holder if a message is addressed to them. Token management is handled at layer 2 of the OSI model. This layer controls the features of the token application with respect to how long a token can be held, the sequence of addresses that is to take place, and the amount of time that is allowed for retrying communications before failure is assumed. If the logical ring is broken at some point, for example, if one equipment unit is no longer able to operate, the other nodes will wait a certain length of time and then will re-form the token passing scheme. They will do this

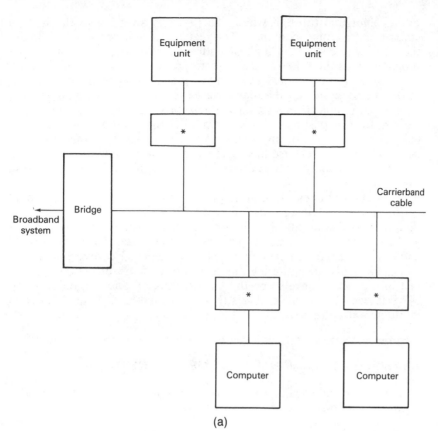

Figure 7-20.
Carrierband system and token passing. (a) Carrierband MAP network. (b) Token passing, given a logical sequence of addresses.

(a)

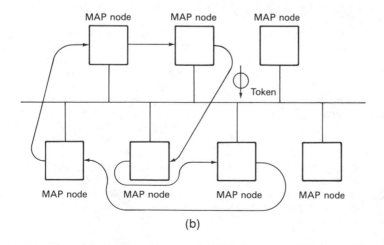

(b)

by an algorithm through which the token is awarded to the highest station address that contends. The rest of the stations on the ring are then determined by the highest address successors. This process is repeated until the token ring is re-formed.

Layer 1, the physical layer, involves the encoding and modulation of the message so that the digital data are transferred into analog and digital communications signals. Each MAP application requires a modem for this purpose. The modem takes the extended message that has been developed at higher layers and modifies it so that it can be used to provide an electronics signal onto the communications medium. The medium itself provides the means for transferring the signal from user 1 to user 2.

MAP continues to be an important protocol approach for application in CIM environments. For very large factories, the broadband option is available, and for smaller factories a carrierband system is also available. A number of vendors now produce the hardware and software necessary to establish a MAP network. However, such networks typically are quite high in cost and, because of the complexity of the protocol, can also be difficult to develop and maintain. Thus, MAP is one of several solutions that are available to the planning and implementation teams.

EXAMPLE 7-5. MAP-BASED COMPUTER NETWORKS

Discuss the advantages and disadvantages of MAP-based computer networks with respect to the network strategies discussed in Example 7-4.

Results

MAP provides a full set of protocols for all layers of a communications system, and has been designed for specific use in demanding factory environments. MAP allows the creation of LANs having a guaranteed maximum waiting time for access and the ability to achieve high rates of data exchange on a deterministic basis. In terms of performance, there is no question that MAP is much more sophisticated than the other approaches discussed in Example 7-4.

In selecting a communications strategy, the objective is typically to decide on the level of performance that is required and the minimum costs associated with achieving that performance. MAP can be an expensive "overkill" solution in simple settings. Thus, MAP should always be considered, but often only in demanding circumstances will it be the best solution.

Several other full protocols represent alternative strategies that may be used to create a full LAN based on commercial products. The selection of an optimal computer network strategy for a given manufacturing setting is not a simple matter, and decision making in this area requires support from those who are knowledgeable about computer networks. Evaluation efforts as to which network strategy should be followed must include contributions by experienced project managers who can discuss the advantages and disadvantages associated with implementing the different strategies, the technical problems that can arise in each case, and the likely costs to be associated with achieving a fully operational system.

EXAMPLE 7-6. COMPUTER NETWORK SIMULATION USING NETWORK II.5.

As discussed in Chap. 4, computer network requirements must be considered as an integral part of the system design activity. Figure 4-5 specifically addresses information handling requirements as steps 11 and 12 in the system design procedure. Chapter 5 discusses the advantages of various types of modeling to evaluate system performance (step 10 in the design process). What modeling capabilities exist to evaluate the performance of computer networks?

Results

(A) General Features: One software product for the assessment of network performance is NETWORK II.5®, offered by CACI Products Company. This software was introduced in 1983, and has passed through several evolutionary stages to date. NETWORK II.5 provides the means for simulating a computer network in order to determine its operational characteristics under specified conditions. Both hardware and software models are employed in the simulation.

In developing a NETWORK II.5 simulation, the user begins by defining the building-block hardware elements of the system—processors, transfer devices, and storage devices. Processors, or processing elements, perform the information processing tasks that are typically associated with a computer and are characterized by the cycle time (which describes the speed of the processor) and the instructions used (which describe the activities that can be undertaken by the processor). A wide range of functional processing instructions may be used. Transfer devices include the buses or methods of interconnection in the network. The data storage devices describe the memory capabilities in a computer system.

The characteristics of these three building blocks are illustrated in Fig. 7-21a. The processing elements are described by the types of instructions they can perform and the cycle time required for these instructions. Data transfer devices are characterized by the ways in which data can be transferred and the protocol that is used. Data storage devices are characterized by the amount of data they can store, the time it takes to access the storage, and the method of access (such as a READ or WRITE operation). As used in this context, the list of instructions being considered is not typically the actual detailed instruction set for the machine being simulated, but is a higher-level functional set of instructions that can be associated with typical operating requirements.

Once the hardware system has been defined, the characteristics of the software must also be described. The software is defined in terms of a number of modules, where each module may contain a series of instructions or operations to be performed. Each module is functionally described in terms of the operations it must perform, the running time required for these operations, and interactions with other software modules. In the simulation, each software instruction or operation can be represented by a time delay. Interactions among modules can control the sequence of operations and can introduce delays in moving among operations. The instructions being performed provide a link between the hardware and the software, as hardware processing speed will determine the associated software delays. During a simulation, a set of logical rules is established to define the input to each program, the nature of the program operations, and the output of the program. The logical relationships are then implemented through hardware capability, which will define software performance.

(B) MAP Application: An interesting application of NETWORK II.5 involves application to a MAP-based local area network. Figure 7-21b shows such a network that has been considered in an example prepared by CACI Products Company. Each automated equipment unit has its own operating software, and the control software resides in the computer. When required, the computer sends control messages to one or more of the three units. These units provide the appropriate response according to their own programming. In addition, each unit may request guidance from the computer by sending a request message. Thus, the three equipment units interact with the computer by both receiving commands and requesting guidance. Each equipment unit performs an operational task, as instructed and within system constraints.

The performance evaluation of such a system involves a statistical analysis of the data flow in the system. This data flow will depend on the nature of the software modules and on the token bus aspects of the MAP network. Each of the software programs is described in terms of its operating characteristics, run time, and probability that a control or response request must be dealt with during a particular period of time. A set of logical requirements can be developed to describe the message priorities and message implementations among the various software modules. The simulation allows the different modules to operate and interact together, using the hardware configuration as a constraint. This allows utilization rates to be established for each element of the system and for communications delays to be predicted on a statistical basis. The characteristics of the controlling computer, the bus, and the equipment units may all be established.

The results of the simulation are printed out in tables of utilization statistics for all elements of the system. Figure 7-21c shows sample output reports. A wide range of information is obtained from such reports.

Figure 7-21.
NETWORK II.5™ simulations (courtesy of CACI Products Company). (a) Characteristics of the three building blocks used to develop a NETWORK II.5 simulation. (b) Application of a NETWORK II.5 to a MAP-based local area network. (c) Sample output reports. (d) Typical input screen. (e) A sample utilization plot that might be associated with bus utilization for a particular configuration.

Element	Descriptive Parameters
Processing element	Types of (functional) instructions Cycle time for instruction
Data transfer device	Methods of data transfer Protocol in use
Data storage device	Amount of data that can be stored Access time Method of access

(a)

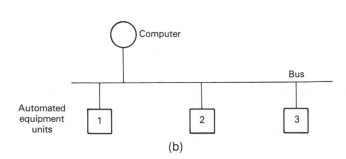

(b)

Figure 7-21. continued

TRANSFER DEVICE UTILIZATION STATISTICS
TO SIMULATED TIME 1. SECONDS

0 0

TRANSFER DEVICE NAME	BUS 1	BUS 2	BUS 3
TRANSFER REQUESTS GRANTED	88	1	40
INTERRUPTED REQUESTS GRANTED	0	0	0
AVG REQUEST DELAY	253.357	0.	0.
MAX REQUEST DELAY	10342.552	0.	0.
STD DEV REQUEST DELAY	1315.136	0.	0.

STORAGE DEVICE UTILIZATION STATISTICS
TO SIMULATED TIME 1. SECONDS

(ALL TIMES REPORTED IN MICROSECONDS)

0

40
1132.864
1500.000
437.538

STORAGE DEVICE NAME	RADAR DISH	MEM 1
STORAGE REQUESTS GRANTED	20	69
INTERRUPTED REQUESTS	0	0
AVG REQUEST DELAY	0.	7.672
MAX REQUEST DELAY	0.	529.368

0.
1.000
0.

63.265

4.531

COMPLETED INSTRUCTION REPORT
TO SIMULATED TIME 1. SECONDS

9

60
0
0.
0.
0.
0.

INSTRUCTION TIME	COUNT	INSTRUCTION TIME	COUNT
PE 1			
GET RADAR DATA	20	GET MESSAGE BUFFER	8
STORE RADAR DATA	20	RADAR DATA AVAILABLE	20
STRING COMPLETE	28	FLIGHT CONTROL SUPPORT	20
CHECK MESSAGE BUFFER	8	RADAR PROCESSING	20

0

68
5028.850
11528.468
4261.783

1
20
9

84000.
8.400

PROCESSING ELEMENT UTILIZATION STATISTICS
TO SIMULATED TIME 1. SECONDS

(ALL TIMES REPORTED IN MICROSECONDS)

20
20
20

72041.194
84000.
6938.917

34.233

PROCESSING ELEMENT NAME	PE 1	PE 2
STORAGE REQUESTS GRANTED	49	40
INTERRUPTED REQUESTS	0	0
AVERAGE WAIT TIME	0.	13.234
MAXIMUM WAIT TIME	0.	529.368
STD DEV WAIT TIME	0.	82.648
GEN STORAGE REQUESTS	29	0
FILE REQUESTS GRANTED	20	40
INTERRUPTED REQUESTS	0	0
AVERAGE WAIT TIME	0.	0.
MAXIMUM WAIT TIME	0.	0.
STD DEV WAIT TIME	0.	0.
TRANSFER REQUESTS GRANTED	69	60
INTERRUPTED REQUESTS	0	0
AVERAGE WAIT TIME	292.744	34.935
MAXIMUM WAIT TIME	10342.552	1269.700
STD DEV WAIT TIME	1472.667	192.434
INPUT CONTROLLER REQUESTS	20	20
DEST PE REQUESTS GRANTED	0	0
AVERAGE WAIT TIME	0.	0.
MAXIMUM WAIT TIME	0.	0.
STD DEV WAIT TIME	0.	0.
RESTARTED INTERRUPTS	0	0
AVG TIME PER INTERRUPT	0.	0.
MAX TIME PER INTERRUPT	0.	0.
STD DEV INTERRUPT TIME	0.	0.
MAX INTERRUPT QUEUE SIZE	0	0
AVG INTERRUPT QUEUE SIZE	0.	0.
STD DEV INTERRUPT QUEUE	0.	0.
MAX MODULE QUEUE SIZE	1	1
AVG MODULE QUEUE SIZE	0.	0.
STD DEV MODULE QUEUE	0.	0.
PER CENT PE UTILIZATION	9.556	33.906

(c)

Figure 7-21. continued

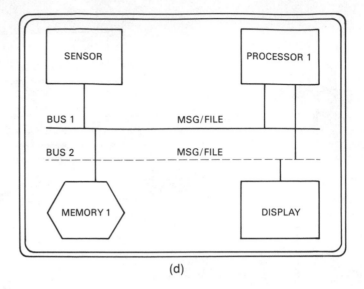

(d)

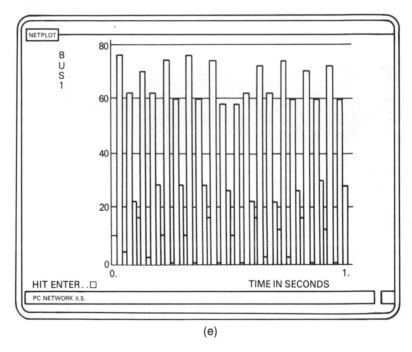

(e)

More recent versions of the software make use of NETANIMATION, which allows a more user-friendly interaction with the system simulation. The application of NETANIMATION allows the user to construct graphically the types of networks of interest without detailed line-by-line programming. Figure 7-21d is a typical input screen that might be associated with application of NETWORK II.5 using the animation feature. A menu-driven editor prompts the user for the necessary information needed to describe the network. The animation feature may be used to produce user-friendly graphics of many different kinds to describe the output. Figure 7-21e shows a sample utilization plot that might be associated with bus utilization for a particular configuration. Another option provided by NETANIMATION is the ability to observe, through event-driven animation, the activities and status of the network as the simulation runs.

NETWORK II.5 is a useful means for evaluating the performance requirements of computer networks. A computer network can be described for each proposed system configuration, and the processing requirements and performance associated with this system can be determined by simulation. In this way, each potential system design can be evaluated in terms of data processing requirements and capabilities. In turn, these results can be used to estimate the costs that would be associated with the development and implementation of such systems.

7-9. The Function Bridge Strategy

The function bridge strategy (see Fig. 7-16) provides a different approach to developing an interface between equipment manufacturers and a system integrator and in achieving integrated computer network operation.* Figure 7-22 illustrates the function bridge concept. A functional interface is defined between each equipment unit and the system integrator. This functional interface takes place in a microcomputer that is utilized by both the equipment vendor and the system integrator to achieve the desired communications.

The function bridge approach provides a reasonable meeting ground for the system integrator and the individual equipment vendors. The system integrator and equipment vendors must work together to define the functions that will be required to implement the desired types of factories. It is the responsibility of the equipment manufacturer, with input from the system integrator regarding needed applications, to develop a set of function calls or subroutines and associated software drivers that can be used to operate the specific equipment. The function calls or subroutines enable the particular equipment unit to be operated remotely from a microcomputer. Each associated subroutine performs the necessary communications function from the microprocessor to the equipment unit.

The function bridge approach requires that the equipment manufacturer specify the hardware environment and software environment to be used for the microcomputer and provide to the system integrator the names of the subroutines that have been prepared and the operations that are associated with these subroutines. The vendor provides to the system integrator the equipment unit, the specifications for the

The function bridge approach to creating computer networks is based on a flexible partnership between equipment vendors and system integrators (users). If adequate communications can be developed among all affected groups and reasonable choices made and adopted, the flexibility of this strategy brings many advantages to participating groups. On the other hand, this same flexibility can result in disagreements and struggles unless an adequate information exchange capability is in place.

* Functionally-based computer communication systems have been studied in a variety of settings. *Direct translators* (as discussed in Chap. 9) can fall into this category. The "function bridge" approach that is described here is closely related to a "meet in the PC" strategy that has been discussed in meetings of the Ad Hoc Advisory Panel (AHAP) for the Integrated Facility for Automated Hybrid Microcircuit Manufacturing (IFAHMM) program, sponsored by the U.S. Naval Ocean Systems Center (NOSC). The "meet in the PC" concept was presented by R. F. Unger and R. A. Unger under NOSC sponsorship; this concept and related approaches to computer communication have been further explored by F. Rybczynski of the National Institute of Standards and Technology. See Example 7-8 for additional background information.

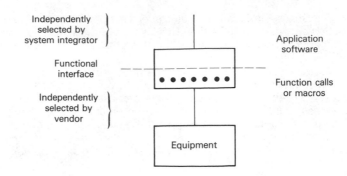

Figure 7-22.
Function bridge
strategy.

Independently
selected by
system integrator

Functional
interface

Independently
selected by
vendor

Application
software

Function calls
or macros

Equipment

hardware and software environments to be used in the microcomputer function bridge, and a computer disk that contains the subroutines that can be used to communicate with the equipment unit. When the system integrator buys the equipment unit, the system integrator then proceeds to implement the necessary microcomputer environment to achieve the function bridge, inserts the computer disk with the function calls into the microcomputer, and then writes application layer software for the microcomputer to achieve the desired operating mode for the equipment.

Thus, by using the function bridge approach, the vendor does not have to be concerned with the manufacturing system in which the equipment is being used, but only with the functional requirements that must be reflected in the system. Once an agreement exists between the vendor and the system integrator about the functional operations required, the vendor is only required to develop subroutines that will enable these functions to be achieved from a microcomputer environment. With specification of this environment and handover of the subroutines, the vendor is not required to learn more about the system itself or to become actively involved in the system design. At the same time, the system integrator does not have to develop a detailed knowledge about the equipment unit itself, but only how to use the subroutines that are available in the function bridge, to achieve the desired system operation.

The function bridge approach provides the means for segmenting the computer communications network for CIM systems. Each equipment manufacturer can restrict activities to the portion of the system interface below the dotted line in Fig. 7-22, and each system integrator can restrict activities to above the line. A useful functional interface is then defined between system integrators, who use the equipment, and the vendors, who manufacture the equipment.

As shown in Fig. 7-23, one approach to developing a CIM network is to make use of the functional system design approach (discussed in Chap. 4) in combination with the function bridge approach. Such a method can be developed by starting out with a cooperative study activity between the system integrators (equipment users) and the original equipment manufacturers (equipment vendors). If these two groups work together in a design activity that makes use of the functional system design strategy, they can agree upon the functional equipment

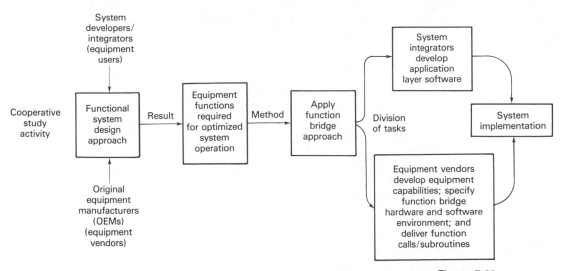

Figure 7-23.
Linking functional system design with the function bridge computer network strategy.

operations that are most appropriate for the systems of interest during the time frame of emphasis. They can develop a mutual understanding of the system and equipment operations that are most appropriate to achieve the highest benefit-cost ratios for the particular product areas and market strategies of interest. The functional system design developed by the two groups can then be used to produce the equipment functions required for optimized system operation.

The function bridge approach to system integration is quite distinct from the others shown in Fig. 7-16. The development of standards and protocols, either for all layers or for just the lower layers in the network, starts with a focus on hardware and software requirements. This is a computer-oriented approach, working toward standard hardware and software. The objective is to achieve setting-independent or application-independent operations that include all possible variations. The function bridge approach is quite different. It requires starting with application issues in order to define the types of functions that will be required. The function bridge approach involves specifying only the top layer of the ISO model, and on solving the particular problems of the given industry group. The emphasis is on the flexible use of hardware and software environments to make the best use of available resources. The function bridge approach is setting-dependent and leads to specialized solutions, as opposed to the protocol focus on generalized solutions.

Trade-offs can develop between (standard) protocol approaches to computer communications and (custom) function bridge approaches. In evaluating both types of systems, it is important to consider not only the original purchase or developmental costs but also the full life-cycle maintenance costs. For simple system configurations, purchase and development costs may be reduced by using custom configurations. On the other hand, the maintenance costs will be highest for custom approaches, because every system change will require intervention by

The function bridge approach is closely related to the functional manufacturing system design concepts introduced in Chap. 4. In both cases, emphasis is on using functional requirements to drive system design. Difficulties may be encountered when the hardware constraints that act on these functional viewpoints prevent the type of coherent functional flow that is desired. Functional approaches to problem solving can provide potential insight into solutions, but can also reveal many real-world constraints and problems in implementing the planned system.

an experienced programmer. At the other extreme, the use of protocols may result in the highest purchase costs and the lowest maintenance costs. Thus, it is important to look at the full operating cost of the computer network when considering these trade-offs.

A major parameter that enters into trade-offs is the number of equipment units or nodes that will be required and the degree of stability that will be obtained for the network. For large networks, custom linkages may become unworkable owing to the many individual computer programs to be prepared. On the other hand, for more limited system sizes, a degree of customization may be the most effective approach. If the system is going to be stable, then the maintenance costs for the custom linkages will be reduced. On the other hand, if the system will be continually evolving and passing through transition stages, custom changes will be required on a continuing basis. A decision as to whether to make sole use of standards and protocols or to introduce customized linkages in the computer network is an important one. The preferred approach will depend on the nature of the system that is being considered.

It should be emphasized that in both cases interoperability problems can ensue. For the function bridge approach, difficulties may occur between the system user and the equipment manufacturer as they attempt to define the nature of the functional interface. And, despite the best effort at developing protocols and standards, it is still not unusual to find interoperability problems between standard products that are intended to work together.

Similar interface communications issues develop with respect to linkages between multiple computer-aided design systems and between elements of integrated design and manufacturing operations. As discussed in Chap. 9, one strategy makes use of *neutral interfaces* or protocols (such as IGES), which correspond to the protocols discussed here. An alternative strategy uses *direct translators,* which correspond to the custom function bridge approach described here. Whether the challenge is to link between equipment units and a computer network or between design computers and the manufacturing system, the issue of standard versus custom software will continue to involve major decisions during system design. Given the high costs associated with software development and maintenance, decisions in this area may have a significant effect on the benefit-cost performance of the final system. Therefore, this aspect of system operation must be carefully and appropriately treated during the design process.

In summary, standards and protocols contribute to setting-independent, all-inclusive solutions. The standard/protocol approach can be tedious, complex, fairly rigid, and costly because of its setting-independent nature. The function bridge approach provides an alternative that may be the low-cost strategy in some circumstances. Whether or not the function bridge approach is appropriate for a given application, or whether systems based on standards and protocols should be developed, depends on the particular application. The appropriate network design approach is to consider system objectives, evaluate the alternatives, and select the highest benefit/cost solution.

EXAMPLE 7-7. COST-EFFECTIVENESS OF ALTERNATIVE NETWORKING STRATEGIES

Consider a situation for which the total cost in developing a CIM system follows the type of curve shown in Fig. 7-24. Discuss how this curve can be interpreted, and the implications associated with such relationships.

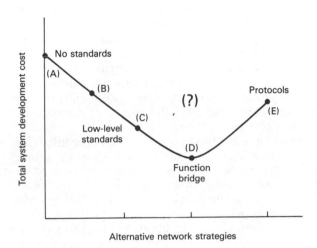

Figure 7-24.
Cost data for one possible system design setting.

Results

The costs of approach A are high because of the costs incurred by the system integrator. The costs at point E are also high because it is necessary for all participants, both the system integrators and equipment vendors, to develop a detailed understanding of all aspects of the protocol that is being applied. The costs at B and C are less than those of A because a lower-level communications interface has been applied, but a significant cost must be incurred by the system integrator to make use of the various equipment units. Finally, it may be in some cases that the function bridge approach associated with position D will be the lowest-cost solution and most adaptable solution when considered by all interested groups.

7-10. Integrating Information Requirements into System Design

Figures 7-6 and 7-9 illustrate how the system design concepts discussed in Chap. 4 can be used in combination with the equipment automation concepts of Chap. 6 to describe the information flow that should be associated with the individual building blocks of the system, and how the system can be configured to make appropriate use of related technology. The system design concepts and equipment and automation concepts can be used to define the required information flow that must take place in the CIM system. By looking at the system functions and the equipment functions, it is possible to describe the information flow that must result.

A complete functional design can be prepared for the information system by combining the top-down system design with the bottom-up equipment design and defining the information flow that must take place in the resulting system. This equipment can be evaluated in terms of the input required by each equipment unit, the data that will be generated and provided by each equipment unit, and how all equipment units must integrate to produce the desired level of system performance. The functional design of the information system can be used as a basis for developing the computer network that will be needed to implement the system. This information system must include both the human information network and the computer network that will be needed to implement the system.

Once the complete information network is specified, it is possible to estimate whether there are any technological limitations to such systems and to determine the costs that will be associated with system implementation. In extreme cases, there may be technological limitations. In many cases, the costs associated with information system implementation may make it necessary to cycle through the system design and the equipment specifications in order to modify functions and reduce the computer network cost. As discussed above, computer network simulations such as NETWORK II.5 can be used to assess the detailed technical performance of alternative network configurations. These simulations can help determine the preferred information strategy for a manufacturing system.

As part of estimating network costs, it is necessary to make decisions regarding the detailed network design strategies for procurement of necessary hardware, installation of the hardware, and approaches to be used for development and installation of the necessary software. In many cases, the development, installation, and maintenance costs of the software will be the dominant determinant in producing an information system.

If the software does not adequately achieve the functions that are planned, then the entire system will fail to work. The development of an effective software system is one of the most difficult tasks facing any CIM planning group. An extensive evaluation must be made of the enterprise capabilities to produce, install, and maintain the required software system. This may well require the use of outside consultants to evaluate any plans that are developed. It may also be necessary to hire system integrators and other software consultants to produce, maintain, and operate the system. It is almost impossible for a manufacturing enterprise to manage such a software development and implementation procedure unless there are strongly qualified in-house personnel who have the necessary skills to monitor and direct the system development. If the entire system development process is turned over to an outside party, the planning group often loses control over the design process, is unable to assess progress that is being made, and cannot appropriately anticipate difficulties and solve problems effectively. An adequate structure must be in place for management and development of the software and to create the required information system.

If an enterprise is evolving toward a CIM system through evolutionary stages, one possible strategy is to continue side-by-side operations of the *as-is* system and the *to-be* system until the *to-be* system is completely debugged and functions effectively. A number of manufacturing concerns have followed this procedure and have been careful never to abandon the *as-is* information flow until the *to-be* system has shown itself to operate correctly. On the other hand, on some occasions, this type of parallel operation is not feasible. In any case, care must be taken to make sure that the information system is developed according to the functional specifications that emerge from the system design and that satisfactory operation is obtained.

As an enterprise evolves through transition stages defined by the planning group, it may be useful to continue side-by-side operation of the old and new systems until the new strategy has been shown to operate as planned. Particularly with respect to information flow, it may be necessary to maintain the old system while the operation of the new system is being evaluated. On the other hand, the physical and organizational costs of parallel operation can be high and must be considered as part of the design trade-off.

EXAMPLE 7-8. IBM AND TELEDYNE MICROELECTRONICS SOFTWARE SYSTEMS FOR FACTORY INTEGRATION*

One of the major difficulties in developing a highly integrated manufacturing system is obtaining the necessary software to link the components of such a system. Because of the multiplicity of possible approaches and the lack of one generally accepted standard, the software development problem can be difficult and expensive to address. Discuss one commercial strategy for dealing with this difficulty.

Results

In response to this situation, International Business Machines (IBM) has developed a software product called Distributed Automation Edition (DAE™). The objective of this product is to simplify the development and implementation of integrated manufacturing systems.

The strategy for developing this product is illustrated in Fig. 7-25a. At the upper level are the multiple applications that must be developed for a CIM factory, including production monitoring, distribution of software programs throughout the system, inventory tracking, quality analysis, scheduling and dispatching of work, tool management, and other functions. At the bottom are the various hardware elements of the system. IBM has created an *application program interface* and *system enabler software* that can link the application development and the manufacturing hardware. The result is a software product that conceals from the user the details of the computer system itself. This product is designed to let the manufacturing engineer define the applications that are to be implemented and to communicate with specific equipment units. The system allows use of a high-level language to achieve

* The author has participated in the development of the facility operated by Teledyne Microelectronics and has received funding from Teledyne for related R&D projects.

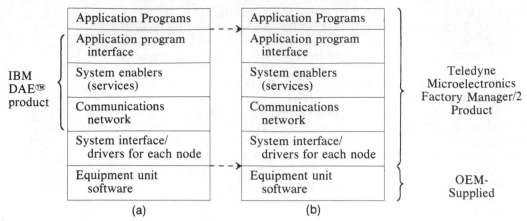

IBM DAE™ product	Application Programs		Application Programs		Teledyne Microelectronics Factory Manager/2 Product
	Application program interface		Application program interface		
	System enablers (services)		System enablers (services)		
	Communications network		Communications network		
	System interface/ drivers for each node		System interface/ drivers for each node		OEM- Supplied
	Equipment unit software		Equipment unit software		

(a)   (b)

Figure 7-25.
Computer integration functions associated with the IBM® DAE® software product and the Teledyne Factory Manager/2 software product (courtesy of IBM Corporation and Teledyne Microelectronics; IBM and DAE are registered trademarks of International Business Machines Corporation).

the desired application. The enabler reduces the complexity and cost of the application task and allows integration to be implemented more easily.

The DAE product achieves a distributed system with a single system viewpoint. The application programs are separated from the hardware/operating system to simplify the application development. The system can utilize MAP, a PC network, or a token ring for LAN communications.

IBM has developed the ARTIC (*A Real-Time Interface Co*-processor) Computer *Card** to be a subsystem for use in IBM industrial computers. This card can be added to a PC or the IBM 7552 Industrial Computer (also called a gear box™) to provide multiple linkages between the computer network and the individual equipment units. The ARTIC Card provides a real-time, multitasking capability in an open architecture for linking many hardware units together. This card facilitates attachment to a wide range of programmable controllers and industrial equipment, and can be coded to support many different communications protocols. IBM has thus worked toward providing hardware and software tools for cost-effective CIM system development.

The application program interface provides isolation between the desired system functions and the detailed implementation of these functions, with the potential for improving programmer productivity. The system enablers provide support in terms of network and device communications, device management, data management and supply, operator interface, and security.

Teledyne Microelectronics was one of the first users of the DAE product. From 1986 to 1988, Teledyne Microelectronics was under contract to the U.S. Naval Ocean Systems Center (NOSC) to develop an Integrated Facility for Automated Hybrid Microcircuit Manufacturing (IFAHMM) program. The objective of the program was to create the requirements, plans, and computer-integrated systems for a demonstration facility for the flexible, cost-effective, and quick turnaround manufacture of a variety of reliable hybrid microcircuit assemblies. The goals of the program were to improve yields, reduce costs, improve production capacity, reduce cycle time, and provide an acceptable return on investment for the government.

Teledyne Microelectronics is the largest U.S. noncaptive hybrid manufacturer. The company produces 15,000 to 20,000 hybrid assemblies per

---

* A computer card is typically a printed circuit board that achieves a specific functional operation when inserted into a complete computer unit.

month, with over 500 circuit types being delivered to over 100 customers on a concurrent basis. A multiplant manufacturing environment of over 245,000 square feet is operated for this purpose. In developing the computer networks to support the IFAHMM demonstration of capability, Teledyne selected the IBM DAE product as an implementation strategy. This product has allowed existing IBM hardware to be linked to a variety of shop-floor equipment units to achieve the desired operating system.

The emphasis of the IFAHMM program was to replace the paper flow in the factory with a computer-based documentation and control system. The IBM DAE product was applied to enable Teledyne to link its existing management computers with the factory hardware and thereby to achieve an integrated information system. The combined IBM and Teledyne developmental efforts are summarized in Fig. 7-25b. IBM provided the hardware and software that enabled the development of an integrated manufacturing system, while Teledyne developed the application programs and device drivers to implement the software in a specific manufacturing setting. The Teledyne Microelectronics version of the software system shown in Fig. 7-25b is called Factory Manager/2 (FM/2).

By use of the FM/2 software, Teledyne has achieved unique data collection system features. The ongoing data exchange between each equipment unit, factory operator, and the system data base results in an interactive operating environment with on-line validation.

At the same time, control instructions and software programs can be uploaded and downloaded at will. A wide range of data is collected and analyzed to achieve the integrated operations. FM/2 makes use of bar-code terminals for the collection of shop floor data. The system tracks the steps associated with the building of each product by serial number and lot identifier. A complete history of each employee operation, part, and equipment unit is recorded from the release of work order to completion. The ongoing tracking of equipment and product allows a detailed knowledge of the system status at any time.

The initial implementation of the information system made use of a MAP LAN supported by 7552 Industrial Computers (gear boxes) and ARTIC Cards used to link the individual equipment units into the network. Teledyne has migrated to a token ring network, which is now also supported by the 7552.

As expressed by Teledyne, the three biggest challenges during system development involved connecting to factory devices in real time, managing a complex programming environment, and achieving program implementation within available budget.

Teledyne's Support Services organization markets this plant floor automation system to the manufacturing industry at large. Teledyne has developed a highly integrated and automated manufacturing system for hybrid microelectronics, and is using this experience base to develop new software products that are available for CIM applications in a wide range of industry settings. Teledyne intends to serve as a "test bed" for new automation strategies and to engage actively in technology transfer activities.

The FM/2 system illustrates the sophistication of the hardware and software capabilities now available for integrating a wide range of equipment types and information-gathering strategies into a single operating system. The advent of low-cost, high-performance computers and related products has provided the opportunity to handle the equipment needs of such systems effectively. Many other similar types of software systems are being developed around the world. This example provides a brief insight into one effort, and also illustrates how the hybrid microelectronics emphasis used in this text can be generalized to a broad range of manufacturing settings.

The experience achieved by IBM, Teledyne Microelectronics, and the IFAHMM program indicates the type of private sector–government cooperation that is taking place today in an effort to address the complexities of CIM system design and implementation. Many other efforts, involving other computer and manufacturing companies, are taking place as industry and government strive to develop more competitive manufacturing systems.

ASSIGNMENTS

7-1. In order to achieve maximum educational insight from Chap. 7, it is helpful to stimulate class discussion of the concepts that have been included, develop critical reviews that probe the advantages and limitations of the concepts, and explore alternative ways for approaching the topic. Select one or several topics from the bold-faced headings that introduce sections or examples in this chapter. Come to class prepared to explain this topic, critically describing the text materials and suggesting ways the topic discussion might be strengthened.

7-2. Identify a particular type of manufacturing equipment that is of interest. Gather information on several different versions of this equipment having varying degrees of automation and computer communications capability. Perform a comparative analysis of the communications requirements of these different equipment units, following the model developed in Example 7-1. Describe the advantages and disadvantages of each equipment unit in terms of information flow.
(a) Come to class prepared to describe your family of equipment units and to discuss information flow requirements associated with each.
(b) Based on class discussion, prepare a brief paper summarizing your understanding of the information flow requirements.

7-3. Design a work cell that makes use of several equipment units. Develop an understanding of the relationships that must exist among these equipment units, the work cell, and the surrounding factory. Develop one information matrix with an emphasis on more manual operations and another with an emphasis on more computer-based operations. The information for this assignment may be obtained by collecting data sheets from original equipment manufacturers, by visiting a manufacturing setting in which equipment units are available and ex-

tending this information base with use of data books and OEM materials, or by discussing with individuals their particular experiences that would be relevant to the types of work cells of interest.

Develop the two information matrices, following Example 7-2. Discuss the relative advantages and disadvantages of each. To the degree that your discussion is based on real-world settings, what have been the organizational experiences with the types of work cells described? What improvements do you think could be made to achieve more cost-effective overall system design?
(a) Come to class prepared to share your results and participate in discussion of the issues raised.
(b) After this discussion, prepare a brief paper on your insights.

7-4. Consider different ways in which information flow might be established for the work cells described in Assignment 7-3. Following Example 7-3, and applying other insights, describe some of the information flow options that are available and the strengths and weaknesses of each. One way to gain insight in this area is to have an experienced information system analyst look at the alternative work cell configurations and to suggest the alternative strategies that might be appropriate in each case.
(a) Come to class prepared to present your finding.
(b) Based on class discussion, prepare a brief paper describing your understanding of the information flow requirements in each case and the relative advantages and disadvantages of each.

7-5. Obtain information from the International Standards Organization (ISO) and other resources on the OSI model. Come to class prepared to give a presentation on these materials, including the advantages and limitations of the model.

7-6. By researching available material and conducting discussions with computer network experts, develop further in-depth knowledge regarding a commercially-available computer communications protocol. Come to class prepared to give a presentation and to answer questions regarding the selected system protocol.

7-7. By using available resources, obtain further information regarding RS-232, Ethernet, or TCP/IP to extend the discussion in the text. What additional features do you think are important to bring out as part of this discussion? Come to class prepared to present and discuss your materials.

7-8. Contact someone who has experience with implementing one of the computer communications strategies discussed in this text. Discuss with that person the reasons for the original se-lection, the person's experience with this protocol, and any learning insights that can be shared with you.
(**a**) Come to class prepared to share your results and participate in discussion of the issues raised.
(**b**) After this discussion, prepare a brief paper on your insights.

7-9. To explore the cost-effectiveness of alternative networking strategies, discuss with several individuals their experiences with creating computer networks of any kind. Ask them to indicate the basis on which they made their original decision, the original cost projections, the final costs incurred, and the perceived effectiveness of the final system. Was the cost-effectiveness of the system greater or less than that originally expected? Would these individuals engage in a different type of cost-effectiveness analysis prior to implementing another system?
(**a**) Come to class prepared to share your results and participate in discussion of the issues raised.
(**b**) After this discussion, prepare a brief paper on your insights.

8

ORGANIZATION AND MANAGEMENT

Chapters 6 and 7 develop relationships between the CIM system design process and equipment and information flow issues. Of equal importance is understanding the ways in which the CIM system design process and organizational and management strategies are related. A CIM planning group must consider organizational and management issues in terms of both constraints and opportunities. It is important to be able to understand how organizational concerns fit into the larger CIM system design process.

This chapter provides an overview of the CIM system design process as related to organizational and management issues. The chapter describes some of the major concerns that arise in achieving organizational change and discusses the types of behavior that can be encountered in the organizational setting.

As noted in previous chapters (see Secs. 3-11 and 4-6), successful implementation of a CIM system depends on the active support and cooperation of the people who are involved in the system. Support must be obtained at every level—from upper management, to those who are responsible for product design, to factory management, and to the individual workers who interact with the manufacturing equipment on a direct basis.

One of the unique features of a computer-integrated system is that each person who is part of the supporting organizational structure must relate directly with the technology involved. Those in upper management must become accustomed to and understand the type of information available and how to use it, must deal directly with computer data bases, and must act to direct the system through interaction with computer networks. Those who are responsible for product design must learn how to use the computer-aided design tools, must be able to interact closely with the supporting design software, must learn to be concerned with design for manufacturing, and must gain an appreciation for the manufacturing capabilities of the factory. Managers in the factory itself and workers who interact directly with equipment must become accustomed to a quite different environment in which their main function is often to serve the technology, make sure that it operates correctly, and problem-solve when the technology is not working correctly.

In a CIM-oriented manufacturing system, every person in the organization is affected by the technology. The ways in which people perceive their own activities and relate to one another and to the organization will be fundamentally changed.

Thus, in all areas of a CIM system, the functioning of the organization is shaped by the way in which the technology has been implemented. The importance of relationships between the social/organizational aspects of an enterprise and the technology that the enterprise uses has been studied in a number of settings under the description of *sociotechnology* (Woodward 1980).

Figure 8-1a demonstrates how the organizational opportunities and constraints (of layer 3) become a screen that selects out only those system configurations that are compatible with the organizational setting. Configurations that cannot be implemented must be discarded from the planning process.

A CIM system is basically different from a more typical manufacturing organization that does not make extensive use of computers and integration. In a traditional setting, the social/organizational structure forms the focus of the enterprise, and the equipment is used as a means to accomplish the tasks and functions associated with the enterprise (Fig. 8-1b). In contrast, in a CIM system, the structure and function of the enterprise change so that the organizational focus is determined by the technology, and the human participants in the system are often oriented toward servicing the technology in order to make maximum effective use of it (Figure 8-1c).

In an advanced manufacturing system, the technology in use often determines the activities of the organization. Relationships and work patterns are defined by the need to service the technology. This shift in organizational control can have significant impact on all members of the organization. The design of a CIM system must reflect a concern with human needs and relationships in this technology-driven setting.

Obviously, a change from a typical organization to a CIM environment can create many difficulties. In the first place, the transition can be quite disorienting and unsatisfactory. All individuals involved may need extensive reeducation to make effective use of the new system. At the same time, a high level of management concern must be maintained if the technology is to be adequately serviced and supported by the individuals involved and the system is to function in a way that

Figure 8-1.
Overview concepts for
Chap. 8. (a) Discussion of Chap. 8 related
to the design model of
Fig. 4-5. (b) Relationship between technology and the social organization for
classically oriented
manufacturing. (c) Relationship between
technology and the
social organization for
CIM-oriented systems.

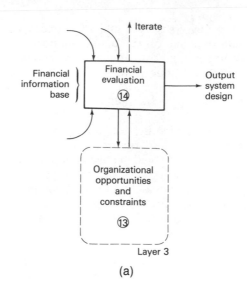

Iterate

Financial
information
base

Financial
evaluation
⑭

Output
system
design

Organizational
opportunities
and
constraints
⑬

Layer 3

(a)

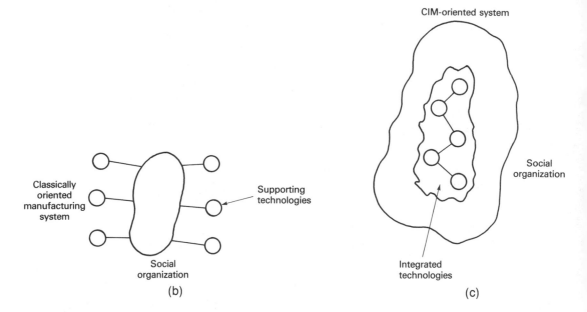

CIM-oriented system

Classically
oriented
manufacturing
system

Supporting
technologies

Social
organization

Social
organization

Integrated
technologies

(b)

(c)

*During implementation
of an advanced manufacturing system, careful, planned efforts
must take place to assist all members of the
organization in adapting to the new system
strategy. Without adequate preparation, alienation and rejection
of the strategy can
occur, leading to system failure.*

is satisfactory to the enterprise. Thus, the design of the CIM system
must reflect a concern with the human aspects of the enterprise, with
how well the CIM "core" can provide a focal point for the entire
enterprise to achieve its objectives, and with how well the human members of the enterprise can make use of the technology in order to obtain
the desired optimized system.

In order to implement an effective CIM system, it is necessary
to understand the present *as-is* organization and behavioral aspects of
the people in the organization and at the same time have a good grasp
of the *to-be* organizational structure and the desired behavioral char-

acteristics for the new setting. Once these two aspects are clearly understood, then a strategy must be developed, and put in place, to help guide the organization from the *as-is* to the *to-be* system through the required intervention processes. This transition may be extremely difficult. If it is not accomplished successfully, the people in the enterprise may be alienated by the technological system focus to which they must relate, will not make effective use of the system, and will obviously produce an unsuccessful outcome for the CIM implementation effort.

As part of the design of any CIM system, it is necessary to have sufficient insight to be able to understand organizational structures and the behavior of individuals within such structures. Further, it is necessary to have a grasp of the intervention methods that may be used to transform an organizational structure and function with one set of behavior patterns into another structure with a revised set of behavior patterns. The objective of this chapter is to provide background information that will be useful in producing a system design that recognizes the essential importance of organizational and management issues in a highly automated and integrated environment. This chapter also provides an overview of intervention strategies that can be used to help guide the system from one design approach to another.*

The implementation group for a CIM system has to view itself as a *change agent* in terms of modifying the organization and changing the behavior of individuals in the organization. It may often be necessary to obtain the help of outside consultants and organizations in order to develop sufficient skill for implementing the desired social change. As has been demonstrated in the United States during the past 30 years, efforts to change social systems can often lead to quite unexpected results (Bardach 1977). Program outcomes can be quite different from those expected. Certainly, the best use should be made of available historical insights regarding change processes for organizations in order to guide organizational change toward a planned CIM system. At the same time, limitations in these insights must also be recognized.

8-1. Alternative Perceptions of Organizations

In order to direct the planning and implementation of organizational change for manufacturing systems, it is necessary to have an understanding of the nature of the problems that exist in transforming the *as-is* enterprise to the *to-be* enterprise. It is necessary to gain an appreciation for the ways in which organizations must change during this transformation process.

An organizational setting will appear different to different individuals. An organization is not a specific physical object that can be

* Rouse (1986) has described a methodological framework that may be used to study the human role in complex, technology-driven systems. The objective of this approach is to identify potential problems and likely solutions, and the approach represents a possible extension of the foundation provided in this chapter.

viewed by many individuals with agreement on the nature of that object; rather, an organization is a concept that exists within the minds of those who are familiar with all or part of the organization. The perceived nature of an organization will vary, depending on the perceptions of an individual. Thus, a hands-on worker involved in a manufacturing process in the factory will not have the same view of the organization as will the engineers designing the product, or the chief executive officer of the corporation.

In order for the planning group to guide the enterprise from an *as-is* to a *to-be* configuration, it is necessary for the members of the planning group to understand the different types of perceptions that exist in the current manufacturing system and to develop a realistic strategy for working with or modifying these perceptions to achieve the desired manufacturing setting. There is a significant difference between individual perceptions in settings for which a social organization functions as the core of the organization (Fig. 8-1b) and settings for which the technological core forms the focus of the enterprise (Fig. 8-1c).

To develop an understanding of the wide range of perceptions that can exist in a manufacturing setting, it is useful first to view some of the relevant literature. This material is introduced in Secs. 8-2 through 8-4. Then some of the types of organizational structures that can exist in manufacturing organizations, and the behavior of individuals that can result in such structures, are considered in Secs. 8-5 through 8-18. Finally, against this background, Secs. 8-19 through 8-23 discuss leadership and organizational change.

An organizational setting appears different to the various individuals who work in the setting. Efforts to prepare the members of an organization for major change must be customized to the perceptions of the groups and individuals involved. Otherwise, the intervention efforts may turn out to be inappropriate and even destructive.

8-2. Perceptual Filtering

Simon (1976) addresses the nature of selective perception in the organizational setting. He writes that "an important proposition in organization theory asserts that each executive will perceive those aspects of the situation that relate specifically to the activities and goals of his department" (p. 309). He continues by referring to Bruner (1957) and noting that

> the proposition that we are considering is not peculiarly organizational. It is simply an application to organizational phenomena of a generalization that is central to explanation of selective perception: presented with a complex stimulus, the subject perceives in it what he is "ready" to perceive; the more complex or ambiguous the stimulus, the more the perception is determined by what is already "in" the subject and the less by what is "in" the stimulus. (p. 309)

Bruner (1957) has further stated: "I would propose that one of the mechanisms operative in regulating search behavior is some sort of gating or filtering system" (p. 141).

Selective perception may thus be viewed as involving a "filtering" process that links each individual to the outside world. The individual's ways of "seeing" can strongly affect the input information that is "received" and can allow the stimulus "in" or block it "out."

Perceptual limitations may restrict the data that are usefully available to the subject.

Writing on this topic, Allison (1971) has contrasted the effects of three viewpoints on approaches to addressing the circumstances and events related to the Cuban Missile crisis. He has concluded that:

> spectacles [or ways of "seeing"] magnify one set of factors rather than another and thus not only lead analysts to produce different explanations of problems that appear . . . but also influence the character of the analyst's puzzle, the evidence he assumes to be relevant, the concepts he uses in examining the evidence, and what he takes to be an explanation. (p. 251)

Holzner and Marx (1979) have discussed the filtering functions by observing that ". . . human sensory experience is . . . affected by shadings of light, distances and the internal state of the organism's moods or needs." Sensory experience is also the result of attention or inattention to particular aspects of the environment (Bruner 1957). Newell and Simon (1972) have recognized the importance of perceptual filters by noting that the input representation formed by a problem statement is "not under control of [the] inputting process" (p. 89). Sheps (University of Pennsylvania 1980) has noted how selective perception can particularly affect professionals: "All of us suffer from what the French call 'deformation professionelle,' the warping that is produced by the special point of view of a person with specialized background and experience" (p. 105).

8-3. The Coding Process

Once the information has been received, the individual may develop different degrees of understanding of its meaning, depending on his or her ability to structure and arrange the data into patterns. Katz and Kahn (1978) have discussed this as a "coding" process:

> Any system that is the recipient of information, whether it be an individual or organization, has a characteristic coding process, a limited set of coding categories to which it assimilates the information received. The nature of the system imposes omission, selection, refinement, elaboration, distortion and transformation on the incoming communication. (p. 433)

And quoting Lippman (1922), they note: ". . . We tend to perceive that which we have picked out in the form stereotyped for us by our culture" (p. 31).

Anthony (1965) has put it another way:

> To be useful, material dealing with any broad subject needs to be organized within a framework of topics and subtopics. If the topics and subtopics are well chosen, the available material can be so arranged as to make it possible for one to reach conclusions generally applicable to each classification but not applicable to other classifications. Such conclusions or generalizations, furthermore, will have validity and significance impossible in the absence of such a framework. (p. 1)

As noted earlier (Sec. 1-2), the coding process has also been explored by Kuhn (1962), who has emphasized the paradigms through which scientific investigators seek to structure the world. He writes that paradigms tend to dominate research and define the "correct" way of thinking at a given time in history.

8-4. Perceptions of Reality

The two concepts of filtering and coding are intertwined, but the linkage of concepts seems more useful (in terms of explanatory power) than either alone. The filtering process deals with the blocking out of data; the subject is restricted in access to the information and is unable to perceive portions of the potentially available materials. The coding process deals with the sorting of data; the subject has full potential use of the data and is seeking to interpret the available materials. This distinction is an important one, for those who have broad access to incoming information may profit by learning new coding principles or models; the willingness to "see" is there and the problems involve complexity and interpretation. On the other hand, those who have limited access to incoming information may gain little from new methods for coding; if the ability to "see" the world is directed and limited, the complexity remains unseen and interpretation, although achieved more simply, lacks a broad base.

Figures 8-2a through 8-2e are heuristic models that may be used to explore these concepts. Figure 8-2a represents an individual with narrow perceptual filters who accepts only one way of seeing based on past conditioning. Figure 8-2b shows an individual with broad perceptual filters who accepts many different ways of seeing a situation. Individuals can draw on explanatory models (or codes) to interpret the information. In the broad-filter case, such models provide an organizing structure for the data being received (Fig. 8-2c); in the narrow-filter case, however, the organizing model is of little value because the individual is unable to see the range of viewpoints (Fig. 8-2d). Therefore, teaching new organizing structures or conceptual files will be useful only when the filters can be opened so that the individual develops an ability to perceive. If two individuals have very different perceptions of a situation or event, it is reasonable to expect that communications, interchange, and understanding will be limited (Fig. 8-2e).

In trying to understand the nature of organizations and their problems and to design appropriate problem-solving methods, a broad perceptual approach is needed. Strategies based on a limited ability to assimilate are likely to be less effective than those based on a broadened information base. This point has been noted by Ramos (1982) who argues that the realities we see are shaped by our past conditioning, training, and experience. He notes that the languages we use affect how our filtering and coding processes develop and function (although he does not use this particular vocabulary). Ramos goes on to argue that the ability to broaden our perceptions is critical to the development of social science and the solving of contemporary social problems.

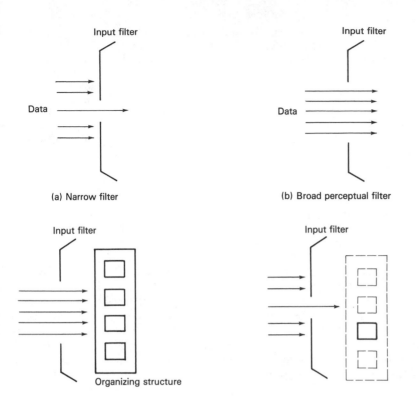

Data

Input filter

(a) Narrow filter

Data

Input filter

(b) Broad perceptual filter

Input filter

Organizing structure

(c) Using codes to interpret the input

Input filter

(d) Impact of limited filters on the use of codes

Figure 8-2.
Heuristic models to illustrate filtering and coding as part of the perceptual process.

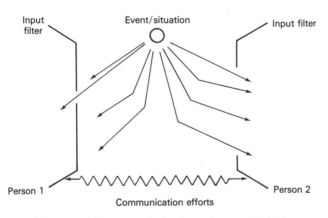

Input filter

Event/situation

Input filter

Person 1

Communication efforts

Person 2

(e) Impact of filters on understanding and communications

The problems that can arise from narrow, inappropriate perceptions have been noted by Bruner (1957). He has observed that

> Perhaps the most primitive form of perceptual unreadiness for dealing with a particular environment is the case in which the perceiver has a set of categories that are inappropriate for adequate prediction of his environment. . . .

Some people are characteristically tuned for a narrow range of alternatives in the situations in which they find themselves. . . . [S]hould the environment contain unexpected events, unusual sequences, then the result will be a marked slowdown in identification and categorizing. . . . [P]eople who are not able to shift categorization under gradually changing conditions of stimulation tend also to show what [Klein] describes as "over-control" on other cognitive and motivational tasks. (pp. 142–144)

It is reasonable to expect that conflict and poor adaptation will ensue if individuals operate from narrow perceptions and are unable to access multiple viewpoints.

The discussion here suggests that in studying organizations, multiple perspectives of reality may be expected to coexist and that interventions are more likely to be successful if a broad approach to understanding the organization is achieved. Attempts to redirect organizations must consider the realities and perceptions of the organizational members. Attempts to help organizations adapt to new situations (such as a CIM environment) must be responsive to the overall context of the membership.

Individual "filtering" and "coding" processes determine how people relate to their surroundings. These processes shape behavior in the as-is *system and help determine the types of change processes that are feasible. New, different ways of "thinking about" the organization may be needed, requiring careful planning of educational efforts that are appropriate for the existing perceptions.*

The perceptions of a given organization will vary among different organizational members and different external observers, in terms of their perceptions of organizational existence, goals, needs, and operations. There may be strong common threads based on common conditioning and experiences or substantial variation among members. Multiple realities can produce adaptive organizations or destructive conflict.

The organization may consciously develop and communicate multiple perceptions of itself. As Wilson (Wilson and Rafkind 1980) has noted

In "selling" itself . . . [a complex organization] presents its specially tailored "pictures" to . . . external and internal groups. Individual managers may come into contact with five, ten or even fifteen different pictures of the organization and what it stands for and where it is headed. (p. 21)

Through the filtering and coding processes, participants have an active role in constructing the organization; actions based on perceptions help produce the composite organizational reality. Any one of the ways of seeing may dominate. The organizational reality is thus socially constructed (Berger and Luckman 1966) and management processes must be actively involved in the development and functioning of this reality.

8-5. Organizational Structure and Function

In order to understand the *as-is* organizational setting and to produce a *to-be* organizational setting that best meets enterprise needs, it is useful to examine a variety of models that describe the structure and function of organizations (see, for example, Carzo and Yanouzas 1967;

Evan 1976; Mouzelis 1978; Presthus 1978; Perrow 1979; Cummings 1980; Katz, Kahn, and Adams 1980). These various models represent limiting cases, and real-world organizations tend to be combinations of these models. Nevertheless, these models can help the planning group and management understand the tasks they face in creating a CIM-oriented organization.

8-6. Classical Bureaucracy

The most famous organizational reference model may be one of the oldest, put forward by Max Weber in the early 1900s. Weber wrote extensively on the rise of formal bureaucracy and attempted to define its essential elements. As translated by Gerth and Mills (1978), Weber described bureaucracies as having the following characteristics:

- There is the principle of fixed and official jurisdictional areas, which are generally ordered by rules.
- The principles of office hierarchy and of levels of graded authority mean a firmly-ordered system of super- and subordination.
- The management of the modern office is based on written documents.
- Office management . . . usually pre-supposes thorough and expert training.
- When the office is fully developed, official activity demands the full working capacity of the official, irrespective of the fact that his obligatory time in the bureau may be firmly delimited.

The bureaucratic form was further defined by Luther Gulick (1937), who stated that such organizations should conform to the following "proverbs":

- Work should be divided into specialized processes and skills.
- Work should be coordinated through organizations.
- The span of control of any organization should be limited by the type of work.
- Unity of command should be observed.
- The functions of the executive are as follows: Planning, Organizing, Staffing, Directing, Coordinating, Reporting, Budgeting.

The bureaucratic model is generally concerned with fitting individuals into the classical organizational structure to produce efficient composite activity. The organization is designed to restrict individual behavior to the scope of the office in order to be effective and productive. Bureaucratic organizations tend to develop in and to function most effectively in stable environments. They tend to be stable places of employment with dependence on formal rules and policies. If the environment shifts away from stability, the bureaucracy can suffer severe functional and structural strains.

A manufacturing organization may be understood from a variety of viewpoints. From an initial perspective (viewpoint 1), many manufacturing organizations can be viewed as bureaucracies in which a formal definition of organizational roles is made within a hierarchical structure of responsibility and authority, as shown in Fig. 8-3. In such a management setting, the objective of management is to "fit" individuals into the formal structure, so that they function according to the desired job descriptions. In many ways, this model is often the starting point for the development of an organizational setting, and in many cases it may be misunderstood that an organization actually functions in this way. No organization functions completely as a bureaucracy, which is an "idealized" reference viewpoint.

From the bureaucratic point of view, management problems arise when individual behavior does not conform to the formal job description. The management solution to this problem is typically to impose a tightened bureaucratic structure in order to achieve the desired behavior. Sometimes this strategy will work, but many times it does not.

Particularly in the design and implementation of CIM systems, it is required to have highly trained individuals who are able to work in many creative ways with the technological system that is being developed. If an effort is made to fit these individuals into highly structured job descriptions with a bureaucratic authority structure, the result may be a major conflict between the organizational structure and the technology demands of the system. In such a case, inefficient operations may be expected.

Figure 8-3.
Bureaucratic model.

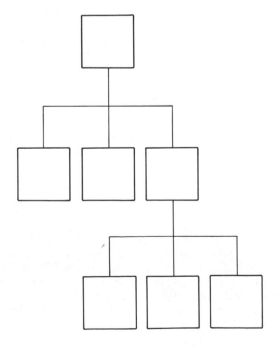

8-7. Informal Organizations

Over the years, observers have noted that people do not fit smoothly into organizationally defined niches, and that important nonformal interactions take place among organizational members. In a famous series of studies called the Hawthorne experiments (introduced in Chap. 2), Roethlisberger and others showed that informal (group) relationships among employees could have broad impact on behavior and productivity (1941). Concerning informal organizations, Barnard (1938, p. 48) noted:

- They establish essential attitudes, understanding, customs, habits and institutions.
- They create the conditions under which formal organizations may arise.
- They are essential to organizational communications.

The informal organization may strongly shape employee behavior and can affect the efficiency and productivity of the formal organization. Based on this organizational model, the manager must be concerned with both the formal behavior and organizational characteristics of employees and the informal group behavior that reflects nonformal characteristics. On the other hand, organizations that have only informal components may be weak in several ways. Such groups depend on the specific relationships among individuals to bind the members together and for organizational continuity. An informal organization can be constantly in a period of change, which is appropriate for some settings but not for those where effectiveness and productivity are valued.

Many types of interactions take place outside the formal job description and organizational structure. A major disruption of a manufacturing system can occur if an effort is made to change only the formal organizational structure and to neglect other aspects. Organizational viewpoint 2 (shown in Fig. 8-4) illustrates how nonformal or informal interactions always take place in any organizational setting. Informal organizations always develop (outside of the formal job descriptions) among the individuals involved in bureaucratic organizations. In fact, in many cases nonformal interactions will determine whether or not the formal structure is able to achieve enterprise objectives.

From this point of view, management problems exist when the informal organization does not support the desired enterprise objectives and conflict develops between the informal and formal organizations. One solution is for management to attempt to shape and harness the informal organization. This requires a very different set of actions than those required to enforce the bureaucratic model of viewpoint 1. In many organizational settings, both efforts must be attempted simultaneously.

Formal and informal organizations will always exist together. Human behavior in an organizational setting results from combined interactions with both organizational features. A focus on either feature alone will often lead to an inadequate understanding of the organization and an inability to achieve the desired change.

Figure 8-4.
Combined bureau-
cratic and informal
organizations.

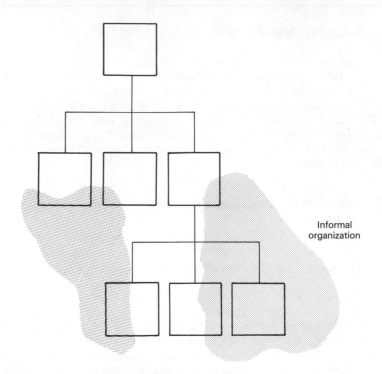

Informal
organization

8-8. Collegial Organizations

As noted by Hefferlin (1969), "Collegiality . . . indicates a community of equals." Hefferlin further describes collegiality in terms of three main characteristics: "(1) Initiative is dispersed throughout the organization rather than being concentrated in any one portion. Most members participate equally in decisions and policy formation. . . . (2) High status is achieved through renown rather than ascribed through seniority. . . . (3) Positions of high status rotate among the members of the organization on a periodic basis. . . ."

Viewpoint 3, representing collegial organizations, is of particular importance in the CIM setting (Fig. 8-5). With many highly trained professionals involved in the system, it may be expected that these individuals will have strong relationships to other professionals with the same training. This collegial relationship can actually be more important than the bureaucratic structure or the informal organizations. It is essential to understand the nature of the collegial relationships that will exist in the new system, and to harness these relationships to the purposes of the organization.

From this point of view, the diagnosis of management problems arises when collegial perceptions and actions have unwanted impact. This is particularly true if the computer hardware and software specialists feel that certain approaches to program implementation should be followed, and if these conflict with the system design as envisioned by the planning group. Such conflicts can often arise. Management

Figure 8-5.
Collegial organization.

problem solving can involve an attempt to influence and redirect the collegial groups. This can be accomplished in many cases, but requires a capability for understanding this type of organizational perception and how to deal with it.

8-9. Matrix Organizations

As open system concepts began to be applied to organizational systems, a number of authors began to explore the relationships between organizations and their environments. These authors noted that bureaucracies seemed to be well designed for stable environments but that they did not seem to function well in rapidly changing or turbulent environments. A matrix organization-design model was put forward as an appropriate response to such turbulent settings (Schon 1973; Kaufman 1979; Biller 1982).

The matrix organization is designed to combine stable and adaptive organizational elements into a single structure (viewpoint 4, as shown in Fig. 8-6). In changeable environments, the matrix organization is able to adapt and survive where the bureaucracy might develop "fractures" and collapse. The matrix organization combines ele-

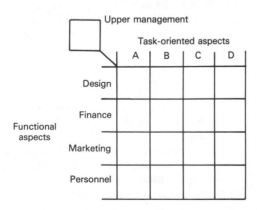

Figure 8-6.
Matrix organization.

ments of the bureaucratic and informal organizational forms with an adaptive "task-oriented" structure. The functional (permanent-structure) portions of the organization are for purposes of stability, whereas the task-oriented (temporary-structure) portions are for purposes of responding to environmental changes. Resources are allocated among the permanent and temporary structures so that each group has to negotiate with the other to function successfully. Upper management becomes involved only to resolve conflicts that are appealed for resolution.

A matrix organization is developed by having each individual assigned to a "home base" functional area (such as design, finance, marketing, or personnel) and also to a number of multifunctional tasks or project teams. The functional areas remain stable over time, whereas the task or project teams are formed and dissolved on a continuing basis, in order to address current organizational issues. The various functional areas and project teams are in a constant process of negotiation to match available people (from the functional areas) to the needed project teams.

Matrix organizations have become common in a variety of settings. Each individual has multiple responsibilities within the organizational setting. Many different teams may be set up to represent the various aspects of the organization. Each team includes representatives from different organizational units and is assigned to a particular task.

Matrix organizations can work well within the CIM environment. At the same time, the management of matrix organizations can be quite difficult. The functional areas and task teams must resolve most of their conflicts without appeal to higher management. The limited intervention role of upper management (as indicated Fig. 8-6) will be viable only if the matrix negotiations take place smoothly, with few appeals in cases of conflict. If significant unresolved conflict develops at the team level, then upper management is drawn into constant decision making and the system does not work. Similarly, if the task groups do not understand how to harness the organizational structure, problems will ensue.

Difficulties may be encountered in making the transition from a more bureaucratic model (as introduced in viewpoint 1) to a more matrix organization (viewpoint 4) as part of the CIM system. Extensive education may be necessary to enable the participating individuals to understand the way the new system is supposed to work and to perceive the advantages (both personal and to the enterprise). Management problems exist when the task teams fail to function smoothly, the teams do not link effectively to the surrounding organizational environment, or individuals do not function effectively in the team setting. Possible management problem-solving strategies are to employ educational efforts and perhaps to apply a relocation of resources to reenforce the desired behavior. It may be difficult to move to this type of matrix organization without adequate preparation.

By definition, the objective of CIM is to create an integrated facility in which the many aspects of the enterprise work together closely. The matrix organizational form is quite compatible with the concepts embodied in computer-integrated manufacturing. At the same

Support by the professionals in the organization, who often form strong collegial relationships, is essential for the effective operation of a matrix organization. It is necessary to create an organizational environment in which professional relationships and matrix operations are compatible and mutually reinforcing. Conflict in matrix operations due to collegial linkages can result in dissatisfaction and resistance among participants.

time, the meshing of a matrix organization with the technological CIM setting can require extensive skill.

8-10. Review of Organizational Models

The four viewpoints that have been discussed above are only a few of the many that can be considered. These illustrations indicate some of the perceptual problems that can be experienced in implementing a CIM system. It is essential to design the organizational structure to complement the technology used in the CIM system design. There must be a close relationship between technical function and organizational function. If the organization and the technology do not match smoothly, extensive conflict can be expected to result, bringing with it inefficient operations. On the other hand, when an effort is made to change the organizational design, major educational and motivation efforts must be undertaken so that employees can function effectively in the new setting. Design of a CIM system thus involves not only the management of technical change but also the management of social change in such a way that the resulting sociotechnical system can effectively address the enterprise objectives.

EXAMPLE 8-1. FORMAL REORGANIZATION

The management of ABC Company decides to introduce new technologies into the manufacturing system and applies bureaucratic concepts to develop a revised formal organizational plan. The activities to be performed by each new organizational unit are carefully defined in a logical way, and new job descriptions are applied throughout the organization. However, six months after implementation of the new organizational plan, it is observed that the organization is not functioning as desired. Upper management is faced with taking corrective action. What types of insights are provided by the four organizational models that have been discussed?

Results

Management has attempted to achieve reorganization by introducing a new formal organizational structure and revised job descriptions. The individuals involved are being treated completely from this point of view. No allowance is being made for the effects of the informal organizational relationships that have developed over the years. If essential attitudes, customs, and habits that have developed in past years are in conflict with the new organizational structure, and if this conflict is not specifically addressed, then a tug-of-war exists between the new formal structure and the continuing customs of the old informal structure. It is essential in this case for organizational management to gain an understanding of the informal relationships that have evolved in the organization and to develop a strategy for harnessing these relationships to the new formal structure being implemented.

It also may be necessary to study carefully the roles of the professionals involved in this system, to determine whether these professionals are functioning as desired within the new organization or whether collegial, professional linkages are serving to bypass or modify the desired organizational function. This may particularly be true in the case of the computer engineering specialists and design engineers. If these groups do not function as anticipated or as desired within the formal organization structure, significant dysfunctional performance of the organization may be noted. Management must thus address the roles of these collegial groups and develop a plan of action for dealing with any problem areas.

The revised formal bureaucratic system may not be well matched to the new manufacturing setting. If an advanced manufacturing setting is intended to adapt very rapidly to a changing environment, and to respond to market opportunities and difficulties as rapidly as possible, then a bureaucratic organization is typically not very effective. It may be that a more matrix-oriented organization would be appropriate. It may be necessary to develop further functional organizational areas to act as the stable basis for the organization and a transitory team structure that is constantly solving the problems of the moment. It may be that the formal organization does not encourage sufficient communication among the different functional areas, so the functional areas continue to try to do business as usual and solve their own problems. At the same time, the necessary multidisciplinary communications among all portions of the manufacturing system may be neglected. In this case, management must consider whether the new design is so mismatched to the type of operation desired that success cannot be obtained without further organizational redesign.

This example provides simple insights into the ways in which reference models can help management evaluate the organizational setting. The example indicates how critical it is for management to develop the perceptual capabilities required to look at the organization from many points of view and to understand any difficulties that arise. If any one point of view is allowed to dominate, and that point of view is not appropriate for the actual situation, then efforts to solve the problem will actually make things worse. If the formal bureaucratic organizational structure is not an appropriate strategy for the new enterprise, then efforts to force this structure on the enterprise can result in chaos. On the other hand, if managers employ a multiviewpoint strategy for dealing with problems, then a learning experience will ensue in which management can determine which viewpoint or combination of viewpoints is preventing accomplishment of the desired results. A wide range of possible strategies must be considered in order to have the maximum likelihood of moving the organization in the desired direction.

It is important to recognize that the organizational form must be developed to allow the enterprise to achieve its objectives. If the organizational structure and function are inappropriate to the system that is being developed, this mismatch will limit the ability of the enterprise to function as desired.

8-11. Organizational Behavior

In order to understand the behavior of the members of an organization, it is useful to become familiar with a variety of different behavioral models that exist in the literature. In general, people do not always behave according to any one model; rather, people are complex and typically behave according to a combination of models. However, it is beneficial to explore a number of specific behavioral viewpoints in order to have a basis for understanding the type of behavior that is observed (Behling and Schriesheim 1976; Hall and Lindzey 1978; DuBrin 1978; Lawlers 1979; Hellriegel and Slocum 1979; Herbert 1981).

8-12. Operant Conditioning Behavior

Operant conditioning theory, proposed by B. F. Skinner (Hall and Lindzey 1978), assumes that people are completely molded by their environmental conditioning (positive and negative reinforcement). People do not initiate behavior on their own, but only react to stimuli, producing acognitive behavior. People are motivated to respond to stimuli in the environment—to avoid negative stimuli and seek out positive stimuli. Advocates of operant conditioning theory are concerned with developing environmental conditions that will reward desired behavior and punish unwanted behavior. Behavior is assumed to be a function totally of the external setting and conditioning.

Figure 8-7 shows viewpoint 1, with behavior based on the operant conditioning model. The behavior of an individual is assumed to be shaped by combinations of positive and negative reinforcements. Under this assumption, a person's behavior is completely shaped by the environment; in order to produce a new type of behavior, it is only necessary to find the appropriate positive reinforcements and negative reinforcements required to produce the desired change. From this point of view, if management problems exist, it may be assumed that the existing reinforcements are not appropriate to produce the desired behavior. Management problem solving involves changing the reinforcements. This viewpoint of human behavior provides useful insights. At the same time, it is restricted in scope and must be viewed as only one consideration in obtaining a full understanding of individual behavior.

In general, the behavior of individuals cannot be interpreted according to any one behavioral model. People are complex, and typically seem to act according to combinations of models. The six behavioral models presented here provide a starting point for understanding the types of actions that are observed, and can serve to expand the perceptual frameworks of organizational members.

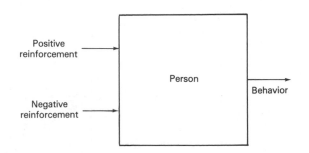

Figure 8-7.
Operant conditioning model.

8-13. Trait Model

Viewpoint 2, as illustrated in Fig. 8-8, is based on a trait model and assumes just the opposite of viewpoint 1. Instead of behavior being completely determined by external reinforcements, in this case behavior is completely determined by internal traits. Traits are assumed to be ingrained and not easily changed (Hall and Lindzey 1978). This is another extreme that is sometimes used as an interpretive framework. Management problems are seen to exist when individuals do not have the desired behavior traits, and one management action for this situation is to replace the person with someone who does have the desired traits.

This viewpoint is a narrow one that does not recognize human value, the full complexity of people and the capability of individuals to learn and adapt to new settings. Situations in which this model is applied may be more an indicator of limited perceptual filters on the part of management than an indicator of the characteristics of an individual. Managers must often be on guard not to let a trait viewpoint prevent the development of other, effective problem-solving insights.

8-14. Rational/Goal-Oriented Models of Behavior

The rational model of behavior assumes that people process information in a logical, orderly way and arrive at optimum decisions. In the rational model, perceived needs constitute the driving motivational dynamic (Behling and Schriesheim 1976). Based on these needs, the individual assigns values for action options; if all possible consequences and values are known, the individual will be motivated to pursue the consequence of highest value. Behavior is externally driven to the degree that needs are formed and shaped by the environment and internally driven to the degree that needs are formed internally.

The optimizing model treats individuals as calculating "computers" that seek optimal solutions before acting. People attach values to their options and select the highest-value option. Economists have often treated human behavior from an "economically optimizing" point of view, resulting in behavior that is "ideal" in the economic sense. Managers who behave according to the rational model try to gather maximum data, study the situation carefully, and make an optimum solution.

Figure 8-8.
Trait model.

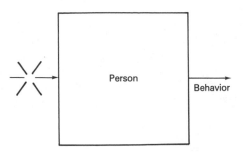

If people do not behave in the desired way, problems are interpreted in the following framework:

- The decision-making patterns being used are not rational.
- The planning function is inadequate and not thought out.
- Cost-benefit analyses and trade-offs need to be more clearly defined.
- There is a lack of task knowledge that requires an educational process.

The rational model may be criticized from many points of view, based on real-world observations of behavior. The optimizing process itself is in doubt, since it

- Requires complete knowledge of the consequences that will follow each choice
- Requires the individual to attach a (numerical) value to each choice
- Requires consideration of all alternative choices (not just a few that happen to come to mind)
- Requires parallel processing of information (where the human mind is basically serial, since it uses short-term memory for processing)

In response to critics, it may be argued that perhaps the rational model is adequate to describe average composite human behavior in groups, even if it does not satisfactorily relate to individual behavior.

Viewpoint 3 is illustrated in Fig. 8-9. It is assumed that people will attempt to optimize their situations based on the data they have. This model has its roots in economic studies, in which it is assumed that people will behave as rational beings in order to optimize their own situations. From this point of view, people behave like computers. There is data input and resulting behavior based on optimizing calculations. Based on this viewpoint, management problems exist when decisions are not rationally correct, and the strategy for management problem solving is to provide better data input and to teach better decision-making methods to the individual.

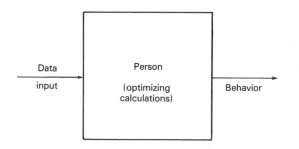

Figure 8-9.
Optimizing rational model.

8-15.   Satisficing Model

A "bounded" rational model of behavior was proposed by Simon (Newell and Simon 1972; Simon 1976), based on his study of the ways in which human beings process information. He concluded that decision making is basically a serial process because data must be drawn from long-term memory into short-term memory for analysis.

This model assumes that people search among a limited number of options to find the first satisfactory decision; they start with obvious solutions and then expand as necessary to other considerations. Each option is weighed against standards of acceptability and the first satisfactory solution is accepted. The initial starting point (and therefore the solution) is biased by individual characteristics. The model assumes that standards of acceptability are known and requires a comparative process with the ability to know when an option is satisfactory. This model suggests that managers may develop predispositions to certain obvious solutions, and will often vary as little as possible from these starting points of decision making. Simon's view of behavior is pragmatic, emphasizing the limits of humans in trying to solve problems.

Viewpoint 4, Simon's *satisficing* model, is represented in Fig. 8-10. It is assumed that when a decision must be made, the individual forms a limited list of the internal options that are available (based on limited available information) and accepts the first satisfactory solution that comes to mind. This model is based on the experimental observations that individuals cannot realistically absorb the quantity of data and perform the calculations required to optimize a solution, as represented in viewpoint 3. Rather, individuals will search for the first satisfactory solution, and this first satisfactory solution will drive behavior.

From this point of view, management problems exist when a person is restricted to a given behavior pattern and keeps looking for satisfactory solutions that are far removed from the optimum for the

Figure 8-10.
Satisficing model.

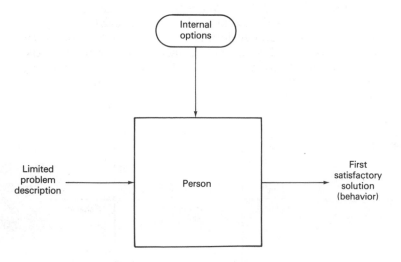

organization. Appropriate management actions in this case might be to provide new information and new settings for the individual. This model suggests that efforts to implement CIM systems may encounter established decision-making patterns that are no longer appropriate. Educational efforts may be required to broaden insights regarding problem-solving approaches.

8-16. Human Relations Model

The human relations model emphasizes the process of social interaction as the primary cause of behavior. The model is based on the assumption that human beings have an innate need to achieve, or actualize, and that social interactions with the environment determine whether or not this need is satisfied (Hall and Lindzey 1978). Thus, this model is an interactive one that is concerned with individual/environment communications. Individuals are seen as developing an inner tension that can be released and satisfied only through social interaction. If this communication exchange is distorted, the drive to achieve will be restrained and tension will develop in the individual. This tension may produce alienation and/or frustration-instigated behavior (see below). The human relations model is concerned with developing positive emotional feelings and reducing negative ones.

Undesired behavior is often interpreted in terms of the flaws inherent in problem solving through the rational approach. It is assumed that problems should be solved by all persons affected by the problem and solution, using a group process, not through top-down management decision making. Problems are defined in the following framework:

- Present group processes are not adequate to create the necessary problem-solving environment.
- Participation in problem solving is inadequate by group members.
- Group communications must be improved.

The human relations model assumes that motivation is present in everyone, and that an adequate social environment will enable needs to be achieved. The most important approach to decision making is through healthy social communication, which will result in a positive emotional state.

Viewpoint 5 assumes that behavior is based on social interaction with other members of the immediate group (Fig. 8-11). The person has an innate need to achieve, and positive and negative emotional feelings are developed by social interactions with the group. It has been shown that the beliefs and values of the members of the group can be important in determining the behavior of the individual. From this point of view, management problems exist when negative emotional feelings produce undesired behavior, and management treatment might be to produce positive emotional feeling through communication and participation.

Figure 8-11.
Social interaction
model.

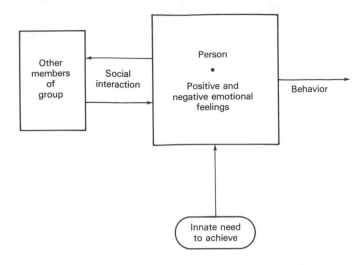

8-17.   Frustration-Instigated Behavior

If individual needs are not being met, irrational behavior can result (Behling and Schriesheim 1976). This may occur if goal-driven behavior is not satisfied (for the rational model) or if subjective communications needs are not satisfied (for social relations model). If goal-driven behavior is thwarted, there may be a suspension of rational judgment; if social communications behavior is thwarted, the tension buildup may be released through negative (irrational) actions. In the former circumstance, "rational thought" fails to produce the expected, reasonable outcomes, so frustration results as this analytic attempt to problem solving does not succeed. In the latter circumstance, behavior is not designed to achieve goals but to reduce internal tension.

Frustration-instigated behavior may be destructive, stereotyped, and rigid. The frustration may produce aggression, regression, or resignation. In general, such behavior patterns can be changed only if the individual changes his or her perceptions or if environmental circumstances change, and catharsis of some type is often necessary before nonfrustrated behavior can resume.

Viewpoint 6 represents the frustration model in which the person behaves in a certain way because past actions have produced failure (Fig. 8-12). This is particularly critical with individuals who have been unsuccessful in dealing with high technology, including computers and automated machinery. If the individual has a history of effort and failure, then his or her behavior is likely to be frustration-driven and therefore not rational but destructive and rigid. In order to produce the desired behavior, management must provide a means for developing a positive experience that will eliminate or reduce the frustration. If many individuals in an enterprise have experienced frustration in dealing with high-technology settings, then extensive effort must be made to produce settings in which the individual can experience success. Related

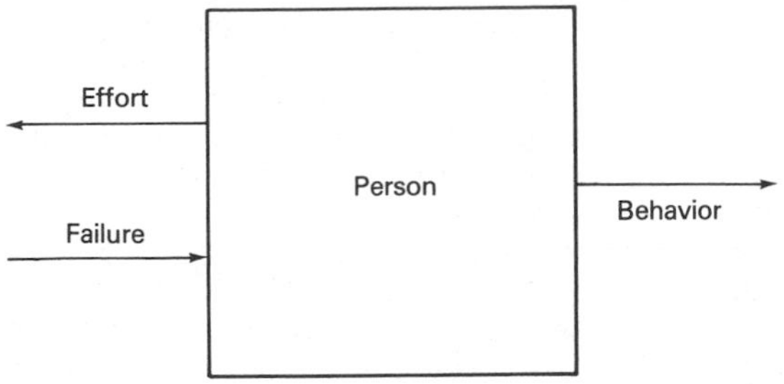

Figure 8-12.
Frustration model.

settings then can provide a learning environment in which the individuals can achieve the desired behavior.

8-18. Review of Behavioral Models

The six models discussed here provide different viewpoints of human behavior and describe how behavior can be linked to the internal and external characteristics of the individual. As noted, people are sufficiently complex so that all of these types of behavior can be observed at various times. With certain individuals, various aspects may dominate in a given setting and at a given time. The models indicate the complexity of the environment in which the CIM planning group and enterprise management are functioning. The best strategy for achieving the desired organizational change will be to understand the complexities of human behavior. By constructing a range of models such as the ones discussed here, the planning group and management can assess whether or not implementation plans will be effective in dealing with all the various behavioral points of view that exist in the organization.

The planning group is faced with deciding whether implementation plans will be effective for achieving the desired system state. By considering combinations of the types of behavioral models introduced here, and by evaluating on a firsthand basis the specific organizational setting of interest, the planning group can improve its efforts to achieve the desired system changes.

EXAMPLE 8-2. INDIVIDUAL BEHAVIOR

During the development of a new CIM-oriented manufacturing facility, the management of ABC Company observes that the people involved are not functioning as anticipated or in an effective way. Management must intervene in order to achieve the appropriate individual behavior. Based on the six behavioral models discussed above, what alternative perceptions of the system might be developed, and what actions might be taken?

Results

One common approach to achieving the desired employee behavior applies operant conditioning concepts. It is assumed that if individuals are behaving in the wrong way, then rewards should be offered to achieve the desired behavior and punishment should be introduced to discourage existing behavior. The assumption is that if people are not behaving as wanted, the solution is to change their environment.

It is interesting to note, based on past experience, that such an approach can either succeed or fail. Particularly, financial reward systems can produce desired behavior or may produce quite unexpected behavior, based on the perceptions of the individual. Financial reward systems can result in conflict and internal struggle in an organization that is not compatible with the type of organizational operations that are intended. For example, if a matrix organizational form is being implemented and each individual is provided with financial rewards (such as bonuses) that are keyed to individual performance, then the task teams, developed as part of the matrix approach, may be dysfunctional because the individual team members are more concerned with how their own accomplishments are going to be measured rather than with how well the team performance will achieve organizational goals. On the other hand, if individual rewards are keyed to team performance, then conflict may be developed among those who think they are contributing most strongly to the team and those who are hostile to other team members. Efforts to guide organizational change by operant conditioning are usually an essential element of such a plan but may not solve the problem that is faced.

From the trait model point of view, an appropriate problem-solving approach is to replace people who do not behave in the desired way. This viewpoint probably says more about limited management insight than about employee capability, and it is a narrow model that can be extremely expensive for a company. Experienced employees are a valuable organizational resource. Removing such employees in order to hire individuals who are perceived to have different traits is likely to be an unproductive problem-solving strategy.

The rational model of behavior assumes that the individuals who are behaving in an undesirable way simply need to be provided with more education. This requires explaining to everyone in the organization why they should behave a certain way, explaining why it is to their advantage to behave in such a way, and using this approach to produce the desired results. The difficulty here is in attempting to explain to each person why it is to his or her rational advantage to behave in a certain way. This strategy should certainly be a part of any such solution, but it may be difficult to conduct these educational efforts in such a way that the individuals will perceive the rational advantages that are being explained. Since the values of the organization and the values of the individual will often be different, and what appears like rational analysis to one person will not appear like rational analysis to another, it may be difficult to develop educational programs that address all issues. At the same time, such rationally oriented educational programs should often be a part of the solution strategy.

The satisficing model suggests that perhaps individuals in the system have developed predispositions to certain types of solutions and need to move away from these types of solutions to other problem-solving approaches. This might be experienced in a CIM system where members of the organization are accustomed to established problem-solving approaches that no longer will work in the new setting. In this case, it is important to be able to develop focal points for decision making that will produce satisficing around a new set of perceptions rather than around the old set of organizational perceptions.

The human relations model suggests that problems can be solved only by improving communication and developing more employee participation in the organization. This has become an important approach for producing improved manufacturing systems during recent years. Based on the experiences of other countries (such as Japan and Sweden) and on the experiences of smaller start-up companies, many large companies have attempted to introduce a spirit of cooperation and participation into the manufacturing environment. The development of *quality circles* (Melcher 1988; Blinder 1989; Hoerr 1989) has been a widespread effort to involve all employees in discussions about the operations and ways to improve them. This model assumes that the type of behavior that is desired in a CIM system can only be obtained if good communications exist, and that by producing this positive environment, the system will achieve its desired performance objectives.

Finally, frustration-instigated behavior may be part of any problem area. If individuals have had frustrating experiences with the types of technology that are being used in the CIM system, they may behave in a way that is not related to problem solving but is destructive. If such frustration-instigated behavior develops, it may be difficult to solve without replacing the individuals involved. One way to avoid this conflict is to provide maximum opportunity for the individuals in the organizational system to work with the technology and become experienced and comfortable with it prior to placing them in a new environment. Familiarization with the technology and the development of comfort with the technology can be critical ingredients in avoiding frustration behavior.

The six behavioral models introduced above provide different interpretations of behavioral problems that may exist in an organizational setting. As the CIM planning group moves the enterprise through a series of transition stages, learning and evolving toward the desired *to-be* system, it is essential that all people in the enterprise go through the learning process. It is not adequate only for upper management or the design team members to learn over time; it is essential that every person who must participate in the system constantly engage in the learning process in order to achieve the desired operations.

In many cases, the type of learning that is involved will require intense effort and emotional commitment. This is a high price to be paid by individuals in order to function in the new environment. It may be necessary for management to make use of all possible strategies to encourage employees to learn and change with the technology in the system. The behavioral models introduced here and others can support the development of perceptions regarding the types of problems that exist and the types of actions that can be taken. Producing a setting in which people function as an integral part of the new CIM system is a difficult task. It is essential that management make the most appropriate use of organizational design and the widest efforts to produce the desired behavior within this structure in order to create a CIM system that will achieve enterprise objectives.

Many manufacturing systems today are basically nonlearning environments in which individuals are not expected to learn and change. As a manufacturing enterprise begins to move toward CIM operations and to a different environment in which learning and change are valued, it will be necessary to work with all the people in the system, to help them move into learning activities that will lead in the desired direction. The types of perceptions introduced here can be useful to management in understanding the problems that are being faced and the strategies that can be used to address these problems.

8-19. Management of Organizational Change

In attempting to guide organizational change, managers often have to recognize that the existing organization is not a learning environment. Individuals may be expected to perform assigned tasks without adaptation or innovation. In this circumstance, the planning group will have to enable and stimulate a change away from standardized performance to learning performance.

The previous sections provide a basis for understanding the types of organizational change that must take place as a result of planning group efforts. It is not enough simply to design a technical strategy for the *to-be* enterprise; it should be clear at this point that the group must also be able to produce an organizational setting that will match the technology. It will be necessary to help create organizational structures and individual perceptions that produce the type of behavior that is desired. Against this background, the following sections address issues associated with managing the necessary change processes to implement a CIM system.

8-20. The Management Literature

Numerous approaches to the management task are described in the literature (see, for example, Kaufman 1971; Huse and Bowditch 1973; Basil and Cook 1974; Toffler 1980; Drucker 1980). In some of the early writings, trait theorists emphasized the desirable characteristics of any manager that were necessary for success (Hall and Lindzey 1978). Selected individuals were seen to contain the perceptions and skills appropriate to directing organizations and their membership. From another perspective, Taylor (1912) explored ways in which managers could enhance the productivity of employees to the mutual benefit of individual and organization; he advocated that a "best way" existed for every task, and through scientific management all organizational tasks could be optimized (refer to Sec. 2-4). Gulick's (1937) step-by-step description of management activities considered that there were a specific set of tasks facing the manager. He concluded that managers could be most effective if they mastered each of these specific activities (see Sec. 8-6).

These approaches to management (and a variety of others) generally emphasize the existence of a single, preferable reality for the manager. Depending on the reality or paradigm adopted, the able manager would have ingrained traits, an optimizing approach to task performance, or a patterned approach to defining appropriate activities.

This "best way" approach to perceiving organizations and management has continued to be an important perceptual framework for many analysts and practitioners. However, these earlier perceptions were followed by a second wave that began to see the management task in terms of multiple parameters. One of the best known approaches to management was put forward by McGregor (1960), who argued that there are Theory X and Theory Y managers (refer to Sec. 2-5). Theory X managers tend to view people as basically unmotivated; with such employees, managers must focus on control and impose rational behavior to achieve organizational goals. On the other hand, Theory Y managers tend to view people as desiring to achieve; with such employees, managers seek to provide an open, supportive environment and to draw on group process. Two perceptions of the organizational reality (related to control versus enabling) were now being conveyed and contrasted, illustrating management choice as a possible behavior.

Another management topology was developed by Blake and Mouton (1980). Managers were ranked in terms of their task orientation and people orientation. These two aspects of management, or ways of perceiving, were enriched by considering that managers could have different degrees of either orientation; by scaling the two orientations a range of possible behavior patterns could be identified. The two descriptors are used to define several management types (based on scales of low, medium, and high for each variable).

In his early work, Maccoby (1976) distinguished between involved managers with fixed perceptions (including the company manager who is captured by organizationally defined reality; the "jungle fighter" who sees the organization as a battlefield, and the "crafts person" who is locked into the perceptual limits of a specific profession or craft) versus those who could remain uninvolved from perceptual commitments and see the organization-management interaction as a "game."

This second wave of perceptions regarding management involves the development of two (or several) alternative realities that can be contrasted and explored as an alternative to the single-approach perception of organizational reality. A related approach to management has developed as an extension of the above. Preferred organizational problem solving may be seen as involving the application of management methods and models based on the setting and circumstances. This contingency approach by Fiedler (1977) and others emphasizes the need for congruence between the management reality in use and the application. Management becomes more an inventory of coping methods that can be drawn upon when needed. The limited choice and contingency approaches to management are currently in the mainstream of organizational study and practice.

8-21. Alternative Concepts

A third wave of perceptions regarding management has begun to be developed. Maccoby (1981) has written that there is a need for a new integrative approach to organizational problem solving. He suggests that the perceptions of managers must be expanded to include the environment outside the organization and a broadened view of people; he calls this new role that of the organizational "leader." Peters and Waterman (1982) have proposed that "the real role of the chief executive is to *manage the values* of the organization" (emphasis added). As reviewed in *Time* (1982), the authors conclude that successful organizations are able to find ways to help people be part of a team and to be recognized individually. They also note that "the best managers value action above else, a spirit of 'do it, fix it, try it.' . . . They solicit their employees' ideas and 'treat them like adults,' allowing talented people 'long tethers' for experimenting" (p. 68). The Peters-Waterman value management model is related to earlier work by Selznick (1957) involving the conversion of organizations to institutions by the "infusing of value."

The "culture management" model is currently being widely discussed. High-technology organizations often experience problems in persuading managers to work together as a team; in order to address strategic and operational needs, the manager may have to "reshape . . . [the organizational] culture" (*Business Week* 1982a, p. 124). It has been noted (*Business Week* 1982b) that "[the] merging of alien corporate cultures requires a strong leader who can push [the component organizational parts] . . . to pool their strengths" (p. 50). The culture management model links to efforts by managers to consciously shape the ways in which organizational members (and outsiders) perceive the organization.

As reported in another summary article, Bennis has proposed the concept of the "superleader" (*Sacramento Union* 1982). He has concluded that "people would rather dedicate their lives to a course they believe in than to lead lives of pampered idleness." Bennis finds that the most effective managers have combined

> . . . [1] vision . . . the capacity to create a compelling picture of the desired state of affairs which inspire[s] people to perform; [2] communication . . . the ability to portray their vision clearly and in a way that enlist[s] the support of their constituencies; [3] persistence . . . the ability to stay on course regardless of the obstacles encountered; [4] empowerment . . . the ability to create a structure which harness[es] the energies of others to achieve the desired result; and [5] organizational ability . . . the capacity to monitor the activities of the group, learn from mistakes and use the resulting knowledge to improve the overall performance of the organization.

Finally, much has been written about the meaning and application of Theory Z management (refer to Sec. 2-7) (Ouchi 1981; Schein 1981; Shortell 1982; Bruce-Briggs 1982). Shortell (1982) has stated:

> The basic premise of Theory Z is that involved workers are the key to increased productivity. . . . Theory Z implicitly recognizes that individuals are composed of elements of both X and Y, and that the central challenge lies in *creating a common culture* in which the organization realizes it can meet its objectives only through the needs and objectives of its participants. (pp. 7–8) [Emphasis added]

8-22.   Context Management

The third wave of management concepts involves conscious efforts by managers to understand that (1) multiple perceptions of an organization exist among organizational members and among external observers, (2) organizational reality can be changed, and (3) effective management methods will depend on the prevailing organizational reality. As presented here, conceptualizing the management task has proceeded from seeing there is "one best way," to perceiving topologies of contrasting choices regarding behavior matched to the setting, to an understanding that the nature of organization reality itself must be molded and directed as part of the management process.

In the following discussion, these third wave perceptions are merged into the concept of *context management*. There is a purposeful double meaning contained in this definition of management; it refers to both management *of* the context and management *in response* to the context. Context management involves three interrelated activities: (1) a continuing effort by the manager to expand personal perceptual files and coding in order to broaden the scope of understanding and possible action; (2) continuing action to shape the reality of the organization; and (3) continuing action to achieve organizational (and individual) objectives within the current reality. The context manager both responds to the environment *and* helps create it, is concerned with actively redefining organizational reality *and* utilizing management methods that are effective in the organizational setting. The context manager is concerned with redirecting organizations and systems by changing the perceptions of organizational members and observers and by acting effectively within the context that is developed.

The context manager has much more potential responsibility and potential power. A conscious awareness of the ability to shape organizational reality may be essential to successful organizational redirection from an *as-is* to a *to-be* environment. At the same time the risks involved may be much greater. The stakes may well be organizational survival and/or survival of the manager within the organizational structure.

Misperceptions of the organizational environment can lead to context changes that once were perceived as desirable but are later found to be destructive. The manager is always a prisoner of personal perceptual limits and must attempt to cope effectively despite these limits.

Redirection efforts by the context manager may also produce significant resistance by organizational members and outside observers. People may be reluctant to allow their reality to be reshaped; they may prefer to continue with familiar perceptions that (to the manager) are destined for failure. The context manager has a much greater potential for successful intervention and for failure, for reward and rejection, than those choosing more limited redirections.

Yet, as noted by several authors, management success in today's setting seems to require a context management approach. The manager is thus caught between competing forces: from one direction, the need to grow and intervene on a basic perceptual level; from the other, the risks involved in such perceptual interventions. It is not yet clear how these competing forces can best be resolved. However, it does seem likely that educational programs for managers will have to be substantially reshaped if they are to meet the needs for a contextual approach to redirecting organizations.

Context management refers both to the management of the organizational context and management in response to the context. In this framework, managers must be committed to expanding their own perceptual capabilities, shaping organizational reality, and achieving objectives in today's reality. Application of such a management strategy requires ongoing learning by managers; resistance may often be expected in response to the encouragement of such efforts by the planning group.

8-23. Application to CIM System Design

As may be concluded from the above sections, the development of transition stages toward the desired *to-be* system can be most effectively accomplished by understanding the perceptions of the organizational members and observers and by acting effectively within the existing organizational perceptions. The planning group and enterprise management must be able to understand the present situation and to assess how effectively organizational reality can be reshaped in order to produce transition stages toward the desired system.

Redirection efforts may produce significant resistance unless care is taken in the change process. An educational effort must be mounted so that the participants in the organization are motivated to participate in a reshaping of reality and changing of perceptions. To be most effective, managers must develop personally and intervene on a basic perceptual level. However, at the same time, this requirement should provide a warning note to the planning group because of the difficulty involved in achieving change of this type. Despite the difficulties, such change efforts provide the planning group with an opportunity to deal with the actual problems that are faced in the organization and not only with superficial factors.

One of the most important strategies for achieving the desired organizational system is to make effective use of organizational communications, as illustrated in Fig. 8-13. Individually directed messages can be provided over the organizational communications networks in order to shape the individual roles. Such communication involves individual explanations, discussions, and the development of the necessary insights. At the same time, management can also provide nondirected messages addressed to the entire organization to help shape the organizational culture or framework. The objective must be to work with both individuals and the organization as a whole to obtain the actions that are desired.

A problem associated with this approach is that management must develop an understanding as to how to send the messages that will be conveyed and received in order to produce the desired actions. This requires understanding the sending and receiving processes for messages in organizations. In the same way that learning activities must be initiated with respect to the technical aspects of the CIM system, learning activities must also be conducted in terms of organizational communications. As illustrated in Fig. 8-14, a learning loop can be set up for use in conjunction with the directed and nondirected messages that are delivered to the system. The resulting behavior is observed and evaluated, and the messages are repeated or modified as necessary in order to work toward the desired behavior. This learning loop will be unique in many ways to the people involved in the specific organization because of unique organizational features. This organizational-based learning loop must take place in parallel with the technical learning loop.

Management of organizational change can thus be viewed as a *teaching process* in which behavior is observed and a learning activity

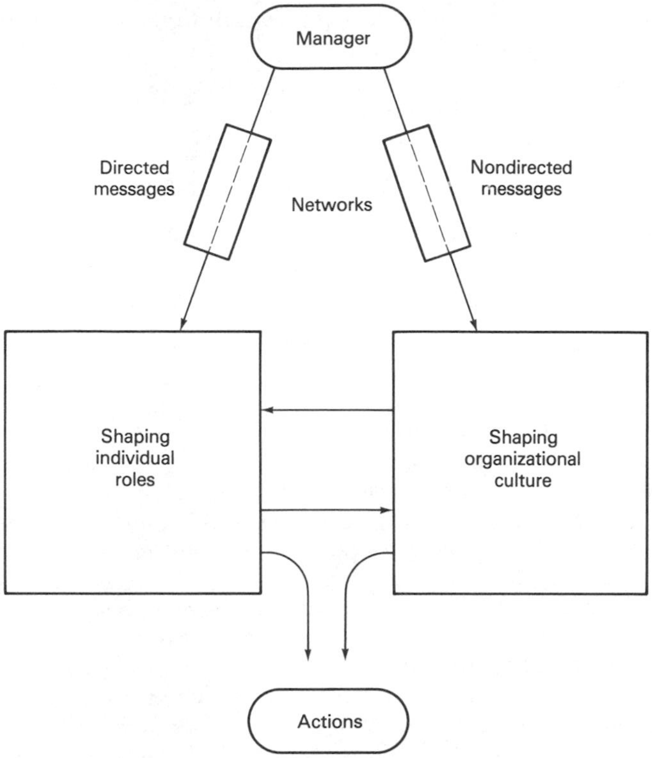

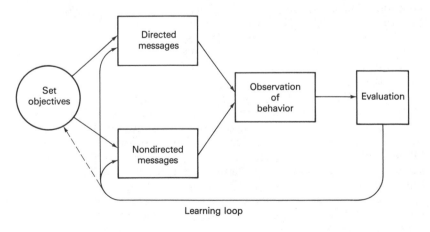

Figure 8-13.
Management through organizational communications.

is used to provide corrected messages until the desired behavior results (Gagne 1975). Teaching is most effective when it follows the following guidelines:

1. It is based on a structured plan (applying the rational concepts of Fig. 8-9).
2. Messages are clear.

Figure 8-14.
System approach to organizational communications.

3. Messages are reinforced through rewards (applying the behavioral concepts of Fig. 8-7).

4. Organizational members participate in the learning process (applying the social concepts of Fig. 8-11).

5. There is continuing encouragement.

6. Maximum use is made of all media possibilities.

In addition, in many CIM settings it has been found effective to make maximum use of hands-on familiarity with the technology that is being applied, to enhance the organizational/technology interface indicated in Fig. 8-1. If bar coders are being used to identify individual products, it is helpful to allow all workers an opportunity to try out these bar coders on their own, in a relaxed setting, and thereby gain familiarity and comfort with the technology involved. This type of familiarization applies to all aspects of the technologies being applied. At the same time, it has been determined that the needs of all participants must be addressed, in terms of how the new technology is going to affect individual functions and opportunities within the organization. It is necessary to deal with all of the individual concerns in order to provide a setting in which people are willing to develop new perceptions and new behavior patterns. A strategy of addressing the individual anxieties and concerns of all participants and providing an opportunity for hands-on familiarization with the technology in use can be a powerful approach to produce the desired type of teaching.

Another strategy that has been successfully implemented in transitioning from *as-is* to *to-be* systems is to run the two systems in parallel for a period of time (to the degree that this is possible). This allows a gradual learning period to take place instead of a sudden transition. If no parallel operation is feasible, then it may be necessary to have hands-on learning sessions well in advance of the implementation of the new system.

The task of the CIM planning and implementation group becomes a learning process from many points of view. There must be learning about the technologies that are available and the best use of these technologies for the particular enterprise. There must be learning about the organization and its members and how to best work with groups and individuals to achieve integration between the technological and human aspects of the system. Finally, there must be a learning process regarding the most effective ways to develop and manage CIM-oriented systems.

The design, implementation, and operation of an advanced manufacturing system can require substantial investment of time and energy by all organizational members. Ultimately, it is this widespread resource investment that may determine the success or failure of the effort. If individuals are unwilling to undergo the stress and demands associated with change, the desired results will not be obtained.

As emphasized in previous chapters, the development of a CIM system is basically an educational activity. Teaching and learning involve all members of the organization on a continuing basis. It is this necessity of entering into a learning environment that often meets significant resistance from participants. The development of new perceptions and new methods of problem solving requires an investment of time and energy that must be strongly motivated. For an organization to be successful as it implements a CIM system, it must be able to break down barriers to change and learning. The enterprise must produce an environment in which all members of the organization are willing to enter into the necessary learning efforts, to understand the nature of the change that must take place, and to learn individually how to operate effectively in the new environment.

EXAMPLE 8-3. MANAGEMENT PLAN

As part of the learning process for implementation of a CIM system, the management of ABC Company participates in a number of courses provided by consultants and learns about perceptions regarding organizations and human behavior in organizational settings. Based on these courses, management develops a plan for implementation of a desired CIM system. Given the insights developed in the above discussion, what type of plan might management develop?

Results

The management literature has evolved from narrow viewpoints regarding the type of organizational change efforts that can be initiated to broad efforts that address the most basic perceptions and orientations of the individuals in the organization. The concept of context management involves efforts by management to expand its own understanding of the situation (by attending consultant courses as discussed here) and to shape the way in which organizational members view the organization and its objectives. Continuous action must be taken to achieve organizational and individual goals within the current reality.

It is important during implementation of the CIM system for management to understand that the organization and its function must be as much a subject of redesign as the technology used to form the core of the new system. This will require efforts by management to understand the nature of the organizational structure and function that will be most appropriate for the new setting and to evaluate the range of problems that can be expected when members of the organization make the transition from the *as-is* system to the *to-be* system.

As emphasized in Chap. 3, transition stages must be robust and able to stand alone because they may remain in place for a long time and may turn out to be the basis for starting a new (unexpected) growth direction. This means that the technology and the organization for each transition state must produce a viable enterprise that continues to grow and change. Management must expand its own perceptions in order to be able to develop the best possible understanding of the situation that presently exists, the type of organization and behavior that are desired, and the strategies that will be necessary to move from the *as-is* to the *to-be* system. This will require management to accept the need for self-education.

A difficult situation will evolve if management decides that the employees of the organization must learn and evolve and change, but that upper management can continue to function as it always has. The learning and change efforts must spread throughout the organization in order to achieve the needed impact. This will require everyone in the organization to apply sufficient commitment and emotional energy to undertake the type of learning and change processes that are required.

Many of the most effective strategies will involve educational efforts. As part of the implementation of any CIM system, there must a structured plan to produce the desired teaching and learning environment. The messages that are distributed through the system must be clear and must be reinforced through rewards. There should be broad participation in the learning process and continuing encouragement. In order to achieve the desired change, it will be necessary to harness all of the perceptions in order to succeed in the task at hand.

Management must first develop the capability to understand its own needs as to the type of change that must take place and then must develop an environment in which this type of change occurs. This will require significant participation and involvement by all levels of management and a willingness by all members of the organization to undertake a learning process. Once this learning environment is established, it may feed upon itself naturally so that all members of the organization will attempt to move from transition step to transition step toward improved operating systems without as much resistance as initially displayed. However, the development and maintenance of a learning environment requires constant attention and effort. If this effort stops, learning will stop. The desire to learn to change will also stop. One of the hidden costs associated with the development and implementation of CIM systems is the need for all people who are part of the system to invest more time, energy, and effort into constant learning and change. This may be difficult for the people involved, but unless this type of willingness to learn and change can be achieved the designed system is not likely to perform as desired.

ASSIGNMENTS

8-1. In order to achieve maximum educational insight from Chap. 8, it is helpful to stimulate class discussion of the concepts that have been included, develop critical reviews that probe the advantages and limitations of the concepts, and explore alternative ways for approaching the topic. Select one or several topics from the bold-faced headings that introduce sections or examples in this chapter. Come to class prepared to explain this topic, critically describing the text materials and suggesting ways the topic discussion might be strengthened.

8-2. The objective of this assignment is to explore the alternative perceptions of organizations. Pick an organization and discuss with members of the organization and outside individuals familiar with it on how they perceive the structure and function of the organization. Try to identify the ways in which the organizational setting appears different to these individuals. Develop a short description of the organization as seen by each individual, including a description of the organization as you perceive it. Then, by looking at the various descriptions you have compiled, consider the nature of the perceptions in each case. Discuss the ways in which per-

ceptual filtering and the coding process may have affected each case. How can you use the concepts of filtering and coding to understand the nature of the information you have received?
(a) Come to class prepared to share your results and participate in discussion of the issues raised.
(b) After this discussion, prepare a brief paper on your insights.

8-3. Identify someone who works in an organizational setting and is willing to discuss past experiences with change in the organization. Discuss the ways in which the change was initiated, and how the members in the organization responded. What was the outcome of the effort? Based on this experience, can you draw conclusions regarding the perceptual filters and coding processes that were in place during this period of time? To what degree was the outcome determined by the perceptions of the participants in the organization, and to what degree was it due to other factors?
(a) Come to class prepared to share your results and participate in discussion of the issues raised.
(b) After this discussion, prepare a brief paper on your insights.

8-4. By reading the references for this chapter at the end of this book or other references, explore an alternative organizational model that might be used to describe organizational settings. Prepare a description of the model as it might likely be experienced in a manufacturing-oriented setting. How does the model relate to those organizational models discussed in the text? What additional insight can be obtained by applying this model to help interpret the nature of an organization?
(a) Come to class prepared to share your results and participate in discussion of the issues raised.
(b) After this discussion, prepare a brief paper on your insights.

8-5. Explore in further depth one of the reference models for organizations discussed in this chapter. By reading reference materials, considering your own experience base, and discussing the issue with others, what conclusions can you draw regarding the relevance and usefulness of the reference model?
(a) Be prepared to present your insights and discuss them in class.
(b) Prepare a short paper on your insights.

8-6. Discuss with someone who is familiar with a manufacturing system and environment the nature of the organization in which this individual functions or has functioned. Explore how the various organizational reference models discussed in this chapter might be used to describe the nature of the setting as perceived by this individual. What other types of reference models might be called upon to produce a broader understanding of this organization?
(a) Come to class prepared to present your materials and to discuss this topic.
(b) Following the class discussion, prepare a brief paper describing your insights in this area.

8-7. By using the references in this text or other reference materials, further explore the types of organizational behavior that are described in the text. To what degree are the behavioral models introduced here incomplete and in need of extended discussion? What other aspects of each of these models can you identify? What are the limitations associated with the models?
(a) Come to class prepared to discuss your findings.

(b) After class discussion, prepare a brief paper on your insights.

8-8. By drawing on the references in this text and other reference resources, identify another organizational behavior model in addition to those presented in the text. Develop a brief description of the nature of this model as it might be most usefully applied to a manufacturing setting. What relationships might exist between this model and the models presented in the text? In what ways does this model strengthen the ability to interpret organizational behavior? To what degree does it conflict with other models provided?
(a) Come to class prepared to discuss your materials and participate in discussion.
(b) Following class discussion, prepare a brief paper describing your understanding of the topic.

8-9. Identify an individual who has had significant experience in the management of organizational change in a real-world setting. Discuss with this person the nature of the change efforts that were both most and least successful. Can you relate the methods that were used to the concepts described in the text and/or to other reference materials? Develop a description of the management change methods that have been tried and a brief description of their results. Try to interpret the reasons for selecting the particular method and the reasons for the outcome that resulted. What conclusions can you draw?
(a) Be prepared to present your materials and discuss them in class.
(b) After class discussion, prepare a brief paper describing your insights and findings.

8-10. Discuss the approach to management change that is described in the text. Develop a critique in which you identify those strengths and weaknesses that might be associated with such an effort. In what ways do you think the management of organizational change and the establishment of a learning environment might be more effectively achieved? Make use of available reference materials, your own experiences, and the experiences of others. Come to class prepared to participate in a discussion of your studies.

9

PRODUCT DESIGN AND MANUFACTURING

System development activities must attempt to achieve a balanced approach to design and manufacturing. Since product design determines manufacturing requirements, and manufacturing determines design opportunities and constraints, an integrated viewpoint must be taken to achieve the most competitive products.

The emphasis in Chaps. 4 through 8 has been on methods for designing and implementing improved manufacturing systems, with the primary focus on the manufacturing activity itself. However, as discussed in Chaps. 1 through 3, the development of competitive manufacturing systems depends in essential ways on the methods that are used for product design. There must be an effort to apply product design activities that make maximum effective use of the manufacturing capabilities, and a determination to develop manufacturing capabilities that are appropriate for the designs in use. It is not rewarding to maintain rigid design strategies while attempting major changes in manufacturing, because the design function drives the manufacturing function. If product design is not integrated with manufacturing, the enterprise will not achieve an effective problem-solving strategy.

This chapter provides an overview of the types of relationships that may be developed between design and manufacturing. Several design principles can be established in this context and used to guide the manufacturing system into an appropriate strategy. This chapter also discusses a number of commercially available software products that can be used to support design activity. However, as systems become further integrated, such software must often be customized to reflect the particular enterprise environment. Design and manufacturing must be considered as parallel areas of emphasis by the planning group, and a balanced approach must be taken at each step to obtain a system that functions as effectively as possible.

As discussed in Chap. 8, the management/organizational aspects of a manufacturing system must be matched to the technological "core" of the system (Fig. 8-1). Otherwise, a mismatch may develop to prevent effective operations. It is also essential that the product design functions of the enterprise be matched to the technology and organization if system potential is to be realized. Product design and product manufacturing must become closely integrated functions in the CIM setting. The designers and the producers must link together the design and manufacturing activities to achieve optimum CIM operations.

Figure 9-1 shows a range of relationships that can exist between product design and manufacturing. At issue is the degree and type of linkage that should exist between these functions. Figure 9-1a shows manual design and the manual transfer of information to the factory.

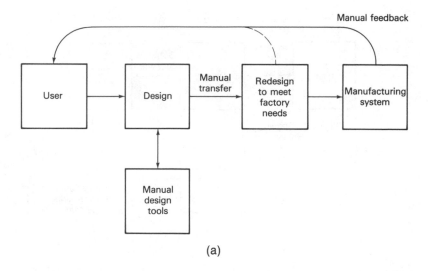

(a)

Figure 9-1.
Alternative relationships between design and manufacturing. (a) Manual design and manual transfer of information to the factory. (b) Combined use of manual and software transfer of information. (c) Use of computer-aided design and computer-aided engineering to strengthen product design. (d) A highly integrated design and manufacturing system configuration.

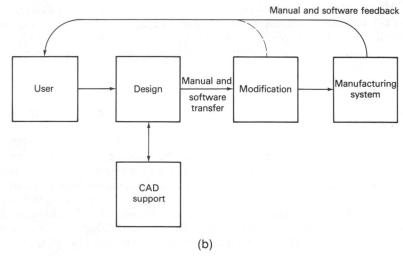

(b)

Figure 9-1. continued

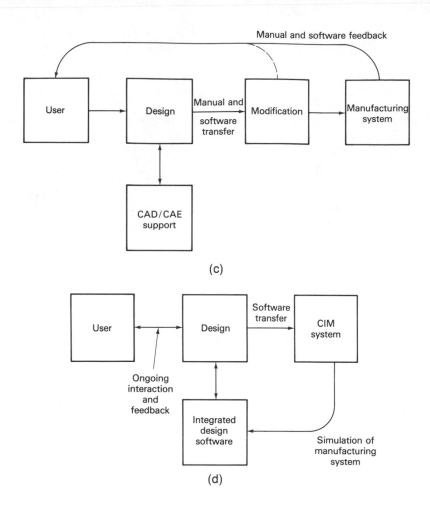

(c)

(d)

Many different types of relationships can exist between design and manufacturing. Isolation between these two aspects can result in products that are not competitive; on the other hand, activities to integrate them will result in widespread change for the technical and organizational features of the system. The degree to which integration is desirable or achievable will depend on a combined evaluation of the competitive needs and the implementation issues that are faced.

This type of system often results in inefficient operations, independent of the level of automation and integration that is applied to the factory itself. In such decoupled settings, the factory may even have to redesign the product for manufacturing. In this environment, the constraints introduced by manual design, manual transfer of information, and lack of communication will likely prevent effective operations.

Figure 9-1b shows how manual design may be replaced by computer-aided design with combined use of manual and software transfer of information to the factory itself. This is a definite improvement, because the designer is now obtaining computer support to improve productivity and efficiency, and the potential for limited software transfer of information develops. This results in a more compatible relationship with a typical CIM system. Figure 9-1c shows how computer-aided design and computer-aided engineering (both introduced in Chap. 3) can further strengthen product design by allowing the designer to understand better how the product will function. Additional computer linkage between design and manufacturing will become particularly useful in this setting. Finally, Fig. 9-1d shows a highly integrated design and manufacturing system configuration in which product design is shaped by manufacturing capabilities, and product design information can be downloaded to the factory through a computer network.

In order to implement the highly integrated configuration shown in Fig. 9-1d, the design-support software in the CAD system must contain detailed information regarding the capabilities of the manufacturing system. The CAD software is basically required to include a simulation of the manufacturing system. When using this software, the designer can anticipate how the manufacturing system will respond to certain designs and can be guided toward the most effective cost-benefit strategy for product design. Further, once the design is complete, it is assumed in this arrangement that the design can be completely transferred to the factory through a software download over a computer network.

This design approach conforms to the design for manufacturing concept, since the capabilities of the manufacturing system are explicitly considered in the design process. It may be possible in such a system to include (with the design information being passed to the manufacturing system) detailed data that will describe exactly how the product should be manufactured. Thus, a feedback loop exists between product design, the manufacturing system, and the integrated design software so that all elements interact to achieve the interfaces that are required for closely linked design and manufacturing operations.

In order to guide the evolution of an enterprise from a manually oriented system (Fig. 9-1a) toward an integrated network system (Fig. 9-1d), the following requirements may be noted:

1. An enterprise orientation toward CIM system concepts, as presented in previous chapters

2. An orientation toward design for manufacturing, which emphasizes linking design and manufacturing capabilities

3. Use of a sophisticated design software that incorporates a manufacturing system simulation and can be used to guide the user toward the most cost-effective product designs

4. Use of sophisticated CIM system that can produce different products based on the software downloaded from the design activity

5. Development of the required software linkages between design and manufacturing

The CIM system capabilities discussed in previous chapters can be effectively harnessed only if an appropriate level of integration is introduced into the design process and into the design-manufacturing interface.

This chapter describes a number of CAD software products and indicates their capabilities and limitations. These software tools are a useful resource for the development of a CIM-oriented enterprise. However, they only partly satisfy the requirements for producing the type of system shown in Fig. 9-1d. In order to simulate a manufacturing system, custom programming is required. This custom programming must be linked to the existing design tools so that they work together.

Such coordination can require extensive software development. Further, the design process must take place in such a way that a software download can occur from design to the CIM manufacturing system. This requires the necessary software interface and a manufacturing system that can adapt from product to product by making use of the information that is provided over the computer network. This will typically require a system with high levels of integration and automation.

9-1. Introduction to Computer-Aided Design and Engineering

The classic approach to product design emphasizes manual activities by engineers and technical staff to develop the documentation that is necessary to build the desired product. In this setting, the purpose of all design activity is to produce the paperwork needed to provide instructions to the factory for product manufacturing. The creation of this documentation is a labor-intensive task; a large, expensive infrastructure often develops in manufacturing organizations to transfer the paper-based information from design to manufacturing. At the same time, because of the complexity of manufacturing processes, additional quantities of paper record keeping are developed during the manufacturing activity. This method of information handling results in an inefficient and ineffective approach to controlling and monitoring production.

With the introduction of computer capabilities, a widespread effort has resulted to determine how the design process can be enhanced by making use of a computer-aided design (CAD). The initial focus of CAD, and still one of the most important areas, has been to replace manual drafting of design documents with an ability to create designs on the computer screen and then to print out the documentation. To this end, many different software programs have been written to enable the engineering designer to use the computer screen to create a drawing. This approach has required the development of software that can be used to draw a variety of shapes and to describe these shapes in sufficient detail that they can be used for the manufacturing process.

The most fundamental requirement involves the designer being able to create straight lines. This can be done by defining beginning and ending coordinates for the line and then having the computer connect the two points (Fig. 9-2a). Another simple drawing aid is to create a circle in which the designer locates the center of the circle and defines the radius, and the computer software then draws a circle around this point (Fig. 9-2b). With more sophistication, it is possible to remove part of a line or part of a circle to produce various types of intersections. These capabilities can be extended to additional shapes until the designer can create on the screen essentially any desired geometric shape.

As these types of drawing aids have become available, typical use has been to have the designer attend a class to learn how to use the computer system, perform geometric design by sitting at the computer, and use the computer to draw the necessary shapes. The computer can then print out the resulting shapes on a large scale printer.

This approach has many advantages for the production of draw-

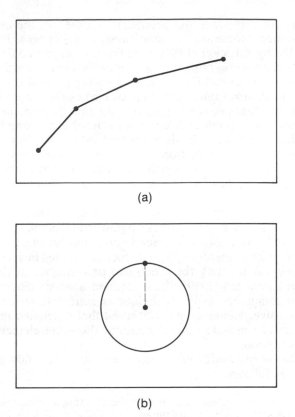

Figure 9-2.
Basic CAD drawing elements. (a) Sequence of straight lines. (b) Defining a circle in terms of its radius.

(a)

(b)

ings, since mistakes or changes can easily be incorporated into the computer data base and a new drawing produced. Thus, one of the earliest advantages of computer-aided design has been to simplify the technology of producing drawings.

The initial mechanical design software was oriented toward two-dimensional drawings. The designer created each drawing as a two-dimensional perspective and mentally "mapped" the views into an integrated three-dimensional composite. More sophisticated versions have allowed the computer to connect the various two-dimensional views of a drawing. A top view might be rotated into a side view, so that the top and side views can be compared with each other for accuracy. Software has been developed to enable two-dimensional perspectives to be combined, resulting in a perspective that relates all three dimensions. Such drawings can be rotated in space to allow the designer to examine them.

The designer can also create *wire frame* drawings, which have the appearance of a three-dimensional (3-D) representation but do not truly provide the designer with three-dimensional insight. The surface of a part can be traced by a number of wire frames to create a three-dimensional illusion. Wire frame CAD uses a set of points and lines to describe a three-dimensional surface. Points above and below the surface and the internal structure of the object are not defined. This type of definition can provide a 3-D geometric "sense" of the object, but can also produce "nonsense" or ambiguous visual images (Jain 1989).

As noted by Voelcker (1988), wire frames "appeared first as simple two-dimensional programs. . . . In the 1970s, the systems' modeling entities—two-dimensional lines and arcs—were generalized to represent segments of three-dimensional space curves that could be linked to represent the edges of solids (hence the name "wireframe")."

Another step in product design has been toward a computer modeling technique, that helps the designer understand the function of the part being described. A transition has been made from looking only at the geometry of the part to understanding how the part functions in its desired setting. This has led to the concept of computer-aided engineering.

One of the standard approaches to creating such models has been to use a *finite element* analysis (FEA), a powerful method for analyzing structures. A physical object is divided into a number of small building blocks called finite elements; a physical model can then be incorporated into the computer to study the functional performance of the object. As noted by Kinnucan (1989), finite element analysis places difficult demands on computers, as the technique is extremely processing and memory intensive. Kinnucan has observed that a detailed model of a large product can include several hundred thousand elements in the product description.

FEA has been a basic tool for engineers since the 1960s. Kinnucan describes it as follows:

> The . . . process begins with an engineer creating a geometric model of a part or structure on a computer. The model is divided into a finite number of small pieces (elements) that are connected to each other at points called nodes. Mathematical equations describe how these nodes respond in terms of deflection, stress and temperature to loads, such as gravity, pressure and heat, applied to the elements. The equations can take into account material composition and other variables as well.* To determine the overall response of an entire structure, the equations for the individual elements are assembled in matrix format to provide a global description. This results in a large set of simultaneous equations to be solved by the computer. (p. 42)

A great many building-block elements are required for FEA. Because of the complexity of the formulas that must be used to represent physical performance, such systems require large computer memories and fast processing speeds. Nonetheless, many complex mechanical products are designed using finite element analysis to assure the correct functional performance. The fierce competition among FEA vendors will continue to lead to new hardware and software products and decreases in price (Kinnucan 1989).

As noted by Hamilton (1989), the growth of mechanical computer-aided engineering (MCAE) is handicapped by a lack of integration capability. The users of MCAE must allow themselves flexibility to ac-

---

* Until recently, such models were limited to two dimensions or surface shells; 3-D solids modeling is discussed later in the chapter.

commodate a rapidly changing market and product environment, while making the best use of available software. As pointed out by Hamilton, the harnessing of computer-aided engineering resources to the enterprise can be accomplished with careful planning and implementation; at the same time, without adequate care, significant problems and poor performance can result.

While mechanical CAD/CAE capabilities were being developed, computer-aided design and engineering were also being applied to many other areas, ranging from electronics products to architecture. In the electronics-products area, the initial application was primarily drafting-related, using a computer screen to draw the locations of the desired thick and thin film layers and components on a printed circuit board or hybrid microcircuit (discussed in Chap. 6).* This type of drafting capability was essential for the development of complex integrated circuits with thousands or millions of elements on a very small surface area.

For all of these electronics products, ranging from integrated circuits to hybrids and printed circuit boards, it became impossible to develop manually the drawings that were necessary to produce the final product. A computer was required to allow the replication of building-block cells for such products; only in this way has it become feasible to produce the electronics systems that are available today.

In the same way that finite element analysis is used to study the performance of mechanical products, it is also applied to electronics products. Finite element analysis can be used to study heating throughout the electronics system, and has been used to study the propagation of microwaves and optical fields within circuits.

Computer-aided design has evolved rapidly to support the drafting functions necessary for a wide range of engineering products and has been extended to computer-aided engineering to allow the designer to perform computer studies of the function of a product as well as its geometric characteristics. Continuing effort is being devoted to producing CAD software that is both more effective at addressing geometric issues and includes strengthened computer-aided engineering models that represent the functions of the products being developed. In a variety of settings, CAD has become the dominant strategy by which companies develop products. Improvement has taken place in product design through the use of CAD resources. On the other hand, the principal function of CAD is still to create the paper documentation that is required by the manufacturing facility to produce a part. There is a significant gap between the CAD function discussed here and the type of integrated design-manufacturing activity that is envisioned as part of a CIM strategy.

9-2. Three-Dimensional Modeling

Major changes are taking place today in an effort to move from combined two-dimensional representations to true three-dimensional

Computer-aided design and computer-aided engineering software can bring many new capabilities to the enterprise, particularly if manufacturing constraints are reflected in the design algorithms included in the software. However, tools of this nature provide the individual designer with much more information regarding product opportunities and constraints than do alternative approaches to design. The designer must be willing to learn how to make use of the improved system design approach and to apply it effectively. Otherwise, efforts will continue with "business as usual," and designers working around and avoiding the tools that have been provided.

* Several CAD systems for hybrid design are listed by *Hybrid Circuit Technology* (1987).

modeling for computer-aided design. In three dimensions, the modeling focus is on defining the complete product geometry and then studying the two-dimensional properties through projections. Three-dimensional modeling enables the designer to consider all aspects of the geometry and to study performance with complex models that couple between perspectives.

Three-dimensional solids modeling defines all aspects of an object, including the observed surface and the internal structure. All ambiguities are removed. This type of definition not only provides an accurate geometric view of an object, but also can be used to perform CAE studies of the entire physical structure (Jain 1989). Three-dimensional solids modeling is often used with complex optical-ray-tracing software to develop use of light reflection and shading appropriate for the object and setting. Solids modeling has the potential to help make the integration of CAD and CIM a reality (Wohlers 1989). A number of 3-D software and supporting hardware systems are available; however, this capability is quite expensive and can require significant retraining of those who are involved in product design.

The 3-D systems present a significant increase in capability over conventional wire frame graphics systems because they allow the complete geometric description of the object to be stored in the computer. Many types of design and analysis can be performed with 3-D representation (Fallon 1986; Skomra 1986; Jain 1989; Wohlers 1989). In moving to complete 3-D representations, errors can be noted early in the design process and performance can be optimized by using a wide range of physical models. At the same time, the ability to develop 3-D models on a computer places new requirements on the designer.

As CAD capabilities become more strongly linked to the manufacturing setting, these requirements will further increase. Thus, product design that draws on 3-D modeling and design for assembly (DFA) manufacturing concepts (optimized for particular factory configurations) will place increasing demands on the designer, who must make use of all these resources.

For both 2-D and 3-D design software, the typical CAE models that are used to study product performance are often oriented toward linear modeling of performance. Nonlinear modeling capabilities are also available, but this step increases both the complexity of the analysis and the difficulty of interpretation (Singampalli 1986; Crosheck and Hite 1989).

CAD/CAE software will often require custom modifications to achieve an adequate linkage to the manufacturing system. Major effort is typically required to understand the nature of the as-is *manufacturing system and to document the manufacturing constraints that should be included in the software. The modification of the software to achieve the type of integration desired can be a demanding task, requiring expert and experienced programmers who can understand and implement management objectives.*

EXAMPLE 9-1. CAD/CAE SOFTWARE PRODUCTS

Many CAD/CAE software products are available. Discuss typical features and illustrate the types of output results that can be obtained for several of these products.

Results

(A) MasterCam Mechanical Design Software: MasterCam®, offered by CNC Software, Inc., is a PC-based product that can be used to design

mechanical parts and to provide data input to CNC milling and lathe machines to produce the part that has been designed. Figure 9-3a shows an application of the milling machine feature. MasterCam is used to develop contours for the cutting tools. The tool moves along the defined surface lines in order to create the desired product shape. MasterCam makes use of dynamic tool display with a tool path increment display set by the user, advanced "look-ahead" compensation to prevent gouging, and automatic feed and speed calculation. The software allows the rapid development of any desired geometry, based on many standard shapes that are available. With a "mouse," the user can select combinations of shapes to produce the desired pattern. An effort has been made to create a product that is simple to use with no complicated programming language to learn.

Figure 9-3b shows how MasterCam can be used to develop designs for lathe production. The CAD drawing illustrates the cutting tool and shows the paths that will be followed by the tool as the part is turned on the lathe. Features of MasterCam include a standard 50-tool library, plus the potential for creating your own tools, "look-ahead" cutter compensation, and an automatic feed and speed calculator that depends on the material being processed.

MasterCam operates on a number of microcomputer hardware hosts. It is a customized product that is intended to help job shops improve their profits, deliveries, and production. Over 3500 job shops use this product today. Such a CAD package can be of significant advantage. The types of machining operations to be performed in such settings can be quite complex, and developing the necessary CNC data to produce such parts would be difficult without adequate CAD capability.

(B) CADKEY Design and Drafting Software: CADKEY® is a computer-aided design and drafting product widely used for mechanical engineering system design. CADKEY 3 is an advanced product that can help the designer integrate all aspects of the manufacturing process into one solution.

Figures 9-4a and 9-4b show CADKEY 3 used as a design tool for product development. This software makes use of a user-oriented menu structure, and prompting guides the user through each function. A history function helps the user recall the present status of the system design process. A number of "immediate mode" commands provide shortcuts through the system and allow the user to bypass the menu and go straight to the desired function. As shown in Figs. 9-4a and 9-4b, a wide diversity of shapes and labels can be created during the design process to define complex mechanical products.

By combining CADKEY 3 with CADKEY SOLIDS, the user can link a wire frame modeler (to create a detailed description of product geometry) with a solids modeler (for visualization and other types of analysis). Figure 9-4c illustrates how a wire frame model of an aircraft can be linked to the solids modeler, resulting in a product description that can be used to support many types of analysis. Figure 9-4d shows another solids modeling result from CADKEY products.

CADKEY is a versatile product that can be used for two-dimensional drafting, three-dimensional wire frames, and three-dimensional solids modeling. The software operates on a variety of workstations and operating system environments.

CADKEY offers a growing line of direct translators to link the software product to other design centers and produces an IGES translator (discussed below) to achieve flexible communications. CADKEY can also be used for reverse engineering and inspection to obtain an accurate blueprint for an object. The CAD Inspector product allows the user to unfold a 3-D product drawing in order to determine the design of the component.

Figure 9-3.
MasterCam® mechanical design software
(courtesy of CNC Software, Inc.). (a) Milling
machine feature. (b)
Lathe production feature.

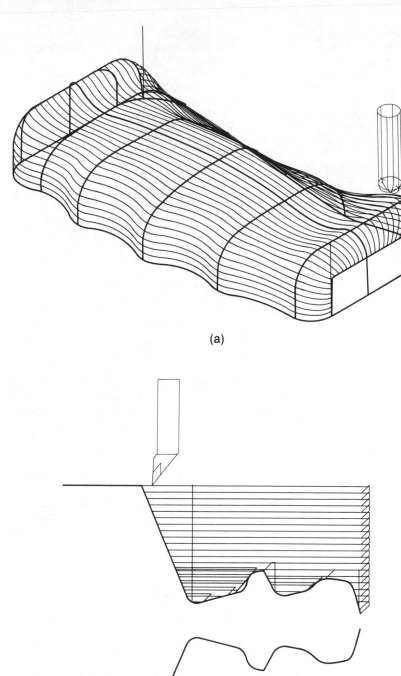

(a)

(b)

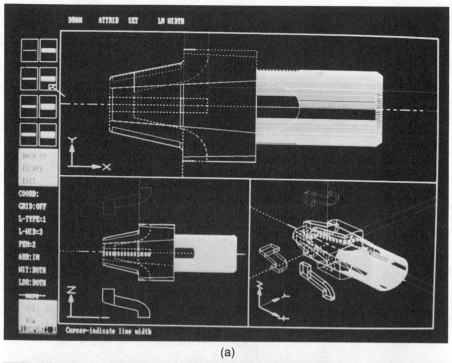

(a)

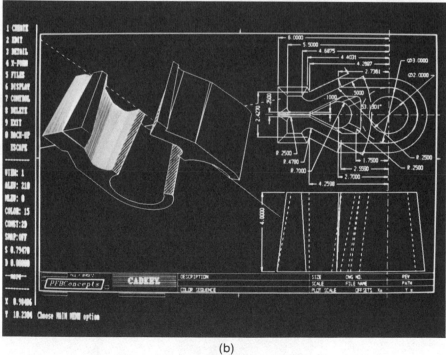

(b)

Figure 9-4. CADKEY® design and drafting software (courtesy of CADKEY, Inc.). (a and b) Creating complex mechanical products. (c) Wire frame and solids model of an aircraft. (d) Solids modeling result.

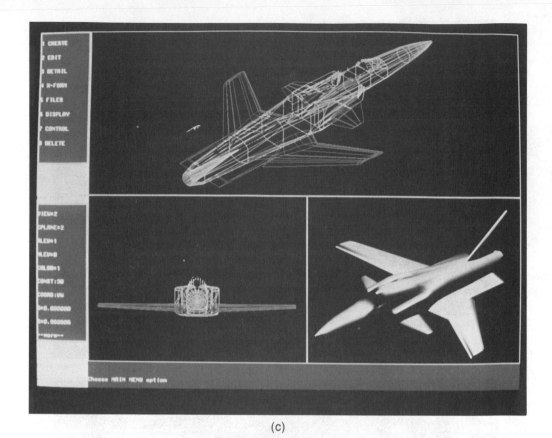

(c)

Figure 9-4. continued

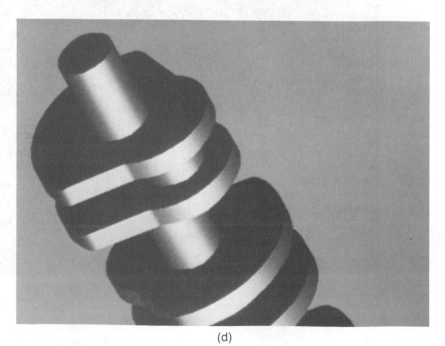

(d)

Since its release in January 1985, more than 55,000 copies of the CAD-KEY product have been installed. Systems are available in a variety of languages and are supported by a worldwide dealer network. Numerous functions, ranging from simple to complex, are available to users.

(C) ToolChest Software System: ToolChest is a CAD/CAM system offered by Battelle that is designed expressly for pattern makers, mold producers, and tool and die shops. The ToolChest software can be used to create models and drawings for complex 3-D shapes and for control of CNC machines to produce the desired product. The direct formation of CNC data from the design process enables a strong design-production link to be established. Specialized capability is available for machining complex cavities, a difficult metalworking task. ToolChest blends design capability, resource data, mechanical drawing, and 3-D picture-making capability with CNC program operations. The software system provides a collision-avoidance feature that can help achieve most effective production of complex shapes. The ToolChest product accepts data from other CAD systems, can be used in networks, and meets IGES and other communications standards.

Figures 9-5a and 9-5b show how the ToolChest software product can be used to design products requiring cavity machining with multiple contoured intersection surfaces. Figure 9-5a shows how one continuous cutter path can be developed by using the CAD system. Figure 9-5b provides a three-dimensional view of the product cavity that will be produced due to the defined machining process. Figure 9-5c shows the design of a connecting rod. Figure 9-5d provides another 3-D surface solids model to illustrate the capability of the ToolChest product; this figure shows the inner trunk lid body panel for an automobile.

Figure 9-6 shows how the ToolChest product can be used to define inspection paths for coordinate measuring machines. The design of such inspection paths can be a labor-intensive task in order to assure that the necessary inspection procedure is being followed. The ToolChest product defines the motion necessary for the coordinate measuring machine and develops the required data to drive the inspection process. Data for the designed paths can be directly downloaded to coordinate measuring machines. The DMIS (Dimensional Measuring Interface Specification) protocol is presently under development to simplify and strengthen the ability to download inspection path information to coordinate measuring machines.

(D) ANSYS Computer-aided Engineering Software: The ANSYS® program, offered by Swanson Analysis Systems, Inc., is a widely recognized large-scale, general-purpose finite element analysis program for engineering. The original versions of the ANSYS program were oriented toward batch processing for mainframe computers and emphasized structural and heat transfer analyses. The program has been extended to include other physical phenomena (magnetic fields and fluid flow), nonlinear effects, on-line documentation, an extended graphics library, solids modeling, light source shading, and many other features.

The preprocessing phase of the ANSYS program is used to create a finite element model and to specify the options needed for the entire subsequent analysis. Static analysis can be performed to determine the displacement, stresses, strains, and forces that occur in a structure or component as a result of applied loads. Dynamic analysis is used to determine the effect of applied loads with a time-changing environment. Buckling and stability analysis can be performed to determine the characteristics of any load-carrying structure. The ANSYS program performs studies of thermal conduction, convection, and radiation; these types of heat transfer can be considered under steady-state or transient, linear or nonlinear conditions.

Figure 9-5. ToolChest software (courtesy of Battelle). (a) Wire frame model of complex surface. (b) Solids modeling of the same surface as in (a). (c) Design of a connecting rod. (d) Solids modeling of automobile body panel (inner trunk lid).

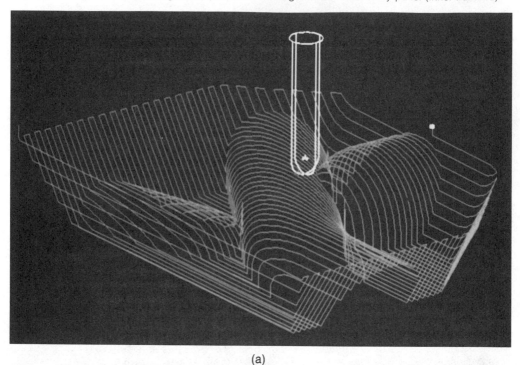

(a)

(b)

Figure 9-5. continued

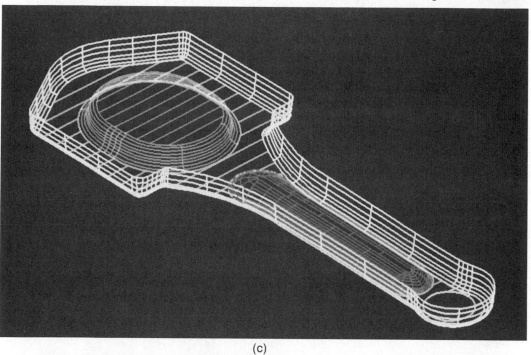

(c)

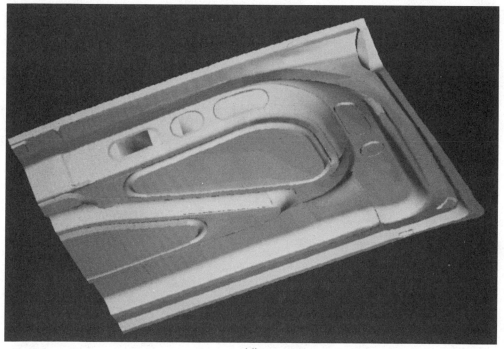

(d)

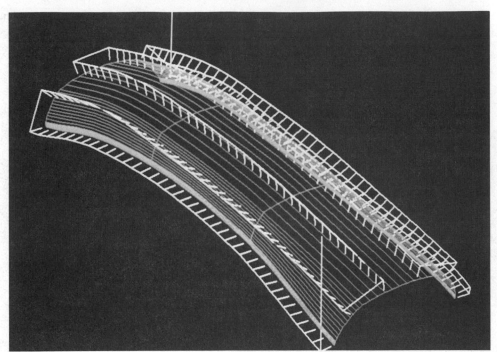

Figure 9-6.
ToolChest application to development of an inspection path for a wordmate measuring machine (courtesy of Battelle).

The magnetic capabilities of the ANSYS program can be used to analyze the different aspects of magnetic fields and are useful for analyzing devices such as solenoids and actuators. The fluid capabilities in the ANSYS program enable the user to study the flow or pressure wave characteristics of a liquid or gas in a given system. The ANSYS program allows various material properties to be specified, in a data base format, through the preprocessor.

The ANSYS program was originally developed for use by the power generation and metals industries, but has since branched out to accommodate the requirements of a wide variety of other industries, from automotive and electronics to aerospace and chemical. The program is now installed in over 1000 sites worldwide and is used by thousands of engineers.

Figures 9-7a and 9-7b illustrate an application of the ANSYS program to thermal-stress analysis of a silicon die (chip).* This analysis of an integrated circuit package shows the stresses generated when the chip is subjected to a 150°C temperature differential. A mismatch in the thermal coefficient of expansion between the IC chip, the frame, and the plastic molding results in stresses. The surface and three-dimensional features of the IC chip are shown, with various shadings used to indicate the temperature gradients.

The types of analyses described here can be of extreme importance in predicting product performance. By understanding the type of performance that can be expected, the designer can more effectively satisfy product design objectives. Combined CAD/CAE capabilities can shorten the design and manufacturing cycle and result in more competitive enterprise operations.

---

* Refer to Chap. 6 for a discussion of electronics components and products.

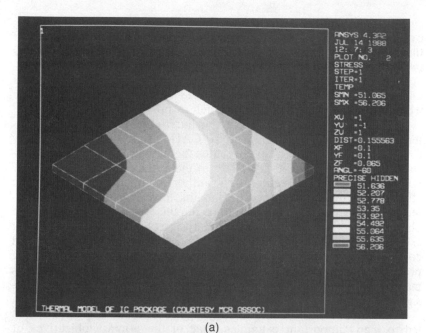

(a)

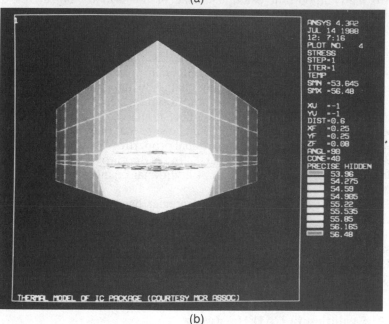

(b)

Figure 9-7.
ANSYS® computer-aided engineering software. (a, b) Application of the ANSYS program to thermal-stress analysis of a silicon die chip (photographs courtesy of MCR Associates, Inc., use of ANSYS courtesy of Swanson Analysis Systems, Inc.).

9-3. Design for Assembly and Manufacturing

Major efforts are continuing to understand the design for assembly (DFA) and design for manufacturing (DFM) concepts as they apply to realistic manufacturing settings. These concepts can be used for single components or entire products and require a substantial learning experience by the enterprise in deciding how to simplify the nature of the manufacturing process. A number of different strategies are being pursued, ranging from the definition of standard building-block components that can be used for product design to a more general method that involves the application of criteria to each part to determine whether it has been simplified as much as possible.

As noted by Boothroyd and Dewhurst (1988), there is strong interest in applying these concepts to CAD systems. However, many CAD applications are begun only after the new product has been substantially defined, and it is too late to make the types of basic DFA/DFM changes that are required. The authors point out how early design efforts must be linked to cost estimating to enable the most cost-effective product design approach to be pursued from the earliest design phases. Results of an early DFA analysis combined with early cost estimation can have significant impact on product manufacturing and cost. The development of this capability early in the manufacturing process can require a major restructuring of the enterprise, as well as major reeducational efforts for those who are engaged in product design.

The importance of simplified design to manufacturing ability has led to many important examples. One of the primary objectives of design for manufacturing is to eliminate screws and other fasteners (Boothroyd and Dewhurst 1988; Port 1989). The objective is to reduce the total number of parts, simplify each of these parts, and simplify the means of assembly. Manufacturing, installation, and maintenance can all be reduced in complexity. Mechanical computer-aided engineering (MCAE) software programs that allow team members to study the three-dimensional properties of a design are critical to the development of such simplified systems. The simulated components can be assembled on a computer screen to make sure that they will fit together properly and have the desired functional properties. Software programs developed by Boothroyd/Dewhurst, Inc. (Boothroyd and Dewhurst 1988), and others can help prompt additional changes to reduce the parts count.

Experience with such design strategies has resulted in a growing awareness that design must be regarded as an integral part of manufacturing. If products are designed and manufactured as efficiently as possible, costs can be reduced and quality can be enhanced. On the other hand, if design remains isolated from manufacturing, continued, unrealistic pressures will be applied to the manufacturing system itself.

9-4. Evolution in CAD Hardware

As discussed in Chap. 7, the introduction of powerful microcomputers and workstations has had a significant impact on product design and manufacturing. With the availability of cost-effective hardware and a wide range of software products, it has become feasible to develop more highly automated and integrated equipment.

The application of computer-aided design to manufacturing has often focused on the use of high-end microcomputers called *workstations*. These workstations have been enhanced to provide the engineering capabilities required for the designing purpose. The rapid growth in workstation capabilities has been widely discussed (Thorell and Hurley 1989; Jain 1989). The workstation market continues to expand and the average price per station continues to drop. Many companies are involved in the field, and systems continue to evolve toward increased levels of capability.

The CAD systems in use today often make use of both workstations and minicomputers (with terminals), with the particular configuration depending on the setting and application requirements. As the capabilities of microcomputers approach those of workstations, PC-based CAD is also growing rapidly.

The hardware and software capabilities of workstations and microcomputers have led to CAD products that are helpful in some ways but limited in others. CAD is broadly applied as a way to replace the pen and paper traditionally used for drafting purposes. The use of a combined computer terminal and printer has many advantages, as revisions and modifications can be quickly and efficiently made. The steady growth in the power of CAD hardware and software has led to efforts to broaden the scope of design support. Computer-aided engineering has been introduced as a way to enable the designer not only to look at the geometry of the product but to explore how the product will function. The effort to increase the ways in which the computer can support the design process has led to progressively more sophisticated hardware and software.

A major difficulty with these applications is that the problems being addressed still often far exceed the workstation capabilities, and the steadily more complex products require a higher investment of user time before they can be effective. Thus, the application of CAD to sophisticated functional design and the linking of design to manufacturing are still rapidly evolving areas. Capability is increasing, but so is an awareness of limitations. This awareness of both potential and limitations is leading to an unstable field characterized by rapid change and splintering of the available products. No one company dominates the hardware and software for CAD, so the user is faced with a bewildering array of products.

9-5. Computer Communication for CAD Integration

The use of sophisticated CAD software can support and enhance design activity. Using the standard software packages that are available on the market, the designer not only can apply the CAD system for support in the drafting process but can also explore functional product performance through various modeling techniques. At first consideration, it may seem that, based on this capability, the design problem has potentially been solved. Unfortunately, the situation is much more complex. Major difficulties can be experienced in trying to link design computers to one another and in linking the design capabilities to the manufacturing operations (Voelcker 1986). As discussed in Chap. 7,

Computer communications difficulties are often encountered when integrating CAD into the manufacturing system. The types of linkages called for by DFA and DFM require that design and manufacturing computers interchange information on a continuing basis. A significant amount of custom software may be needed to achieve an adequate flow of information to support the system integration concept.

continuing effort has been devoted to developing computer communication standards that will allow different types of equipment units and computers to be linked together for automated and integrated manufacturing. In the same way, efforts continue to solve the computer communication problems associated with design information developed through the use of CAD systems.

Two types of communication strategies are shown in Fig. 9-8 (Stark 1988). Figure 9-8a illustrates two CAD systems that require communication with one another, with data flow in either one or both directions. Hardware and software system type A must be able to exchange information with hardware and software system type B. If both computers A and B can use the same hardware and software, then the exchange can be accomplished easily, either through one of the computer communication methods discussed in Chap. 7 or by exchanging floppy disks from one computer to the other. However, the typical situation in a manufacturing setting is more complex.

Unfortunately, the CAD market is highly fragmented. There are many different software products and hardware hosts that are used for these products, and there is no generally available CAD product for *all* purposes that can provide the needed capability for all applications. Typically, subcontractors and contractors will use different CAD systems that are suited for their own design needs and are customized for the types of products of interest. (Because of the diversity of product types in various manufacturing sectors, it is reasonable to expect that a wide diversity of CAD systems will have evolved.) Given the complexity of the design process in each setting, it is difficult to produce customized CAD products that are required for each setting. An effort to merge all of these into a single CAD product seems unworkable.

In any large manufacturing setting, it is often necessary for multiple vendors to be able to work together by the exchange of CAD data. In many cases, the exchange is being achieved by the use of documents.

Figure 9-8.
CAD communication strategies. (a) Communication by making use of a common neutral format. (b) Communication by making use of a direct translator. (c) Communication between a CAD system and a CIM factory using a direct translator. Based on J. Stark, *Managing CAD/CAM: Implementation, Organization, and Integration* (New York: McGraw-Hill, 1988), pp. 124–128.

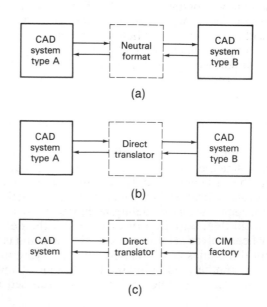

CAD system type A produces documents that are then transported to the second CAD system, and must be fed through a data entry process to convert the documents back to computer data. This is obviously an inefficient procedure that sharply limits the use of CAD to improve productivity.

Given the complexity of the situation, two other approaches are being followed. One is to apply standards of the type introduced in Chap. 7, so that (ideally) every CAD system will be able to transfer its output into a standard or *neutral* format and will be able to receive input from this standard or neutral format. All CAD systems would be able to exchange information by passing through this neutral or standard data base. Unfortunately, this effort has achieved only limited success. Because of the many different ways to represent geometric data and the characteristics of the parts being described by the drawing, transforming complex shapes and specifications from one format to another becomes a difficult process.

In the same way that much effort has been applied to MAP to develop a universal factory standard,* continuing effort has also been expended in developing a uniform CAD standard (*Computer Aided Design Report* 1987; Stark 1988; Wilson 1989). The Initial Graphics Exchange Specification (IGES) was adopted as a standard in 1981. IGES is principally used to exchange information regarding the geometric shape of models. Despite a number of major revisions to the original IGES, this standard still does not provide the universal format that has been sought.

A number of information types are not handled in the IGES format. In order to apply IGES, it is necessary to consider the specific types of CAD systems that are being employed and examine the particular data types that are to be used (which include geometric points, definitions of lines and surfaces, and other symbols). A detailed evaluation process must take place to determine the degree to which available IGES systems will perform the desired translation and the degree to which they must be enhanced or "flavored." (*Flavoring* has been used to describe writing additional software that is necessary to overcome the limitations of vendor-provided IGES products.)

There are many different ways to describe the geometry of a product; unfortunately, a true IGES system will have to accommodate every possible method. This results in such an arduous task that it is likely that IGES will continue to apply only to limited cases; the user must evaluate the applicability of IGES depending on the particular use. Major companies often have to make large investments of funds and expert time in order to achieve a desired level of data interchange using IGES.

Another method of translating the CAD data from system A to system B involves *direct translation* (Fig. 9-8b). It is possible to examine the two particular systems of interest and develop software that will achieve a limited translation. The advantage of this approach is the potential for simplicity, when compared with a universal neutral format. On the other hand, if many different systems are in use, and

---

* Refer to Sec. 7-8.

a direct translator is necessary to link all systems, the number of translators rapidly becomes unworkable. This approach is similar to the function bridge approach introduced in Sec. 7-9.

All of these information exchange processes are being used to link CAD station data. IGES is a widely used method for transferring CAD information from one site to another. IGES has become a de facto standard for geometric data interchange, but must be customized for different applications. Direct translators are available commercially and can also be developed by individual companies.

In addition, a number of other standard-development efforts are being attempted. Some of these have been extended to include nongeometric information about materials properties and manufacturing tolerances. Several different standard-oriented approaches have been taken (Stark 1988, Ch 8). Because of the rapid rate of change in both hardware and software capabilities in computer-aided design, it may be expected that the communication solutions will continue to lag the application needs and that the exchange of CAD information will remain a major problem area.

As shown in Fig. 9-8c, another major area of need is for the transfer of CAD information to the CIM manufacturing setting and the extraction of information from the manufacturing system for use in the design process. If computer numerically controlled (CNC) equipment is being used, then it is desirable to translate the design information into the correct type of computer control information to produce the desired part. In general, this linkage requires custom software. The output of CAD systems is not usually in a format that can be directly or easily used to drive the manufacturing system. However, such capability is an essential ingredient in integrating the design and manufacturing aspects, so effort is being applied to understand the nature of the problems that are being faced and proposing methods for solutions.

As shown in Fig. 9-1, it is also essential that manufacturing data be extracted from the factory and made available to the CAD system, so that the CAD system can best accommodate the requirements and opportunities of the manufacturing system. As the design engineer creates products, it is necessary that a knowledge of the manufacturing facility be included in the design data base. In this way, the designer can choose products that can be most effectively manufactured in the particular manufacturing setting. The withdrawal of data from the factory, the production of the desired modeling framework for the data, the transfer of data to the CAD system, and the inclusion of the data in the CAD data base all require custom software.

The application of CAD/CAE tools can result in a competitive manufacturing system that can rapidly respond to market opportunities and make the best possible use of manufacturing capabilities. At the same time, many difficult implementation problems can be experienced in developing the required software and in helping designers learn how to use the available resources.

Computer-aided design has provided a great many capabilities to the draftsman and designer in support of product development. However, when efforts are made to link companies and to link design and manufacturing, the CAD systems must be evaluated by the CIM design committee. These CAD/CAE systems must be studied in terms of the stand-alone support they can provide to the design function of the particular organization. At the same time, if there is a desire to link CAD systems, either within the organization or between the organization and other groups, or to link the CAD system to the manufacturing system, careful consideration must be given to the state of the art and capabilities (Wohlers 1988; Hays 1989; Metzger 1989) and the cost of achieving the desired transfer.

EXAMPLE 9-2. RELATING CAD AND CAM CAPABILITIES

A number of efforts are being pursued to relate CAD capability and a CIM-oriented manufacturing system. One such project has been conducted by Melkanoff and Bond (1989) of the University of California, Los Angeles.* Discuss the strategies used to develop this linkage.

Results

The UCLA strategy provides an approach for bridging the gap between computer-aided design and computer-aided manufacturing. A methodology and software tools have been developed that permit the automatic download of design information from the CAD to the CAM system to drive manufacturing, and the linking of the manufacturing experience directly to the designer by extracting CAM information and incorporating it in the CAD system. The proposed two-way CAD-CAM bridge can be used to enhance both factory operations and the design process.

Significant strides have been made by the UCLA team in developing and applying the software required for this linkage. Various portions of the CAD-CAM bridge have been achieved and described. Prototype modules have been developed and tested, and UCLA is presently working with several companies to develop the systems further by applying them to actual manufacturing settings.

Along with the many important potential advantages of such a system, a number of difficulties have been observed in thoroughly achieving the project goals. All individuals involved must work in close collaboration to break down the traditional barriers between design and manufacturing. The present system may be appropriate for certain geometries, but must be revised for more arbitrary geometric shapes. Actual drawings and CAD models may be sufficiently complex to demand extensive computer time and high-powered computer resources. The large diversity of existing manufacturing systems may possibly limit the application of such systems because they must be customized for each setting.

The program concludes that the proposed two-way CAD-CAM bridge should yield productivity improvements through significant reduction in manufacturing time and costs and increases in quality and flexibility. However, the linkage requires a major commitment from management as well as future users. The strategy should be planned top-down and implemented bottom-up, and will provide a major step toward CIM.

The type of work being done by the UCLA team and others is important to the integration of product design and manufacturing. However, as is illustrated by this state-of-the-art effort, such integration capabilities cannot be based on commercial, off-the-shelf products.

---

* This example is provided courtesy of the University of California at Los Angeles Manufacturing Engineering Program.

Any efforts to integrate design and manufacturing must be based on customized activities that make use of the available research and development resources.

EXAMPLE 9-3. ACHIEVING DESIGN-MANUFACTURING INTEGRATION

Example 9-3 describes how all aspects of a manufacturing enterprise might be integrated for maximum effectiveness. A step-by-step description of information flow is included to provide insight into the complexities that can exist in such systems, and into the requirements that arise when addressing the as-is *system and creating an adequate information flow to support the planned* to-be *system. Care must be taken to ensure that the information flow associated with transition stages will be stable and effective, as required to achieve the necessary operations capability as the system evolves.*

A manufacturing system to produce hybrid microelectronics products was introduced in Chap. 6 (and will be further explored in Chap. 10). How can the general design and manufacturing concepts discussed in this chapter be applied to such a manufacturing setting to result in a design strategy that makes most effective use of a highly automated and integrated manufacturing capability?

Results

One way to prepare a functional description for a complete hybrid CIM enterprise is illustrated in overview in Fig. 9-9. The following discussion provides further detail.

Figure 9-9 shows the major subsystems and interactions for the complete enterprise, including three sectors relating to computer-aided design and computer-aided engineering, CIM production, and facility management. The result is a system characterized by distributed processing functions and maximum adaptability to allow for continuing system modification.

The description begins with the setting of product objectives and the interaction of the designer with the system (1). The interaction takes place through a user interface (UI). The first step in hybrid design requires the development of a circuit schematic. The design objectives and trial design strategies are input by the user to a schematic performance simulator (SPS) that can model the circuit and provide performance estimates (2). The interactive process between the designer/user and the simulation program can draw on a schematic design library (of past and standard circuits) and a component library that includes all hybrid components to be fabricated or mounted on the hybrid substrate. Both the schematic design library and the component library are resident in the computer-aided design/computer-aided engineering data base (D/E DB).

The schematic design library enables the user to make maximum use of previous design efforts and thus to minimize design costs and delays. The component library is a central feature of the hybrid CIM system. Each component that can be selected for use is assumed to be described in terms of technical parameters associated with the component, the manufacturing steps required to produce or use the component, and the data describing the costs associated with use of the component. The user can access all of this information or can suppress access if convenient during various stages of design.

The circuit simulation program (2) thus can draw on information relating to component technical performance, manufacturing events, and cost, as contained in the D/E DB (3). Trade-offs can be considered to select schematic configurations and components that are best matched to the manufacturing capability of the specific facility and lowest in cost for the performance desired.

After each new circuit design is complete, the schematic library is updated. The component library is updated as new components and processes are added to the CIM factory and new cost data are available.

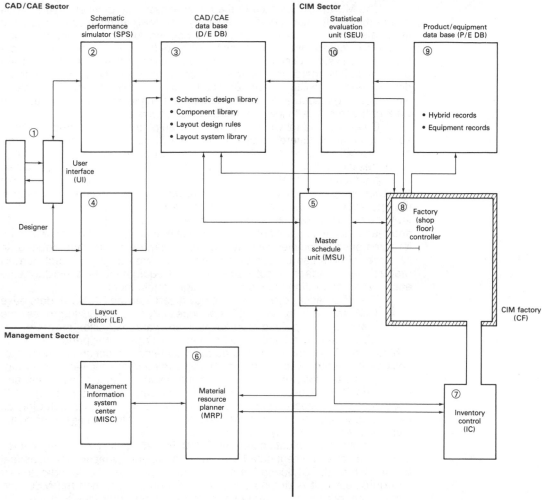

CAD/CAE Sector

Schematic performance simulator (SPS) ②

CAD/CAE data base (D/E DB) ③
- Schematic design library
- Component library
- Layout design rules
- Layout system library

User interface (UI) ①

Designer

Layout editor (LE) ④

Management Sector

Management information system center (MISC)

Material resource planner (MRP) ⑥

CIM Sector

Statistical evaluation unit (SEU) ⑩

Product/equipment data base (P/E DB) ⑨
- Hybrid records
- Equipment records

Master schedule unit (MSU) ⑤

Factory (shop floor) controller ⑧

CIM factory (CF)

Inventory control (IC) ⑦

Figure 9-9.
Overview of a hybrid CIM facility, showing the functional interactions among the CAD/CAE sector, manufacturing (CIM) sector, and management sector.

Once the schematic is complete, the user transfers the circuit data base to the hybrid layout editor (4). The D/E DB contains information on general design rules for all components and past experiences with desirable layout configurations. The user draws on design objectives and design strategies, the circuit schematic, and the layout system library to produce the hybrid topology. Using the extended data base in the component library, alternative configurations are evaluated in terms of technical considerations, manufacturing requirements, and cost. The user is able to choose a topology that will enhance product performance, simplify manufacturing, and minimize costs.

It may be necessary to iterate between the schematic simulation (2) and the layout editor (4) to experiment with the interactions between electronics performance and circuit topology. This is easily achieved through the linked data bases.

Once the layout is complete, it is filed in the layout system library for future reference (3). The master schedule unit (MSU) is notified that the new design is complete, and receives information regarding product quantities required and the due date for the new design (5).

The MSU provides data to the material resource planner (MRP) that is part of the facility management sector (6). The MRP automatically orders the materials and components that will be needed for hybrid production and monitors inventory through the inventory control (IC) unit (7). The management information system center (MISC) provides management reports on all aspects of facility operation. The MSU causes process/event control data to be downloaded to the CIM factory (8). The factory (shop floor) controller assigns manufacturing resources to work orders based on the priorities provided. The work cell controllers follow the instructions received.

Given the adaptive nature of fully automated (software-controlled) equipment, many different types of hybrids will be in various stages of production at all times, and most efficient use can be made of equipment. It is not necessary to batch similar types of hybrids. The same type of hybrid can be in production at several equipment units, and a given equipment unit can be sequencing many different types of hybrids. Each equipment unit changes the operations it is performing based on the software that is downloaded to it by the factory controller, and thus can work on any hybrid that is available and matches unit capabilities. In this type of manufacturing system, the small lot sizes often associated with hybrid orders can be efficiently produced. In essence, all lots can be combined and considered a single (much larger) product batch. Adaptive, software-controlled equipment units will fabricate each hybrid according to its unique design requirements.

The MSU causes direct software download from the CAD/CAE data base to the CIM factory. In this way, it is not necessary for the MSU to process the high level of throughput required for each software instruction program.

A complete data base is maintained on all operations performed in the CIM factory (9). Hybrid data records are maintained to document every process performed on every hybrid, including part numbers and vendors, equipment units used for each process, and the results of all automatic test and evaluation (ATE). Concurrently, an equipment data record is maintained to document the processes performed by every equipment unit, including all hybrids operated upon and processes performed, control settings, diagnostic (sensor) feedback, and hybrid test results.

A statistical evaluation unit (SEU) (10) is continually performing studies of the product/equipment data base (9). This provides a means for statistical quality control (SQC) during the manufacturing process. If data indicate that an equipment unit is drifting out of quality boundaries, then the MSU and factory controller can be warned to stop assigning processes to the equipment unit until adjustments or repairs are made. More catastrophic equipment failures can be handled internally by the factory controller, with notification to the MSU (5) and SEU (10).

The functional description of Fig. 9-9 is general in nature and is intended to illustrate some of the major concepts that can be associated with a CIM facility. Specific implementation of highly integrated and automated facilities can vary in many details. The emphasis in each case should be on using computer systems and networks to their utmost potential to optimize data management and production activities.

Figure 9-10 provides further detail on the CAD/CAE sector; Fig. 9-11 summarizes the sequence of events performed by the designer as this sector is exercised. These two figures may be reviewed together in the following discussion.

The complete CIM facility is driven by the external specification of product performance objectives and initial (preliminary) circuit/schematic concepts. The designer can either input a specific description of a trial circuit/schematic strategy or draw on the schematic design library (in the D/E DB) for a trial circuit/schematic strategy. Once the trial schematic is selected, the

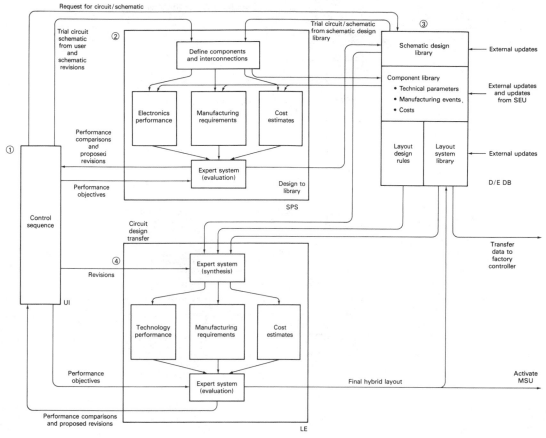

Figure 9-10.
Further detail on the CAD/CAE sector.

schematic performance simulator (SPS) is activated. By making use of the CAD/CAE data base (3), the SPS is able to evaluate this design in terms of technical performance, manufacturing advantages and disadvantages, and cost. The SPS then provides the designer/user with comparisons between the input performance objectives and predicted circuit performance, and suggests possible circuit revisions. The user selects those revisions of interest and iterates through another SPS cycle.

After a circuit/schematic is selected, the layout editor (LE) is used to prepare the topography for the hybrid (4). The circuit performance is evaluated, and further iteration through the SPS and LE continues until a final layout is approved. The user then activates the MSU (5).

Continuing with Figs. 9-10 and 9-11, the flow of information in the CAD/CAE sector can be further defined. The user initiates a circuit design sequence by providing a trial circuit/schematic (1 → 2) or requesting that a trial circuit/schematic be provided from the schematic design library (1) → (3) → (2). In either case, the schematic is defined in terms of the required values for all components and all interconnections.

The schematic is then evaluated in three different ways. The electronics performance is simulated by a program, with artificial intelligence (AI) enhancements. The output of this simulation will determine how well the circuit satisfies the electronics-parameter performance requirements. The selected

Figure 9-11.
Functional operation of
the user interface (UI).

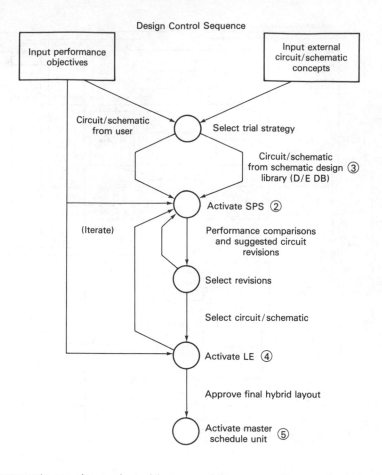

components are also evaluated in terms of the composite manufacturing requirements. The objective is to determine whether the resulting circuit will be easy or difficult to produce in the specific CIM factory associated with the facility. This evaluation is driven by the manufacturing events list associated with each component (to be fabricated on or attached to the hybrid substrate) as maintained by the D/E DB (3). The cost estimate for the circuit is prepared by combining the component cost data from the D/E DB with algorithms that involve circuit complexity and the specific combination of components to be used.

These three evaluations serve as input to an (evaluation) expert system. The expert system will compare the evaluation inputs with the performance objectives defined by the designer/user. Based on the rules incorporated into the expert system (using the best heuristics available from experienced circuit designers), the expert system will provide both performance comparisons and proposed circuit revisions to the user.

In order to perform its functions, the expert system may access the schematic design library and component library in the D/E DB and perform a series of trial and error studies without interaction with the designer/user. The output to the user reflects the results of such exploration. The user will assess the output and decide on the revisions to be tried. The objective is to produce a circuit schematic that is matched to technical performance objectives, the manufacturing capability of the CIM factory, and minimum-cost objectives.

After a design is selected, it is stored in the schematic design library (in the D/E DB) and is transferred to the layout editor (LE). The synthesis expert system draws on the layout system library (of past and reference layouts) and general layout design rules to prepare a trial hybrid topology. Three evaluations are then conducted to assess the technical (electrical and thermal) hybrid performance, manufacturing requirements, and costs. The (evaluation) expert system then measures these results against stated performance objectives. This expert system can also explore variations on a layout by accessing the layout system library and layout design rules, without user interaction.

Based on these results, performance comparisons and proposed revisions are provided to the user. These may result in directed revisions to either the schematic or the layout. This cycle continues iteratively until a final hybrid design is achieved. The final schematic and layout are filed in the D/E DB and a message is sent to the MSU to activate scheduling and production.

Once production begins, the needed manufacturing data are transferred directly from the D/E DB to the factory controller. Since all manufacturing steps have been previously identified and incorporated into the design, no further consideration of the manufacturing events is required.

The functions of the master schedule unit (MSU) are further illustrated in Fig. 9-12. As shown, the MSU maintains a file of all hybrid designs in the

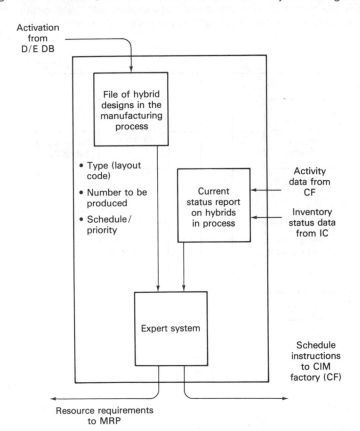

Figure 9-12.
Further detail on the master schedule unit (MSU) functions.

manufacturing process by type/layout code, quantity to be produced, and scheduling/priority information. The MSU also maintains a current status report on hybrids in process and inventory status. Based on available data files, the MSU expert system provides information on resource requirements to the MRP unit and schedule instructions to the CIM factory (CF). To decide on appropriate use of the CIM factory, the factory (shop floor) controller (8) makes use of a current model of the factory processes, which defines the available manufacturing capability, and current status report information on how the available factory capacity is being utilized. As a result of the above activities, scheduled instructions are provided to the CIM factory on a continuing basis. This ensures the most effective use of all manufacturing resources while maintaining high levels of quality control (QC).

The product/equipment data base (P/E DB) and statistical evaluation unit (SEU) function are shown in Fig. 9-13. The P/E DB maintains detailed records on each hybrid that is manufactured, based on both process data and automatic testing that is used to validate these processes. In addition, the P/E DB maintains a detailed record for each equipment item in the factory, tracking process performance over time.

The statistical evaluation unit (SEU) analyzes the P/E DB files to achieve the quality assurance function. The hybrid records are evaluated to develop positive control over the fabrication of each individual hybrid and to look for patterns that indicate external corrective action is required. These records are also used to update the component library in the D/E DB to most effectively identify the manufacturing constraints associated with each circuit component (for use during circuit design). The equipment records are evaluated in

Figure 9-13.
Further detail on the P/E DB and SEU functions.

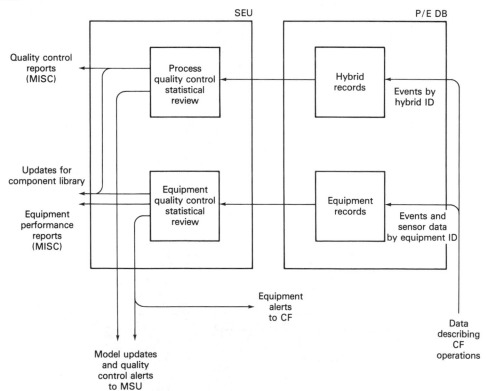

order to track equipment performance and identify any potential statistical quality control (SQC) problem before flawed circuits are produced. The objective of the SEU is to achieve a positive quality assurance function that will eliminate the necessity for rework.

The output of the SEU is also used to update the factory controller models, provide SQC alerts to the MSU, and provide equipment (failure) reports to the CIM factory (CF). The CF itself has been discussed earlier. The inventory control (IC) center consists of an automated entry, storage, and retrieval subsystem integrated with a data base that can identify all items in stock and cause them to be automatically ordered and transferred to the factory (for incoming materials and components) or packaged and shipped (for outgoing products).

The management sector includes the material resource planner (MRP) and the management information system center (MISC). The MRP is driven by the production schedules and projections developed by the MSU. The MRP converts the productions plans into supply requirements for the factory and causes the appropriate orders to be placed by inventory control (IC). The MISC is the focal point for all management input to the system and for all reports provided to describe system operation. The dominant management interaction with the MISC takes place through interface with computer terminals that are networked to the system. Hard copy is produced only to satisfy legal and other carefully defined corporate requirements.

Example 9-3 applies integrated design and manufacturing concepts to a particular production environment. The two-way linkages between design and manufacturing are explicitly explored. Preparation of this type of functional description, involving both manufacturing activities and information flow, is a necessary step toward understanding the implementation of systems that use advanced design support software.

ASSIGNMENTS

9-1. In order to achieve maximum educational insight from Chap. 9, it is helpful to stimulate class discussion of the concepts that have been included, develop critical reviews that probe the advantages and limitations of the concepts, and explore alternative ways for approaching the topic. Select one or several topics from the bold-faced headings that introduce sections or examples in this chapter. Come to class prepared to explain this topic, critically describing the text materials and suggesting ways the topic discussion might be strengthened.

9-2. Visit an individual who is experienced in a particular setting that involves product design and manufacturing. Develop an understanding of the ways in which the design activity is performed and the means for transferring design information to manufacturing for the setting being discussed. Identify the areas in which information transfer takes place by paperwork, and any areas in which computer resources are being applied. Discuss the strengths and weaknesses of the system as you understand it. What suggestions would you have to improve enterprise operations by changing the way in which design takes place?

(a) Come to class prepared to describe the facility you have selected and to discuss the detailed features of the design-manufacturing interface. Draw a flowchart showing the flow of information that takes place in this setting.
(b) Based on class discussion, prepare a brief paper summarizing your understanding of the information flow requirements that relate design and manufacturing for the selected setting.

9-3. Discuss with a product designer the ways in which his or her particular products reflect the design for assembly (DFA) and the design for manufacturing (DFM) concepts. In what ways does the use of DFA/DFM improve the integration of design and manufacturing activities for the application and location? What difficulties are experienced with the type of design that is taking place, and what types of improvements are being considered? Develop a description of an idealized design activity that might be most effectively applied in a highly automated and integrated enterprise. Compare the specific setting with the more idealized description you have developed. Consider the difficulties that might be experienced in moving from the present method to the more idealized situation. Can you recommend various transition stages that might be useful for a change in procedure?

(a) Come to class prepared to describe both the product design efforts at your study site and your reference model design efforts that would represent a maximum of automation and integration. Discuss the issues raised and possible transition stages that could lead from one design method to another.

(b) After this discussion, prepare a brief paper on your insights.

9-4. Obtain access to documentation and software for a CAD system. Become familiar with the operation of this software on a firsthand basis. Use it to develop several different types of product designs that illustrate the features of the CAD software. Determine the limitations you experience in using this software.

(a) Come to class prepared to describe your CAD application experience and to compare your experience with those of others in the class who have had similar access to other types of CAD software.

(b) Following class discussion, write a paper describing your learning experiences about the specific CAD product and contrast this product with those reported on by other members of the class.

9-5. Develop firsthand access to a computer-aided engineering software product. Based on the available documentation, discuss the advantages and disadvantages of this product and the areas in which it can be most effectively applied. Describe the ways in which the CAE analysis is performed. Identify the inputs required from the user and the types of outputs produced. Identify limitations associated with this product.

(a) Be prepared to describe your CAE product and to participate in class discussions regarding a variety of CAE products.

(b) Following the class discussion, write a brief paper analyzing your CAE insights and those of other members of the class. Based on the combined class experience, what insights have you gained into the CAE field? What kinds of CAE products would be helpful to the design process, but are not available today? What limitations prevent the availability of such a product?

9-6. Obtain documentation for several 2-D and 3-D modeling systems for computer-aided design. Evaluate the documentation in terms of how effectively it will assist the user in applying the software products. What questions do you have about the product application that are not easily answer by the materials you have been provided? How would you revise the software documentation to make it more appropriate for a new user? Discuss the problems associated with documenting these systems, and estimate the amount of user time that would be required to become effective at product application.

(a) Summarize your findings for presentation to the class. Show the class a variety of screens that illustrate the operation of the software products and the types of outputs that can be obtained. In what ways might the type of capability represented in each program be of use to the designer? What difficulties might be encountered in this use?

(b) Following your analysis and class discussion, turn in a brief paper describing each of the software products for which you have documentation, with an evaluation of their strengths and weaknesses.

9-7. Discuss with a knowledgeable person the range of computer hardware that is available for CAD/CAE applications. Determine the trends that have been associated with the capability and costs of these systems so that you can show an historical analysis of past performance and costs. By extending these data, draw some conclusions about the likely capabilities and costs during the next five years. Describe the advantages and disadvantages of each of the computer hosts that might be used for CAD application. Come to class with your historical and projected data and describe the basis for your results.

9-8. Meet with an individual experienced with computer communications to discuss the specific requirements for CAD and CAE information exchange. What problems are most familiar to this individual, and how are these difficulties being addressed? Discuss the ways in which IGES and other standards could be used and the possible roles of direct translators. Based on your discussion, describe a particular setting and type of information flow that might be a useful model for enterprise design. Describe why you have chosen this particular configuration and be prepared to deal with questions regarding the decisions you have made. Come to class prepared to describe your computer network and respond to questions regarding its strengths and weaknesses.

9-9. Examples 9-2 and 9-3 describe possible ways in which the gap may be bridged between CAD and a CIM-oriented manufacturing facility. Based on a familiar manufacturing setting, develop your own flowchart and information strategy that can be used to achieve effective integration of design and manufacturing for the given setting. Discuss your system design with a person knowledgeable about the particular setting and obtain a reaction.

(a) Come to class prepared to draw your proposed information system design and to explain the reasons for your configuration.

(b) Based on class discussion, contrast the several different information system designs that have been produced by the class. What common themes can you identify, and what significant differences exist? Prepare a paper on these topics.

10

SYSTEM DESIGN
AND IMPLEMENTATION

Chapter 10 describes a system design effort for a particular type of factory and environment in order to illustrate how the concepts described in earlier chapters can be linked together. The discussion takes the reader through the activities of a typical planning group, illustrating decisions and the application of design tools. This chapter provides a first-cut look at the types of system design efforts that can support the planning group. Numerous iterations and improvements would be required in a real-world setting.

Chapter 10 brings together many of the ideas that have been presented earlier in the text. The focus is on illustrating how a step-by-step system design activity might take place, following the strategy introduced in Chap. 4. The intent is to illustrate how such strategies can be implemented. The resulting system design is a function of available cost-benefit options and provides insights into possible future configurations for an enterprise.

As discussed throughout the text, system design efforts must be considered as supporting the planning group. The objective is to perform step-by-step activities that support the learning and adaptation processes of the manufacturing system.

This chapter integrates many elements of the system design strategy, but these supporting analyses must be viewed at all times as "feeding into" the system planning process. The chapter summarizes one way in which CIM planning groups can begin to address the problem of system change. The information contained here and in similar studies may be drawn upon as input material by the planning group in order to strengthen understanding of the choices that are available and the impact associated with these choices. The planning group must draw upon such design efforts and combine them with a broader range of insights about the nature of the particular system enterprise and its environment. Conclusions may then be formed regarding the types of activities and learning experiences that are most appropriate for system improvement.

The purpose of Chap. 10 is to address the design and implementation of CIM systems and to integrate the insights and strategies that have been developed in previous chapters. This chapter illustrates how the preceding material can be used to proceed through a system design procedure.

Emphasis is again on the product area introduced in Chap. 6. It is not helpful always to treat manufacturing strategies in general terms, because manufacturing systems vary so widely that many real issues are not adequately addressed by applying a survey approach. On the other hand, it is important to focus on an area that will illustrate many of the important general issues. As discussed in Chap. 6, the manufacture of small electronic products has associated with it many of the features and issues that are experienced in a wide variety of other manufacturing sectors.

The material presented in this chapter applies the general methods of analysis of Chap. 4, particularly with respect to Fig. 4-5. All of the concepts developed in Chaps. 1 to 3 and the more detailed discussions of Chaps. 5 to 9 are applied to the design of manufacturing systems for hybrid microelectronic products. The objective is to walk through the various design considerations that might be encountered in the development of a manufacturing facility for the production of these products. The materials that are included can be directly transferred for application in other manufacturing sectors.

This chapter again considers small electronic circuit assemblies as the manufacturing focus. Earlier discussions of this product area covered the functional manufacturing requirements and many of the different types of equipment available for system implementation. This background affords a better understanding of the issues and decisions presented here.

10-1. Market Environment (Step 1)

As indicated in Fig. 4-5, the CIM design process must begin with an understanding of the market environment. The market setting will determine the fiscal viability of the proposed manufacturing facility. It is essential, therefore, to gain an understanding of the ways in which this market is presently being addressed and to define areas of opportunity that are available to the facility being designed. The design strategy for CIM systems that is being applied here involves the definition of *categories* and *types* of products. The design requirements for the facility will thus be driven by decisions regarding the products that are to be produced.

There are many ways in which the hybrid microelectronics market environment can be described. Typical data are often collected to describe sales by market component, as shown in Fig. 10-1. These data indicate the size of the market and indicate how sales are divided into market segments. In order to couple this market information to factory design, it is necessary to have an understanding of the types of products that are sold into each of these markets. For this discussion, seven product categories are defined, with the distribution of sales as shown in Fig. 10-2.

In studying the relationships between product categories and the hybrid market today, it is estimated that the dominant dollar volume of sales is typically associated with categories 1 to 4. In addition, it is determined that these categories might be defined as forming the mainstream or foundation product areas for the hybrid industry, and that

Figure 10-1. Hybrid market information. (a) By functional application. (b) By market segment. Reprinted by permission from Robert V. Allen and Thomas W. Matt, "Industrial Commercial Applications Pace Hybrid Market Growth Through 1992," *Hybrid Circuit Technology*, March 1986, p. 26. Copyright © 1986 Lake Publishing Corporation, 17730 W. Peterson Road, Libertyville, IL 60048.

Functional Application	Merchant			Captive			Total		
	1986	1989	1992	1986	1989	1992	1986	1989	1992
Digital, total	407	598	877	1065	1681	2724	1472	2279	3601
(Compound growth rate)	—	13.7%	13.6%	—	16.4%	17.5%	—	15.7%	16.5%
Linear, total	1287	2013	3110	1592	2394	3680	2879	4407	6790
(Compound growth rate)	—	16.1%	15.6%	—	14.6%	15.4%	—	15.2%	15.5%
Optoelectronic, total	256	382	599	152	256	431	408	638	1030
(Compound growth rate)	—	14.3%	16.2%	—	19.0%	19.0%	—	16.1%	17.3%
Microwave, total	174	238	324	289	381	497	463	619	821
(Compound growth rate)	—	11.0%	10.8%	—	9.7%	9.3%	—	10.2%	9.9%
All hybrids, total	2124	3231	4910	3098	4712	7332	5222	7943	12242
(Compound growth rate)	—	15.0%	15.0%	—	15.0%	15.9%	—	15.0%	15.5%

Note: All figures in millions of dollars.

(a)

Market Segment	Merchant			Captive			Total		
	1986	1989	1992	1986	1989	1992	1986	1989	1992
Automotive	56	72	97	111	144	193	167	216	290
Communications	515	833	1298	621	1013	1664	1136	1846	2962
Consumer	142	200	282	143	208	302	285	408	584
Data processing	463	708	1078	798	1308	2177	1261	2016	3255
Government	488	687	960	874	1174	1600	1362	1861	2560
Industrial	188	312	521	224	350	548	412	662	1069
Test and measurement	272	419	674	327	515	848	599	934	1522
Total hybrid market	2124	3231	4910	3098	4712	7332	5222	7943	12242
Compound growth rate	—	15.0%	15.0%	—	15.0%	15.9%	—	15.0%	15.5%

Note: All figures in millions of dollars.

(b)

	Compatible Products				Product Variations		
Product Category	1	2	3	4 most complex product	5	6	7
Percentage of Industry Sales	10	15	20	30	10	10	5

Figure 10-2.
Summary of market data.

they can form a compatible product area through which to develop the functional design for a manufacturing facility. Thus, a preliminary decision is made to guide the design process toward a facility that can address these four selected categories. In the study of categories 1 to 4, it is determined that a factory capable of producing category 4 products, which are the most complex, will also be capable of producing categories 1 to 3. Category 4 products thus have been selected as the design focus for the hybrid manufacturing facility.

10-2. Enterprise Objectives (Step 2)

Given an understanding of the market setting, it is essential for an enterprise to understand its organizational objectives (step 2). The enterprise must decide on what will typically be a range of objectives and on a relative weight to be given to these objectives in the system design process. The objectives must reflect the financial realities of the market and the capabilities of the enterprise. The objectives will also reflect the particular interests of the enterprise and its members and the specific opportunities that are available.

For the facility design being developed here, it is decided that a major enterprise objective will be to develop a manufacturing setting that can achieve the lowest-cost hybrid microelectronic products as defined by categories 1 to 4. Further, a decision is made to consider factory strategies that are five to ten years beyond the current state of the art in the field to determine if such strategies can provide a significant competitive advantage. A decision is made at the earliest design stages to evaluate technologies and manufacturing strategies that extend beyond the current industry standard. Consideration of both manually oriented and flexible manufacturing system work cells develops from these objectives.

An emphasis on becoming the low-cost producer for a given product is fundamental to most system design efforts. While many other emphases must also be considered, a concern with cost is always necessary in a highly competitive market environment.

A decision is made to seek a manufacturing configuration that will have maximum potential for being in the forefront of industry capability over the next five to ten years. The alternative work cell strategies described below are studied as means for achieving program objectives.

10-3. State-of-the-Art Technology (Step 3)

In general, the enterprise will have available a range of equipment capabilities varying with price and opportunities for upgrading present equipment based on recent research and development and changes in the field. In each case, the design team must have an understanding of the equipment that is available, including performance and cost. The study of equipment items must link closely to the categories and types of products that are to be produced, which in turn (as discussed above), relate to the market environment.

Once a set of product categories has been defined, then it is possible to define the manufacturing functions and the types of manufacturing equipment that must be available to produce the desired categories. Based on the categories of hybrid products that are chosen, the manufacturing functions of Fig. 10-3 are defined as necessary for the planned CIM facility. These manufacturing functions and equipment units were introduced in Chap. 6, and are assumed here to be adequate for the production of the types of hybrid products being considered. In general, multiple visits may be required to any process station during product manufacturing. For example, multiple printing and firing cycles are often conducted to produce multiple layers on a substrate. Multiple cleaning and inspections may also occur, as may a wide variety of packaging and electrical tests.

In order to simplify the following discussion, several assumptions have been made. It is assumed that the manufacturing system to be designed here does not include the package sealing, testing, and shipping functions shown in Fig. 10-3. The objective of the manufacturing system is to produce a complete hybrid prior to these operations. In addition, it is assumed for simplicity the desired hybrids can be produced by passing the work-in-progress sequentially through the (remaining) steps shown in Fig. 10-3 with only one visit required to each station. Variations that require multiple passes through any station can be accommodated and studied without difficulty; however, such additional complexity can obscure some of the more important insights into an initial system design and so are excluded here.

It is necessary to select a number of scenarios that will span the various degrees of automation. As proposed for this study, the scenarios range from very manually oriented facilities (scenario 1) to extensively automated facilities (scenario 10). The objective in selecting these scenarios is to determine the level of automation that will be most cost-effective for the facility design. Varying equipment properties are

Figure 10-3.
Manufacturing functions/equipment for product category 4.

Screen printer or direct write unit	Package assembly unit
Drying oven	Curing oven
Firing furnace	Cleaning unit
Laser trimmer	Wire bonder
Cleaning unit	Package seal unit
Inspection unit	Package leak test unit
Component assembly unit	Electrical test unit
Curing oven	Package for shipping unit

associated with each scenario, depending on the level of automation selected.

In order to satisfy the design objectives for this study, it is decided to use a work cell approach to defining a range of manufacturing system capabilities. The work cell approach is selected as having the potential for improving product quality and reducing product cost through varying degrees of integration and automation.

Scenarios 1 to 10 make use of the general equipment and work cell concepts that were introduced in Chap. 6. The scenarios make use of varying levels of automation and integration, but employ a common grouping strategy to create the work cells.

As shown in Fig. 10-4, scenarios 1 to 10 vary in terms of the degree of automation and integration that is applied to each work cell. It is assumed for this particular study that these are the most effective equipment geometries. As discussed in Chap. 6 (Sec. 6-9), a variety of other work cell geometries could be considered as part of the system design. The ones shown here were selected for the present study.

The ten scenarios discussed here vary in the degree of automation and integration applied to the work cells under discussion. By evaluating scenarios that range from manually oriented operations to the maximum application of automation and integration, the design group can understand the trade-offs involved in system design and can seek to identify a system concept that is best matched to the specific enterprise and environment.

EXAMPLE 10-1. HYBRID MANUFACTURING

Discuss how the work cell concepts of Fig. 10-4 relate to historical manufacturing strategies for hybrid products. Describe further the features of the limiting (scenario 10) FMS work cell concept.

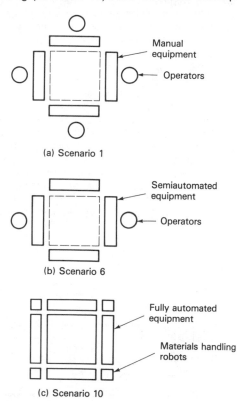

(a) Scenario 1

Manual equipment

Operators

(b) Scenario 6

Semiautomated equipment

Operators

(c) Scenario 10

Fully automated equipment

Materials handling robots

Figure 10-4.
Alternative work cell configurations. The arrangements represent a grouping strategy with varying levels of automation and integration.

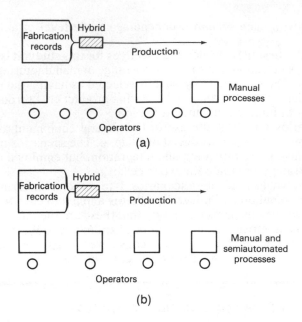

Figure 10-5.
Alternative hybrid manufacturing system designs. (a) Manual system. (b) Isolated semiautomated equipment.

Results

Figures 10-5 to 10-7 illustrate ways in which the evolution of hybrid industry manufacturing strategies can be described.

Figure 10-5a is a manual-intensive system with stand-alone process stations, manual equipment, and standard documentation and tracking based on linear product flow. This category represents a type of operation that is not competitive by today's standards. Figure 10-5b is an "islands of semiautomation" system with linear flow in which some of the stand-alone process stations use semiautomated equipment. Handover among stations is manual, as are documentation and tracking. Increased quality and throughput are obtained. Systems in this category might be considered the norm of today.

Figure 10-6 is another approach to achieving higher levels of automation and integration. Equipment becomes fully automated and linearly interconnected by robots of varying types. Computer networks link these automated stations together and to a host controller. Hybrids and components are individually tracked by the computer system, limiting the need for extensive paperwork. And the design system is linked to manufacturing in order to simplify and improve the design-manufacturing interface. Several large companies are experimenting with different versions of this system in part or whole (refer to Example 6-24).

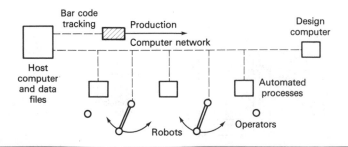

Figure 10-6.
Automated equipment robots.

To optimize industry-wide manufacturing strategies, other system design concepts also need to be pursued. The highly automated facility in Fig. 10-7 represents the limiting work cell approach associated with scenario 10 (as introduced in Fig. 10-4c) for this geometry. As noted in Secs. 3-7 and 6-9, an FMS is a multipurpose equipment system that can perform a variety of manufacturing tasks based on the use of computer-driven instructions. For the configuration shown, each FMS work cell consists of several subsystem equipment units and a core coupling unit. The subsystem units are designed to have interchangeable mechanical and computer interfaces to the coupling unit. Equipment units in each FMS cell share a pattern recognition unit and computer control unit to allocate the cost of these units among the several subsystems. The automatic distribution system can make use of standard carriers and magazines, vacuum pickup and handling systems, or other distribution techniques. The computer-assisted design operations

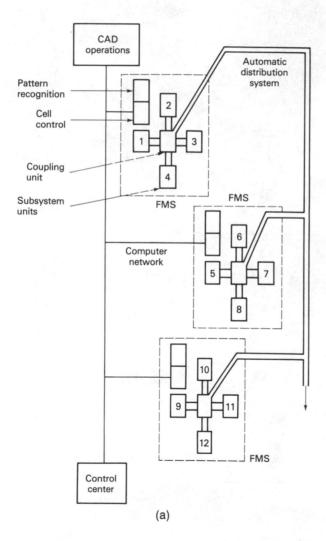

Figure 10-7.
Flexible manufacturing system (FMS) CIM design. (a) Functional overview. (b) Artist's drawing of the manufacturing facility discussed in this chapter. (Drawing by Tan, Boyle, Heyamoto, Architects, Spokane, WA)

(a)

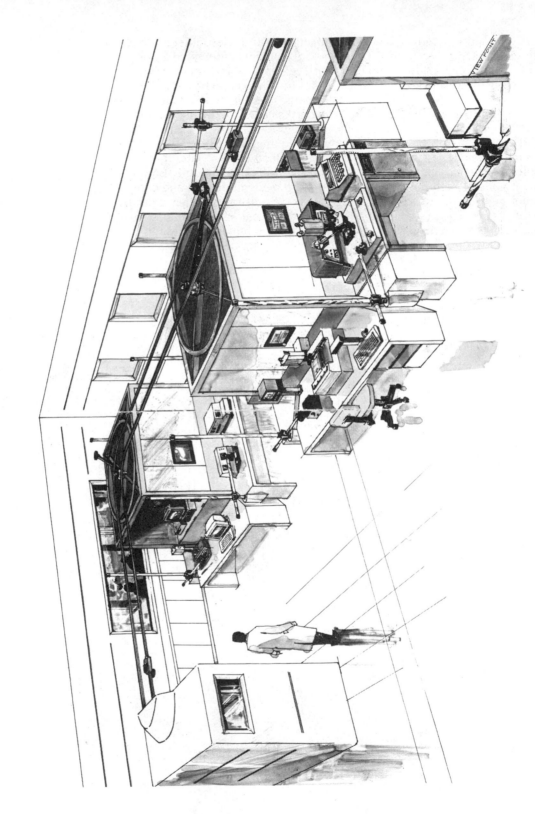

Figure 10-7. continued

(b)

are connected directly into the control center, allowing the design software to reflect the manufacturing system's opportunities and constraints.

This system concept is based on using many existing equipment subsystems with varying levels of modification. As intended, the coupling unit will make use of mechanical and computer interfaces that will easily accommodate upgraded versions of present equipment. The elements of this FMS modular strategy are the following:

1. A building-block approach to system assembly.

2. A low-cost, software-controlled materials handling and inspection coupling unit that can link a set of four manufacturing subsystem units (as shown in Fig. 10-7).

3. A coupling unit design that will accommodate any subsystem unit with any interface and will provide product flow among all interfaces.

4. A subsystem unit design based on software control and standard coupling unit interfaces for all manufacturing processes (to be accomplished through upgraded versions of existing products).

5. Multifunctional subsystem units to take advantage of software flexibility.

6. Application of group technology concepts and specific facility objectives to configure subsystem unit combinations for each coupling unit. Every combination will be viewed as an FMS. The building-block subsystem units for each FMS can be removed and replaced over time, depending on facility needs.

7. A shared pattern recognition control system and programmable coupling unit control system that can be used as building-block units for each FMS.

8. An adaptable control center and computer network to link together the different FMS work cells.

9. A CAD unit that will provide the capability to optimize hybrid design for the FMS/CIM facility.

10. A configuration that will readily allow the incorporation of artificial intelligence for process planning and control and application of a common data base for all aspects of design and manufacturing.

Figure 10-7b shows an artist's sketch of the FMS modular strategy as described above.

By applying the work cell concept to the list of functions in Fig. 10-3 (and deleting the packaging, testing, and shipping functions from the factory), it is determined that 12 workstations are required. Given the particular cell configuration being applied here, this will lead to a minimum of three work cells that will be required for the product category selected (category 4). The degree of automation and integration for the cells will vary with scenario, as seen in Fig. 10-4. As shown in Fig. 10-8, each manufacturing equipment function can be associated with a manufacturing unit, and these units can be grouped into three

Figure 10-8.
Definition of equip-
ment units for each
cell.

#1: *Print and Trim Cell*
Screen printer or direct write unit
Drying oven
Firing furnace
Laser trimmer

#2: *Assembly Cell*
Cleaning unit
Inspection unit
Component assembly unit
Curing oven

#3: *Interconnect and Packaging Cell*
Package assembly unit
Curing oven
Cleaning unit
Wire bonder

*A simplified set of
functions is used here
to define the manufac-
turing system under
discussion. The se-
lected functions pro-
vide an adequate vari-
ety of processes and
apply a diverse group
of equipment families
to illustrate the system
design trade-offs that
are experienced. Ex-
tensions of this set can
be developed without
difficulty to study
other issues of inter-
est. Similarly, a shift
in design focus to
other product areas
can be accomplished
by following a step-by-
step procedure similar
to the one provided
here.*

work cells. These cells provide the framework in which the study design process can proceed.

The twelve processes and three work cells in Fig. 10-8 define the scope of the manufacturing system considered here. In the following discussion, it is assumed that each hybrid passes once through each station during the manufacturing process. (In many situations, work-in-progress is required to make multiple passes through the same cell and may also make returns from cell to cell. While these additional complexities are essential in a complete manufacturing system design, they are not pursued further here because of the unnecessary complexity.) The manufacturing system thus consists of a print and trim cell with four sequential functions, followed by an assembly cell with four sequential functions and an interconnect and packaging cell with four sequential functions. The study of more complex sequences, both within cells and between cells, can further evaluate the utility of the particular equipment grouping in each cell and the advantages and disadvantages of the four-unit cell structure that has been assumed here. Linear manufacturing systems and other cell combinations thus could be productively compared with the one shown here. On the other hand, the functional groupings shown in Fig. 10-8 do provide a logical strategy for many aspects of hybrid manufacturing.

For purposes of system evaluation, the operating parameters of each work cell are assumed to be dominated by the equipment unit having the most restrictive level of performance. The equipment units thus can be considered in composite to predict cell performance (given the assumption of sequential product flow). Supporting studies can be performed to relate the individual equipment units to cell performance; the composite cell performance can be used in the system design trade-offs. Each FMS cell is thus viewed as a single complex equipment unit, with several sequential component operations. In this way, an integrated multipurpose manufacturing cell is created.

A number of simplifying assumptions regarding hybrid manufacturing and processes are made in this discussion. The objective is not to provide an exhaustive and detailed introduction to hybrid technology, but to use this manufacturing area to illustrate CIM design methods in a specific application. This approach provides a way to understand the key aspects of the methodology and the associated learning opportunities.

Figure 10-8 lists the manufacturing functions of the particular factory being considered here. Substrates from vendors arrive at the print and trim cell. Screen printing or direct write processes are used to apply a thick film paste to the substrate. Subsequently, a drying oven and a firing furnace complete the processing of the thick film material. The laser trimmer is used to trim the material as required.

The substrate then moves to the assembly cell, where it is cleaned and inspected. The components are then assembled onto the substrate and the epoxy is cured to achieve the desired bond.

The substrate then moves to the interconnect and packaging cell. The package assembly unit attaches the substrate to the package with epoxy, which is then cured. Another cleaning operation is performed, followed by wire bonding to attach the components to the circuit and the circuit to the connector pins on the package assembly. It is assumed at this point that additional factory operations would be performed to achieve package sealing, package leak test, electrical test, and packaging for shipping.

Figure 10-9 shows the strategy that has been developed to relate the equipment units in each work cell to the ten proposed scenarios. As noted, each scenario is described in terms of the degree of automation that is applied for the processing equipment, materials handling equipment, and information flow. Scenarios range from manually oriented systems to fully automated versions that make maximum application of possible technology (as described in Fig. 10-7). Figure 10-9 thus provides a framework in which more detailed assumptions can be made about the relationships between equipment properties and the scenarios.

The processing equipment ranges through ten levels of automation. The manual equipment description (scenario 1) implies that an operator must always be at the equipment unit and is required to intervene in processing on an almost continuous basis. The operator is responsible for supplying process information, correcting any process errors, providing operational control, and loading and unloading all parts. For scenario 2, it is assumed that one-fourth of the processing equipment is semiautomated. Machine operation at this level involves having one-fourth of the process equipment link into a computer controller. This transition allows a computer-driven product with faster processing speeds and the ability for future upgrading.

The transition to scenarios 3 to 5 involves an evolution to equipment that has computer-supported operations for half of the equipment in place. For scenarios 6 to 7, the transition is to complete use of semiautomated equipment.

Scenarios 8 to 10 involve fully automated equipment with only remote operator support required. A "fully automated" designation implies that the equipment is networked into an FMS work cell con-

Scenario	Processing Equipment	Materials Handling	Information Flow
1	Manual	Manual	Manual
2	One-quarter semiautomated	Manual	Manual
3	One-half semiautomated	Manual	One-quarter automated
4	One-half semiautomated	One-quarter automated	One-half automated
5	One-half semiautomated	One-quarter automated	Fully automated
6	Semiautomated	One-quarter automated	Fully automated
7	Semiautomated	Fully automated Remote operator	Fully automated
8	Fully automated Remote operator (best use of commercial technology and available enhancements)	Fully automated Remote operator	Fully automated
9	Fully automated Remote operator (extended technology, partial application of state-of-the-art R&D)	Fully automated Remote operator	Fully automated
10	Fully automated Remote operator (limiting technology, maximum application of state-of-the-art R&D)	Fully automated Remote operator	Fully automated

Figure 10-9.
Definition of scenarios.

figuration and the amount of operator intervention is reduced to a minimum. Scenarios 8 to 10 involve increasing use of the most available state-of-the-art technology to produce equipment that has maximum capability for computer-driven operation. For these cases, the operator intervenes only occasionally, since the sensors and actuators associated with the automatic equipment are continually monitoring the system.

The materials handling scenarios vary from manual to fully automated. At the manual level, all products are picked up by operators and delivered to the appropriate equipment units. At the one-quarter automated level, conveyor belts or a similar transport method have been added to the factory. The automated system moves products to and from various equipment units. At the level of full automation with a remote operator, the materials handling system utilizes the materials transport approach that has been described above for the FMS configuration. With the use of this transport system and a computer acting as a system scheduler, the materials handling process is fully automatic for all equipment units.

The final description relates to information flow. At the manual level all data entry and part verification are done by hand by the operators. The next level is that of one-quarter automation, which decreases the amount of paper flow by the addition of bar scanners for part inventory and product verification. This coincides with the introduction of station managers to control the process equipment.

At the level of one-half automation, paper flow has been reduced further and the processing equipment is controlled by computers. In addition, files now can be uploaded and downloaded from a mainframe, and process information can be accumulated. The final level of full automation associated with scenarios 5 to 10 involves system designs in which paper flow essentially has been eliminated. All information on the product, such as process information, product history, and necessary parts, is distributed to the operations from the computer scheduler.

In Fig. 10-9, scenario 2 describes most small hybrid manufacturers today in the industry. Scenario 4 exemplifies where leading technology companies in the field might be placed. Scenario 5 is associated with, perhaps, the ultimate in existing hybrid factories. It is anticipated that scenarios 6 to 10 represent manufacturing facilities that extend beyond the current state of the art.

In order to perform the functional trade-offs among these different scenarios, it is necessary to associate specific operational parameters with each of the scenarios shown in Fig. 10-9. Further information regarding each constituent equipment unit, the characteristics of each work cell configuration, and the appropriate parameters for each scenario to be associated with each item of equipment and work cell are discussed in Sec. 10-9 (relating to step 6 of the design process illustrated in Fig. 4-5).

Figure 10-9 provides general scenario descriptions. Information about the available families of equipment to implement these scenarios must be drawn upon in order to associate specific equipment parameters with the various levels of automation and integration.

10-4. CIM Design Principles and Reference Models (Step 4)

As introduced in Chap. 3, the design of any CIM system is usually based on a set of design principles and reference models that suggest new ways for achieving a manufacturing function in order to take maximum advantage of available technology. As discussed earlier, the concept of a work cell as a major CIM design principle has been applied to this study. A variety of other design principles have also been factored into the development of the proposed factory. The objective is to design a manufacturing operation that can achieve continuous flow manufacturing and demonstrate the ability to group manufacturing equipment into effective work cells using group technology concepts. Computer-aided design software will drive the manufacturing system, requiring close linkages between the CAD and CIM facilities. The concept of continuous flow information is applied to achieve the most effective information-handling capability.

10-5. Design of Manufacturing Operations

The four input areas to the system design process described in steps 1 to 4 (Fig. 4-5) are introduced above. These inputs drive the formation of product categories and types. In turn, the categories and types can be used to define the manufacturing functions for each category and as a basis for selecting the appropriate parameters to describe the manufacturing processes for each category.

10-6. Product Definition in Terms of Functional Manufacturing Operations

As discussed above, seven categories of hybrids are defined for this study (Fig. 10-2). A decision has been made to focus all modeling and evaluation efforts on product category 4, since this product category includes categories 1 to 3. Thus, a system design that is intended to address category 4 products in an optimum way also addresses categories 1 to 3. Given product category 4 as the emphasis, 12 manufacturing functions have been defined as needed to produce this category (Fig. 10-8).

Figure 10-10.
IDEF format and application (courtesy of Meta Software Corporation).

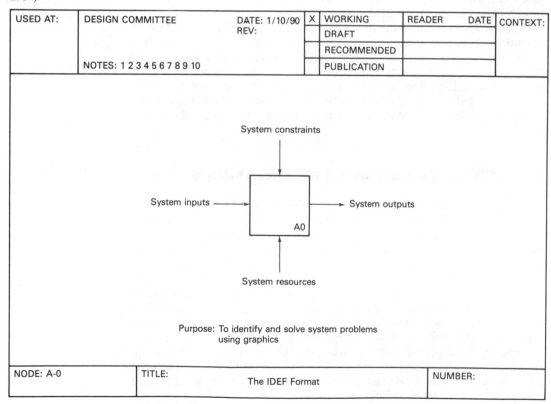

USED AT:	DESIGN COMMITTEE	DATE: 1/10/90 REV:	X	WORKING	READER	DATE	CONTEXT:
				DRAFT			
				RECOMMENDED			
	NOTES: 1 2 3 4 5 6 7 8 9 10			PUBLICATION			

System constraints

System inputs → | A0 | → System outputs

System resources

Purpose: To identify and solve system problems using graphics

NODE: A-0	TITLE: The IDEF Format	NUMBER:

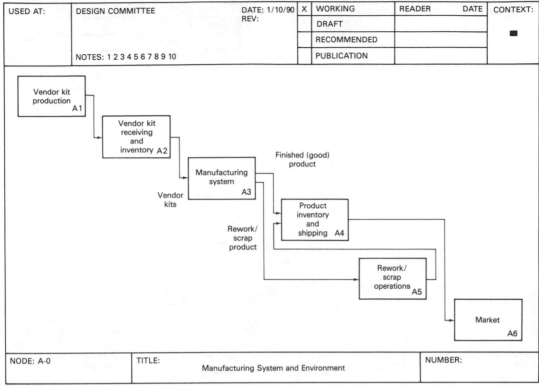

USED AT:	DESIGN COMMITTEE	DATE: 1/10/90 REV:	X	WORKING	READER	DATE	CONTEXT:
				DRAFT			■
				RECOMMENDED			
	NOTES: 1 2 3 4 5 6 7 8 9 10			PUBLICATION			

NODE: A-0	TITLE: Manufacturing System and Environment	NUMBER:

Figure 10-11.
Manufacturing system and environment using Design/IDEF.

As also discussed above, the system design emphasizes a work cell approach to applying CIM principles to hybrid manufacturing. Thus, it has been decided to incorporate the 12 manufacturing functions for product category 4 into a number of such cells. Analysis of the manufacturing functions has produced a decision that three different types of building-block work cells are adequate to manufacture product category 4.

In order to develop descriptions of the functional operations required for each product category and for composite functional operations, the Design/IDEF software package has been used.

Figures 10-10 through 10-15 provide an overview of the manufacturing system and its environment using the Design/IDEF software, and provide definitions of the manufacturing system in terms of three work cells related to print and trim, assembly, and interconnect and packaging. The generic manufacturing overview of Figs. 10-10 and 10-11 has been introduced in Chap. 1 (Figs. 1-9 and 1-10) and further discussed in Chap. 3, where it is used to support development of the SES model (Fig. 3-9). Figures 10-12 through 10-15 provide further definition of each work cell for the particular application being discussed. These three block functions, or cells, include the materials transport system. The Design/IDEF module for the print and trim cell is shown

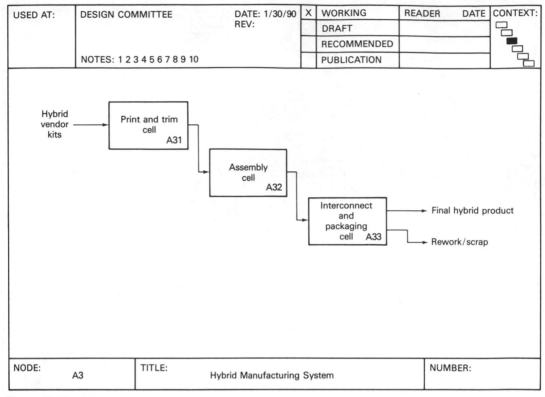

USED AT:	DESIGN COMMITTEE	DATE: 1/30/90 REV:	X	WORKING	READER	DATE	CONTEXT:
				DRAFT			
				RECOMMENDED			
	NOTES: 1 2 3 4 5 6 7 8 9 10			PUBLICATION			

Hybrid vendor kits → Print and trim cell **A31** → Assembly cell **A32** → Interconnect and packaging cell **A33** → Final hybrid product / Rework/scrap

NODE: A3	TITLE: Hybrid Manufacturing System	NUMBER:

Figure 10-12.
Hybrid manufacturing system.

in Fig. 10-13; the assembly cell, in Fig. 10-14; and the interconnect and packaging cell, in Fig. 10-15.

10-7. Composite Manufacturing Functions for the Entire Product Line (Step 8)

The emphasis on product category 4 has resulted in the choice of a composite category that includes several other product categories. By incorporating categories 1 through 3 into category 4, a composite has previously been defined. The function list of Fig. 10-8 and the Design/IDEF drawings of Figs. 10-10 to 10-15 provide a detailed foundation for understanding the composite manufacturing functions required for the facility.

10-8. Functional Parameter Set (Step 9)

As discussed in Sec. 6-1, the following equipment parameters can be used for the functional design of a CIM system:

1. Scope of operations
2. Mean time between operator interventions (MTOI)
3. Mean time of intervention (MTI)
4. Product yield
5. Processing time

The scope of operations parameter has been addressed by associating manufacturing functions and equipment units, as shown in Fig. 10-8. As discussed below, parameters 2 to 5 form the basis for the system trade-off studies.

10-9. Manufacturing Equipment Information Base (Step 6)

Equipment for the manufacture of hybrids is introduced in Chap. 6. Figure 6-32 summarizes a range of processes and equipment strategies for the manufacture of thick film hybrids. Sections 6-12 through 6-24 provide details regarding the equipment families that can be considered during system design. The following sections build on the

Figure 10-13.
Print and trim work cell.

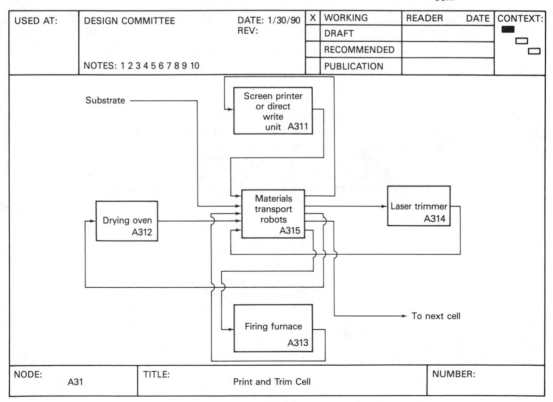

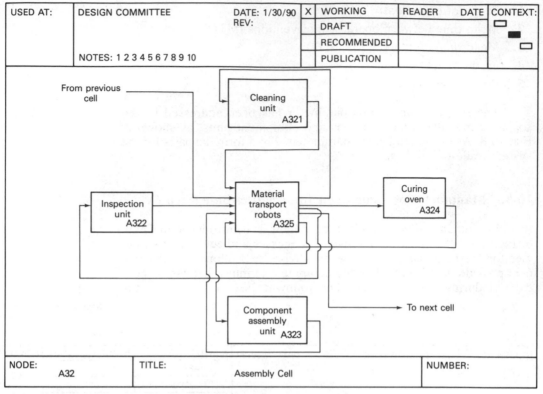

USED AT:	DESIGN COMMITTEE	DATE: 1/30/90 REV:	X	WORKING	READER	DATE	CONTEXT:
				DRAFT			
				RECOMMENDED			
	NOTES: 1 2 3 4 5 6 7 8 9 10			PUBLICATION			

From previous cell

Cleaning unit
A321

Inspection unit
A322

Material transport robots
A325

Curing oven
A324

To next cell

Component assembly unit A323

NODE: A32	TITLE: Assembly Cell	NUMBER:

Figure 10-14.
Assembly work cell.

Figure 10-16 summarizes the functional parameters used to describe the equipment families considered for this study. These choices are compatible with the general scenario concepts introduced in Fig. 10-9 and apply the equipment families introduced in Chap. 6. The following discussion describes the 12 manufacturing operations that are included and explains and justifies the parameter values that have been selected. A similar type of analysis is required for all system design efforts in order to link equipment properties to the scenario definitions.

equipment introductions of Chap. 6. Decisions are made regarding the specific equipment alternatives to be considered for the system design discussed in this chapter.

Figure 10-16 summarizes the functional parameters used to describe the hybrid manufacturing equipment in this study. The following discussion explains how these parameter choices (which are consistent with the general framework of Fig. 10-9) relate to the equipment discussed in Chap. 6. (The equipment selected represents only one set of choices from a wide range of available performance characteristics. The purpose is to illustrate a typical set of choices; in a complete system design, other choices would also be evaluated.)

The print and fire work cell (Fig. 10-8) can apply, dry, fire, and trim only a single layer of paste during a cycle. Multiple layers require multiple cells (operated in parallel) or multiple cycles through the same cell. For simplicity, the discussion considers only a single print and fire work cell and only a single printing cycle for each substrate. Other cases can be considered as alternative studies to the particular case presented here.

The costing of manufacturing systems has been discussed in Chap. 3 (in terms of fixed and variable costs), in Chap. 5 (with respect to commercial software products that support costing efforts), and in

Chap. 6 (with respect to the capital costs experienced in manufacturing systems). The discussion in this chapter builds on these earlier considerations. The capital equipment costs in Fig. 10-16 include both commercial and customized equipment, but without the R&D costs associated with development of the customized equipment. The costs used in this chapter are based on scaling model concepts (introduced in Chap. 6). These first-cut estimates can be extended to more detail in future iterations.

10-10. Screen Printer or Direct Write Unit

Two principal substrate printing processes are currently in use. The silk-screen printing process makes use of a photographic screen that defines the specific areas of the substrate that are to be printed. Operation of a screen printer must be supported by appropriate equipment to produce photographic reductions of the desired circuit geometry (or topography) for each interconnection layer, allowing preparation of the screen masters. This process is most effective in settings in which a large number of identical substrates must be printed. This type of system cannot change the pattern produced based on software

Figure 10-15.
Interconnect and packaging work cell.

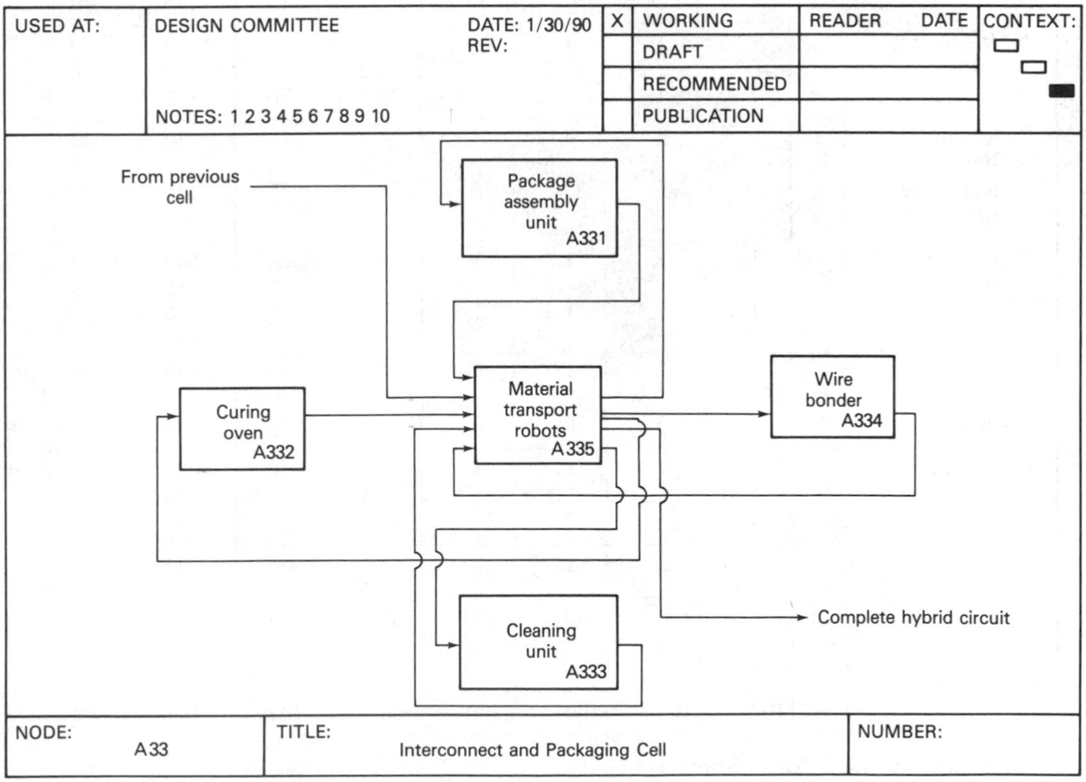

USED AT:	DESIGN COMMITTEE		DATE: 1/30/90 REV:	X	WORKING	READER	DATE	CONTEXT:
					DRAFT			
					RECOMMENDED			
	NOTES: 1 2 3 4 5 6 7 8 9 10				PUBLICATION			

NODE: A 33	TITLE: Interconnect and Packaging Cell	NUMBER:

Figure 10-16. Equipment parameters, combining Figs. 10-8 and 10-9 to provide further detail, by scenario, on the equipment units.

MTOI (min)
MTI (min)
Cost ($000)

Equipment Parameters
Scenario

Cell	Equipment	1	2	3	4	5	6	7	8	9*	10*
CELL 1	1. Printer/writer	1	1	1	1	1	4	4	10	20	60
		.5	.5	.5	.5	.5	.5	.5	2	3	5
		3	3	3	3	3	135	135	185	400	600
	2. Drying oven	1	1	1	1	1	4	900	1000	1500	3000
		.5	.5	.5	.5	.5	.5	2	2	3	5
		1	1	1	1	1	35	35	35	100	150
	3. Furnace	2	2	2	2	2	2	4000	5000	6000	7000
		.5	.5	.5	.5	.5	.5	2	2	3	5
		4	4	35	35	35	35	35	35	200	300
	4. Laser trimmer	1	10	10	10	10	10	10	20	30	60
		.5	2	2	2	2	2	2	2	3	5
		1	230	230	230	230	230	230	280	500	600
CELL 2	5. Cleaning unit	1	1	2	2	2	2	4000	5000	6000	7000
		.5	.5	.5	.5	.5	.5	.5	2	3	5
		1	1	15	15	15	15	15	20	50	200
	6. Inspection unit	1	1	1	1	1	4	4	5000	6000	7000
		.2	.2	.2	.2	.2	.5	.5	2	3	5
		1.25	1.25	1.25	1.25	1.25	70	70	110	350	450
	7. Component assembly unit	1	4	4	4	4	4	10	10	20	60
		.3	.5	.5	.5	.5	.5	2	2	3	5
		2	75	75	75	75	75	75	135	300	600
	8. Curing oven	1	1	1	1	1	4	4000	5000	6000	7000
		.5	.5	.5	.5	.5	.5	2	2	3	5
		1.3	1.3	1.3	1.3	1.3	15	15	20	50	200
CELL 3	9. Package assembly unit	1	1	1	1	1	4	2000	4000	8000	10000
		.5	.5	.5	.5	.5	.5	.5	2	3	5
		10	10	10	10	10	150	150	250	650	1000
	10. Curing oven	1	1	1	1	1	4	500	4000	8000	10000
		.5	.5	.5	.5	.5	.5	.5	2	3	5
		2	2	3	3	3	35	35	35	50	200
	11. Cleaning unit	1	1	2	2	2	2	1000	2000	3000	4000
		.5	.5	.5	.5	.5	.5	.5	2	3	5
		1.25	1.25	100	100	100	100	100	100	200	450
	12. Wire bonder	1	2	2	2	2	2	4	10	20	60
		.5	.5	.5	.5	.5	.5	.5	2	3	5
		11	100	100	100	100	100	100	185	400	600

* Based on additional R&D, but such costs not included.

482   *System Design and Implementation*

control, and does not allow direct information handling between product design and manufacturing.

As defined here, scenarios 1 to 5 apply the silk-screen printing process. The functional parameters and costs of this method are shown in Fig. 10-16. These values imply that for the equipment selected here, an operator will be needed to participate in and monitor the process at all times (given the low value for MTOI).

An alternative strategy for substrate preparation is to use a direct write unit (DWU). The direct write unit does not use a silk-screen process; the thick film paste is pressed out the nozzle of a special computer-controlled dispenser system. Thus, the system can write with various pastes. Because of its configuration, the DWU can produce a hybrid circuit pattern directly from a computer-aided design download of design information. The DWU system can be an effective match to a hybrid manufacturing system that is intended to be flexible and highly automated. However, current direct write units are limited in terms of speed so that high-volume operations can require many equipment units.

Estimated functional parameters and costs for a direct write unit are shown in Fig. 10-16 (scenarios 6 and 7). The value for the MTOI is significantly better for the direct write unit than for the screen printer because the former is intended to achieve a much more automated process (given a variety of product types). An operator will still be present in scenario 6, but will only be responsible for loading and unloading parts. The printing process will be completed by the semiautomatic DWU. Scenario 7 is basically the same as scenario 6, but instead of having the operator load and unload, a robot is employed. The value for the MTOI will not change since only the materials handling system differs between these two scenarios and the same level of operator participation in unit operation is required. The MTI value is the suggested time for an operator to set up or correct any errors.

For scenarios 8 to 10, a DWU unit is still suitable, but the equipment available today would need many modifications. The addition of robots for materials handling means that an operator will no longer be needed to load and unload parts. The DWU process would have to be improved to increase throughput and to improve automation (so that no operator is needed to monitor the machine and so that the downtime of the machine will be reduced). Extended computer control would have to be introduced into the system by means of new sensors, actuators, and effectors that could direct the accurate printing of the substrate pattern.

As additional technology is incorporated in the higher order scenarios, the MTOI will increase. Since the DWU will fail less often, it will not require as many operator interventions. The MTI will increase slightly since the equipment is more complex. These modifications are represented by the parameters shown in Fig. 10-16.

As before, the cost of the equipment increases with the level of technology. In scenario 8 this increase is modest, but the cost rises dramatically for scenarios 9 and 10. Scenarios 9 and 10 require a high level of state-of-the-art technology based on additional research and development.

10-11. Drying Oven

Once the hybrid thick film layer has been printed on the substrate, the layer must be dried in an oven. In scenarios 1 to 5, the drying process is manual, with operators inserting the substrates into the oven and removing them at the appropriate time. The MTOI is the amount of time between operations, and the MTI is the amount of time it takes to perform the operation. A low value for the MTOI applies, so the operator will always have to be present to participate in the process. This type of drying process is relatively inexpensive. Functional performance and cost data are provided in Fig. 10-16.

For scenarios 6 and 7, the drying oven will be computer controlled and will make use of a conveyor belt running through the heated area. The MTOI value for scenario 6 is determined by the time required for insertion and removal of substrates, and the MTI is the required operator time. The MTOI value for scenario 7 increases significantly because the substrates will be loaded and unloaded from the oven by a robot. For scenarios 8 to 10, the drying unit is assumed to be fully automated with extended technology. This will increase the cost for such a system, as shown in Fig. 10-16, but with these increased costs come increasing values for the MTOI. The values shown for MTOI reflect the maintenance schedule. The MTI will also increase because the equipment is more complex and more effort will be needed to repair it.

10-12. Firing Furnace

Once the hybrid thick film layer has been printed and dried on the hybrid substrate, the layer must be fired in a furnace. All ten scenarios use a firing furnace, but employ equipment with different levels of sophistication. Scenarios 1 and 2 use a small zone furnace with a conveyor belt. The parts are assumed to be loaded by an operator. The MTOI value is chosen so that an operator would have to load and unload the substrates from the conveyor belt as needed. The MTI is low because these furnaces are typically reliable and only a small part of operator time would actually be spent in substrate handling.

In scenarios 3 to 6, the firing furnace is assumed to be operated with computer control. The MTOI and MTI are low because these scenarios still include an operator to load and unload the parts from the conveyor belt. These values take into account a monthly maintenance schedule to keep the furnace in optimal working order.

In scenario 7, the furnace is the same, but robots have been added for materials handling. The MTOI value has increased significantly because the addition of robots means that an operator is no longer required to load and unload parts from the belt. The only operator intervention in this scenario is monthly maintenance. The MTI will increase slightly.

For scenarios 8 to 10, the firing furnace will be completely automated with extended technology. All materials handling will be done by robots. Since the furnace is completely automatic, the MTOI will increase with each ascending scenario, as will the cost of such equipment. To achieve the level of technology required for scenarios 8 to 10, the furnace must make use of extended technology, new sensors and actuators, and additional research and development. The functional parameters and costs for these various furnace versions are shown in Fig. 10-16.

10-13. Laser Trimmer

The next step in the manufacturing process is to trim any resistors printed on the substrate to the correct values. For scenario 1, an abrasive trimmer might be used. This process is typically slow and operator intensive. The MTOI value is based on fairly continuous operator involvement. The MTI reflects the time it would take the operator to perform the trimming function.

A much better method for resistor trimming makes use of a laser trimmer. All scenarios above the first use a laser trim system. Most modern laser trimmers are computer controlled so they can link to a computer network.

In scenarios 2 to 6, a semiautomatic laser trimmer is assumed. In these scenarios, operator intervention is needed to load and unload a part from the equipment, which will affect the functional parameters, as shown in Fig. 10-16. Another factor that affects the MTOI is the daily maintenance schedule. In scenario 7, robots are added for materials handling, but trimmer function will continue to limit the MTOI.

For scenarios 8 to 10, the laser trimmer is assumed to be fully automatic. As before, this will increase costs because of the much higher performance level. As improved technologies are used for the laser trimmer, the value for the MTOI will increase due to full automation. The MTI will increase because the equipment is more complex and will require more time to repair or maintain.

10-14. Cleaning Unit

In scenarios 1 and 2, manually oriented cleaning processes are assumed and operator action is necessary to achieve the desired operations. In scenarios 3 to 6, a semiautomated unit reduces the required operator intervention, and therefore enables more efficient operations. Inserting substrates into the cleaning unit, controlling the cleaning process, and viewing the cleaned substrate still require significant operator involvement. In scenario 7, a robot feeds the substrates into the cleaning unit, greatly reducing involvement as this is a major aspect of operator activity for this station. In scenarios 8 to 10, the remaining cleaning operations are fully automated, with lower impact due to automated material handling. For scenarios 8 to 10, there is a steadily increasing MTOI based on improved application of research and development to allow reduced operator involvement.

10-15. Inspection Unit

Once the substrate is complete, it is necessary to clean and inspect the resulting product to make sure that it meets all specifications and to prepare the product for assembly.

In scenarios 1 to 5, inspection consists of an operator looking at each part through a microscope. Although in scenarios 3 to 5 the information flow will no longer be manual, this will not greatly affect the MTOI because the operator still must visually inspect the parts. The MTOI is the amount of time between operations and the MTI is the amount of time it takes to perform the operation. The low value for the MTOI applies so that operators will always have to be present to participate in the processes. Functional performance and cost data are provided in Fig. 10-16.

For scenarios 6 and 7, inspection equipment is assumed to be semiautomatic with combined operator and computer control. Semiautomatic pattern recognition is used to detect any cracks or errors on the substrate. An operator is still needed to load and unload the parts, which drives the MTOI. The MTI is the amount of time it takes the operator to service the unit during scheduled maintenance. The MTOI value for scenario 7 is not significantly affected by automated materials handling because a high level of operator involvement is still required by the inspection process itself.

In scenarios 8 to 10, the inspection unit is assumed to be fully automated with extended technology. This will increase the costs for such a system, as seen in Fig. 10-16, but with these increased costs come increasing values for the MTOI. The values shown for MTOI reflect principally the maintenance schedule. The MTI will also increase because the equipment is more complex and more effort will be needed to repair it.

10-16. Component Assembly Unit

After the substrate is cleaned and inspected, various components can be assembled on the substrate. In scenario 1 this is accomplished by the operator using a vacuum system to pick up and place the parts onto the substrate. The MTOI value is very low because the operator has to intervene at all times, and the value for the MTI is the amount of time it takes the operator to perform the process.

In scenarios 2 to 7, assembly is accomplished by using a commercially available hybrid assembly unit that is essentially a computer-driven pick and place system. Placement of components by such a system is typically fast and reliable. For scenarios 2 to 6, the operator will always have to place the part on to the machine prior to assembly. This drives the value for the MTOI. The MTI value is the amount of time it takes the operator to load and unload the required components. The MTOI and the MTI are low because the assembly unit requires ongoing operator support. The operator has to intervene to support system operations. In scenario 7, the addition of the robot handling system means that an operator is no longer required to place the parts onto the machine. This factor results in an increased MTOI value. In addition, the information flow is completely automatic in this scenario.

In scenarios 8 to 10, the assembly unit becomes fully automatic with extended technology and remote control. For scenario 8, little improvement is noted in MTOI because an operator is still required for monitoring unit operations. The values for the MTOI increase in the ascending scenarios, and in turn these factors increase the operator intervention when the assembly unit requires repair. The cost of systems in the ascending scenarios increases significantly.

10-17. Curing Oven

Once the required components are placed on the substrate, curing is required to harden the epoxy used to attach the components. In scenarios 1 to 5, this task is accomplished in a small curing oven. The operator is responsible for loading and unloading parts from the oven. The value for the MTOI is driven by the time required for operator loading of a batch of parts, and includes some maintenance. Once the batch is complete, the MTI is dominated by the time required to load and unload the batches.

In scenario 6, the curing oven is replaced by a semi-automatic oven with computer control. Information flow is completely automatic. The curing oven basically requires the same maintenance as the furnace. In scenario 7, robot materials handling is added to load and unload parts from the curing oven. This results in a large increase in the MTOI. In scenarios 8 to 10, the curing oven is assumed to be fully automatic. The values for the MTOI, MTI, and costs are similar to those of the firing furnace.

10-18. Package Assembly Unit

The function of the package assembly unit is to apply epoxy to the package mount that is being used, then to place the substrate into the package on this mount. As shown in Fig. 10-16, scenarios 1 to 5 depend on manually intensive operations to apply the epoxy and to put the substrate into the package. The assembly unit used here is manually oriented, requiring continuous operator assistance.

In scenario 6, a semiautomated package assembly unit is introduced. The requirement for materials handling still performed by the operator maintains a low value for the MTOI. In scenario 7, the addition of automated materials handling greatly reduces operator involvement. Scenarios 8 to 10 reflect a steadily increasing application of state-of-the-art research and development and continuing improvement in the MTOI.

10-19. Curing Oven

The package and substrate are now exposed to a curing process to form an epoxy bond between the substrate and the package. As illustrated in Fig. 10-16, scenarios 1 to 5 make use of operator-intensive activities to load the package assembly unit into the oven, monitor the curing process, and remove the assembly from the oven. In scenario 6, a semiautomated oven is introduced, but only a modest increase in the MTOI is noted because materials handling is still a limiting factor. In scenario 7, the addition of automated materials handling causes a large increase in the MTOI. For scenarios 8 to 10, additional application of improved automation continues to improve equipment performance.

10-20.   Cleaning Unit

Once the curing is complete, cleaning is done to prepare for wire bonding. Scenarios 1 and 2 make use of an operator-intensive manual cleaning unit. Scenarios 3 to 7 have a semiautomated cleaning unit that still requires significant operator involvement for materials handling. For scenario 7, there is a large increase in the MTOI associated with the introduction of automated materials handling. In scenarios 8 to 10, again there is a steadily increasing MTOI, reflecting the increased application of research and development capabilities for the cleaning unit.

10-21.   Wire Bonder

In scenario 1, wire bonding is done using a manual wire bonder that is under complete operator control. This is a very tedious operation, as the operator has to locate visually each bond site using a microscope. Functional parameters for the wire bonder are shown in Fig. 10-16. The MTOI value implies that the operator is present at all times, and the MTI value is the amount of time it takes the operator to wire bond the required parts. In scenarios 2 to 7 a semiautomatic wire bonder is assumed, using computer-controlled operations and pattern recognition to achieve high efficiency and quality bonding. The MTOI value for the wire bonder is low because the operator is required to supervise the operations at all times and to load and unload parts. Another reason for the low MTOI is that the wire bonder sometimes fails to shift correctly from one bond to the next, requiring operator intervention. If the bonder is bonding the same part for an extended period, such errors will not occur as frequently because the bond parameters can be maximized over time. Time to correct this error is usually quite short and is given as the value used for the MTI.

In scenario 7, robot materials handling will eliminate the need for an operator to load and unload the parts from the wire bonder. This partial reduction in operator activity will increase the MTOI, but the MTI value will remain similar because the wire bonder still has errors and requires the operator to correct them.

In scenarios 8 to 10, the wire bonder will be a remotely controlled, fully automatic equipment unit with a closed control loop, requiring much less operator intervention. The sensors and actuators/effectors will be able to correct many of the wire bonder errors.

10-22. Materials Transport System

In scenarios 1 to 3, the materials transport system consists of an operator hand-carrying each part as required. Performance parameters for scenarios 1 to 3 are incorporated into the manual descriptions of the process equipment units. In scenarios 4 to 6, a conveyor belt is added to the materials transport system. The associated MTOI values are still low, as an operator will have to be present to move parts on and off the conveyor belt. The amount of time it takes to load and unload contributes to the MTI.

In scenarios 7 to 10, a robot materials handling system is assumed. The system has been designed and described in Sec. 3-3, which also discusses the flexible manufacturing system materials integration concept. In scenario 7, the MTOI and MTI values improve somewhat, depending on the equipment function and the impact of automated materials handling. In scenarios 8 to 10, the MTOI and MTI increase rapidly with application of higher levels of automation. Figure 10-17 summarizes the materials transport system costs.

The materials transport costs are divided into those applied once to the whole factory and those applied on a per cell basis. In the latter case, the costs shown below in Fig. 10-17 have been presented for a three-cell configuration. In scenarios 1 to 3 there are no costs with automated materials handling because parts are handled manually. Scenarios 4 to 6 use a conveyor belt for materials handling, and scenarios 7 to 10 use the FMS work cell configurations. In moving from scenario 7 to scenario 10, there is a rapid increase in costs as the materials handling capabilities of the FMS work cells also increase rapidly.

Materials transport concepts are defined by the general scenario descriptions of Fig. 10-9. These concepts have been incorporated into the final parameter data base of Fig. 10-16. These concepts must be sufficiently described to develop materials handling cost estimates for each scenario.

EXAMPLE 10-2. LINKING GENERAL SCENARIO CONCEPTS TO SPECIFIC EQUIPMENT PARAMETERS

Figure 10-9 described a general strategy that may be used to develop a range of scenarios for the design of a manufacturing system, from manually oriented to highly automated. Figure 10-16 showed the specific parameter values that have been used in the design of a manufacturing system for hybrid microelectronic products. Discuss how the general scenario definitions of Fig. 10-9 are related to the specific parameter values of Fig. 10-16.

Results

As may be noted in Fig. 10-9, scenario 1 involves manual processing equipment, materials handling, and information flow. Appropriately, the values for scenario 1 in Fig. 10-16 are associated with manually oriented operations. Note that the MTOI is 1 to 2 minutes, and the MTI ranges from 0.2 to 0.5 minute. Thus, all of these equipment units require ongoing interaction with an operator.

In Fig. 10-9, scenario 2 involves changing one-fourth of the processing equipment to semiautomated operations. Observe that in Fig. 10-16 equipment units 4, 7, and 12 for scenario 2 have been upgraded to semiautomated operation. For these cases, significant improvement has been noted in the MTOI. Because one out of each four equipment units has been upgraded, the desired one-fourth modification has been achieved.

From Fig. 10-9, scenario 3 involves half of the processing equipment being semiautomated and one-fourth automation of information flow. In Fig. 10-16, for scenario 3, equipment units 3, 5, and 11 are upgraded for improved operation. In Fig. 10-9, for scenario 4, there is additional upgrading to a one-fourth level of automation for material handling (the use of conveyor belts) and the information flow becomes one-half automated. These changes do not directly affect the equipment parameters shown in Figure 10-16, but will affect the type of integration that can be achieved and system cost.

In scenario 5, information flow is completely automated. This does not change the equipment parameters, but provides the foundation for higher levels of system integration. Levels of processing equipment and materials handling automation remain unchanged. For scenario 6, all of the equipment becomes semiautomated. As may be seen in Fig. 10-16, all of the operating parameters have been upgraded by this point.

For scenario 7, materials handling is brought to full automation (with remote operation possible). Once materials handling becomes fully automated, it is possible to make maximum effective use of the semiautomated processing equipment. In moving to scenario 7, there is significant improvement in the parameters shown in Fig. 10-16 because the automated materials handling allows much more effective use of the semiautomated equipment with a reduction in operator involvement.

Scenarios 8, 9, and 10 involve fully automated equipment with varying uses of technology. Scenario 8 involves the best use of commercial technology, scenario 9 involves limited use of state-of-the-art R&D, and scenario 10 involves maximum application of state-of-the-art R&D. For each of these scenarios, materials handling and information flow continue to be fully automated. As may be noted in Fig. 10-16, there is steady improvement in the equipment performance parameters, but also a rapid growth in the costs associated with these equipment units.

Figure 10-16 represents one specific implementation of the scenario definitions shown in Fig. 10-9. For any particular industry setting, general scenario definitions may be prepared and specific equipment parameters then chosen.

EXAMPLE 10-3. PRODUCING COMPOSITE WORK CELL COST ESTIMATES

Figure 10-16 provides costs for each of the equipment units for the three work cells under discussion here. Figure 10-17 shows total equipment costs for three cells (one of each type) with a minimum configuration (called here *version 1*). Show that the data from Fig. 10-16 may be combined to estimate the total equipment costs for three cells, for each scenario, as shown in Fig. 10-17.

Results

The equipment cost data in Fig. 10-17 have been obtained by adding together the costs in Fig. 10-16. For example, for scenario 1, if all of the

Figure 10-17.
Equipment and materials handling costs for version 1.

Scenario	Equipment Cost* ($000) (3 cells) (version 1)	Automated Material-Handling Cost ($000)	
		Whole Factory	Per Cell × 3
1	38.80	0	0
2	429.80	0	0
3	574.55	0	0
4	574.55	30	0
5	574.55	30	0
6	995	30	0
7	995	100	900
8	1390	1000	1800
9	3250	1000	2400
10	5350	1000	3300

* Developed by adding together the costs included in Fig. 10-16.

individual equipment unit costs are combined, a total of $38,800 is obtained. Similarly, if the same columns for scenario 2 in Fig. 10-16 are combined, $429,800 is obtained.

Note that the (manual) equipment for scenario 1 is the least expensive. There is rapid growth in cost associated with the introduction of more semiautomated equipment (moving from scenario 1 to scenarios 2 and 3). The equipment costs for scenarios 3 to 5 are the same, as no equipment upgrading takes place. The costs for scenarios 6 and 7 are the same because the equipment is fully semiautomated for both of these situations. The growth in equipment cost for scenarios 8 to 10 involves fully automated equipment with higher levels of R&D application.

The automated materials handling costs shown in Fig. 10-17 indicate that only modest costs are experienced in scenarios 4 through 6 (which involves a conveyor belt approach to partially automated materials handling). Scenario 7, which is the first fully automated scenario, requires additional factory-level costs of $100,000 plus $300,000 in automated materials handling costs per FMS work cell. Scenarios 8 to 10 show fixed automated materials handling costs at the factory level, with steadily increasing costs for materials handling at the cell level as additional research and development is applied to maximize the level of automation for each equipment unit and FMS cell.

10-23.   Producing Composite Cell Data

The descriptions and information in Fig. 10-16 and 10-17 indicate the functional parameters and costs associated with each equipment unit for each scenario. This information provides the foundation on which functional simulations can be performed.

The modeling here is based on evaluating alternative work cell combinations. Thus, the information in Fig. 10-16 must be used in a synthesis process to estimate the functional parameters and costs that would be associated with the three building-block work cells that have been introduced. Figure 10-18 shows the resulting data base. For each

The performance of each work cell is assumed here to be limited by the equipment unit with the lowest MTOI value. For future iterations, it would be necessary to model each work cell as a combination of equipment units and to arrive at composite functional parameters that would reflect performance associated with the combined equipment units in each cell.

scenario, information is presented on each of three building-block work cells. For each of these cells, cell performance is limited by the equipment unit with the lowest performance in terms of MTOI. The information in Fig. 10-18 represents the output from step 6 that was used to drive the functional modeling of step 10 (Fig. 4-5).

As may be noted in Fig. 10-18, the MTOI and MTI for the composite cells are the same for scenarios 1 to 5. The overall cell performance is dominated by the manually oriented equipment units in the cells. As a gradual transition takes place from manual to semiautomated equipment, the primary effect is on the percentage yield from the cell and the processing time required by the cell.

In scenarios 6 to 10, the composite cell MTOI and MTI increase rapidly and dominate the differences among the scenarios. A slight improvement in the yield and process time continues to be noted. Throughout the entire evolution from scenario 1 to 10, a slow overall improvement in transport time can also be noted. The final column shows the increases in equipment costs for each cell in scenarios 1–10 and composite costs for the three-cell version across all scenarios.

EXAMPLE 10-4. ESTIMATING COMPOSITE CELL PERFORMANCE

Figure 10-16 shows operational parameters for each equipment unit for each scenario that is under consideration. Figure 10-18 shows the resulting cell-level parameters that are assumed for this study. Confirm that the equipment data from Fig. 10-16 result in the cell data of Fig. 10-18, given that the lowest MTOI value dominates cell performance in each case.

Results

The data in Fig. 10-18 have been obtained from Fig. 10-16 by assuming that the lowest MTOI value dominates cell performance in each case. For example, for scenario 1, the lowest MTOI value for each cell is 1 minute (from Fig. 10-16). As shown in Fig. 10-18, for scenario 1, 1 minute has been chosen as the composite cell parameter for all these cells. This 1-minute minimum MTOI extends through scenarios 1 to 5 and thus limits performance for each of these cases. This means that in a work cell that consists of partly manual and partly semiautomated equipment, the manual equipment will continue to dominate cell performance.

An improvement begins to be noted in the minimum MTOI for scenarios 6 and 7. At this point, all of the equipment units are semiautomated. Continuing growth in the minimum MTOI is seen in scenarios 8 to 10, as fully automated equipment is being used with increasing levels of research and development.

The MTI values in Fig. 10-18 are those associated with the equipment unit having the lowest MTOI. The percentage yield and processing times are those estimated for each of the work cell types for each scenario, and the transport times are for the assumed materials handling system. The costs in Fig. 10-18 are the sum of the individual equipment unit costs and materials handling costs that originate in Fig. 10-17.

FMS cell	type 1
	2
	3

Scenario	MTOI (min)	MTI (min)	Yield Rate (%)	Process Time (min)	Transport Time (min)	*Cost ($000)/ Total
1	1	.5	50	3	3	9
	1	.5	75	10	3	5.55
	1	.5	50	20	3	24.25/38.80
2	1	.5	60	2	3	238
	1	.5	80	5	3	78.55
	1	.5	60	15	3	113.25/429.80
3	1	.5	60	2	3	269
	1	.5	80	5	3	92.55
	1	.5	60	15	3	213/574.55
4	1	.5	65	2	2	269
	1	.5	85	5	2	92.55
	1	.5	65	15	2	213/574.55
5	1	.5	80	2	2	269
	1	.5	90	5	2	92.55
	1	.5	80	15	2	213/574.55
6	2	.5	95	2	2	435
	2	.5	95	5	2	175
	2	.5	95	5	2	385/995
7	4	.5	95	2	1	435
	4	.5	95	5	1	175
	4	.5	95	5	1	385/995
8	10	2	97	2	1	535
	10	2	97	3	1	285
	10	2	97	3	1	570/1390
9	20	3	98	2	1	1200
	20	3	98	3	1	750
	20	3	98	3	1	1300/3250
10	60	5	99	2	1	1650
	60	5	99	3	1	1450
	60	5	99	3	1	2250/5350

* Does not include materials handling or information network.

Figure 10-18. Composite cell parameters developed from the equipment unit parameters of Fig. 10-16.

10-24. Functional Process Model (Step 10)

A framework has been constructed in which alternative hybrid manufacturing facilities can be evaluated. Possible hybrid products have been divided into categories and a selection made of a specific category to be emphasized in this manufacturing design. A set of ten scenarios has been defined to span the range from manually oriented to highly automated, integrated systems. A detailed discussion has been presented with respect to the manufacturing equipment capabilities associated with each scenario for all manufacturing processes required to produce the selected product category. The manufacturing strategies applied in this design make use of work cell concepts. It has been illustrated that three building-block work cells are a minimum configuration for the factory under consideration.

The functional modeling that follows emphasizes the performance of these composite cells. The properties of each cell have been determined by examining the equipment units that are constituents of that cell. Thus, the information in Fig. 10-16 has been used to produce the cell parameters shown in Fig. 10-18. The most restrictive equipment unit performance parameters dominate the work cell performance in each case.

Section 10-25 examines alternative work cell combinations using these parameters. Different equipment and cell capabilities are drawn upon for each of the ten scenarios. A minimum configuration is three cells to produce product category 4; thus, a fundamental three-cell configuration (Version 1) has been studied across all ten scenarios. In addition, three additional versions have been studied for each scenario. These additional versions have been developed by extending the total number of work cells to four, five, or six. In each case, each additional cell duplicates one of the basic three cells. First, the work cell in the three-cell minimum configuration with the highest utilization rate is duplicated to improve system balance. Then the cell with the next highest utilization rate is duplicated and so forth until six cells are obtained.

The following study examines how the three building-block cells function across the ten scenarios (or levels of automation) and how the versions with three, four, five, or six cells function across these ten scenarios. A total of 40 modeling efforts is then obtained (four cell versions times ten scenarios, as shown in Fig. 10-19; the specific cell choices shown here were determined from step 10 of the modeling).

Figure 10-19.
Cell definitions for each version/configuration.

Number of each FMS type

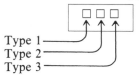

Type 1
Type 2
Type 3

Scenario

Version Number	Total Number of Cells	1	2	3	4	5	6	7	8	9	10
1	3	1,1,1	1,1,1	1,1,1	1,1,1	1,1,1	1,1,1	1,1,1	1,1,1	1,1,1	1,1,1
2	4	1,1,2	1,1,2	1,1,2	1,1,2	1,1,2	1,1,2	1,1,2	1,1,2	1,1,2	1,1,2
3	5	1,2,2	1,1,3	1,1,3	1,1,3	1,1,3	1,2,2	1,2,2	1,2,2	1,2,2	1,2,2
4	6	1,2,3	1,2,3	1,2,3	1,2,3	1,2,3	2,2,2	2,2,2	2,2,2	2,2,2	2,2,2

Each modeling effort describes the number of hybrids that can be manufactured in a given period of time. The scenario versions are contrasted in terms of the productivity levels they can obtain for product category 4. The results of this section can be described in terms of tables and graphs that indicate the total productivity of each scenario and each version.

10-25. Application of MANUPLAN II

The MANUPLAN II software (introduced in Sec. 5-6) has been used to support the evaluation of alternative system designs (step 10). This software makes use of rough-cut analytic methods to estimate the performance properties of manufacturing systems. Further detail on MANUPLAN II is included in Chap. 5.

Three principal input variables are used as independent variables for MANUPLAN II: the mean time between operator interventions (MTOI) and the mean time of intervention (MTI) for each cell and the processing time required per cell. (The mean time to failure [MTTF] and mean time to repair [MTTR] in MANUPLAN II correspond to the MTOI and MTI of this text.) The processing time per product item through a continuous flow automated system is actually the delay observed between the times parts emerge from a particular equipment group. That is, the delay time in MANUPLAN II is not the time it takes for an individual part to move through a process, but the time delay that occurs between parts as they emerge from a given process.

MANUPLAN II makes use of lot size definitions; a lot size of one is chosen for this study because this value reflects a continuous flow manufacturing system. This implies that the setup time per lot must be modeled as zero and accounted for in the MTI input variable.

One of the dominant expenses in hybrid manufacturing is associated with the large percentage of rework required. Although it is possible to model such rework in the MANUPLAN II software by creating a loop whereby the parts are rerouted back through the necessary units, such rework is designated as scrap in this study in order to keep the first approach as simple as possible. It is expected that a trend toward increasing automation will be accompanied by a decrease in the percentage of the product that is routed to scrap, which will eventually remove rework as a consideration.

For the present study, an effective yield rate was established for the entire manufacturing system. The percentage yield rates for each work cell were multiplied together to achieve an effective factory yield rate. The transit time between cells was based on the materials handling strategy being applied for each scenario.

The remaining MANUPLAN II input variables were kept constant in the study and given the following values:

- The equipment utilization limit was set at the maximum allowable 95 percent.
- The variability in arrival time indicates the variability (around

MANUPLAN II is used here to determine the number of products (of the defined type) that can be manufactured by the alternative system configurations in a given length of time. These productivity data can be combined with cost data (associated with the same given length of time) to calculate cost per product, which has been selected as a system function for optimization.

scheduled times) for the arrival of material to the system. This was set at the recommended standard of 30 percent.

- The variability in equipment indicates the variability (around the average time) for the performance of operation. This was set at the recommended standard of 30 percent.

- All operations were performed by dedicated equipment units. The portion of equipment assigned to each operation was set at one.

Figure 10-18 (columns 1 to 6) summarizes the input parameters that were used in the MANUPLAN II–supported analysis. The objective of the study was to determine the maximum achievable production in parts per year for each version of each scenario. However, the maximum achievable production is not provided as a MANUPLAN output. Instead, MANUPLAN provides a qualitative output that designates "desired production achievable" or "desired production not achievable." It was necessary, therefore, to perform successive approximations to obtain the maximum achievable demand. Several runs of each system configuration were required before the limiting production value was found. The primary dependent variable in this study was, therefore, the maximum demand in parts per year that could be achieved under each facility configuration (scenario/version).

Four versions are considered for each scenario (Fig. 10-19). Version 1 makes use of three work cells. Version 2 makes use of four work cells, with the additional cell duplicating the cell with the highest utilization rate in version 1. Version 3 extends this concept to five cells, and version 4 to a total of six cells.

The results of the MANUPLAN II analysis are shown in Fig. 10-20. Shown on the graph along the vertical axis is the maximum achievable production in parts per year for the factory (expressed in thousands). Across the bottom are the scenarios, ranging from 1 (manual) to 10 (maximum automation and integration).

The upgrading of individual equipment units to higher levels of automation and integration may have only limited impact on total system productivity until a certain critical level is reached. Until a balanced system design is obtained, partial improvement may not result in the improved system performance that might be expected. The planning group must prepare the organization to expect this effect in order to avoid disappointment and possible abandonment of the strategy.

Important conclusions may be drawn from this graph. Only a modest increase in production capability is observed in passing from scenario 1 to scenario 5. Above scenario 5, a rapid increase in productivity is noted. This is an important effect that has been discussed in other CIM contexts. That is, upgrading individual items of equipment will have only limited impact on total system performance until a certain critical level of automation and integration is obtained. Scenarios 1 to 5 have not provided balanced systems in which the automation and integration features can be optimally linked to produce the highest order of productivity. Instead, partial improvements are being incorporated that do lead to some improvement but not to the large improvement that might be hoped for. These results provide a cautionary note to anyone moving toward a more automated and integrated facility, in the sense that only modest payoffs should be expected during initial upgrading activities.

It is essential to perform the type of system design study shown here as part of any CIM implementation effort in order to anticipate the types of results that can be expected on a system-wide basis. Otherwise, discouragement and frustration may set in when specific subsystems are improved and the system only shows a modest improvement. Above scenario 5, the production capability of the system increases rapidly with broadened application of CIM strategies.

Based completely on Fig. 10-20, it might seem optimum to work immediately toward scenario 10. However, scenario 10 involves extensive state-of-the-art technology, which can cause implementation problems. The more automated scenarios may also lead to such high costs that they are not the preferred solution from a cost-performance point of view. Thus, in order to proceed now to the selection of a preferred hybrid factory, it is necessary first to review some of the information processing and organizational issues associated with these more automated scenarios and then look at the costs associated with these scenarios.

EXAMPLE 10-5. ACHIEVABLE PRODUCTION FOR ALTERNATIVE SCENARIOS AND VERSIONS

The graph in Fig. 10-20 describes the maximum achievable production for each of four system versions for the range of scenarios being considered here. From Fig. 10-20 find the maximum achievable production for versions 1 to 4 for scenario 8. What work cell configurations are associated with each value?

Results

Figure 10-20 shows the results of the MANUPLAN II simulation for the maximum achievable production of parts per year for each scenario for the

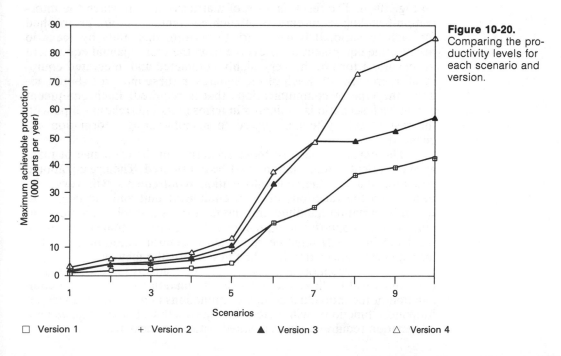

Figure 10-20. Comparing the productivity levels for each scenario and version.

Maximum achievable production (000 parts per year)

Scenarios

□ Version 1   + Version 2   ▲ Version 3   △ Version 4

four versions being considered. By examining this graph it may be observed that for scenario 8, the maximum achievable production is about 37,000 parts/year for versions 1 and 2, about 48,000 parts/year for version 3, and about 73,000 parts/year for version 4.

These data describe the levels of production that can be achieved for each of the scenarios and versions. However, they do not provide enough information on which to base a choice of the preferred production strategy. Figure 10-20 provides only an estimate of system performance, not of system cost. Each of the data points must also be evaluated in terms of implementation cost in order to choose the preferred solution.

10-26. Information Processing

As part of the design of a manufacturing facility, it is essential to understand the information processing requirements and the costs associated with implementation of the required information-handling networks. In addition, it is necessary to anticipate whether or not highly integrated configurations will be limited by the availability and cost of information-handling equipment. Therefore, the scenarios must be evaluated in terms of their implementation viability with respect to information processing.

10-27. Functional Information Models

A starting point is to apply the information matrix approach introduced in Sec. 7-1. The result is illustrated in Fig. 10-21, which builds on Fig. 10-16. The result is a set of matrices through which the information-handling requirements of each equipment unit are established for each scenario. It is necessary to perform this study by scenario because the equipment units evolve from the more manual equipment in scenario 1 toward the very highly automated and integrated equipment in scenario 10. Each of the squares in these matrices should address the type of communications that is required. Each equipment unit in each scenario is evaluated in terms of its interrelationships with the rest of the system in regard to manufacturing information exchanged.

The impact of each of these scenarios on the exchange of management-oriented information must be considered. The organizational structure and management information requirements will vary, depending on the functions of each equipment unit and the degree to which information handling and the factory as a whole become computer-based as opposed to human-based. The information shown in the results of Fig. 10-21 must be combined with information on the entire management information-handling system for the factory.

For each scenario, a decision must be made as to how the equipment units will be linked to achieve information exchange. The combination of the individual equipment functions and the systemwide information functions will determine the technical and management information requirements associated with each scenario.

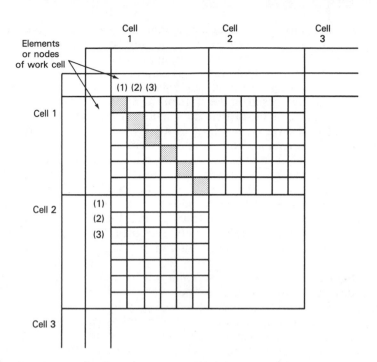

From the above analysis, a description can be prepared of all information exchanges that will be taking place on the computer network. A list of required functional exchanges can then form a basis for the design of computer networks and protocols to implement the system. By combining the individual equipment requirements with the systemwide network requirements, a complete list of information functions can be prepared. The information functions can then be considered as input to the computer network designers.

For the present study, it was concluded that all scenarios would be feasible and that equipment selection issues would be associated with cost. In terms of information-handling capability, more detailed studies of all configurations, particularly scenarios 8 to 10, would be required to make a final determination on this point. Much of the high cost for the equipment in scenarios 8 to 10 is due to the information-handling requirements of the equipment units. It was assumed for purposes of the study that the technology could be achieved if the defined resources were made available.

10-28. Information Processing Equipment Data Base (Step 12)

In order to develop adequate cost models for the manufacturing equipment units, it is necessary to have an information processing equipment data base. This information base is required in order to understand the performance capabilities that are achievable and the costs associated with these varying performance levels. In the same way that the manufacturing equipment data base must be prepared (step

6), the information processing equipment data base must be available (step 12) so that only realistic capabilities are considered and the costs associated with this performance are appropriate.

For the present study it was assumed, after a brief review, that the data processing performance requirements for all ten scenarios could be achieved based on the state of the art in computer technology. That is, it was concluded that no version of any scenario would be eliminated based on an inability to achieve desired data processing functional parameters. Initial estimates were made of the approximate costs that would be associated with upgrading equipment units for the required information capabilities, and these costs have been incorporated into the equipment cost estimates shown in Fig. 10-16. However, much more detailed analyses of information system performance and cost are needed for further iterations of this study. It is also important to have an estimate of the total network costs that will be required in addition to those information-handling costs for specific equipment units.

Figure 10-22 provides a summary of the network costs that have been developed for this initial study version. The costs in the figure for the information/computer network apply to the whole factory and therefore are a one-time expenditure. The information/computer costs that are included are scaling estimates of the costs that would be required to achieve the desired level of integration for each scenario and include both hardware and software development. (In a sensitivity study, the impact of R&D costs on the various scenarios is also considered below. These R&D costs are necessary to produce the first unit of any kind of equipment, work cell, or factory, and in large part are driven by estimates of the software integration requirements necessary to achieve the functional equipment objectives.)

Figure 10-22 also shows the total capital costs of the three-cell factory configuration, or version 1. Component capital costs are the equipment costs for three cells, the automatic materials handling costs, and information/computer network costs.

Figure 10-22.
Total capital costs.

| Scenario | Equipment Cost (3 cells, version 1) | Automated Material* Handling | | Information/ Computer Network Cost (whole factory) | Total Capital Cost (3 cells, version 1) |
		Whole Factory	Three Cells		
1	38.80	—	—	—	38.80
2	429.80	—	—	—	429.80
3	574.55	—	—	260	834.55
4	574.55	30	—	300	904.55
5	574.55	30	—	500	1104.55
6	995	30	—	635	1660
7	995	100	900	700	2695
8	1390	1000	1800	1000	5190
9	3250	1000	2400	1000	7650
10	5350	1000	3300	2000	11650

* From Fig. 10-17.

10-29. Organizational Opportunities and Constraints (Step 13)

It is necessary to assess the organizational impact of each of the proposed scenarios. As indicated in Fig. 10-23, all labor in the proposed facility can be divided into contact labor and indirect labor. *Contact labor* is associated with hands-on product manufacturing, and *indirect labor* is associated with the information handling and management of the system.

In scenario 1, there is a large amount of contact and indirect labor and a small amount of equipment support provided for each of these labor groups. In scenario 10, operations are performed by integrated and automated equipment and the function of labor then becomes to support the equipment. This is an essential change between scenarios 1 and 10 in the role of the factory labor. In the lower-numbered scenarios, individuals will typically function in a formal setting with well-specified jobs that define how they support the processing of materials or information. In the higher-numbered scenarios (8 to 10), individuals typically will be required to have skills that are necessary to support and assist in the operation of the integrated and automated equipment.

These two extremes can require quite different skills, organizational settings, and worker attitudes toward the manufacturing system. If a decision is made to shift from one scenario to another, substantial reeducation can be expected, and significant effort will be required to prepare all workers in the organization for a changed working environment. It may be necessary to hire and include in the organization individuals with many new skills such as computer programming and equipment repair as the system evolves toward the higher scenarios.

The difference among scenarios will also produce an evolution in organizational structure, as indicated in Fig. 10-24, which shows a shrinking of upper management and indirect labor and a large reduction in contact labor, which is replaced by integrated and automated equipment. Evolution from one organizational structure to the other will require significant educational and culture development efforts. In addition, in the shift from scenario 1 toward scenario 10, the emphasis of the organization will change from one in which rework is common and the organizational philosophy is to keep the manufacturing system

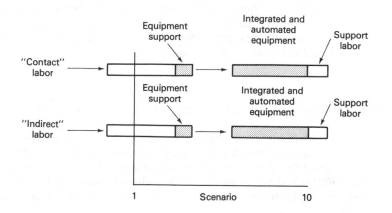

Figure 10-23.
Evolution of labor requirements.

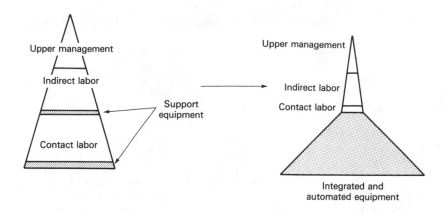

Figure 10-24.
Changes in organizational structure with low and high levels of automation and integration.

running and "fix" the product at the end of the system, to a very different one in which quality of each manufacturing step is essential and the intent is to eliminate rework completely. A significant amount of reeducation may be required in order to produce a change in perception and organizational culture to accommodate this emphasis on high-quality production and continuous flow manufacturing.

It has been assumed here that it would be possible to implement each of the scenarios discussed in this study. It has been further assumed that sufficient educational and reorganizational efforts will be made possible to accommodate these scenarios. No effort has been made to cost out individually the educational programs and retraining programs that would be part of this evolution. Staffing patterns in each case have been shown based on an assumption that education and training programs have been completed. Therefore, for all changes, some additional temporary costs would have to be included in a final budget analysis.

Figure 10-25 shows the organizational staffing per scenario that has been assumed for the three-cell factory configuration of version 1. The staffing is divided into the number of people required for direct, contact participation in the manufacturing process and the number of indirect people required to support the direct manufacturing process. The staffing estimates provided are based on the MTOI and MTI values associated with the equipment units and cells. By using these functional parameter values, it is possible to estimate the number of operators who must be available for system operation for each scenario. The total staffing requirements generally decrease as the degree of semiautomation and automation increases from scenario 1 to scenario 10.

10-30. Financial Evaluation (Step 14)

Various CIM facility configurations and capabilities are described in the preceding sections. All these data, however, will not completely satisfy the decision-making requirements of a profit-oriented enterprise that must identify the best scenario for hybrid manufacturing. A definition of the cost-benefit relationships for each scenario is necessary

	Staffing (people)		
Scenario	Direct	Indirect	Total
1	10.8	12.0	22.8
2	10.8	12.0	22.8
3	13.3	9.0	22.3
4	12.47	9.0	21.47
5	12.8	6.0	18.8
6	12.01	6.0	18.01
7	1.27	10.73	12.00
8	1.89	7.11	9.00
9	1.89	3.17	5.05
10	1.30	4.70	6.00

Figure 10-25.
Staffing for version 1 (3 cells).

to maximize the profitability of the factory. To this end, information generated by both the equipment analysis and MANUPLAN II analysis must be incorporated into a financial model.

There are many different ways to compare scenarios and factory configurations on a financial basis. The strategy applied here is a simple one that emphasizes the cost per (good) unit produced as an indicator of factory competitiveness (Lane 1986). Based on this point of view, the driving consideration in the design of a hybrid manufacturing facility is to obtain the lowest cost for hybrid production. Many other factory characteristics enter into a complete evaluation, including market responsiveness, the ability to intermix many different products on a continuing basis, and the ability to adapt as the market changes. In the present analysis, all of these indicators or selection criteria are considered secondary to the cost per unit calculations.

The financial evaluation performed here is based on a differential analysis of each scenario and version as generated by MANUPLAN II. This means that the various choices are evaluated by comparing only selected financial costs. This approach simplifies the analysis, without compromising the results, by excluding those costs that are dependent on the specific product but are relatively independent of the production technique. If this analysis needs to be extended to define all costs, only minor modifications would be required for the model. The specific items excluded from the costing analysis are the following:

Material costs: It is assumed that the same quantity and grade of input materials will be used in all cases. This is not fully accurate because the usage of materials will increase with the degree of scrap and rework incurred with each scenario.

Buildings and occupancy: The physical plant requirements and the number of upper-management personnel utilized for each scenario are assumed to be constant.

Marketing costs: Since marketing costs are product-specific and this study is a general analysis, it is assumed that these costs will be covered as a fixed percentage of the selling price of the product and thus will be independent of the production method.

Additional assumptions made to simplify the model further are as follows:

Rework: Rework costs are excluded from the analysis. It is assumed that all unsatisfactory product becomes scrap. This assumption is a conservative one, in that it will lead to underestimating costs for the manual (higher scrap rate) scenarios.

Labor rate: A labor rate of $18/hour is used for all scenarios (including salary and benefits). Direct labor requirements (given in Fig. 10-25) are derived from the properties for the equipment MTOI and MTI and work cell properties.

Discount rate: A discount rate of 18 percent was applied, following a strategy of using two times the prime rate for discounting an investment.

Figure 10-26.
Equations for financial calculations.

$$CC = \frac{\dfrac{SUM_{1\,to\,x}\,[PVann_x\,(Cost,\,Life,\,Disc)]}{TOH}}{Units/Hour}$$

where CC   = Capital costs
       $PVann_x$ = Present value annuity calculation per individual component
       Cost   = Initial capital cost of the component x
       Life   = Economic life of the component x
       Disc   = Discount rate
       TOH   = Total operating hours
       x   = Equipment number designation

$$DLHC = \frac{\dfrac{SUM_{1\,to\,x}\,[(TOH_x/MTOI)\,MIT]}{TOH}}{Units/Hour}$$

where DLHC   = Direct labor hourly cost
       TOH_x = Total operation hour for each fabrication step
       TOH   = Total fabrication time
       MTOI   = Average time between operator interventions
       MIT   = Average time for the operator to function at the station
       x   = Equipment number designation

$$ILC = \frac{\dfrac{HC \times P \times H}{TOH}}{Unit/Hour}$$

where ILC   = Indirect labor hourly costs
       HC   = Hourly rate
       P   = Number of personnel
       H   = Hours facility staffed

Three basic equations have been used to determine the unit costs for each scenario and version, relating to capital costs, direct labor hourly costs, and indirect labor costs. These equations are summarized in Fig. 10-26. The equations build on the concepts introduced earlier in this chapter and in previous chapters. The capital and staffing requirements have been previously estimated for each scenario. Given estimates for these values, it is possible to develop the production cost estimates for each scenario as shown.*

The results apply only to version 1 of the four possible factory versions introduced. Version 1 consists of three work cells, one of each type, and provides a minimum configuration for the manufacture of the desired product category 4. The financial data for version 1 provide an initial insight into the factory scenario that will be most cost-effective. Additional studies can be performed to explore versions 2 to 4 and to identify the version that provides an optimum number of work cells for each scenario. Figure 10-27 shows the labor requirements for each scenario for version 1. Figure 10-28 shows the capital requirements associated with each scenario for version 1.

EXAMPLE 10-6.   LABOR REQUIREMENTS

Figure 10-27 shows a graph of the labor requirements for each scenario, and Fig. 10-25 provides a table in which these labor requirements are summarized. Show that the labor requirements for the graph of Fig. 10-27 can be derived from the data of Fig. 10-25.

Results

Figure 10-25 gives for each scenario the staffing required in terms of numbers of direct, indirect, and total people. The staffing level ranges from 22.8 people for three cells (scenario 1, version 1) down to a low of 5 to 6 people for scenarios 9 and 10, version 1. (The increase from 5 to 6 people between scenarios 9 and 10 occurs because of the increased number of indirect personnel necessary to operate this state-of-the-art system.)

The data in Fig. 10-25 have been directly translated to the graphical form of Fig. 10-27. For each scenario, the direct labor is shown on the bottom half of each bar and the indirect labor on top, so that the sum represents the total staffing. Figure 10-27 is a visual aid to understanding the way in which the staffing requirements have dropped in going to higher levels of operation. Not only is the total staff decreasing to scenario 9, but a shift is occurring from direct to indirect labor. The functions of labor are also changing, from support of individual equipment units to support of the integrated technology. Therefore, in addition to changes in number of people, there are qualitative changes in the skill levels that are required in moving from scenario 1 to scenario 10.

---

* The method of financial analysis presented here describes only one approach; many groups have discussed the need for improved cost accounting methods to deal with the complex manufacturing equipment of today (Capenttini and Clancy 1987).

Figure 10-27.
Required staffing levels by scenario.

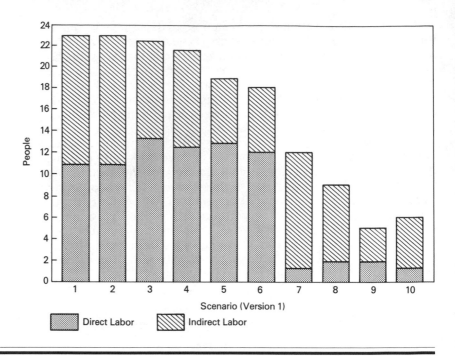

EXAMPLE 10-7. CAPITAL REQUIREMENTS

Figure 10-28 shows a graph of the capital requirements for each scenario, and the table in Fig. 10-22 summarizes data for capital requirements. Show that the capital requirements for the graph in Fig. 10-28 can be derived from the data of Fig. 10-22.

Results

The total capital cost for three cells (version 1) shown in Fig. 10-22 has been calculated by combining the individual equipment costs, the automated materials handing costs (for the whole factory and three cells), and the information-computing network costs for the whole factory. These combined costs (for a version 1 system) range from $38,800 for scenario 1 to $11,650,000 for scenario 10.

The total capital cost data shown in Fig. 10-22 have been graphed by scenario, as shown in Fig. 10-28. It is immediately obvious that a rapid growth in capital is associated with moving from scenario 1 to scenario 10. Scenarios 1 to 5 show relatively low levels of total capital costs because of an emphasis on combinations of manual and semiautomated equipment.

Scenarios 6 and 7 show significant increases in capital costs. For scenario 6, all the processing equipment becomes semiautomated, and for scenario 7 the materials handling system becomes fully automated (refer to Fig. 10-9). Scenario 7 thus consists of semiautomated equipment, fully automated materials handling (allowing remote operation), and fully automated information flow.

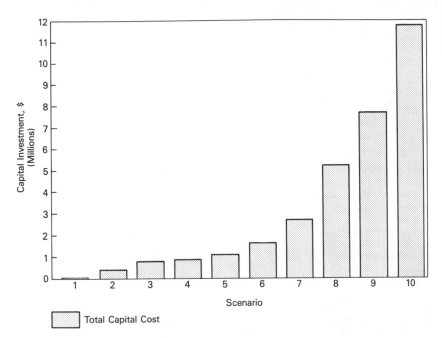

Figure 10-28.
Required capital costs
by scenario.

Scenarios 8, 9, and 10 move to fully automated equipment, which allows remote operator involvement, combined with fully automated (remote operator driven) materials handling and fully automated information flow. The differences among 8, 9, and 10 are the degree of state-of-the-art application of research and development. As may be noted, the increasing dependence on state-of-the-art technology rapidly increases capital costs.

Figure 10-29 shows the total manufacturing cost per unit for each scenario for version 1. This is an extremely interesting result. This graph indicates how higher-performance equipment can offset much larger capital costs. Also, it indicates how a manual system may not be the most cost-effective solution; low capital cost may also correspond to low performance. In general, the manual scenarios show high cost per unit values and the very highly automated and integrated scenarios show a slow increase in cost per unit values. Based on this figure, scenarios 7 to 9 provide a range for further study of optimal operations. It is also interesting to note that between scenarios 2 and 3, the addition of more automated equipment can actually increase unit production cost temporarily until an acceptable critical level is reached for each group of equipment.*

Based on Fig. 10-29, the study predicts that for the three-work-cell configuration, an optimum facility should make use of scenario 7, 8, or 9. Further implications of this result are discussed below.

Figure 10-29 provides manufacturing cost per product data as a function of system automation and integration levels. For the parameters selected here, a preferred system design can be selected for further study. The planning group is thus provided with an information base that can be used to help guide system evolution toward lower-cost approaches to manufacturing and an improved competitive capability.

* As noted by Lardner (1986), CIM can have "measureable effect . . . only when it has reached some minimum critical mass."

Figure 10-29.
Results of CIM financial analysis (without R&D costs).

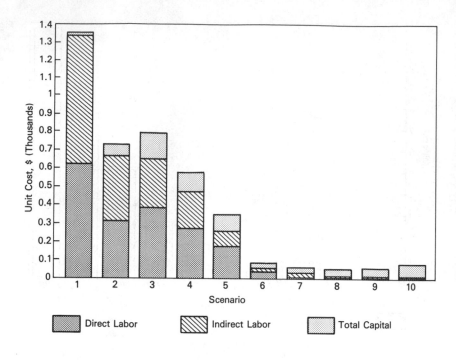

Figure 10-30.
The effect of R&D costs on scenarios 9 and 10.

Equipment	Scenario 9 Without R&D ($000)	Scenario 9 With R&D ($000)	Scenario 10 Without R&D ($000)	Scenario 10 With R&D ($000)
1	400	1,200	600	6,000
2	200	600	300	3,000
3	500	1,500	600	6,000
4	100	300	150	1,500
5	350	1,050	450	4,500
6	300	900	600	6,000
7	50	150	200	2,000
8	50	150	200	2,000
9	50	150	200	2,000
10	400	1,200	600	6,000
11	200	600	450	4,500
12	650	1,950	1,000	10,000
Automated materials handling	1,000 (whole factory 2,400 (per cell × 3)	3,000 7,200	1,000 3,300	10,000 33,000
Information/ computer network	1,000	3,000	2,000	20,000
Total	7,650	22,950	11,650	116,500

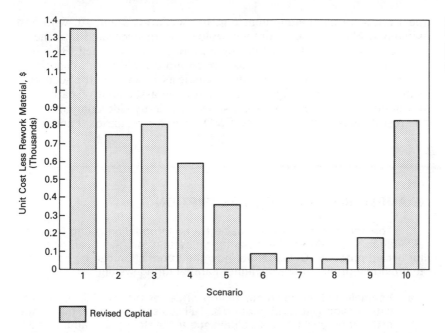

Figure 10-31.
Results of the CIM financial analysis (with R&D costs included for scenarios 9 and 10).

It is important to recognize that scenarios 9 and 10 in Fig. 10-29 do not include any allowance for the research and development costs that would be incurred with the use of such highly automated and integrated equipment. Therefore, an additional sensitivity study has been performed in which the capital costs of these scenarios are increased in order to include R&D. Figure 10-30 shows the total equip-

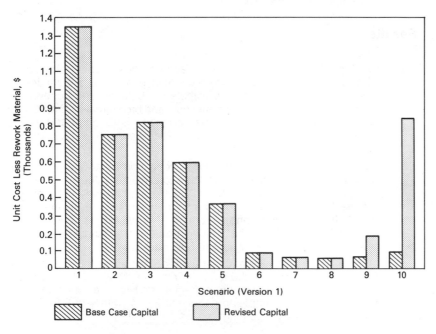

Figure 10-32.
Comparison of unit costs with and without R&D costs.

ment costs for each equipment unit for scenarios 9 and 10, with and without R&D. If these equipment values are incorporated into the financial cost model, the results in Fig. 10-31 are obtained. The unit manufacturing costs now increase much more rapidly for scenarios 9 and 10. This graph provides a more realistic assessment of the scenarios if the implementing group has to pay for the associated research and development costs. Figure 10-32 shows a side-by-side comparison of the results with and without the R&D costs for scenarios 9 and 10.

EXAMPLE 10-8. FACILITY DESCRIPTION

The above study provides background information giving the characteristics of a preferred manufacturing system that has been designed for hybrid products. Demonstrate how the information can be combined to provide a profile of the resulting manufacturing system.

a. Estimate the approximate selling price for the hybrids associated with version 1, scenarios 7 and 8, if all costs in addition to differential production costs and an allowance for system profit add an additional 100 to 200 percent to production cost.

b. What will be the total approximate market value of the hybrids produced by the factory described by version 1, scenarios 7 and 8? What will be the percentage system yield and rate of production?

c. What will be the approximate capital equipment costs and number of people employed (excluding marketing and upper management) for version 1, scenarios 7 and 8? How could higher production levels be obtained for these scenarios?

Results

a. From Fig. 10-31, a differential manufacturing cost (capital plus labor) of $50 to $100 per hybrid may be determined for optimum system design (version 1, scenarios 7 and 8). If material, building and occupancy, marketing, and upper management costs and profit are assumed to add $50 to $200 per hybrid to the sales price (an addition of 100 to 200 percent), then the hybrids being discussed here will sell for $100 to $300 per unit.

b. From Fig. 10-20, the facility (version 1, scenarios 7 and 8) will produce 20,000 to 40,000 hybrids per year. Thus, total market value of the hybrids produced will be ($100 to $300) × (20,000 to 40,000) = $2 to 12 million. As expected, this corresponds to a small manufacturing facility. From Fig. 10-18, system yield ranges from $(.95)^3 = .86$ (86 percent) to $(.97)^3 = .91$ (91 percent) and the limiting process time (the time spacing between hybrids) ranges from 3 to 5 minutes.

c. From Fig. 10-22 or 10-28, a total capital equipment investment of $3 to $6 million is required for the facility. From Fig. 10-27, 10 to 14 people will be employed for direct and indirect labor. Higher production levels can be achieved by employing versions 2 to 4 (with 4 to 6 FMS work cells) or simply by replicating version 1 capability.

10-31. Summary of the *To-Be* System

The system design discussed here has illustrated a step-by-step procedure to determine a preferred CIM factory design for the given objectives and within the given constraints. The study began with an introduction to the nature of the design problem and with a discussion of the flowchart of Fig. 4-5 as the design method to be followed.

The study has discussed the anticipated market environment, enterprise objectives associated with the planned manufacturing facility, the state-of-the-art technology available for use in the facility, and CIM design principles and reference models that can be used in conceptualizing the type of facility planned. These input variables have been used to define product categories to span the planned market. From these categories, product category 4 was selected as a desirable design focus since it also includes product categories 1 to 3, provides a mainstream focus in keeping with the hybrid industry, and is appropriate to the enterprise objectives.

A functional parameter set was used to describe the manufacturing activities of the system. A manufacturing equipment information base was developed to drive the functional process modeling efforts, which used the MANUPLAN II software in this particular study. Functional modeling has been performed for a set of ten scenarios, with four configurations or versions for each scenario. A total of 40 potential configurations were then analyzed in terms of the number of hybrid units that could be produced in a given period of time. A preliminary evaluation has been made of the functional information requirements associated with these scenarios and the discussion has outlined the need for a more detailed data processing equipment information base. Computer network equipment costs have been estimated (Fig. 10-22). Organizational opportunities and strengths associated with the different scenarios were briefly introduced. Staffing levels have been estimated for each scenario for version 1 (Fig. 10-25).

Finally, financial evaluations have been used to estimate a cost per unit time associated with version 1 manufacturing configurations. The productivity and cost data were then combined to estimate the cost per unit for each scenario for version 1 (Fig. 10-29). This productivity analysis has suggested the importance of reaching a certain critical phase in the development of a CIM facility to achieve significant impact, and the cost per unit analysis has indicated an optimum scenario within the study constraints.

The analysis indicates that scenario 7 or 8 is an optimum factory configuration (within the present study constraints). An optimum scenario then results from a mix of semiautomated and fully automated equipment, making effective use of the flexible manufacturing work cell strategy that has been introduced for equipment and information-handling integration. This preferred solution represents a more automated and integrated hybrid factory than presently exists within industry or government.

EXAMPLE 10-9. SYSTEM IMPLEMENTATION

The discussion above indicates the nature of an advanced manufacturing system for the production of hybrid microelectronic products. Given the study results, what actions might be taken to begin implementation of the manufacturing system described above?

Results

It is appropriate at this point to perform further detailed studies in order to gain additional insight into some of the issues raised in this study. A more detailed look is required at the product categories that can be defined to span the market to be addressed and the specific strategies to be used to obtain the manufacturing equipment parameters that have been assumed for the preferred configuration. It will be necessary to develop detailed plans for upgrading available equipment in order to have equipment that will operate with the parameters that have been assigned to each scenario. This will require significant detailed work with the original equipment manufacturers.

In preparation for implementation of the results, strategies must be developed for all 12 building-block equipment units listed in Fig. 10-8. In addition, the FMS building-block strategy must be completely implemented. This will require completion of the materials handling system (discussed in Example 10-1), as well as the modification of equipment units to be incorporated into the FMS structure.

Additional work is needed to extend the discussion of information-handling requirements for individual equipment items, for each FMS work cell, and for the system as a whole. A more detailed evaluation must be performed of the computer automation and integration requirements associated with each equipment unit to achieve the desired functional parameters. In addition, the functional information-handling requirements identified in Fig. 10-21 must be obtained for each equipment unit. Based on these functional requirements, the computer network for each FMS and for the whole system can then be evaluated in further detail. As part of this analysis it will be necessary to evaluate the data processing information base to make sure that the resulting information system can be implemented with available equipment and within estimated costs.

As discussed in Chap. 4, the results of the modeling activity shown here must be considered as input to a planning group. The models do not provide an answer but support the decision-making process. The results of such models cannot be viewed as "solving the problem"; rather, they can be used to help establish a learning environment that will lead to successful evolution of the enterprise.

Further organizational studies are required to determine an operational plan for the new facility in terms of organizational structure, the types of job descriptions that will be developed, and the overall costs associated with organizational function and training. The financial analysis must be extended in order to consider factors that have not been considered in detail in this first study. Cost estimates must be improved, and, as iterations through the study procedure of Fig. 4-5 are continued, additional modeling efforts will provide steadily improved estimates of the proposed system performance.

Finally, it must be remembered that the modeling efforts discussed in this chapter are performed in support of a planning group (as indicated in Fig. 4-1). The input data and output results must be evaluated for realism by the planning group, and the insights gained from the modeling must be combined with a wide range of qualitative assessments in order to determine the most appropriate plans. The types of analyses described in this chapter can support the planning group, but they cannot give the final answer. The most appropriate learning environment must be established and efforts initiated toward the robust transition stages that are needed.

10-32. Overview: The Design and Manufacture of CIM Systems

This example of a system design effort is intended to strengthen understanding of the linkages that exist among the many aspects of CIM system design. Chapters 1 through 3 have provided a problem-solving foundation, Chap. 4 has introduced the system design strategy employed here, and Chaps. 5 through 9 have addressed specific elements of any problem-solving strategy. This chapter has linked these materials together to show how various considerations may be combined in a design activity in support of a planning group.

The intent of this text is to demonstrate how a wide range of considerations associated with CIM system design can be integrated and applied in a reasonable problem-solving approach. At the same time, a wide range of specific interest areas has been considered. These specific interest areas can be further examined by those with professional and academic interest by using this text as a bridge between overall system design issues and specific areas of emphasis.

ASSIGNMENTS

10-1. In order to achieve maximum educational insight from Chap. 10, it is helpful to stimulate class discussion of the concepts that have been included, develop critical reviews that probe the advantages and limitations of the concepts, and explore alternative ways for approaching the topic. Select one or several topics from the bold-faced headings that introduce sections or examples in this chapter. Come to class prepared to explain this topic, critically describing the text materials and suggesting ways the topic discussion might be strengthened.

10-2. Select several class members to role-play the planning group for the organizational setting described in this chapter. The planning group is to present the results of this study to corporate management in order to obtain approval for adoption of the plan that has been developed. At the same time, form a second group to identify areas in which the system design has been inadequate. This group should develop the most critical possible review of the system design and express the full range of concerns regarding management action based on this plan. Form a third group to represent corporate management.

To the planning group: Working as a group, prepare materials that enable you to describe to corporate management the reason for the study, the way the study has been performed, the results of the study, and your interpretation of the actions that should now be taken to implement the study. Come to class prepared to present these materials to corporate management and to propose the most viable action at this point.

To all groups: Come to class prepared to role-play a setting in which the system design and suggested strategies for implementation are presented by the planning group to corporate management. The critical review group can be treated as a staff team that is invited to provide corporate management with a critical review of the design.

Develop a setting in which the planning group, the critical review staff team, and corporate management interact to arrive at an action plan. The study described in this chapter is assumed to be the result of a planning group effort and thus has this group's full support. Additional step-by-step strategies for implementation should be put forward by the group. The role of the staff group is to bring out all the critical points that might challenge the design and implementation plan. The function of corporate management is to consider the information provided by both groups and to decide on the most appropriate action.

10-3. Based on written materials, personal opportunities, or the experiences of others, gather data on system design and implementation efforts that have taken place in a specific setting of interest to you. As much as possible, describe the way in which the system design and implementation activity took place and the outcome of the effort. Contrast the system design method used in your selected setting with that described in this chapter. What are the strengths and weaknesses of each approach? If you were to conduct a second system design of the nature described in your example, what design and implementation strategy do you think would be most advantageous? Be prepared to present your materials in class so that all students can understand your setting, design, implementation strategies, and outcomes.

CLASS PRESENTATIONS AND WRITTEN REPORTS

CLASS PRESENTATIONS

When making a presentation, it is important to convey the maximum amount of information in the time available. This will typically require the use of various types of visual aids and handouts. Visual aids are essential to improve the impact of a presentation. In general, they should be used to structure the discussion and achieve the maximum communication of ideas among all class members.

Visual aids can be prepared and presented in a number of ways. Charts may be drawn on large poster board or easel paper and then propped up on or hung from an easel or hung on the wall for viewing by the class. Another simple method of presentation is to make use of overhead transparencies. These can be prepared by outlining the material on several sheets of paper and using a copy machine with transparent "pages" intended for this purpose. A projector is used to display these materials onto a wall or screen.

In general, visual aids should guide the discussion, not provide the full details of the material being presented. Typically, the high points can be illustrated with a few figures or a few key words. An outline format can be used effectively in enabling the class to follow the flow of discussion. Figures A-1a to A-1h show typical presentation charts in an effective format.

Another method of strengthening discussion is to provide handout copies of materials for all members of the class. Typically, a handout might include a step-by-step outline of an entire presentation, or it might describe specific materials to be emphasized in discussion. With an outline handout, the class can follow the flow of the presentation and make notes as they listen to comments. With handouts that emphasize certain points, the class can use them for reference and for in-depth explanations.

Assignment:	1 = 3
Topic:	Developing system models
Setting visited:	ABC Manufacturing Co.
Personal contacts:	Ms. Kathy Jones
	Mr. Al Smith

(a)

Figure A-1.
Typical visual aid format for presentation.

Figure A-1. continued

Presentation Outline
- Description of manufacturing setting
- System elements
- System relationships
- System behavior and operation
- Areas of information not available
- Insights gained and conclusions

(b)

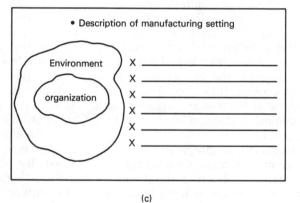

(c)

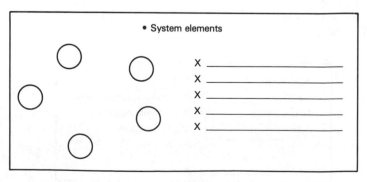

(d)

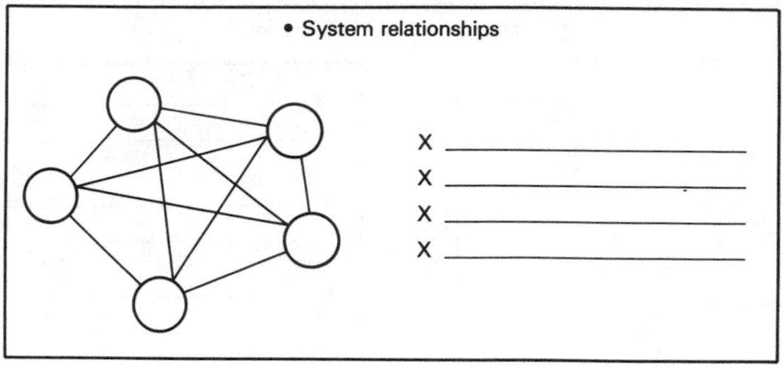

• System relationships

X _____
X _____
X _____
X _____

(e)

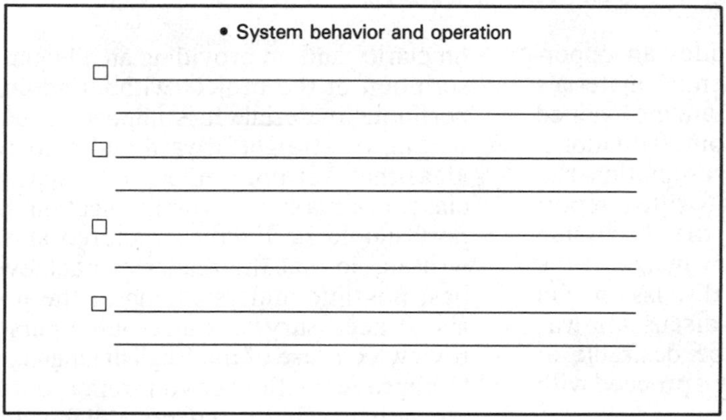

• System behavior and operation

☐ _____
  _____
☐ _____
  _____
☐ _____
  _____
☐ _____
  _____

(f)

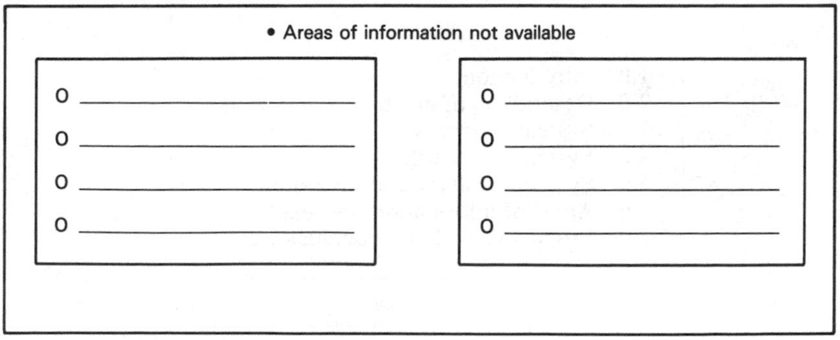

• Areas of information not available

O _____   O _____
O _____   O _____
O _____   O _____
O _____   O _____

(g)

• Insights gained and conclusions

O _____
  _____
O _____
  _____
O _____
  _____

O _____
  _____
O _____
  _____
O _____
  _____

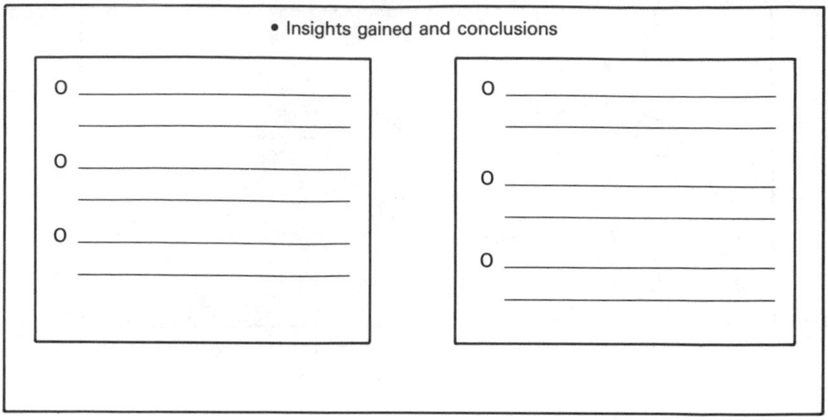

(h)

WRITTEN REPORTS

A written report provides an opportunity to combine the presented materials with insights and understanding gained from class discussion. In some situations, the initial outline for the presentation can be effectively used for the written report (Fig. A-2). The content of the discussion can be expanded to take advantage of the insights gained from class discussion. On the other hand, if the class discussion was sufficiently useful, it may be desirable to prepare a new outline and then proceed with writing the report.

In report writing, emphasis should be on clarity and on providing an adequate description of the project without adding superfluous material. It is important that the writing be straightforward and convey the ideas that were presented and discussed in class. For maximum communication the report should be tightly structured and well written, so that the reader can achieve the best possible understanding of the materials. If necessary, ask a resource person to review your use of the English language and to improve the flow of your report until you are satisfied that it reflects a high standard of quality.

Figure A-2.
Typical outline for
written report.

Developing System Models
by _____

Outline

1.0 Introduction
2.0 Description of manufacturing setting
3.0 System elements
4.0 System relationships
5.0 System behavior and operation
6.0 Areas of information not available
7.0 Insights gained and conclusions

APPENDIX B

OVERVIEW OF THE SYSTEM-ENVIRONMENT SIMULATION (SES) MODEL

The operation and application of the SES model has been introduced in Chap. 3. This appendix provides further information on the design and use of this software product. The executable code and source code for this product are provided on a floppy disk with this text.

Figure B-1 shows a flowchart for the SES model, and Fig. B-2a to B-2r illustrate the screens that will be encountered when using the product. As shown in Fig. B-1 and Fig. B-2a, use of the model begins with an introductory screen. After reading the screen, the user should hit any key on the keyboard to proceed.

The next series of screens allows the user to specify various parameters that determine the performance of the program. Screen 2 is used to specify the maximum output of the factory (denoted by EMAX*U). Values between 50 and 1000, in increments of 50, can be selected by pressing keys 1 through 0 on the keyboard for values 50 to 500 and by pressing the keys SHIFT 1 through SHIFT 0 for the values 550 to 1000. The program will remember the last value used and will default to that value if RETURN is pressed. The default value of 450 is given in brackets.

As shown on screen 3, the user is asked to specify the percentage rework (denoted by R) for the factory as a whole. Values ranging from 0 to 10 percent can be selected by pressing the number key for that percentage and SHIFT 0 for 10 percent. For example, 2 percent is selected by pressing the 2 key and 6 percent is selected by pressing the 6 key.

After these two parameter values have been entered by the user, market performance must be determined. As shown on screen 4, the user is asked to specify market performance in terms of point-by-point data, data for a function that is provided, or a random number generator. If the user selects point by point, the program will prompt the user to enter the values for each time point, as indicated on screen 5. If the user chooses to specify the market performance by function, screen 6 will appear, listing all of the available functions (these are further discussed below). If the user decides to use a random number generator, the program will prompt for the upper and lower bounds of the random variation (as per screen 12). For this latter case, the program will calculate a series of random numbers between the boundaries that are set and provide these values in a market equation array, MT.

As may be noted from Fig. B-1, screens 7 to 11 provide further information on the function approach to defining MT. Five different functions are provided: a constant MT (screen 7), MT changing from a higher constant to a lower constant value (screen 8), MT changing from a higher to a lower value and back to the higher vaslue (screen 9), MT varying between higher and lower levels (screen 10), and MT changing

Figure B-1.
Flowchart for SES model.

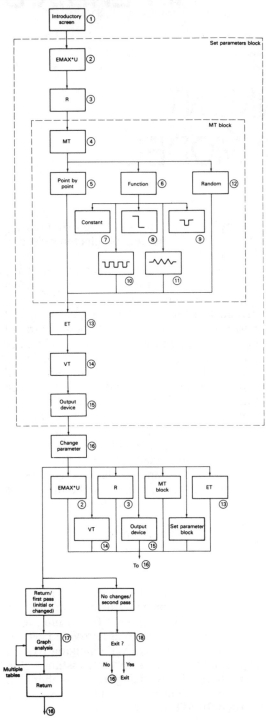

Figure B-2.
User interface screens for SES model.

```
                                          ①
Welcome to the System-
Environment Simulation
(SES) Model to be used with
the text:              •

   CIM Systems: An
   Introduction to Computer
   Integrated Manufacturing
   By F. H. Mitchell, Jr.

   Copyright (C) Prentice
   Hall, Inc.

Press any key to continue
                      • • •
```
(a)

```
                                          ②
Define the maximum output
of the factory (EMAX * U)
as a constant value: [450]
     1) 50      SH1) 550
     2) 100     SH2) 600
     3) 150     SH3) 650
     4) 200     SH4) 700
     5) 250     SH5) 750
     6) 300     SH6) 800
     7) 350     SH7) 850
     8) 400     SH8) 900
     9) 450     SH9) 950
     0) 500     SH0) 1000
Choose 0-9, Shift 0-9 or
RETURN!
```
(b)

```
                                          ③
Define the percentage
rework (R) for the factory
as a whole: [10.000000]
     0) 0
     1) 1
     2) 2
     3) 3
     4) 4
     5) 5
     6) 6
     7) 7
     8) 8
     9) 9
   Shift 0) 10
Choose 0, 1, 2, 3, 4, 5, 6,
7, 8, 9, Shift 0 or RETURN!
```
(c)

④

```
Select the market
performance (MT[]):
   1) Point by point
   2) By function
   3) Randomly
Choose 1, 2, or 3!
```

(d)

⑦

```
Constant value of MT[]:
```

(g)

⑤

```
Point by point:
   Enter value for point 1:
   Enter value for point 2:
   Enter value for point 3:
                      .
                      .
                      .
   Enter value for point
100:
```

(e)

⑧

```
First constant value of
MT[]:
Second constant value of
MT[]:
Time point at which MT[]
changes:
```

(h)

⑥

```
By function:

   1) MT[] constant

   2)              4)


   3)              5)


Choose 1, 2, 3, 4, or 5!
```

(f)

⑨

```
First constant value of
MT[]:
Second constant value of
MT[]:
Time point at which MT[]
first changes:
Time point at which MT[]
changes again:
```

(i)

Figure B-2. continued

⑩
```
First constant value of
MT[ ]:
Second constant value of
MT[ ]:
Time point at which MT[ ]
changes:
Time point at which MT[ ]
changes:
```
(j)

⑪
```
First value of MT[ ]:
Second value of MT[ ]:
Time point at which MT[ ]
changes:
Time point at which MT[ ]
changes again:
Time point at which MT[ ]
changes again:
```
(k)

⑫
```
Randomly:
    Upper bound of MT[ ]:
    Lower bound of MT[ ]
```
(l)

⑬
```
Define the algorithm to
select the level of factory
production (ET[ ]):
    1) Constant at EMAX*U
    2) ET starts out at
       maximum value,
       EMAX*U.
       If Q2 + N3 > 2*MT, ET
       is reduced by 50%.
       If Q2 = 0, ET returns
       to maximum value.
       ET changes occur 3
       days after conditions
       are met.
    3) ET starts out at
       maximum value,
       EMAX*U.
       If Q2 >= 2.0*EMAX*U,
       ET is reduced by 50%.
       If Q2 <= 0.2*EMAX*U,
       ET returns to maximum
       value.
       ET changes occur 3
       days after conditions
       are met.
Choose 1, 2, or 3!
```
(m)

⑭
```
Define algorithm to select
the level of vendor kit
orders (VI[ ]):
    1) VT constant at
       maximum production
       value EMAX*U
    2) VT starts out at
       maximum value,
       EMAX*U.
       If Q2 + N3 > 2*MT, VT
       is reduced by 50%.
       If Q2 = 0, VT returns
       to maximum value.
       N1=VT with 5-day
       delay.
    3) VT starts out at
       maximum value,
       EMAX*U.
       If Q2 >= 2.0*EMAX*U,
       VT is reduced by 50%.
       If Q2 <= 0.2*EMAX*U,
       VT returns to maximum
       value.
       N1=VT with a 5-day
       delay.
Choose 1, 2, or 3!
```
(n)

```
                              ⑮
Print to:
   1) Screen
   2) Printer
Choose 1 or 2!
```

(o)

```
                              ⑰
Output variable:
      1) MT[ ]
      2) VT[ ]
      3) ET[ ]
      4) N1[ ]
      5) N2[ ]
      6) N3[ ]
      7) N4[ ]
      8) Q1[ ]
      9) Q2[ ]
      0) R[ ]
      RETURN) Output finished.
Choose 1, 2, 3, 4, 5, 6, 7,
8, 9, 0 or RETURN!
```

(q)

```
                              ⑯
Change parameters:
    1) Maximum output
       (EMAX*U)
    2) Percentage rework (R)
    3) Market equation
(MT[ ])
    4) Vendor kits produced
       (ET[ ])
    5) Kit ordered (VT[ ])
    6) Output device
    7) Completely new model
RETURN) Change
parameters finished
Choose 1, 2, 3, 4, 5, 6, 7,
or RETURN!
```

(p)

```
                              ⑱
Are you finished with the
simulation? (y/n)
```

(r)

between lower and higher levels with ramping between the higher and lower levels (screen 11).

Once the MT block has been completed, screen 13 appears to enable the user to select an algorithm that will determine the level of factory production (ET). ET can be selected either as a constant or to respond to the changing market conditions in two different ways. The conditions shown for options 2 and 3 correspond to Examples 3-4 and 3-5 in the text. The user must next specify the desired VT algorithm. The op-

tions shown on screen 14 are the same as those available for factory production (ET) but with changes in the delay times. (These choices are the ones applied in Examples 3-4 and 3-5.)

This completes the data input to the program itself. The user must now decide on the type of output device to be used (screen 15). If the output is sent to the computer screen, only 70 time points can be shown because of the constraints of the computer screen. If the data are sent to the printer, 100 time points can be shown.

Screen 16 is displayed to allow the user to change any selection previously made or to specify a revised model for use. A list of possible change parameters is shown on screen 16. If any of these options is selected, the program returns to the appropriate prior screen, allows the change to be made, and then returns to screen 16. A return to screen 2, 3, 4, 13, 14, or 15 will provide an opportunity to change a single parameter. Return to screen 2 (through option 7) will allow the setting of a completely new parameter set.

On the first pass through any application of the model (with initial or changed parameters), screen 17 will appear. This screen allows the user to print graphs of parameters associated with the simulation. These parameters are selected by pressing keys 1 through 0 on the keyboard. The user can continue to print graphs until the RETURN key is pressed, returning to screen 16 and allowing changes to be made in the parameters.

If the user exits from screen 16 without having made any changes in a parameter (so that this is the second pass through a particular set of parameters), the user will be routed to the exit screen (screen 18), which provides an opportunity to exit from the program or to return to screen 16 and further change the parameters.

The program is written in the C language, and will operate on IBM-compatible computers using a monochrome monitor. To change to other types of platforms, a few modifications in the source code are required. The source code for the SES program is included on the floppy disk provided with this text, and a hard copy of the source code may be obtained on request from Prentice Hall.

The following discussion provides further detail on the SES model applications in Examples 3-4 and 3-5. The model is designed so that modifications can be incorporated. Consideration of multiple product lines (as discussed in Chap. 3) and other types of system-environment interactions can be obtained by working within the provided program structure and changing only selected parts of the structure.

DETAILED DISCUSSION OF EXAMPLE 3-4

Figure 3-14 illustrates a market change (MT) from a level of 500 parts per day (for which the market demand is greater than the factory output) to a level of 350 parts per day (for which the market demand is less than the factory output). This drop occurs on day 50, as shown.

The behavior of the manufacturing system is controlled by the sum of the factory output (N3) and the product inventory (Q2). Note that prior to the market drop, the level of factory output (N3) is 405 parts per day, which is less than the market maximum but greater than the market level after the drop. The factory production drop occurs when the factory output (405), plus product inventory (297), exceeds twice the market demand (2 × 350).

Inventory begins to build on day 50 when the market drops. On day 55, five days after the market drop, inventory plus production reaches the critical decision level. Vendor kit orders (VT) drop on day 55, and vendor kit arrivals (N1) drop five days later, on day 60. A three-day delay produces a drop in desired production (ET) on day 58. At this time, the factory production is cut to 50 percent of its prior maximum. Once the factory level is cut back, the product inventory Q2 decreases (after the last day of production is completed) because the factory is no longer overproducing for the market.

By day 63, the total inventory Q2 has been reduced to zero. This is the trigger level to turn the manufacturing system back on. Vendor orders return to the maximum level. ET increases back to the maximum level on day 66, three days after the inventory is reduced to zero. After a delay of five days (from day 63) the newly ordered vendor kits will be received by the manufacturing system.

The vendor kit orders are reduced on day 55, when the critical inventory level is reached. Five days after this reduction of vendor orders (on day 60), the shipments of vendor kits (N1) is reduced to 225. The return in kit shipments N1 to the level of 450 occurs five days after the vendor kit orders are increased back to the original level.

On day 58, the factory cutback results in fewer kits being taken in by the factory. Then on day 63 the Q2 = 0 point is reached, at which time factory production increases back to the level of 450. N2 increases back to the higher level on day 66 (three days later) in order to meet the return to higher factory production. However, the higher level of production can be sustained only for a short period because there is not adequate kit inventory to maintain the higher level, and the new orders for inventory are being delayed by five days. Thus, after temporarily resuming a higher production level, the shipment of vendor kits drops to a lower level due to an inadequate kit inventory. Finally, on day 68, the increased number of vendor kits begins to arrive, enabling the higher level of operations to be sustained.

The kit inventory Q1 varies over this period of time, as shown. The inventory begins to build on day 58, when the factory is cut back, and then levels off when the vendor supply is reduced, two days later. On day 63, when Q2 = 0, the critical decision point is reached. The factory is told to begin production, and after a three-day delay (day 66), the factory increases to the higher level of production. This higher production level quickly draws down the kit inventory.

Once kit inventory reaches zero, the factory must cut back until two days later, when the increased delivery of kit orders is received. ET represents the factory production level that is being attempted. The graph for N3 shows the results of the factory

operations. The figure illustrates the effect of the factory cutback, with a reduced product shipment after a delay. When an attempt is made to return to full production level, there is a temporary drop in shipment again because of the lack of kit inventory.

The graph for N4 shows the shipments that actually make it to the market. On day 50, there is a drop in market demand. The shipment of product drops from 405 to 350, which is fulfilling the reduced market. Output stays at 350 until the effect of the factory cutback is felt. The factory cutback takes place on day 58. On day 63 the decision point Q2 = 0 is reached, and the factory begins to go back to the higher level of production. From day 63 to 66 there is a three-day delay before factory production level is increased; factory production then shifts back, so that by day 67 the higher level of production is obtained and the full market of 350 parts is shipped.

The temporary dip shown in day 68 is the effect of the kit inventory Q1. A small product inventory Q2 has been built up previously, but there is still not sufficient Q2 to fulfill the market opportunities until vendor parts are available at the factory.

The resulting behavior of the product inventory Q2 is shown. Once the market drops, there is a buildup of inventory. When the critical point is reached (day 55), the factory cuts back (day 58). One day later the reduction in inventory begins. Inventory begins to build again on day 66, as the desired factory production level ET increases. During the turn-on phase, inventory Q2 drops again (on day 69) because the factory has inadequate vendor kits to achieve full production. There is a temporary drop in the product inventory (Q2) buildup as a factory cutback is forced by the lack of adequate vendor kits. Once the number of vendor kits returns to the upper level, inventory Q2 continues to rebuild due to overproduction.

When the critical level is reached, the decision is made to reduce the order of vendor kits and to reduce factory operations; after a delay, the same process is repeated.

As shown in Fig. 3-15, a one-time permanent reduction in the market (MT) is noted on day 20. The market drops from 500 to 350 parts per day. Factory production is related directly to the product inventory level Q2. As soon as the market drops, inventory level of the product begins to build. The level of Q2 builds until it is two times the maximum factory production level of 500. When the product inventory reaches 1000 parts, a message is sent to reduce factory production. However, three days must pass before factory production is reduced; the inventory Q2 continues to build to 1100 as shown. As soon as the factory cutback to 50 percent is achieved, the market is greater than the factory production level and the inventory begins to drop.

The inventory becomes equal to 20 percent of production on day 45. A five-day delay is experienced in increasing vendor kit orders and a three-day delay is noted in bringing the factory back up to full capacity again. The inventory level thus continues down to zero.

The vendor kit orders (VT) are cut back on day 37, which corresponds to the inventory level for which the critical two times factory production level is reached. The vendor kit orders remain at the reduced level until day 48, at which time the inventory reaches the 20 percent level. The vendor kit orders are then increased on day 48, but the response (of N1) is not noted until five days later, on day 53. The relationship between the vendor orders VT and the receipt of vendor kits by the factory shows that ET and N1 develop similar patterns, displaced only by the five-day time delay between the time the vendor orders are placed and the vendor orders are received.

The shipment of vendor kits from inventory Q1 is shown as N2. The shipment of vendor kits to the factory rises and falls, depending on the level of desired factory operations, ET. The rise and fall of kit inventory Q1 is indicated as a result of the interplay between vendor orders and factory production levels. As may be seen, inventory Q1 starts to rise each time the factory cuts back operations and builds to a maximum level of 675 parts. Once the factory production is reduced to the 50 percent point, the inventory no longer rises, and once the market expands again, the inventory is worked off as again the market exceeds the production capacity.

The output of goods being shipped to output inventory is N3. Again, this parameter follows the basic production pattern established for the factory.

Parameter N4I shows the shipment of the final product to the market. Prior to the market cutback, output is stable at the level of 405 products per day. Following the cutback on day 20, output is cut back to a reduced level and continues at that level until such time as the inventory is worked down and the factory is brought back to full production level. The drop in output on day 48 is associated with the fact that the product inventory is reduced to 0 between days 48 and 50 (refer to Q2). During the time that there is no inventory, there is not adequate production from the factory because of waiting for vendor kits to arrive. Once the full level of vendor kits starts arriving, then factory production increases again up to its maximum level, and sales to the market are achieved at full market potential.

When the market drops, the product inventory Q2 begins to build. Once it reaches the 200 percent point, messages are sent out to reduce vendor purchases and factory operations. After delay times, these changes are implemented, and after the cutback, inventory starts to fall. When inventory reaches the 20 percent level, messages are sent out again to turn on the factory and begin a higher level of vendor orders, but due to a delay in the system, the inventory falls to zero for a two-day period before starting to repeat the pattern.

A comparison of the results of Examples 3-4 and Examples 3-5 shows how the different algorithms have produced different system responses. As may be noted for Example 3-4, the maximum kit inventory

Q1 is 450 parts, and this level is maintained for six days. Comparing that with the results of Example 3-5, note that Q1 reaches a maximum of 675 parts and remains at that level for eight days. Product inventory Q2 for Example 3-4 reaches a peak of 495 parts per day, whereas the completed product inventory Q2 for Example 3-5 reaches a maximum of 1100 parts. In comparing the system sales to the market for Examples 3-4 and 3-5 (parameter N4), note that different average numbers of product are provided to the market in the two cases. For Example 3-4 an average of 318 parts is provided after the reduction in market demand, whereas in Example 3-5 the number changes to 340. Thus, a trade-off exists between the advantages and disadvantages of the algorithms used. The nature of this trade-off is such that the preferred decision-making approach will depend on the cost structure of the manufacturing system in each case.

REFERENCES

PREFACE AND CHAPTER 1

Abernathy, W. J., K. B. Clark, and A. M. Kantrow. 1983. *Industrial Renaissance: Producing a Competitive Future for America.* New York: Basic Books.

Air Force Systems Command. 1981a. "Integrated Computer-Aided Manufacturing (ICAM) Dynamics Modeling Manual (IDEF$_2$)," Wright-Patterson Air Force Base, Ohio.

————. 1981b. "Integrated Computer-Aided Manufacturing (ICAM) Function Modeling Manual (IDEF$_0$)," Wright-Patterson Air Force Base, Ohio.

Bermond, J. C., and G. Memmi. 1985. "A Graph Theoretical Characterization of Minimal Deadlocks in Petri Nets," in Y. Alavi et al., *Graph Theory with Applications to Algorithms and Computer Science.* New York: John Wiley.

Bowler, T. D. 1981. *General Systems Thinking: Its Scope and Applicability.* New York: Elsevier North-Holland.

Chiantella, N. A. 1986. "A CIM Business Strategy for Industrial Leadership," in *Management Guide for CIM,* ed. Nathan A. Chiantella. Dearborn, Mich.: Society of Manufacturing Engineers.

Clark, K. B., R. H. Hayes, and C. Lorenz. 1985. *The Uneasy Alliance: Managing the Productivity-Technology Dilemma.* Boston: Harvard Business School Press.

Cohen, S. S., and J. Zysman. 1988. "The Myth of the Post-Industrial Economy," *Siemens Review,* 55 (March/April).

D. Appleton Co. 1985. "IISS: Integrated Information Support System (Information Modeling Manual IDEF$_1$—Extended)," Manhattan Beach, Calif.

Forrester, J. W. 1969. *Urban Dynamics.* Cambridge, Mass.: MIT Press.

————. 1973. *World Dynamics* (2nd Ed). Cambridge, Mass.: Wright-Allen Press.

Gleick, J. 1987. *Chaos: Making a New Science.* New York: Viking Penguin.

Harary, F. 1972. *Graph Theory.* Reading, Mass.: Addison-Wesley.

Jonas, N. 1987. "Can America Compete?" *Business Week,* April 20.

Katz, D., and R. L. Kahn. 1978. *The Social Psychology of Organizations.* New York: John Wiley.

Krooss, H. E., and C. Gilbert. 1972. *American Business History.* Englewood Cliffs, N.J.: Prentice Hall.

Kuhn, T. S. 1970. *The Structure of Scientific Revolutions* (2nd ed.). Chicago: University of Chicago Press.

Lardner, J. F. 1986. "Computer Integrated Manufacturing and the Complexity Index," *The Bridge* (National Academy of Engineering), 16, 1, Spring.

Mitchell, F. H., and C. C. Mitchell 1980. "Development, Application and Evaluation of an 'Action-Reaction' Planning Method," *Academy of Management Review,* 5, 1.

National Academy of Engineering. 1985. *Education for the Manufacturing World of the Future.* Washington, D.C.: National Academy Press.

National Research Council. 1986. *Toward a New Era in U.S. Manufacturing: The Need for National Vision.* Washington, D.C.: National Academy Press.

————. 1987. *Management of Technology: The Hidden Competitive Advantage.* Washington, D.C.: National Academy Press.

Perrow, C. 1972. *Complex Organizations* (2nd ed.). Glenview, Ill.: Scott, Foresman.

Peterson, J. L. 1981. *Petri Net Theory and the Modeling of Systems.* Englewood Cliffs, N.J.: Prentice Hall.

Porter, M. E. 1980. *Competitive Strategy: Techniques for Analyzing Industries and Competitors.* New York: Free Press.

Ralston, A., ed. 1983. *Encyclopedia of Computer Science and Engineering* (2nd ed.). New York: Van Nostrand Reinhold.

Simon, H. 1976. *Administrative Behavior* (3rd ed.). New York: Free Press.

Solberg, J. J., et al. 1987. "Design and Analysis of Integrated Manufacturing Systems," supplement to *The Bridge,* (National Academy of Engineering), 17, 2 (Summer).

Van Gigch, J. P. 1978. *Applied General Systems Theory* (2nd ed.). New York: Harper & Row.

von Bertalanffy, L. 1968. *General System Theory: Foundations, Development, Applications.* New York: George Braziller.

Yeomans, R. W., A. Choudry, and P. J. W. TenHagen. 1985. *Design Rules for a CIM System.* New York: North-Holland.

CHAPTER 2

Abernathy, W. J., K. B. Clark, and A. M. Kantrow. 1983. *Industrial Renaissance: Producing a Competitive Future for America.* New York: Basic Books.

Carey, A. 1967. "The Hawthorne Studies: A Radical Criticism," *American Sociological Review,* 32, 3 (June).

Chandler, A. D., Jr. 1977. *The Visible Hand: The Managerial Revolution in American Business.* Boston: Belknap Press.

Clark, K. B., R. H. Hayes, and C. Lorenz. 1985. *The Uneasy Alliance: Managing the Productivity-Technology Dilemma.* Boston: Harvard Business School Press.

Cyert, R. M., and D. C. Mowery, eds. 1987. *Technology and Employment: Innovation and Growth in the U.S. Economy.* Washington, D.C.: National Academy Press.

Derry, T. K., and T. I. Williams. 1961. *A Short History of Technology.* London: Oxford Press.

Groner, A. 1972. *The American Heritage History of American Business and Industry.* New York: American Heritage Publishing Co.

Guile, B. R., and H. Brooks, eds. 1987. *Technology and Global Industry: Companies and Nations in the World Economy.* Washington, D.C.: National Academy Press.

Harrington, J., Jr. 1973. *Computer Integrated Manufacturing.* New York: Industrial Press.

Hawke, D. F. 1988. *Nuts and Bolts of the Past: A History of American Technology 1776–1860.* New York: Harper & Row.

Hayes, R. H., and S. C. Wheelwright. 1984. *Restoring Our Competitive Edge: Competing Through Manufacturing.* New York: John Wiley.

Kuttner, R. 1988. "U.S. Industry Is Wasting Away—But Official Figures Don't Show It," *Business Week,* May 16, p. 26.

Lebergott, S. 1984. *The Americans: An Economic Record.* New York: W. W. Norton.

Lindblom, C. E. 1959. "The Science of Muddling Through," *Public Administration Review,* 19 (Spring), as reprinted in *Classics of Public Administration,* ed. Jay M. Shafritz and Albert C. Hyde. Oak Park, Ill.: Moore, 1978.

McGregor, D. M. 1957. "The Human Side of Enterprise," *Management Review,* November, as reprinted in *Classics of Public Administration,* ed. Jay M. Shafritz and Albert C. Hyde. Oak Park, Ill.: Moore, 1978.

Marcus, A. I., and H. P. Segal. 1989. *Technology in America: A Brief History.* New York: Harcourt Brace Jovanovich.

Maslow, A. H. 1943. "A Theory of Human Motivation," *Psychological Review,* 50

(July), as reprinted in *Classics of Public Administration,* ed. Jay M. Shafritz and Albert C. Hyde. Oak Park, Ill.: Moore, 1978.

Mishel, L. 1988a. "Of Productivity Numbers: Of Manufacturing's Mismeasurements," *New York Times,* November 27.

———. 1988b. "Manufacturing Numbers: How Inaccurate Statistics Conceal U.S. Industrial Decline." Washington, D.C.: Economic Policy Institute.

National Academy of Sciences. 1987. *Balancing the National Interest: U.S. National Security Export Controls and Global Economic Competition.* Washington, D.C.: National Academy Press.

Ouchi, W. G. 1981. *Theory Z: How American Business Can Meet the Japanese Challenge.* New York: Avon.

Owen, J. V., and J. F. Entorf. 1989. "Where Factory Meets Faculty," *Manufacturing Engineering,* (February), 48.

Pennar, K. 1988. "The Factory Rebound May Be More Fantasy Than Fact," *Business Week,* December 12, p. 98.

Ratner, S., J. H. Soltow, and R. Sylla. 1979. *The Evolution of the American Economy: Growth, Welfare and Decision Making.* New York: Basic Books.

Roethlisberger, F. J. 1941. "The Hawthorne Experiments," in F. J. Roethlisberger, *Management and Morale.* Cambridge, Mass.: Harvard University Press, as reprinted in *Classics of Public Administration,* ed. Jay

M. Shafritz and Albert C. Hyde. Oak Park, Ill.: Moore, 1978.

Shrensker, W. L. 1985. "A Brief History of CIM," in *A Program Guide for CIM Implementation,* ed. Charles M. Savage. Dearborn, Mich.: Society of Manufacturing Engineers.

Simon, H. A. 1946. "The Proverbs of Administration," *Public Administration Review,* 6 (Winter), as reprinted in *Classics of Public Administration,* ed. Jay M. Shafritz and Albert C. Hyde. Oak Park, Ill.: Moore, 1978.

Simpson, J. A., R. J. Hocken, and J. S. Albus. 1982. "The Automated Manufacturing Research Facility of the National Bureau of Standards," *Journal of Manufacturing Systems,* 1, p. 17.

Smith, A. 1776. "Of the Division of Labour," from Adam Smith, *The Wealth of Nations,* as reprinted in *Classics of Organizational Theory,* ed. Jay M. Shafritz and Albert C. Hyde. Oak Park, Ill.: Moore 1978.

Taylor, F. W. 1916. "The Principles of Scientific Management," *Bulletin of the Taylor Society* (December), as reprinted in *Classics of Organization Theory,* ed. Jay M. Shafritz and Albert C. Hyde. Oak Park, Ill.: Moore, 1978.

U.S. Department of Commerce. 1975. *Historical Statistics of the United States: Colonial Times to 1970* (Bicentennial Edition).

———. 1987. *Statistical Abstract of the United States.*

CHAPTER 3

Alic, J. A. 1984. Statement of John A. Alic, Project Director, Office of Technology Assessment, Before the Subcommittee on Science, Research and Technology, Committee on Science and Technology, U.S. House of Representatives, Washington, D.C.: Office of Technology Assessment, July 13.

Appleton, D. S. 1984. "The State of CIM," *Datamation Magazine,* December 15, 72.

———. 1985. CIM—A Program, Not a Project," *CIM Review,* Winter, 4.

Bertain, L., and L. Hales, eds. 1987. *A Program Guide for CIM Implementation;* 2nd ed. Dearborn, Mich.: Society of Manufacturing Engineers.

Burggraaf, P. 1985. "Semiconductor Factory Automation: Current Theories," *Semiconductor International,* October, p. 88.

Chiantella, N. A., ed. 1986. *Management Guide for CIM.* Dearborn, Mich.: Society of Manufacturing Engineers.

Cousins, S. A., ed. 1988. *Integrating the Automated Factory.* Dearborn, Mich.: Society of Manufacturing Engineers.

Fuchs, J. H. 1988. *The Prentice Hall Illustrated Handbook of Advanced Manufacturing Methods.* Englewood Cliffs, N.J.: Prentice Hall.

Gould, L. 1986. "Where Is the Quality in Quality Control?" *Managing Automation,* September.

Gustavson, R. E. 1986. "Choosing Manufacturing Systems Based on Unit Cost," in *Automated Assembly,* ed. Jack D. Lane. Dearborn, Mich.: Society of Manufacturing Engineers.

Hayes, R. H., S. C. Wheelwright, and K. B. Clark. 1988. *Dynamic Manufacturing: Creating the Learning Organization.* New York: Free Press.

Hyer, N. L., and U. Wemmerlov. 1984. "Group Technology and Productivity," *Harvard Business Review* (July–August), 140.

———. 1988. "Assessing the Merits of Group Technology," *Manufacturing Engineering* (August), 107.

Hutchinson, G. K. 1985. "Information: The Unifying Force in Production Systems," paper presented at the APMS-COMP-CONTROL 85 Conference, Budapest, Hungary, August 27–30, University of Wisconsin, Milwaukee.

International Business Machines. 1988. "CIM: More Than Robots Dancing in the Dark," *IBM Directions* (October).

Jaikumar, R. 1986. "Postindustrial Manufacturing," *Harvard Business Review,* 86, 6 (November–December), 69.

Klippel, W. H., ed. 1984. *Statistical Quality Control.* Dearborn, Mich.: Society of Manufacturing Engineers.

Kutcher, M. 1983. "Automating It All," *IEEE Spectrum* (May), 40.

Lardner, J. F. 1986. "Computer Integrated Manufacturing and the Complexity Index," *The Bridge* (National Academy of Engineering), 16, 1 (Spring).

National Research Council. 1984. *Computer Integration of Engineering Design and Production: A National Opportunity.* Washington, D.C.: National Academy Press.

———. 1986. *Toward a New Era in U.S. Manufacturing: The Need for a National Vision.* Washington, D.C.: National Academy Press.

———. 1987. *Management of Technology: The Hidden Competitive Advantage.* Washington, D.C.: National Academy Press.

Piszczalski, M. 1987. "NSF Restructures for Greater Research Impact," *Managing Automation* (December), 17.

Ryan, T. P. 1989. *Statistical Methods for Quality Improvement.* New York: John Wiley.

Savage, C. M. ed. 1986. *Management Guide for CIM.* Dearborn, Mich.: Society of Manufacturing Engineers.

Suri, R., and S. DeTreville. 1986. "Getting from 'Just-in-Case' to 'Just-in-Time': Insights from a Simple Model," *Journal of Operations Management,* 6, May, 295.

Votapka, T. 1988. "Kitting Eases Inventory Control," *Purchasing Management,* May 16, p. 7.

Warndorf, P. R., and M. E. Merchant. 1986. "Development and Future Trends in Computer-Integrated-Manufacturing in the USA," *International Journal of Technology Management 1,* 1–2, 162.

Wittry, E. J. 1987. *Managing Information Systems: An Integrated Approach,* Dearborn, Mich.: Society of Manufacturing Engineers.

CHAPTER 5

Air Force Systems Command. 1981a. "Integrated Computer-Aided Manufacturing (ICAM) Dynamics Modeling Manual (IDEF$_2$)," Wright-Patterson Air Force Base, Ohio.

———. 1981b. "Integrated Computer-Aided Manufacturing (ICAM) Function Modeling Manual (IDEF$_0$)," Wright-Patterson Air Force Base, Ohio June.

CACI Products Co. 1988. "SIMFACTORY with Animation."

Cheung, S., S. Dimitriadis, and W. J. Karplus. 1988. "Introduction to Simulation Using NETWORK II.5." La Jolla; Calif.: CACI Products Co.

D. Appleton Co. 1985. "IISS: Integrated Information Support System (Information Modeling Manual IDEF1—Extended)."

Goldberg, J. 1987. "Rating Cost-Estimating Software," *Manufacturing Engineering* (February), 37.

Grant, F. H. 1988. "Simulation in Designing and Scheduling Manufacturing Systems," in *Design and Analysis of Integrated Manufacturing Systems,* ed. W. Dale Compton. Washington, D.C.: National Academy Press, p. 134.

Law, A. M., and C. S. Larmey. 1984. "An Introduction to Simulation Using SIMSCRIPT II.5." La Jolla, Calif.: CACI Products Co.

Law, A. M., and M. G. McComas. 1988. "How Simulation Pays Off," *Manufacturing Engineering* (February), 37.

Meta Software Corp. n.d. "Design/IDEF Operations Manual."

Mills, R. 1988. "Telecommunication Network Analysis with COMNET II.5." La Jolla, Calif.: CACI Products Co.

Network Dynamics, Inc. n.d. "MANUPLAN II Operations Manual."

Pritsker, A. A. B. 1986. *Introduction to Simulation and Slam II,* 3rd Ed. New York: Halsted Press.

Pritsker Corp. 1988a. "SLAMSYSTEM: Total Simulation Project Support."

————. 1988b. "XCELL+ Manufacturing Simulation System."

Standridge, C. R., and A. A. B. Pritsker. 1987. *TESS: The Extended Simulation Support System.* New York: Halsted Press.

Suri, R. 1983. "Robustness of Queuing Network Formulas," *Journal of the Association for Computing Machinery,* 30, 3 (July), 564.

————. 1985. "An Overview of Evaluative Models for Flexible Manufacturing Systems," *Annals of Operations Research,* 3, p 13.

————. 1988. "RMT Puts Manufacturing at the Helm," *Manufacturing Engineering* (February), 41.

————, and G. W. Diehl. 1986. "A Variable Buffer-Size Model and Its Use in Analyzing Closed Queuing Networks with Blocking," *Management Science,* 32, 2 (February), 206.

————. 1987. "Rough Cut Modeling: An Alternative to Simulation," *CIM Review* (Winter), 25.

Suri, R., and J. W. Dille. 1985. "A Technique for On-Line Sensitivity Analysis of Flexible Manufacturing Systems," *Annals of Operations Research,* 3, 381.

Suri, R., and C. K. Whitney. 1984. "Decision Support Requirements in Flexible Manufacturing," *SME Journal of Manufacturing Systems,* 3, 1, 61.

Talavage, J., and R. G. Hannam. 1988. *Flexible Manufacturing Systems in Practice: Applications, Design and Simulation.* New York: Marcel Dekker.

Vig, M., K. Dooley, and P. Starr. 1989. "Simulating Cell Activities in the CIM Environment," *Manufacturing Engineering* (January), 65.

Wild, W. G., Jr., and O. Port. 1987. "This Video Game Is Saving Manufacturers Millions," *Business Week,* August 17.

CHAPTER 5 COMPANY ADDRESSES

CACI Products Co.
3344 North Torrey Pines Court
La Jolla, CA 92037

D. Appleton Co., Inc.
1334 Park View Avenue
Suite 220
Manhattan Beach, CA 90266

E-Z Systems, Inc.
1433 West Fullerton Avenue
Suite M
Addison, IL 60101

Meta Software Corp.
150 Cambridge Park Drive
Cambridge, MA 02140

MiCAPP, Inc.
16956 230th Avenue
Grand Rapids, MI 49307

Network Dynamics, Inc.
128 Wheeler Road
Burlington, MA 01803

Pritsker Corp.
8910 Purdue Road
Suite 500
Indianapolis, IN 46268

CHAPTER 6

Acarnley, P. P. 1982. *Stepping Motors: A Guide to Modern Theory and Practice.* Stevenage, U.K., and New York: Peter Peregrinus.

Andeen, G. B., ed. 1988. *Robot Design Handbook.* New York: McGraw-Hill.

Anthony, J. 1985. "Leak Detection of Hermetically Sealed Devices," *Test and Measurement World* (January), 68.

Ayres, R. U., et al. 1985. *Robotics and Flexible Manufacturing Technologies: Assessment, Impacts and Forecast.* Park Ridge, N.J.: Noyes.

Bader, M. 1989. "U.S. Hybrid Circuit Markets, Applications, Technology," *Solid State Technology/Hybrid Supplement* (February), 51.

Blache, K. M. 1988. *Success Factors for Implementing Change: A Manufacturing Viewpoint.* Dearborn, Mich.: Society of Manufacturing Engineers.

Computer Design. 1989. "Hybrids Offer Alternative to Higher Chip Densities," January 1, p. 63.

Crowley, D. F. 1988. "Economic Justification for Hybrid Substrate Testing," *Hybrid Circuits,* 17 (September).

Dickinson, D. 1989. "Reliable and Economical Assembly of Hybrid Microelectronic Packages," paper presented at 1989 Southern California Symposium of the International Society for Hybrid Microelectronics.

Edson, D. V. 1986. "Bar Coding the Factory," *Managing Automation* (September), 48.

Fuchs, J. H. 1988. *The Prentice Hall Illustrated Handbook of Advanced Manufacturing Methods.* Englewood Cliffs, N.J.: Prentice Hall.

Gayman, D. 1987. "Ramping Up Automation with Machining Centers," *Manufacturing Engineering* (October), 41.

Gould, L. 1987. "Technologies for Machine Perception," *Manufacturing Engineering* (December), 32.

Hall, E. L., and B. C. Hall. 1988. "Focus: Machine Vision Research," *Manufacturing Engineering* (July), 50.

Hoska, D. R. 1988. "FLAM: What It Is. How to Achieve It," *Manufacturing Engineering* (April), 49.

Hybrid Circuit Technology. 1986. "Report Predicts Hybrid Growth" (June), 6.

Institute of Electrical and Electronics Engineers. 1989. "Worldwide Production of Electronic Products," *IEEE Institute,* 13, 3 (March), 5.

Kenjo, T. 1984. *Stepping Motors and Their Microprocessor Controls.* New York: Oxford University Press.

Lane, J. D., ed. 1986. *Automated Assembly.* Dearborn, Mich.: Society of Manufacturing Engineers.

Larin, D. J. 1986. "Vision's Next Steps," *Manufacturing Engineering* (December), 35.

————. 1989. "Cell Control: What We Have, What We'll Need," *Manufacturing Engineering* (January), 41.

Lizari, J. J., and L. R. Enlow. 1988. *Hybrid Microcircuit Technology Handbook,* Park Ridge, N.J.: Noyes.

Malone, R. 1987. "Tooling for End of Arms," *Managing Automation* (December), 65.

Martin, J. M. 1989. "Cells Drive Manufacturing Strategy," *Manufacturing Engineering* (January), 49.

National Academy of Sciences. 1984. *The Competitive Status of the U.S. Electronics Industry.* Washington, D.C.: National Academy Press.

————. 1986. *Toward a New Era in U.S. Manufacturing: The Need for a National Vision.* Washington, D.C.: National Academy Press.

National Institute of Standards and Technology. 1988. "Manufacturing Technology Centers Program." Gaithersburg, Md.

Pirocanac, D., ed. 1986. "Automation, Sooner or Later," *Hybrid Circuit Technology* (April), 4.

Robotics World. 1987. "Vision Complements Modular System" (January), 23.

Schrelber, R. R. 1988. "Whither Sensors?" *Manufacturing Engineering* (February), 54.

Stauffer, R. N. 1988a. "Robotics Research: What the Labs Might Hold in Store," *Manufacturing Engineering* (June), 58.

———. 1988b. "Robots and Vision: Tying the Technologies Together," *Manufacturing Engineering* (August), 103.

Talavage, J., and R. G. Hannam. 1988. *Flexible Manufacturing Systems in Practice: Applications, Design and Simulation*. New York: Marcel Dekker.

Tummala, R. R., and E. J. Rymaszewski, eds. 1989. *Microelectronics Packaging Handbook*. New York: Van Nostrand Reinhold.

Wick, C. 1987. "Advances in Machining Centers," *Manufacturing Engineering* (October), 24.

CHAPTER 6 COMPANY ADDRESSES

Affiliated Manufacturers, Inc.
U.S. Highway 22
P.O. Box 5049
North Branch, NJ 08876

Amistar Corp.
237 Via Vera Cruz
San Marcos, CA 92069

BEI Motion Systems Co.
2111 Palomar Airport Road
Suite 250
Carlsbad, CA 92009

CerProbe Corp.
600 South Rockford Drive
Tempe, AZ 85281

Chicago Laser Systems, Inc.
4034 North Nashville
Chicago, IL 60634

Cobehn, Inc.
Airport Road
Route 1, Box 206V
Winchester, VA 22601

Cognex Corp.
15 Crawford Street
Needham, MA 02194

DEK USA, Inc.
100 Corporate Drive
Suite 102A
Lebanon, NJ 08833

Electro Scientific Industries, Inc.
13900 N.W. Science Park Drive
Portland, OR 97229-5497

Giddings & Lewis FMS/Cellular Systems
142 Doty Street
P.O. Box 590
Fond du Lac, WI 54936-0590

Grieve Corp.
500 Hart Road
Round Lake, IL 60073-9989

Hughes Aircraft Co.
Industrial Products Division
6155 El Camino Real
Carlsbad, CA 92009

ITERE, Inc.
323 Sinclair Frontage Road
Milpitas, CA 95035-5443

Kearney & Trecker Corp.
11000 West Theodore Trecker Way
Milwaukee, WI 53214-1127

Kulicke and Soffa Industries.
2101 Blair Mill Road
Willow Grove, PA 19090

LazerData Corp.
2400 Diversified Way
Orlando, FL 32804

Lindberg Co.
304 Hart Street
Watertown, WI 53094

Lord Corp.
118 MacKenan Drive
Suite 300
Box 8200
Cary, NC 27512-8200

Manufacturers Technologies
59 Interstate Drive
West Springfield, MA 01089

March Instruments, Inc.
125-J Mason Circle
Concord, CA 94520

Micromanipulator Co. Inc.
2801 Arrowhead Drive
Carson City, NV 89701

Micropen, Inc.
3800 Monroe Avenue
Pittsford, NY 14534

Mikron Instrument Co. Inc.
445 West Main Street
Wyckoff, NJ 07481

Panasonic Factory Automation Co.
9401 West Grand Avenue
Franklin Park, IL 60131

Pulnix America, Inc.
770 Lucerne Drive
Sunnyvale, CA 94086

Scientific Sealing Technology
9801 Everest Street
Downey, CA 90242

Teledyne TAC
10 Forbes Road
Woburn, MA 01801

Teradyne, Inc.
30801 Agoura Road
Agoura Hills, CA 91301

Universal Instruments Corp.
Box 825
Binghamton, NY 13902-0825

Vanzetti Systems
166 South Victory Boulevard
Burbank, CA 91502

VIATRAN Corp.
300 Industrial Drive
Grand Island, NY 14072

Watkins-Johnson Co.
440 Kings Village Road
Scotts Valley, CA 95066

CHAPTER 7

Ames, J. G. 1988. "Which Network Is the Right One," *Manufacturing Engineering* (May), 56.

Borggraaf, P. 1985. "Semiconductor Factory Automation: Current Theories," *Semiconductor International* (October), 88.

Bux, W. 1981. "Local Area Subnetworks: A Performance Comparison," *IEEE Transactions on Communications,* COM-29, 10, 1465.

Campbell, J. 1984. *The RS-232 Solution.* Alameda, Calif.: Sybex, Inc.

Farowich, S. A. 1986. "Communicating in the Technical Office," *IEEE Spectrum* (April), 63.

Kaminski, M. A., Jr. 1986. "Protocols for Communicating in the Factory," *IEEE Spectrum* (April), 56.

Kleinrock, L., and S. S. Lam. 1975. "Packet Switching in a Multiaccess Broadcast Channel: Performance Evaluation," *IEEE Transactions on Communications,* COM-23, 4 (April), 410.

Metcalfe, R. M., and D. R. Boggs. 1976. "Ethernet: Distributed Packet Switching for Local Computer Networks," *Communications of the ACM 19,* 395.

Shock, J. F., and J. A. Hupp. 1980. "Measured Performance of an Ethernet Local Network," *Communications of the ACM,* 23, 711.

Tanenbaum, A. S. 1988. *Computer Networks,* 2nd ed. Englewood Cliffs, N.J.: Prentice Hall.

Voelcker, J. 1986. "Helping Computers Communicate," *IEEE Spectrum* (March), 61.

CHAPTER 7 COMPANY ADDRESSES

International Business Machines
P.O. Box 1328
Boca Raton, FL 33432

Teledyne Microelectronics
12964 Panama Street
Los Angeles, CA 90066

CHAPTER 8

Allison, G. T. 1971. *Essence of Decision: Explaining the Cuban Missile Crisis.* Boston: Little, Brown.

Anthony, R. N. 1965. *Planning and Control Systems: A Framework for Analysis.* Cambridge, Mass.: Harvard University: Division of Research, Graduate School of Business Administration.

Bardach, E. 1977. *The Implementation Game.* Cambridge, Mass.: MIT Press.

Barnard, C. I. 1938. "Informal Organizations," in *Classics of Public Administration,* ed.

J. M. Shafritz and A. C. Hyde. Oak Park, Ill.: Moore, 1978, 48.

Basil, D., and C. Cook. 1974. *The Management of Change.* New York: McGraw-Hill, 1974.

Behling, O., and C. Schriesheim. 1976. *Organizational Behavior: Theory, Research and Application.* Boston: Allyn & Bacon.

Berger, P. L., and T. Luckman. 1966. *The Social Construction of Reality.* New York: Anchor.

Biller, R. 1982. "Matrix Organizations," personal discussions with Robert Biller, professor of Public Administration, University of Southern California.

Blake, R. R., and J. S. Mouton. 1980. *The Versatile Manager: A Grid Profile.* Homewood, Ill.: Dow Jones-Irwin.

Blinder, A. S. 1989. "Want to Boost Productivity? Try Giving Workers a Say," *Business Week,* April 17, p. 10.

Bruce-Briggs, B. 1982. "The Dangerous Folly Called Theory Z," *Fortune,* May 17, p. 41.

Bruner, J. S. 1957. "On Perceptual Readiness," *Psychological Review,* 64, 123.

Business Week. 1982a. "TRW Leads a Revolution in Managing Technology," November 15, p. 124.

Business Week. 1982b. "How They Manage the New Financial Supermarkets," December 20, p. 50.

Carzo, R., Jr., and J. N. Yanouzas. 1967. *Formal Organization: A Systems Approach.* Homewood, Ill.: Irwin-Dorsey.

Cummings, T. A., ed. 1980. *Systems Theory for Organization Development.* New York: John Wiley.

Cunniff, J. 1982. "Why Corporate Realignments Become So Messy," *Sacramento Bee,* May 20, p. C9.

Drucker, P. 1980. *Managing in Turbulent Times.* New York: Harper & Row.

DuBrin, A. J. 1978. *Fundamentals of Organizational Behavior,* 2nd ed. Elmsford, N.Y.: Pergamon Press.

Evan, W. 1976. *Organization Theory: Structure, Systems and Environments.* New York: John Wiley.

Fiedler, F. E. 1977. *Improving Leadership Effectiveness: The Leader Match Concept.* New York: John Wiley.

Gagne, R. M. 1975. *Essentials of Learning for Instruction.* New York: Holt, Rinehart & Winston.

Gerth, H. H., and C. W. Mills. 1978. *From Max Weber: Essays in Sociology.* New York: Oxford University Press.

Gulick, L. 1937. "Notes on Theory of Organizations," in *Classics of Public Administration,* ed. J. M. Shafritz and A. C. Hyde. Oak Park, Ill.: Moore, 1978.

Hall, C., and G. Lindzey. 1978. *Theories of Personality.* New York: John Wiley.

Hefferlin, L. B. Lon. 1969. *Dynamics of Academic Reform.* San Francisco: Jossey-Bass.

Hellriegel, D., and J. W. Slocum, Jr. 1979. *Organizational Behavior,* 2nd ed. St. Paul, Minn.: West.

Herbert, T. T. 1981. *Dimensions of Organizational Behavior,* 2nd ed. New York: Macmillan.

Hoerr, J. 1989. "Is Teamwork a Management Plot? Mostly Not," *Business Week,* February 20, p. 70.

Holzer, B., and J. H. Marx. 1979. *Knowledge Application: The Knowledge System in Society.* Boston: Allyn & Bacon.

Huse, E. F., and J. L. Bowditch. 1973. *Behavior in Organization: A Systems Approach to Managing.* Reading, Mass.: Addison-Wesley.

Katz, D., and R. L. Kahn. 1978. *The Social Psychology of Organizations.* New York: John Wiley.

Katz, D., R. Kahn, and J. Adams, eds. 1980. *The Study of Organizations.* New York: Jossey-Bass.

Kaufman, H. 1979. *The Limits of Organizational Change.* Tuscaloosa: University of Alabama Press.

Kuhn, T. 1962. *The Structure of Scientific Revolutions,* 2nd ed. Chicago: University of Chicago Press.

Lawlers, D. J. 1979. *Organizational Behavior,* 2nd ed. Englewood Cliffs, N.J.: Prentice Hall.

Lippmann, W. 1922. *Public Opinion.* New York: Harcourt Brace Jovanovich.

Maccoby, M. 1976. *The Gamesman.* New York: Simon & Schuster.

———. 1981. *The Leader.* New York: Simon & Schuster.

McGregor, D. 1960. *The Human Side of Enterprise.* New York: McGraw-Hill.

Melcher, R. A. 1988. "What's Throwing a Wrench into Britain's Assembly Lines," *Business Week,* February 29, p. 41.

Mouzelis, N. 1978. *Organization and Bureaucracy.* Chicago: Aldine.

Newell, A., and H. A. Simon. 1972. *Human Problem Solving.* Englewood Cliffs, N.J.: Prentice Hall.

Ouchi, W. 1981. *Theory Z: How American Business Can Meet the Japanese Challenge.* Reading, Mass.: Addison-Wesley.

Perrow, C. 1979. *Complex Organizations.* Glenview, Ill.: Scott, Foresman.

Peters, T. J., and R. H. Waterman, Jr. 1982. *In Search of Excellence.* New York: Harper & Row.

Presthus, R. 1978. *The Organizational Society,* rev. New York: St. Martin's Press.

Ramos, A. G. 1982. *The New Science of Organizations: A Reconceptualization of the Wealth of Nations.* Buffalo, N.Y.: University of Toronto Press.

Roethlisberger, F. J. 1941. "The Hawthorne Experiments," in F. J. Roethlisberger, *Management and Morale.* Cambridge, Mass.: Harvard University Press, as reprinted in *Classics of Public Administration,* ed. Jay M. Shafritz and Albert C. Hyde. Oak Park, Ill.: Moore, 1978.

Rouse, W. B. 1986. "The Human Role in Advanced Manufacturing Systems," in *Automated Assembly,* ed. Jack D. Lane. Dearborn, Mich.: Society of Manufacturing Engineers.

Sacramento Union. 1982. "Superleader Study: A Dynamic Boss Inspires Workers," October 10, p. E4.

Schein, E. H. 1981. "Does Japanese Management Style Have a Message for American Managers?" *Sloan Management Review,* Fall, 55.

Schon, D. 1973. *Beyond the Stable State.* New York: W. W. Norton.

Selznick, P. 1957. *Leadership in Administration.* New York: Harper & Row.

Shortell, S. 1982. "Theory Z: Implications and Relevance for Health Care Management," *Health Care Management Review,* Fall.

Simon, H. A. 1976. *Administrative Behavior,* 3rd ed. New York: Free Press.

Taylor, F. W. 1912. "Scientific Management," in *Classics of Public Administration,* ed. J. M. Shafritz and A. C. Hyde. Oak Park, Ill.: Moore, 1978.

Time. 1982. "How to Be Great," November 15, p. 68.

Toffler, A. 1980. "The Third Wave: The Corporate Identity Crisis," *Management Review,* 69, 8.

University of Pennsylvania. 1980. *Conference Summary: National Conference on Management Development in Health Care.* Philadelphia: Wharton School.

Wilson, M. P., and F. B. Rafkind. 1980. *Developing Management Leaders: An Interview with Marjorie P. Wilson, M.D.* Philadelphia: National Health Care Management Center, University of Pennsylvania.

Woodward, J. 1980. *Industrial Organization: Theory and Practice,* 2nd ed. New York: Oxford University Press.

CHAPTER 9

Boothroyd, G., and P. Dewhurst. 1988. "Product Design for Maintenance and Assembly," *Manufacturing Engineering,* April, 42.

Computer Aided Design Report. 1987. "Exchanging Data with Suppliers" (841 Turquoise Street, Suites D and E, San Diego, CA 92109-1159), 7, 2, February.

Crosheck, J., and T. Hite. 1989. "Large Displacement Dynamics on the PC," *CADENCE,* June, 91.

Fallon, M. R. 1986. "Solid Modeling Software: An Emerging Market," *Managing Automation,* October, 79.

Hamilton, C. H. 1989. "MCAE Enters the 1990s," *Manufacturing Engineering,* June, 80.

Hays, M. 1989. "Networks for CAD/CAM," *CADENCE,* March, 64.

Hybrid Circuit Technology. 1987. "Hybrid Circuit Technology CAD Equipment Guide," March, 10.

Jain, R. 1989. "Methodology of Solid Modeling," *CADENCE,* June, 151.

Kinnucan, P. 1989. "Finite-Element Analysis Software Survey," *Computer Graphics Review,* March, 42.

Melkanoff, M. A., and A. H. Bond. 1989. "Bridging the CAD/CAM Gap." Los Angeles: School of Engineering and Applied Science, University of California.

Metzger, J. 1989. "Moving from a Drafting Table to a CAD/CAM Network," *Manufacturing Systems,* April, 55.

Port, O. 1989. "The Best-Engineered Part Is No Part at All," *Business Week,* May 8, p. 150.

Singampalli, R. 1986. "Using COSMOS/M for Nonlinear Finite Element Analysis," *CADENCE,* August 19.

Skomra, A. R. 1986. "Solid Modeling Round Table," CASA/Society of Manufacturing Engineers, August 18.

Stark, J. 1988. *Managing CAD/CAM: Implementation, Organization and Integration.* New York: McGraw-Hill.

Thorell, L., and K. Hurley. 1989. "Work-Station Vendor Scorecard," *Computer Graphics Review,* March, 50.

Voelcker, H. A. 1988. "Modeling in the Design Process," in *Design and Analysis of Integrated Manufacturing Systems,* ed. W. Dale Compton. Washington, D.C.: National Academy Press.

Voelcker, J. 1986. "Helping Computers Communicate," *IEEE Spectrum,* March, 61.

Wilson, M. 1989. "Creating Realistic Images from IGES Data," *MicroCAD News,* July, 70.

Wohlers, T. 1988. "Linking It All Up," *Computer Graphics World,* November 100.

———. 1989. "Does the 3D Market Exist?" *CADENCE,* June, 113.

CHAPTER 9 COMPANY ADDRESSES

Autodesk, Inc.
2320 Marinship Way
Sausalito, CA 94965

Battelle
CAD/CAM Products Office
505 King Avenue
Columbus, OH 43201-2693

Boeing Computer Services
P.O. Box 24346, MS 7W-05
Seattle, WA 98124-9985

CADAM, Inc.
1935 North Buena Vista
Burbank, CA 91504

CADKEY, Inc.
440 Oakland Street
Manchester, CT 06040-2100

CNC Software, Inc.
45 Industrial Park Road
Suite 1
Tolland, CT 06084

Computervision Corp.
100 Crosby Drive
Bedford, MA 01730

Intergraph Corp.
One Madison Industrial Park
Huntsville, AL 35807-4201

ISICAD
1920 West Corporate Way
P.O. Box 61022
Anaheim, CA 92803-6122

Swanson Analysis Systems, Inc.
P.O. Box 65
Johnson Road
Houston, PA 15342

VersaCAD Corp.
2124 Main Street
Huntington Beach, CA 92648

Wisdom Systems
100 North Main Street
Chagrin Falls, OH 44022

CHAPTER 10

Capenttini, R., and D. K. Clancy, eds. 1987. "Cost Accounting Robotics and the New Manufacturing Environment," Edited Presentations of the First Annual Management Accounting Symposium, American Accounting Association, Vanderbilt University, February 26–28.

Lane, J. D., ed. 1986. *Automated Assembly.* Dearborn, Mich.: Society of Manufacturing Engineers.

Lardner, J. F. 1986. "Computer Integrated Manufacturing and the Complexity Index," *The Bridge* (National Academy of Engineering), 16, 1, Spring, 10.

INDEX

539